Springer Series on Polymer and Composite Materials

Series editor

Susheel Kalia, Waknaghat, India

More information about this series at http://www.springer.com/series/13173

Kishor Kumar Sadasivuni
Deepalekshmi Ponnamma
Jaehwan Kim · Sabu Thomas
Editors

Graphene-Based Polymer Nanocomposites in Electronics

Springer

Editors
Kishor Kumar Sadasivuni
Department of Mechanical Engineering, Center for EAPap Actuator
Inha University
Incheon
Korea, Republic of (South Korea)

Jaehwan Kim
Department of Mechanical Engineering, Center for EAPap Actuator
Inha University
Incheon
Korea, Republic of (South Korea)

Deepalekshmi Ponnamma
School of Chemical Sciences
Mahatma Gandhi University
Kottayam
India

Sabu Thomas
School of Chemical Sciences
Mahatma Gandhi University
Kottayam
India

Springer Series on Polymer and Composite Materials
ISBN 978-3-319-13874-9 ISBN 978-3-319-13875-6 (eBook)
DOI 10.1007/978-3-319-13875-6

Library of Congress Control Number: 2014957476

Springer Cham Heidelberg New York Dordrecht London

Printed on acid-free paper

Springer International Publishing AG Switzerland is part of Springer Science+Business Media (www.springer.com)

Contents

Graphene/Polymer Nanocomposites: Role in Electronics

Deepalekshmi Ponnamma and Kishor Kumar Sadasivuni

Abstract Discovery of graphene nanolayers has made a big bang in nanotechnology and many industrial innovations have come up as a result. Graphene and its derivatives filled polymers also contribute towards the fabrication of numerous electronic and mechanical devices such as sensors, capacitors, tyres, shields etc. In this regard a detailed survey of this topic has utmost importance mainly focusing on the application side. This chapter is aimed at providing a brief introduction about various graphene nanocomposite systems, its major properties especially relevant in electronics and a few applications. We also discuss the thermal, piezoelectric, optoelectric and actuating properties of graphene polymer nanocomposites in addition to the electromagnetic interference shielding and ferroelectric performances.

Keywords Electrical properties · Actuation · Piezoelectrics · Shielding · Ferroelectrics

1 Introduction

Nowadays technology depends mainly on nanoparticles and the two dimensional graphene sheets has emerged as a subject of enormous scientific interest due to its exceptional electron transport, mechanical properties, and high surface area. This one atom thick single layer of graphite was first produced by mechanical exfoliation and is the lightest, thinnest and strongest material ever discovered [1]. Because graphene is only 1 atom thick, it is possible to create other materials by

D. Ponnamma
School of Chemical Sciences, Mahatma Gandhi University,
Kottayam 686560, Kerala, India

K.K. Sadasivuni (✉)
Center for EAPap Actuator, Department of Mechanical Engineering,
Inha University, 253 Yonghyun-Dong, Nam-Ku, Incheon 402-751, South Korea
e-mail: kishor_kumars@yahoo.com

K.K. Sadasivuni et al. (eds.), *Graphene-Based Polymer Nanocomposites in Electronics*, Springer Series on Polymer and Composite Materials,
DOI 10.1007/978-3-319-13875-6_1

interjecting the graphene layers with other compounds, effectively using graphene as atomic scaffolding from which other materials are engineered. The promising applications of graphene as in electronics is due to its high theoretical surface area of 2,630 $m^2\ g^{-1}$ and ability to facilitate electrons or hole transfer along its two-dimensional surface. The only problem with graphene is that high-quality graphene is a great conductor that does not have a band gap (it can't be switched off). Therefore to use graphene in the creation of future nano-electronic devices, a band gap will need to be engineered into it, which will, in turn, reduce its electron mobility to that of levels currently seen in strained silicone films. The exceptional property of graphene overtakes the silicon based electronics and the field is hanging around commercialization. Graphene possesses low cost, two external surfaces, facile fabrication and modification and absence of toxic metal particles when compared to its carbon counterparts [2, 3]. The novel functionalities of it include ultracapacitors, transparent conducting electrodes etc. [4, 5] since because of the exceptional electron mobility of 200,000 $cm^2\ V^{-1}\ s^{-1}$ [6] and room-temperature quantum Hall effects [7].

Generally quasi one dimensional graphene of narrow width and atomically smooth edges called graphene nanoribbons (GNRs) provide field effect transistor (FET) application with excellent switching speed and high carrier mobility at room temperature. The very high electrical conductivity and mechanical flexibility of graphene make its derivatives applicable in light emitting diode (LED), FET, memory and photovoltaic devices. The GNRs, were quite different from those of graphene in terms of carrier mobility and fabrication [8]. It was Geim and Novoselov [9] who first reported the electric field effect of graphene by demonstrating its ambipolar characteristics with electron and hole concentration of 1,013 cm^2 with mobilities up to 10,000 $cm^2\ V^{-1}\ s^{-1}$ at room temperature based on the fabricated FET [8, 9]. Graphene based micro-supercapacitors have use in low energy applications such as smart phones and portable computing devices and could potentially be commercially available within the next 5–10 years. A graphene foam-based battery design bridges the gap between batteries and capacitors based on lithium technology. It can discharge and charge as quickly as a capacitor and can discharge completely in 20 s. It is also flexible, and works perfectly when it is bent.

Graphene and its polymer composites show excellent properties in electroluminescent devices, gate dielectrics, and touch panel devices due to their high flexibility, thermal and electrical conductivity, low temperature processing conditions, and simple fabrication process [10–12]. They have many applications in nanoelectronic devices, chemical and biological sensors, energy storage and biomedical fields and many other applications as shown in Fig. 1. Though many articles discuss about the graphene-based polymer composites and their applications in different fields [13–16], the topic is so hot that an updated picture of its applications in electronics is aimed here at. As the introductory chapter of a book on applications of graphene nanocomposites in electronics, the major characteristic properties of graphene is discussed briefly here along with an introduction towards the various applications which this book focuses on. A glimpse to polymer graphene composites thermal

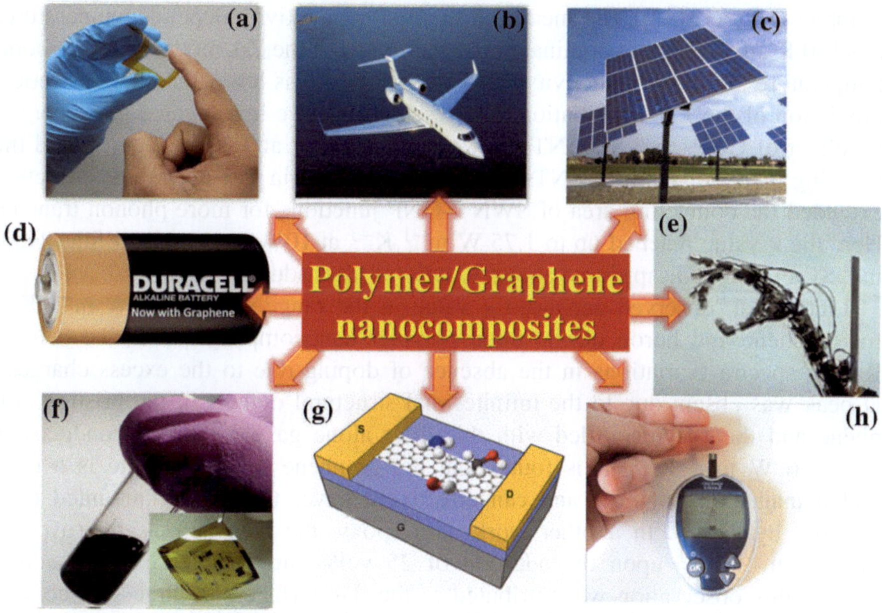

Fig. 1 **a** Flexible transparent electronics, touch screen. **b** Aeronautical field. **c** Solar panels. **d** Batteries. **e** Actuators. **f** Conductive ink, flexible electrodes. **g** Gas sensors. **h** Biosensors

and electrical conductivity, dielectric property, piezoresistivity, optical behavior, actuation and electromagnetic interference shielding (EMI) is made in the beginning. Later on we move to the major applicability of graphene filled polymers in fabricating fuel cells, actuators, energy storage devices and so on.

2 Characteristic Properties of Graphene and Its Nanocomposites

2.1 Thermal Properties

Graphene possesses high thermal conductivity, κ [17] and the electronic devices based on it outperform silicon devices in the record-high electron mobility [18]. The κ values calculated from the temperature rise of a laser heated graphene sheet by Raman spectroscopy [19, 20] are respectively in the range of 1,500–5,800 and ~2,800 W m^{-1} K^{-1} for single and bilayer suspended graphenes. It is reported that the lattice thermal conductivity of freestanding graphene is dominated by contributions from the out-of plane acoustic (ZA) phonons in flexural modes [21] and the damping of which limits the κ of single-layer graphene to ~600 W m^{-1} K^{-1}, considerably lower than the basal-plane values of highly ordered pyrolytic graphite (HOPG) [22]. The measured thermal conductivity depends temperatures at <150 K attributed to a dominant contribution from the ZA modes, but the room-temperature thermal conductivity <225 W m^{-1} K^{-1} was lower than the theoretical prediction of the ZA contribution. The researchers have investigated the synergistic effect of GNPs and SWCNTs as well (10–20 wt% at 1:3 ratio) and found the bridging effect of flexible CNTs with planar GNPs via van der Waals attraction extended the contacting area of SWNT–GNP junctions for more phonon transfer. Here the κ value reached up to 1.75 W m^{-1} K^{-1} at 10 wt% combined filler loading. Synergy was disappeared above 30 wt% of filler addition [21].

Liem et al. observed enhanced κ value in the range of 6.2–9.5 W m^{-1} K^{-1} for graphene and boron nitride filled polymer nanocomposites with the help of Raman spectra (variations in the absence of doping due to the excess charges). D peak was absent due to the infinitesimal structural defects in the pristine graphene and the polymer filled with this filler alone gave a thermal conductivity of ~21.6 W m^{-1} K^{-1}. It is found that the graphene nanocomposite is a thermal management material and can effectively lower the surface mounted temperature by 21 °C. In another work with epoxy, the κ value was improved to 6.44 W m^{-1} K^{-1} upon the addition of 25 vol% multi-layer GNPs. The reason for this observation was attributed to the flat surface and large surface area of high aspect GNPs with which strong interactions were made with epoxy [20]. The silane functionalized thermally expanded graphite at 20 wt% also enhanced the κ value of epoxy from 0.2 to 5.8 W m^{-1} K^{-1}, though the high loading of the nanofillers affect the ductility of polymers seriously [22]. Exfoliated graphite at 20 vol% filled nylon-6 also exhibited a κ value of 4.1 W m^{-1} K^{-1}. All these results highlight graphene sheets of various structures and thicknesses are useful to fabricate highly thermally conductive polymer/graphene nanocomposite materials.

Finally graphene sheets are capable of acting like barriers to prevent the propagation of heat generated from the external environment in polymeric matrices and improve the thermal stability. The thermal conductivity mainly depends on the thermal stability, aspect ratio and dispersion of nanofillers and the latter two are affected by the percolation threshold of nanofillers. In order to achieve good dispersion, functionalization of graphene has also been used and such functionalized graphene exhibit super thermal conductivity; however, low level of functionalities may result in unsatisfactory dispersions of the fillers in polymers [19].

2.2 Electrical Properties

Electrical conductivity of graphene is excellently used in fabricating polymer based sensors, anti-statics, microwave absorbers, and conductive coatings. The conducting mechanism in graphene polymer nanocomposites is explained as the formation of a continuous electrical conductive network at percolation threshold depending on the nature and rate of dispersion and the interaction with polymer matrix [23, 24]. The larger aspect ratio graphenes more effectively decrease the percolation threshold in nanocomposites [25] and thus synthesizing excellent conducting films at low filler loading. For polymer/graphene systems, the high reduction temperature of graphene oxide results towards increased electrical conductivity compared with extending the reduction time at lower temperatures. When reduced thermally, the nanocomposites obtained show low percolation threshold and high electrical conductivity. This is attributed to the removal of surfactants originally present during thermal reduction and thus maintaining the low contact resistance between the interlayers of rGO sheets. Moreover the thermal reduction takes shorter time (15–30 min) than the chemical reduction of graphene oxdie (24 h).

For instance the electrical conductivity of the polyvinyl alcohol (PVA)/LrGO nanocomposites is observed to be 2.5×10^{-6} S/m at 0.3 wt% of LrGO. At 1 wt% LrGO the electrical conductivity again increased to 2.09×10^{-4} S/m indicating the efficiency of LrGOs in improving the conducting network. In this case the percolation threshold was found to be only 0.198 wt% [26]. It is also established that graphene's lowest percolation value achieved is 0.1 vol% and this is in the case of graphene/polystyrene composites.

2.3 Energy Storage

Polymer graphene composites outperform the conventional rechargeable lithium batteries in energy storage applications as they possess superior energy as well as power density. Graphene-based composites are widely reported as supercapacitors [27–32]. Guo et al. [32] used graphene to enhance the electrochemical performance of a polymer cathode at a loading of 60 %. This is really high concentration and to commercially develop such products, enhanced electrochemical performance must be achieved at low graphene loading. Graphene supercapacitors are able to be charged very quickly, yet also be able to store a large amount of electricity and are lightweight. The graphene based lithium ion batteries could be used in higher energy usage applications such as electrically powered vehicles, in smartphones, laptops and tablet PCs but at significantly lower levels of size and weight. The general requirement of an electrochemical double layer material in the capacitor is high specific surface area and high conductivity. The chemically reduced GO in hydrazine is reported to have high conductivity, large specific surface area and chemical stability, with a specific capacitance of up to 205 F/g in aqueous electrolyte [33]. In other works, the chemically modified graphene showed specific capacitances of 135 F/g [34] and 247 F/g [35] in aqueous electrolytes.

There are several reports regarding graphene-based ultracapacitors using metal oxides/graphene [36–38]. The SnO_2-graphene exhibits reversible capacity of 810 mAh/g and enhanced cycling performance compared to bare SnO_2 nanoparticles [39]. Another electrochemically active material out of self-assembled TiO_2-graphene hybrid nanostructure is prepared by Wang et al. [40]. A SnSb nanocrystal/graphene hybrid nanostructure is used as a potential high capacity anode material for lithium ion battery by Xie et al. [41]. Xiao and his coworkers used hierarchical arrangement of functionalized graphene sheets to fabricate a novel air electrode with an exceptionally high capacity of 15,000 mAh/g [42]. GO-MnO_2 composite supercapacitor was reported by Chen et al. [38].

Polymers have low dielectric constants and generally ferro-electric ceramic (e.g., barium titanate or strontium titanate) and/or electrically conductive fillers are incorporated in them to increase the dielectric constant. These fillers cause induced interfacial or space charge polarization and increase the capacitance [43]. Partially chlorinated rGO platelets filled polymer composite films showed an enhanced dielectric constant with a low dielectric loss due to the polar and polarizable C–Cl bonds [44]. The electrical conductivity, the specific surface area of fillers and polymer-filler interactions are strongly related to the dielectric properties of composite films as the fillers increase interfacial polarization at the polymer filler interface [45, 46]. Our group has also developed modified graphene filled polyurethane composites and the filler-polymer interfacial reinforcement was addressed through rheological analysis [47]. The rGO/CNT hybrid structure affects the dielectric performance of cyanoethyl pullulan polymer (CEP) and at 0.062 wt% the composite film showed a dielectric constant of 32 and a dielectric loss of 0.051 at 100 Hz while neat rGO had a dielectric constant of 21 with a dielectric loss of 0.038, and the pristine polymer film had a dielectric constant of 15 with a dielectric loss of 0.036. The increase in dielectric constant of the composite film was attributed to its increased specific surface area and increased density of interfacial polarization. Though this is the case, several problems like formation of the gate dielectrics for FET should be solved to develop graphene-based electronic devices. Graphene is chemically inert and impedes uniform deposition of gate dielectric films [48]. Also the usually used plasma-enhanced chemical vapor deposition and sputtering degrade the already formed graphene layer due to damages caused by the oxygen radicals created in the process [49]. Due to this reason spin coated polymer materials are used as the gate insulator in organic thin-film transistors and such gate shows excellent dielectric properties despite its low-cost and easy-fabrication nature [50, 51].

2.4 Optoelectrical Properties

A material must transmit more than 90 % of light and also offer electrical conductivity exceeding $1 \times 106\ \Omega^1 m^1$ in order to find use in optoelectronics and the highly transparent graphene (transmit up to 97.7 % of light) is widely applied in this field in the form of touchscreens, liquid crystal displays (LCD) and organic light emitting diodes (OLEDs). This nanomaterial matches the properties of the currently used brittle and relatively unstable indium tin oxide (ITO) and also, its optical absorption changes by adjusting the Fermi level. Highly strong and flexible graphene based e-paper display interactive and updatable information and applied in flexible electronic devices including portable computers and televisions. In OLED, graphene acts as anode and also makes it entirely transparent and flexible, expanding the design and application in solar cells as well [52, 53]. In solar cells, graphene in the form of p-doped or n-doped is applied in which photoelectronic reactions can occur [54]. However the surface roughness and uneven distribution of hole density [55–57] are considered to be the shortcomings of graphene which can be solved by regulating the mode of synthesis and doping techniques.

Rice researchers have come up with a graphene based transparent, flexible and eco-friendly electrode possessing higher conductivity than ITO. For the FETs fabricated on PET substrates graphene works as source/drain electrodes and the device showed high optical transmittance and mechanical bendability. In addition to the low contact resistance between graphene and organic materials, the outstanding stability in conductivity even after hundreds of bending cycles make graphene an ideal electrode material [54]. It was Bae et al. who demonstrated the applicability of graphene as touch screen [58] by leaving behind the easy to handle ITOs having low sheet resistance. This has become a basement for the graphene based electronics market and even the single layer graphene shows excellent resilience upon bending and when written on with a touch screen pen.

Graphene is also used as a novel acceptor for bulk hetrojunction polymer photovoltaic cells due to its high hole transport mobility, large surface area and inertness against oxygen. Such cells showed reduced photoluminescence and efficient energy transfer [5]. In short the optoelectronic applications of graphene films include the highly flexible polymer coated films, display devices and solar cells [59]. Graphene filled cellulose nanocrystal is fabricated for proximity sensing in flexible electronics [60] as well.

2.5 *Piezoelectric Properties*

Graphene enhances the piezoelectric properties of the polymer without detriment to the elastic modulus and an optimized composite was reported to design a synthetic muscle fiber with a cross-section of 26 mm^2. This rGO–PVDF (poly(vinylidene fluoride)) composite was actuated at 4.5 V, producing an actuation force of 0.5 N, sufficient to flex a human finger joint. The polymer, PVDF has been used in environmental energy harvesting and in low-cost sensors due to its high ductility, high applicability to curved structures, advantage of free adjustment of sizes etc. [61]. But the low coupling between electrical and mechanical properties and its relatively low generated voltage and force cause polymers to be less applicable in such area and graphene fillers are often employed to clear this issue.

In another study MWNTs and GnPs filled epoxy composites are studied for their piezoelectric properties and the developed hybrid composite film exhibited enhanced mechanical strength, poling behavior and sensitivity to excitation. Here GnPs offered more advantages over MWNTs, as evidenced by the greater improvement in electromechanical properties and poling behavior and the composite is very applicable in pressure sensors. Our group has also developed graphene filled butyl rubber pressure sensors where a construction destruction mechanism was used to explain the piezoresistivity [62].

2.6 Actuating Performance

Actuators undergo large scale shape or mechanical property changes in response to an appropriate external stimulus and optical-induced actuators compared to the actuators driven by other stimuli, offers an alternative means to couple energy remotely into actuator structures [63]. Polymer graphene composite actuators are used in micro-electro-mechanical system (MEMS)-based micro-grippers [64, 65, 69] photonic switches, robotics, plastic motors and adaptive micro-mirrors due to their wireless/remote actuation, displacement control using infrared intensity modulation, low noise, massively parallel actuation of positioner arrays from a single light source and electrical/mechanical decoupling. GNP photo-actuators with an optical-to-mechanical energy conversion factor of 7 MPa/W report reversible actuation, high actuator stresses, saturable absorption, layer and dimensional state dependent mechanical stiffness, low fabrication costs and ability to integrate with existing commercial electronics [66–78].

Graphene based polymer composite actuators include overall mechanical enhancement of the matrix material and novel characteristics such as salient thermal stability and electrical conductivity derived. In graphene/TPU nanocomposite IR triggered actuator graphene works as photoactive energy transfer and TPU as molecular switch units and possesses enhanced mechanical performance. Sulfonated-graphene/TPU nanocomposite was also reported for remarkable light-triggered actuation [69]. Park et al. prepared a GO/CNT bilayer paper showing a curling actuation behavior depending on humidity and temperature. Rogers and Liu [70] also demonstrated the electrochemical actuation of graphene-based paper, the mechanism being quantum chemical expansion due to electrochemical double-layer charging. Zhu et al. [71] also developed an electromechanically driven microactuator with rapid response based on a graphene-PDA (polydiacetylene) film hybrid bimorph. Upon low direct current, the flexible graphene-PDA actuator showed large and controllable actuation bending motion higher than natural muscles (more than 160 MPa/g under dc of only 0.29 A/mm^2). Many other high-performance actuators were also established out of graphene based composites [72–75]. Polymer graphene composites possess flexibility, light weight, biocompatibility and a large displacement under low potential stimuli and are applicable in robotic actuators, artificial muscle, and dynamic sensors [76].

2.7 Electromagnetic Interference Shielding

Interference among the communication devices and instruments in electronics degrade their performance and the tools to shield or block the electromagnetic signals are very necessary. Metals are usually used in this regard, but since they are expensive, heavy and prone to corrosion, light-weight polymeric composites have much significance. Flexible multilayer graphene/polymer composite films achieved high EMI shielding depending on the optimized thickness, skin depth and electrical conductivity in the shielding materials. The shielding effectiveness up to 27 dB was observed and reflection was proposed to be the dominating shielding mechanism for the as fabricated polymeric graphene films [77].

Conducting polymer composites have tremendous opportunity in EMI shielding applications [78]. Calculation based on electrical conductivity was usually applied to suggest the shielding performance in order to understand the shielding mechanism and the target value required for specific EMI shielding applications is 0.1 S m^{-1}. Hu et al. developed a graphene/ultra high molecular weight polyethylene segregated composite with an electrical conductivity of 0.04 S m^{-1} at 0.62 vol% graphene [79]. The graphene/natural rubber composite synthesized by Zhan et al. possessed an electrical conductivity of 0.03 S m^{-1} at 1.78 vol% of graphene loading [80]. Very recently, Li et al. also prepared TRGO/poly(vinylidene fluoride) segregated composites with ultimate electrical conductivity of 0.02 S m^{-1}, still much lower than the target conductivity [81].

PMMA embedded with chemically reduced GOs showed good EMI performance of 13–19 dB [80]. Graphene-PMMA nanocomposite microcellular foams formed using subcritical CO_2 possessed excellent electrical conductivity, EMI shielding efficiency and mechanical properties. Microcellular foams solve this problem by providing increased toughness, improved fatigue life and energy absorption [82, 83]. The tensile strength, ductility, and fracture toughness of such polymer foams are also investigated since the larger pores in the foams can make them brittle. Polymer composite foams of light weight were reported with high electrical conductivity and high EMI shielding efficiency and are applicable in the areas of aircraft, spacecraft and automobiles [84, 85]. The electrical conductivity and thus EMI shielding efficiency of polymer composites depend mainly on the intrinsic conductivity, aspect ratio and filler concentration.

3 Applications of Graphene Nanocomposites

3.1 Fuel Cell

In this section some major applications of graphene filled polymers such as fuel cell, sensor, microwave absorbers and photovoltaics are discussed. For a fuel cell, graphene acts as a catalyst support for oxygen reduction and methanol oxidation [86, 87] and can also work as both window electrode and counter electrode. Fuel cells provide feasible energy conversion due to their high power density and low greenhouse gas emission [88, 89]. Graphene constitutes low cost alternatives to Nafion membranes in fuel cell technology by offering additional advantages of thermal and mechanical stability and high selective permeability [90]. Wang et al. [91] demonstrated the influence of polymer/graphene composites in organic solar devices. Graphene doped conducting polymers such as poly (3,4-ethylenedioxythiophene), poly (styrene sulphonate) and poly(3-hexylthiophene) have shown better power consumption efficiency of 4.5 % compared to the neat polymer counter electrode (2.3 %) [92]. Functionalized GOs exhibit unique ion-exchange capacity, phosphate adsorption ability, electro-chemical detection of H_2O_2, the onset potential, electron transfer number in oxygen reduction reaction process, electrochemiluminescence activity, good solubility and stability. Such functionalized graphenes was used to synthesize low humidifying polymer electrolyte membrane fuel cells with Nafion using the microwave reduction method. Though the fuel cell exhibited quick start-up and high efficiency, the high cost of Nafion with good chemical resistance and high conductivity often limits its widespread usage [93–95].

The electrical conductivity of hydrophilic pure GO is 5×10^{-3} S cm^{-1} at a bias voltage of 10 V and is an electronic insulator [96, 97]. GO based polymer composites using high tensile strength and flexibile poly(ethylene oxide) (PEO) [98] led to high ionic/protonic conductivity as abundant protons are present in the PEO/GO composite membrane. The high conductivity at low temperatures 20–60 °C resulted in the fabrication of low temperature membrane materials for fuel cells. The comparison of GO filled PVA-based membranes with Nafion revealed interesting properties such as improved conductivity and reduced biofouling on the membranes leading to enhanced power generation and higher power densities in single chambered microbial fuel cells [99–101]. The unique structure and high surface area of the GO provide more proton transport channels and hold more water, which could be beneficial for the improvement of the proton conductivity and mechanical properties of the membranes [102]. It is found that the variation in GO loadings and functionalization of GOs affect the current generation and biofouling of the membranes.

Microbial fuel cells (MFC's) of graphene electrochemically deposited on carbon cloth [103] as well as polyaniline hybridized three-dimensional graphene were also made [104]. The membrane materials in MFCs separate the anolyte from reaching the cathode [105, 106] and at the same time membrane less fuel cell with low cost, simple configuration, and relatively high power density at a small scale have also been found to be promising. However the absence of a membrane increases the oxygen and substrate diffusion and thus decreases the catalytic activity making the MFC less efficient. Moreover the electrode spacing is about 2–4 cm (possibility of short circuit) and the design is also limited in this case [107, 108]. All these facts necessitate the use of membranes in fuel cells and for this, other than reducing the manufacturing costs, the internal resistance of the membrane and biofouling on the membrane [109] should be optimized. Economically viable membrane fuel cells are possible if the manufacturing costs are reduced by simplifying the design [102]. In addition these composite membranes show good performance at high temperature and at low relative humidity conditions.

3.2 Photovoltaics and Photodetectors

Conversion of solar energy to electricity generally takes place by the mechanism of light absorption exciton generation, exciton diffusion, dissociation and charge formation and charge transport and collection. The separation of the photoexcited electron-hole pair is achieved in conjugated polymers by creating a heterojunction with an acceptor material with larger electron affinity than the polymer. It is this heterojunction supplies the driving force to dissociate the tightly bound electron-hole pairs into separate charges. The acceptor material should have a relatively lower HOMO level than the conjugated polymer to transfer an electron to it and to retain the hole in the polymer valence band [110]. The efficiency of the planar heterojunction device depends on the exciton diffusion length [110] and various polymer composites of graphene satisfy the essential criteria for good photovoltaic devices [111–113].

Graphene based materials are used as window electrodes for bulk heterojunction polymer organic photovoltaic cells with ease of processing and flexible substrate compatibility [114]. Direct spin-casting graphene solutions as electrodes make low fabrication cost and simple processing. Electrically conductive graphene electrodes were made by chemical reduction of spin-coated functionalized graphene solution [115]. The obtained graphene film showed increasing red shift of 52 nm in its UV-Vis spectra and the absorption increased significantly, indicating the restoration of electronic conjugation within the graphene sheets [116]. It is found that the chemically reduced graphene films at temperatures of 300, 500 and 700 °C respectively give transmittances of 75, 73 and 69 % at a wavelength of 550 nm. This darkening of the reduced material is due to the partial restoration of the p-electron system in the graphene structure [117] and such films find application for transparent conductors.

A thin film of GO and rGO is used for fabricating photovoltaics with the device performance depending on the thickness of the insulating GO [118–121]. The very low levels of light absorption (at around 2.7 % of white light) along with high electron mobility of graphene offer itself an alternative to expensive silicon or ITO in the manufacture of photovoltaic cells. The advantage of graphene lies in the fact that during photon absorption it generates multiple electrons thereby reducing the heat energy loss observed for conventional silicon devices. Moreover graphene is able to work on all wavelengths and is flexible and thin; making it applicable in photovoltaic cells in clothing, retro-fitted photovoltaic window screens or curtains etc.

In a bulk heterojunction solar cell, the basic requirements include strong light absorption over the whole solar spectrum, appropriate energetic distance between HOMO and LUMO of the donor/acceptor for a high open-circuit voltage, and high and balanced electron and hole mobilities in the active layer. Hsu and his coworkers also reported a layer-by-layer molecular doping process on graphene for forming sandwiched graphene/tetra cyano quino dimethane stacked films for polymer solar cells [122]. Such a cell is made of solution processable conjugated p-type semiconducting and environmentally stable material poly (3-hexylthiophene) with extended absorption in the red region up to 650 nm, high hole mobility in regioregular state and high degree of crystallinity blended with a soluble fullerene derivative. The solar cell performance depends on the intrinsic electronic characteristics of the individual materials, processing conditions, morphological issues and device structure [123]. Other structural characteristics like molecular weight, polydispersity, region regularity and the purity of the polymer also affect solar cell performances.

3.3 Microwave Absorbers

Microwave absorptive fabrics avoid pollution of EM radiation and widely applied in military field [124]. Microwave absorbing materials absorb electromagnetic waves and dissipate it by converting to other forms of energy [125]. Generally dielectric loss, magnetic loss and impedance matching characteristics affect the electromagnetic absorption property [126, 127] and such dielectric or magnetic materials are often incorporated [128, 129]. Because of the non magnetic nature of graphene, magnetic particles are doped with it to make it good microwave absorber [130]. Liu et al. [131] found the absorption bandwidth of FeCo alloy particles/graphite nanoflake composites of thickness <2 mm as 5 GHz. The magnetic particles have the advantage of controllable size and shape [132, 133] and Fe_3O_4 particles enhance the electromagnetic absorption property of RGO to great extent [134–136]. The semiconducting Co_3O_4 filled composites are also used as microwave absorbing materials [137]. A successful combination of RGO and Co_3O_4 particles for microwave absorption with a narrow absorption bandwidth exceeding −10 dB was practiced by Liu and his coworkers.

Polymer based absorbers are flexible and light weight compared to metals but the too narrow bandwidths cause awful absorption effects. For instance carbon fiber fabric impregnated with conductive polymer and the fabric coated with barium ferrite doped composite exhibit a bandwidth of only 4 and 6–14 GHz respectively. Researchers synthesized EG/novolac phenolic resin composite and an attenuation of 13 dB were obtained in the range of 8–12 GHz for 3.7 mm thickness. It is also observed that the bandwidth of flake graphite prepared by milling reach 6 GHz and for graphite nanoplatelet after acid treatment 14–18 GHz for a sample thickness of >2 mm. Another composite expanded graphite/polyaniline/$CoFe_2O_4$ [138] reports a maximum reflection loss of −19.13 dB with a thickness of 0.5 mm. The shielding effectiveness of a hybrid polyaniline/graphene/Fe_3O_4 was found to be 26 dB [139]. The graphite nanosheets/PU nanocomposites (30 wt%) coated vinylon fabrics with thinner coatings exhibited good shielding effectiveness with a wet coating thickness of 0.39 mm and dry coating areal density of 130 g/m^2. An electromagnetic attenuation mechanism is attributed to the better impedance matching and the maximum reflection loss of the polymer—RGO–Co_3O_4 was − 51.1 dB at 10.7 GHz and the bandwidth below −10 dB was 3.1 GHz at a thickness of 2 mm.

3.4 Sensors

The superior electrical conductivity of graphene is the major factor contributing towards its use in sensing technology. A sensor detects the variations in physical property and converts the changes to measurable signal responses (like resistance, capacitance etc.). An evenly distributed graphene nano ripple was created to fabricate highly sensitive sensors by applying graphene strips to pre-strained polymer substrate [140]. Here the sensor works based on the increase in the sheet resistance of graphene due to the electron scattering that occurs at the ripples when the ripples were disturbed by physical change. Monolayer graphene finds excellent use in sensing gases and biomolecules [141, 142] where the charge transfer between the adsorbed molecule and graphene regulate the chemical response. While Schedin et al. demonstrated sensing of graphene towards NO_2, NH_3, H_2O and CO [141] Fowler et al. reported graphene sensitivity to dinitrotoluene as well [143]. By electrodeposition of Pd nanoparticles on graphene surface Sundram et al. fabricated H_2 sensors [144]. At the same time gold nanocomposite of graphene nanosheets was utilized for electrochemical sensing by Hu et al. [145]. High performance NO_2 gas sensors were reported from few-layer graphene [146] and reduced graphene oxide (rGO)-conjugated Cu_2O nanowire mesocrystals [147]. Graphene derivatives are used as a super quencher for organic dyes and quantum dots and also in measuring nanoscale distance and change [148]. Graphene carbon nanotube hybrid filler combination was used by our group to synthesize a natural rubber solvent sensor [149]. Since graphene possess excellent capability in fluorescent resonance energy transfer (transfer of energy from a donor fluorophore to an acceptor fluorophore), it is applied in fluorescent detection also.

One of the important sensing properties of graphene is attributed to its capability to detect biomolecules. It was Lu et al. [148] who first reported a graphene-based biosensor consisting of a dye-labeled as DNA probe that could be bound and quenched by GO, resulting from the fluorescent energy transfer between the dye and GO. In addition the sensitive charge carrier modulation of chemically modified graphene has allowed the development of biodevices that can detect a single bacterium/sense DNA. Jang et al. [150] developed a novel GO-related assay to investigate the helicase-mediated duplex DNA unwinding activity. The molecular beacons are elaborately designed DNA hairpin structures with more sequence specificity compared to linear probes, and have been widely used in genetic screening biosensors and biochips, detection of single nucleotide polymorphism and mRNA monitoring in living cells [151]. The GO quenched beacons detect DNA with higher sensitivity and single-base mismatch selectivity [152–155]. Graphene fluorescent bio optical sensors to recognize DNA are widely reported [156, 157]. Several other studies reveal the use of graphene and its derivatives in cellular target monitoring (used for DNA and protein analysis and intercellular tracking) [158], protein micropatterning (glucose sensors and cell sensors) [159] and in identifying amino acids, oligonucleotides, dopamine, adenosine triphosphate (ATP) thrombin etc. [160–162]. Nickel and silver nanoparticles coupled with rGO was used as non-enzymatic glucose sensor and amperometric immunoassay for thrombomodulin respectively by Wang et al. [163] and Yang et al. [164]. The graphite oxide and graphene modify electrodes in electrochemical detection of hydroquinone and ascorbic acid and also works as uric acid sensor (graphene doped chitosan as functional matrix and uric acid as template) [165, 166]. Graphene based drug delivery has combined with other techniques for cancer therapy as well.

The recent discovery of graphene's ability to facilitate the differentiation of stem cells without interference of growth or alteration of the growth environment of the cells [167] have instigated more interest in graphene in the biomedical community. Graphene can be fast and efficient bioelectric sensory devices, with the ability to monitor glucose levels, haemoglobin levels, cholesterol and even DNA sequencing. Finally we can observe in the recent future that the applicability of 'toxic' graphene in preparing antibiotics or even anticancer treatment. Also, due to its molecular make-up and potential biocompatibility, it could be utilised in the process of tissue regeneration. Though graphene derivatives are widely used in biological applications we still need to understand its biocompatibility.

In short the graphene-based polymer nanocomposites are applied in the fields of transparent flexible electrodes, mechanical parts, energy storage, memory storage, sensors and organic electronics. The charge mobility of RGO is considerably higher than that of the amorphous silicon and existing semi-conducting conjugated polymers, enlightening the probable path for the application in electronics. Better understanding of physics and chemistry at the surface of graphene and interaction of chemicals and biomolecules at the interface of graphene will play an important role in applying graphene as nanoscaffold in catalysis, chemical/biosensing, imaging and drug delivery [168, 169].

3.5 Ferroelectrics

Graphene based ferroelectric systems have the unique ability of ferroelectric polarization and form distinct Schottky junctions to win over conventional semiconductors. Such materials have potential applications in old non-volatile memories, transparent flexible electrodes, solar cells, photodetectors, etc. Jie et al. narrated the graphene integrated ferroelectric lead zinc niobate (PZT), lead magnesium niobate-lead titanate (PMN-PT) and organic PVDF as promising materials for future electronics and optics. Since the hybrid systems of graphene-functional materials are useful components for widespread potential applications, heterostructures are developed based on atomically thin graphene films. This facilitates the development of ultrathin and flexible devices for future nanoelectronic and nanophotonic applications [170]. A flexible graphene-based nonvolatile memory device was made using PZT ferroelectric material by Lee et al. During the fabrication step the graphene and PZT ferroelectric layers were respectively deposited using CVD and sol–gel methods. The graphene–PZT memory device on a plastic substrate displayed an on/off current ratio of 6.7, a memory window of 6 V and reliable operation [171]. However a comprehensive understanding of the nonlinear, hysteretic ferroelectric gating and its effective control are still needs to be investigated.

Ni et al. applied nonvolatile ferroelectric polymer gating in engineering large-scale CVD monolayer graphene having low sheet resistance. The graphene ferroelectric transparent conductors (GFeTCs) exhibited more than 95 % transmittance from the visible to the near-infrared range owing to the highly transparent nature of the ferroelectric polymer. The ferroelectric polymer acts as an excellent mechanical supporting layer, making GFeTCs easily transferable and adaptable with flexible electronics, optoelectronics, and photonics platforms [172]. It is really important to know about the graphene ferroelectric interface to effectively manufacture advanced ferroelectrics and this was addressed by Yusuf et al. They have observed that the intrinsic charge trap from the ferroelectric superlattice $PbTiO_3/SrTiO_3$ surface defects adversely affect the graphene channel hysteresis and can be controlled by sample processing. The authors have derived a charge trapping mechanism for the graphene-superlattice system from which the reduction of interfacial adsorbates has come out by the absence of slow charge trapping processes. The extent of this intrinsic charge trapping on the graphene channel as well as the nature of the channel resistance-voltage response is identified from the growth conditions or surface morphology [173]. In another work, a prototype of graphene/(Ba, Sr)TiO_3 hybrid thin film field effect sensor was developed by Rajapitamahuni et al. They analyzed the dielectric constant, pyroelectric coefficient, and ferroelectric polarization of the 100–300 nm thin films and observed that the interface chemistry regulates the device performance. Moreover the ferroelectric hysteresis in resistance and carrier density below 100 K cause an antihysteresis behavior activated by the combined effects of electric field and temperature [174]. Finally graphene based metal oxides have many applications as excellent ferroelectrics and its study needs to be further investigated for future electronics.

4 Conclusion

In summary, graphene and polymer graphene composite combines several functionalities into a single material and thus applicable in a wide variety of technological devices. However certain points restrict its applications in electronics and these technical challenges are summarized below.

1. Though CVD makes bulk manufacture of graphene layers possible, the cost and quality needs to be normalized. The implementation of graphene based electronics and consumer products mainly lies in the cost effective fabrication process and reduction in damages during etching and control in graphene's electrical properties.
2. Though the exceptional properties of graphene replace silicon in electronics its inherent semi-metal characteristic has to be modified to realize the applications in transistors. However, the chemical and thermal modification lowers its electrical and mechanical properties, along with its chemical reactivity.
3. The health risk associated with graphene and their derivatives needs to be evaluated through the investigation of the toxicity and biocompatibility.
4. In order to fabricate the composite, homogeneous distribution of individual graphene platelets, their orientation, connectivity, and interface bonding with matrix still require more investigation.

5. More research is needed to develop a simple detection method of pristine graphene that is independent of support material. Although the advanced deposition technique of single layer and bilayer made it possible to fabricate large area devices, creating band gap in a controlled and practical manner is still challenging for the application in logic devices. Several methods aiming at tuning the substrates properties and nanoribbon dimension have been proposed to introduce tunable band gap essential for nanoelectronics.
6. Cutting graphene into nanoribbon has shown great promise for FET logic applications, but associated with the electron scattering at the rough edges and disorder from the back substrate. On the other hand, FETs based on graphene bilayer still require large on/off ratio for logic applications. However, the electronic band structure is not clearly understood due to the nanoscale inhomogeneity.
7. The several challenges existing in each application field must be solved. For example, taking electrochemical sensing into consideration, there is an urgent need in this area to fabricate reliable, reproducible, and low-cost sensors with high detection sensitivity using well-defined graphene.

References

1. Sadasivuni K K, Ponnamma D, Thomas S, Grohens Y (2014) Evolution from graphite to graphene elastomer composites. Progress in Polymer Science 39: 749–780.
2. Lu C C, Lin Y C, Yeh C H, Huang J C, Chiu P W (2012) High mobility flexible graphene field-effect transistors with self-healing gate dielectrics. ACS Nano 6: 4469-4474.
3. Szafranek B N, Fiori G, Schall D, Neumaier D, Kurz H (2012) Current saturation and voltage gain in bilayer graphene field effect transistors. Nano Letters12: 1324-1328.
4. Banks C E, Crossley A, Salter C, Wilkins S J, Compton R G (2006) Carbon nanotubes contain metal impurities which are responsible for the "Electrolysis" seen at some nanotubesmodified electrodes. Angewandte Chemie International Edition 45: 2533-2537.
5. Liu Z, Liu Q, Huang Y, Ma Y, Yin S, Zhang X, Sun W, Chen Y (2008) Organic photovoltaic devices based on a novel acceptor material: Graphene. Advanced Materials 20: 3924-3930.
6. Bolotin K I, Sikes K J, Jiang Z, Klima M, Gudenberg G, Hone J, Kim P, Stormer H L (2008) Ultrahigh electron mobility in suspended grapheme. Solid State Communication 146: 351–355.
7. Zhang Y, Tan Y W, Stormer H L, Kim P (2005) Experimental observation of the quantum Hall effect and Berry's phase in grapheme. Nature 438: 201–204.
8. Obradovic B, Kotlyar R, Heinz F, Matagne P, Rakshit T, Giles M D, Stettler M A (2006) Analysis of graphene nanoribbons as a channel material for field-effect transistors, Applied Physics Letters 88: 142102-142104.
9. Geim A K, Novoselov K S (2007) The Rise of Graphene. Nature Mater 6: 183-191.
10. Barber P, Balasubramanian S, Anguchamy Y, Gong S, Wibowo A, Gao H, Ploehn H J, Zur Loye H C (2009) Polymer Composite and Nanocomposite Dielectric Materials for Pulse Power Energy Storage. Materials 2: 1697–1733.
11. Kim J Y, Park S H, Jeong T, Bae M J, Kim Y C, Han I, Yu S (2012) High electroluminescence of the ZnS:Mn nanoparticle/cyanoethyl-resinpolymer/single-walled carbon nanotube composite using the tandem structure. Journal of Materials Chemistry 22: 20158–20162.
12. Yang S Y, Shin K, Park C E (2005) The Effect of Gate-Dielectric Surface Energy on Pentacene Morphology and Organic Field-Effect Transistor Characteristics. Advanced Functional Materials 15: 1806–1814.

13. Ponnamma D, Sadasivuni K K, Strankowski M, Moldenaers P, Thomas S, Grohens Y (2013) Interrelated shape memory and Payne effect in polyurethane/graphene oxide nanocomposites. RSC Advances 3: 16068-16079.
14. Potts J R, Dreyer D R, Bielawski C W, Ruoff R S (2011) Graphene-based polymer nanocomposites. Polymer 52 : 5 – 25, doi:10.1016/j.polymer.2010.11.042.
15. Sun Y, Shi G (2013) Graphene/Polymer Composites for Energy Applications. Journal of polymer science, Part B: Polymer Physics 51: 231–253.
16. Das T K, Prusty S (2013) Graphene-Based Polymer Composites and Their Applications. Polymer-Plastics Technology and Engineering 52: 319–331.
17. Ramanathan T, Abdala A A, Stankovich S, Dikin D A, Herrera-Alonso M, Piner R D, Adamson D H, Schniepp H C, Chen X, Ruoff R S, Nguyen S T, Aksay I A, Prud'homme R K, Brinson L C (2008) Functionalized graphene sheets for polymer nanocomposites. Nature Nanotechnology 3: 327–331.
18. Bolotin K I, Sikes K J, Jiang Z, Klima M, Fudenberg G, Hone J, Kim P, Stormer H L(2008) Ultrahigh electron mobility in suspended graphene. Solid State Communications 146: 351– 355.
19. Cai W, Moore A L, Zhu Y, Li X, Chen S, Shi L, Ruoff R S (2010) Thermal Transport in Suspended and Supported Monolayer Graphene Grown by Chemical Vapor Deposition. Nano Letters 10: 1645–1651.
20. Faugeras C, Faugeras B, Orlita M, Potemski M, Nair R R, Geim A K (2010) Thermal Conductivity of Graphene in Corbino Membrane Geometry. ACS Nano 4: 1889–1892.
21. Lindsay L, Broido D A, Mingo N (2010) Flexural phonons and thermal transport in grapheme. Physical Review B 82: 115427:1–6.
22. Hooker C N, Ubbelohde A R, Young D A (1965) Anisotropy of Thermal Conductance in Near-Ideal Graphite. Proceedings of the Royal Society London A 284: 17–31.
23. Yu A, Ramesh P, Itkis M E, Bekyarova E, Haddon R C (2007) Graphite Nanoplatelet-Epoxy Composite Thermal Interface Materials. Journal of Physical Chemistry C 111: 7565–7569.
24. Yu A, Ramesh P, Sun X, Bekyarova E, Itkis M E, Haddon R C (2008) Enhanced Thermal Conductivity in a Hybrid Graphite Nanoplatelet – Carbon Nanotube Filler for Epoxy Composites. Advanced Materials 20: 4740–4744.
25. Ganguli S, Roy A K, Anderson D P (2008) Improved thermal conductivity for chemically functionalized exfoliated graphite/epoxy composites. Carbon 46: 806–817.
26. Zhao X., Zhang Q, Chen D, Lu P (2010) Enhanced mechanical properties of graphene-based poly(vinyl alcohol)composites. Macromolecules, 43, 2357–2363.
27. Wu Q, Xu Y, Yao Z, Liu A, Shi G (2010) Supercapacitors based on flexible graphene/polyaniline nanofiber composite films. Acs Nano 4: 1963-1970.
28. Wang Y, Shi Z, Huang Y, Ma Y, Wang C, Chen M, Chen Y (2009) Supercapacitor devices based on graphene materials. The Journal of Physical Chemistry C113: 13103-13107.
29. Wang D W, Li F, Zhao J, Ren W, Chen Z G, Tan J, Wu S, Gentle I, Lu G Q, Cheng H M (2009) Fabrication of graphene/polyaniline composite paper via in situ anodic electropolymerization for high-performance flexible electrode. ACS Nano 3: 1745 – 1752.
30. Wu Q, Xu Y, Yao Z, Liu A, Shi G (2010) Supercapacitors based on flexible graphene/polyaniline nanofiber composite films. ACS Nano 4:1963 – 1970.
31. Yan J, Wei T, Shao B, Fan Z, Qian W, Zhang M, Wei F (2010) Preparation of a graphene nanosheet/polyaniline composite with high specific capacitance .Carbon 48: 487 – 493.
32. Guo W, Yin Y X, Xin S, Guo Y G, Wan L (2012) Superior radical polymer cathode material with a two-electron process redox reaction promoted by grapheme. Journal of Energy and Environmental Science 5: 5221 – 5225.
33. Wang Y, Shi Z, Huang Y, Ma Y, Wang C, Chen M, Chen Y(2009) Super-capacitor devices based on graphene materials. J Phys Chem C 113: 13103 – 13107.
34. Stoller M D, Park S, Yanwu Z, An J, Ruoff R S(2008) Graphene-based ultracapacitors. Nano Letters 8: 3498 – 3502.
35. Yoo J J, Balakrishnan K, Huang J, Meunier V, Sumpter B G, Srivastava A, Conway M, Reddy A L M, Yu J, Vajtai R, Ajayan P M(2011) Ultrathin planar graphene supercapacitors. Nano Letters 11: 1423 – 1427.

36. Zhang Y P, Li H B, Pan L K, Lu T, Sun Z (2009) Capacitive behavior of graphene-ZnO composite film for supercapacitors. J. Electroanal. Chem. 634: 68-71.
37. Li F H, Song J F, Yang H F, Gan S Y, Zhang Q X, Han D X, Ivaska A, Niu L (2009) One step synthesis of Graphene/ SnO2 nanocomposite and its application in electrochemical super capacitors. Nanotechnology 20 : 455602-1 – 455602-6.
38. Chen S, Zhu J, Wu X, Han Q, Wang X (2010) Graphene oxide-MnO2 nanocomposites for supercapacitors. ACS Nano 4: 2822-2830.
39. Paek S M, Yoo E J, Honma I (2009) Enhanced cyclic performance and lithium storage capacity of SnO/graphene nanoporous electrodes with three-diemnsionally delaminated flexible structure. Nano Letters 9: 72-75.
40. Wang D, Choi D, Li J, Yang Z, Nie Z, Kou R, Hu D, Wang C, Saraf L V, Zhang J, Aksay I A, Liu J (2009) Self-assembled TiO2- graphene hybrid nanostructures for enhanced Li-ion insertion, ACS Nano 3: 907-914.
41. Xie J, Song W, Zheng Y, Liu S, Zhu T, Cao G, Zhao X (2011) Preparation and Li-storage properties of SnSb/graphene hybrid nanostructure by a facile one-step solvothermal route. International Journal of Smart and Nano Materials 2: 261-271.
42. Xiao J, Mei D, Li X, Xu W, Wang D, Graff G L, Bennett W D, Nie Z, Saraf L V, Aksay I A, Liu J, Zhang J G (2011) Hierarchically porous graphene as a lithium-air battery electrode. Nano Letters 11: 5071-5078.
43. Nan W, Shen Y, Ma J (2010) Physical Properties of Composites Near Percolation. Annual Review of Materials Research 40: 131–151.
44. Kim J Y, Lee W H, Suk J W, Potts J R, Chou H, Kholmanov I N, Piner R D, Lee J, A kinwande D, Ruoff R S(2013) Chlorination of Reduced Graphene Oxide Enhances the Dielectric Constant of Reduced Graphene Oxide/Polymer Composites. Advanced Materials 25: 2308–2313.
45. Sadasivuni K K, Saiter A, Gautier N, Thomas S, Grohens Y (2013) Effect of molecular interactions on the performance of poly (isobutylene-co-isoprene)/graphene and clay nanocomposites. Colloid Polym Sci 291:1729–1740.
46. Dang Z M, Zhang Y H, Tjong S C (2004) Dependence of dielectric behavior on the physical property of fillers in the polymer-matrix composites. Synthetic Metals 146: 79–84.
47. K K, Ponnamma D, Kumar B, Strankowskie M, Cardinaels R, Moldenaers P, ThomasS, Grohens Y (2014) Dielectric properties of modified graphene oxide filled polyurethane nanocomposites and its correlation with rheology. Composite science and technology doi: 10.1016/j.compscitech.2014.08.025.
48. Kim S, Nah J, Jo I, Shahrjerdi D, Colombo L, Yao Z, Tutuc E,Banerjee S K (2009) Realization of a high mobility dual-gated graphene field-effect transistor with Al_2O_3 dielectric. Applied Physics Letters 94 : 062107 :1-3.
49. Lu X, Huang H, Nemchuk N, Ruoff R S (1999) Patterning of highly oriented pyrolytic graphite by oxygen plasma etching. Applied Physics Letters 75: 193-195.
50. Huitema H E A, Gelinck G H, Putten J B P H V, Kuijk K E, Hart C M, Cantatore E, Herwig P T, Breemen A J J M V, Leeuw M L (2001) Plastic transistors in active-matrix displays. Nature 414 :599.
51. Klauk H, Zschieschang U, Pflaum J, Halik M (2007) Ultralow-power organic complementary circuits. Nature 445: 745-748.
52. Jin S H, Kim J W, Lee C A, Park B -G, Lee J D (2004) Surface-State Modification of OTFT Gate Insulators by Using a Dilute PMMA Solution. J. Korean Phys. Soc. 44:185-189.
53. Ihm K, Lim J T, Lee K J, Kwon J W, Kang T H, Chung S, Bae S, Kim J H, Hong B H, Yeom G Y(2010) Number of graphene layers as a modulator of the open-circuit voltage of graphene-based solar cell. Applied Physics Letters 97: 0321133.
54. Kim B J, Jang H, Lee S K, Hong B H, Ahn J H, Cho J H(2010) High-per-formance flexible graphene field effect transistors with ion gel gate dielectrics. Nano Letters 10: 3464 – 3466.
55. Wu J, Agrawal M, Becerril HA, Bao Z, Liu Z, Chen Y, Peumans P (2010) Organic light-emitting diodes on solution-processed graphene transparent electrodes. ACS Nano 4: 43-48.

56. Chang H, Wang G, Yang A, Tao X, Liu X, Shen Y, Zheng Z(2010) A transparent, flexible, low-temperature, and solution-processible graphene composite electrode. Advanced Functional Materials 20: 2893–2902.
57. Wang Y, Tong S W, Xu X F, Ozyilmaz B, Loh K P (2011) Interface engineering of layer-by-layer stacked graphene anodes for high-performance organic solar cells. Advanced Materials 23: 1514-1518.
58. Bae S, Kim H, Lee Y, Xu X, Park JS, Zheng Y, Balakrishnan J, Lei T, Ri Kim H, Song YI, Kim YJ, Kim KS, Ozyilmaz B, Ahn JH, Hong BH, Iijima S (2010) Roll-to-roll production of 30-inch gra-phene films for transparent electrodes. Nature Nanotechnol 5: 574-578.
59. Kim K S, Zhao Y, Jang H, Lee S Y, Kim J M, Kim K S, Ahn J H, Kim P, Choi J Y, Hong B H (2009) Large-scale pattern groth of graphene films for stretchable transparent electrodes. Nature 457: 706-710.
60. Kang J, Hwang S, Kim J H, Kim M H, Ryu J, Seo S J, Hong B H, Kim M K, Choi J B (2012) Efficient Transfer of Large-Area Graphene Films onto Rigid Substrates by Hot Pressing. ACS Nano 6 : 5360-5365.
61. Lovinger A J (1981) Crystallization of the β phase of poly(vinylidene fluoride) from the melt. Polymer 22: 412-413.
62. Sadasivuni K K, Castro M, Saiter A, Delbreilh L, Feller J F, Thomas S, Grohens Y Development of poly(isobutylene-co-isoprene)/reduced graphene oxide Nanocomposites for barrier, dielectric and sensing applications. Materials Letters 96: 109–112.
63. Thompson B C, Fréchet J M J (2008) Polymer-Fullerene Composite Solar Cells. Angew. Chem. Int. Ed. 47: 58-77.
64. Lu S X, Panchapakesan B (2006) Nanotube micro-optomechanical actuators. Appl Phys Lett. 88: 253107.
65. Lu S X, Liu Y, Shao N, Panchapakesan B (2007) Nanotube micro-opto-mechanical systems. Nanotechnology 18: 065501.
66. Loomis J, King B, Burkhead T, Xu P, Bessler N, Terentjev E, Panchapakesan B (2012) Graphene-nanoplatelet-based photomechanical actuators. Nanotechnology 23: 045501.
67. Loomis J, King B, Panchapakesan B (2012) Layer dependent mechanical responsesof graphene composites to near-infrared light. Appl Phys Lett. 100: 072108.
68. Loomis J, Panchapakesan B (2012) Dimensional dependence of photomechanicalresponse in carbon nanostructure composites: a case for carbon-based mixeddimensionalsystems. Nanotechnology 23: 215501.
69. Stankovich S, Piner R D, Nguyen S T, Ruoff R S (2006) Synthesis and exfoliation of isocyanate-treated graphene oxide nanoplatelets. Carbon 44: 3342-3347.
70. Rogers G W, Liu J Z (2011) Graphene actuators: quantum-mechanical and electrostatic double-layer effects. Journal of the American Chemical Society 133: 10858 – 10863.
71. Zhu S E, Shabani R, Rho J, Kim Y, Hong B H, Ahn J H, Cho H J (2011). Graphene-based bimorph microactuators. Nano Letters 11: 977 – 981.
72. Shin K Y, Hong J Y, Jang J S (2011) Flexible and transparent graphene films as acoustic actuator electrodes using inkjet printing. Chem. Commun. 47: 8527–8529.
73. Xie X J, Qu L T, Zhou C, Li Y, Zhu J, Bai H, Shi G Q, Dai L M (2010) An Asymmetrically Surface-Modified Graphene Film Electrochemical Actuator. ACS Nano 4: 6050–6054.
74. Zhu S E, Shabani R, Rho J, Kim Y, Hong B H, Ahn J H, Cho H J (2011) Graphene-Based Bimorph Microactuators. NanoLett. 11: 977–981.
75. Liang J J, Huang Y, Oh J Y, Kozlov M, Sui D, Fang S L, Baughman R H, Ma Y F, Chen Y S (2011) Electromechanical Actuators Based on Graphene and Graphene/Fe_3O_4 Hybrid Paper. Adv. Funct. Mater. 21: 3778–3784.
76. Shahinpoor M (2003) Ionic polymer–conductor composites as biomimetic sensors, robotic actuators and artificial muscles—a review. Electrochim. Acta 48: 2343-2353
77. Song W L, Cao M S, Lu M M, Bi S, Wang C Y, Liu J, Yuan J, Fan L Z(2014) Flexible graphene/polymer composite films in sandwich structures for effective electromagnetic interference shielding. Carbon 66: 6 7 –7 6.

78. Gelves G A, Al-Saleh M H, Sundararaj U (2011) Highly electrically conductive and high performance EMI shielding nanowire/polymer nanocomposites by miscible mixing and precipitation. J. Mater. Chem. 21: 829–836.
79. Liang Q, Hsie S A, Wong C P (2012) Low temperature solid state microwave reduction of graphene oxide for transparent electrically conductive coatings on flexible polydimethylsiloxane (PDMS). ChemPhysChem 13: 3700–3706.
80. Thomassin J M, Jérôme C, Pardoen T, Bailly C, Huynen I, Detrembleur C (2013) Polymer/carbon based composites as electromagnetic interference (EMI) shielding materials. Mater. Sci. Eng. R 74: 211–232.
81. Kuilla T, Bhadra S, Yao D, Kim N H, Bose S, Lee J H (2010) Recent advances in graphene based polymer composites. Prog. Polym. Sci. 35, 1350–1375.
82. Li D, Muller M B, Gilje S, Kaner R B, Wallace G G (2008) Processable aqueous dispersions of graphene nanosheets. Nat. Nanotechnol. 3: 101–105.
83. Lee L J, Zeng C C, Cao X, Han X M, Shen J, Xu G J (2005) Polymer nanocomposite foams. Compos. Sci. Technol. 65: 2344–2363.
84. Zeng C C, Han X M, Lee L J, Koelling K W, Tomasko D L (2003) Polymer-clay nanocomposites foams prepared using carbon dioxide. Adv. Mater. 15: 1743–1747.
85. Verdejo R, Barroso-Bujans F, Rodriguez-Perez M A, deSaja J A, Lopez-Manchado M A (2008) Functionalized graphene sheet filled silicone foam nanocomposites. J. Mater. Chem. 18, 2221–2226.
86. Xu C, Wang X, Zhu J (2008) Graphene metal particle nanocomposites. The Journal of Physical Chemistry C 112: 19841-19845.
87. Li Y, Tang L, Li J (2009) Preparation and electrochemical performance for methanol oxidation of Pt/graphene nanocomposites. Electrochemistry Communications 11: 846-849. 107.
88. Pant D, Singh A, Bogaert G V, Gallego Y A, Diels L, Vanbroekhoven K (2011) An introduction to the life cycle assessment (LCA) of bioelectrochemical systems (BES) for sustainable energy and product generation: Relevance and key aspects Renew. Sust. Energ. Rev. 15: 1305–1313.
89. Sridhar P, Perumal R, Rajalakshmi N, Raja M, Dhathathreyan K S (2001) Humidification studies on polymer electrolyte membrane fuel cell. J. Power Sources 101: 72–78.
90. Kim J R, Cheng S, Oh S –E, Logan B E (2007) Power Generation Using Different Cation, Anion, and Ultrafiltration Membranes in Microbial Fuel Cells. Environ. Sci. Technol. 41: 1004 – 1009.
91. Wang J, Wang Y, He D, Wu H, Wang H, Zhou P, Fu M (2012) Influence of Polymer/Fullerene-Graphene Structure on Organic Polymer Solar Devices. Integrated Ferroelectrics: An International Journal. 137: 1-9.
92. Hong W, Xu Y, Lu G, Li C, Shi G (2008) Transparent graphene/ PEDOT-PSS composite films as counter electrodes of dye-sensitized solar cells. Electrochemistry Communications 10: 1555-1558.
93. Higashihara T, Matsumoto K, Ueda M (2009) Sulfonated aromatic hydrocarbon polymers as proton exchange membranes for fuel cells. Polymer 50: 5341–5357.
94. Sen U, Bozkurt A, Ata A (2010) Nafion/poly(1-vinyl-1,2,4-triazole) blends as proton conducting membranes for polymer electrolyte membrane fuel cells. J. Power Sources 195: 7720–7726.
95. Sopian K, Daud W R W (2006) Challenges and future developments in proton exchange membrane fuel cells. Renewable Energy 31: 719–727.
96. Dreyer D R, Park S, Bielawski C W, Ruoff R S (2010) The chemistry of graphene oxide. Chem. Soc. Rev. 39: 228–240.
97. Manoratne C H, Rajapakse R M G, Dissanayake M A K L (2006) Ionic Conductivity of Poly(ethylene oxide) (PEO)-Montmorillonite (MMT) Nanocomposites Prepared by Intercalation from Aqueous Medium. Int. J. Electrochem. Sci. 1: 32–46.
98. Wei B, Tokash J C, Chen G, Hickner M A, Logan B E (2012) Development and Evaluation of Carbon and Binder Loading in Low-Cost Activated Carbon Cathodes for Air-Cathode Microbial Fuel Cells. RSC Adv. 2: 12751 – 12758.

99. Tseng C –Y, Ye Y –S, Cheng M –Y, Kao K –Y, Shen W –C, Rick J, Chen J –C, Hwang B –J (2011) Sulfonated Polyimide Proton Exchange Membranes with Graphene Oxide Show Improved Proton Conductivity, Methanol Crossover Impedance, and Mechanical Properties. Adv. Energy Mater. 1: 1220 – 1224.
100. Liu S, Zeng T H, Hofmann M, Burcombe E, Wei J, Jiang R, Kong J, Chen Y (2011) Antibacterial Activity of Graphite, Graphite Oxide, Graphene Oxide, and Reduced Graphene Oxide: Membrane and Oxidative Stress. ACS Nano 5: 6971 – 6980.
101. Vielstich W, Lamm A, Gasteiger H (2003) Handbook of Fuel Cells: Advances in Electrocatalysis, Materials, Diagnostics and Durability, Wiley, New York.
102. Mayer A C, Scully S R, Hardin B E, Rower M W, McGehee M D (2007) Polymer-based solar cells. Mater. Today 10: 28-33.
103. Liu J, Qiao Y, Guo C X, Lim S, Song H, Li C M (2012) Graphene/carbon cloth anode for high-performance mediatorless microbial fuel cells. Bioresour Technology 114: 275-280.
104. Yong Y C, Dong X C, Chan-Park M B, Song H, Chen P (20120 Macroporous and monolithic anode based on polyaniline hybridized three-dimensional graphene for high performance microbial fuel cells. ACS Nano 6: 2394-2400.
105. Kalathil S, Lee J, Cho M H (2013) Gold Nanoparticles Produced in Situ Mediate Bioelectricity and Hydrogen Production in a Microbial Fuel Cell by Quantized Capacitance Charging. ChemSusChem 6: 246 – 250.
106. Oh S T, Kim J R, Premier G C, Lee T H, Kim C, Sloan W T (2010) Sustainable Wastewater Treatment: How Might Microbial Fuel Cells Contribute. Biotechnol. Adv. 28: 871 – 881.
107. Ayyaru S, Letchoumanane P, Dharmalingam S, Stanislaus A R (2012) Performance of Sulfonated Polystyrene – Ethylene – Butylene – Polystyrene Membrane in Microbial Fuel Cell for bioelectricity production. J. Power Sources 217: 204-208.
108. Li W –W, Sheng G –P, Liu X –W, Yu H –Q (2011) Recent Advancesin the Separators for Microbial Fuel Cells. Bioresour. Technol. 102: 244 – 252.
109. Cropper M A, Geiger S, Jollie D (2004) Fuel Cells: A Survey of Current Developments. J. Power Sources 131: 57 – 61.
110. Tang C W (1986) Two-layer organic photovoltaic cell. Appl. Phys. Lett. 48: 183.
111. Alam M M, Jenekhe S A (2004) Efficient Solar Cells from Layered Nanostructures of Donor and Acceptor Conjugated Polymers. Chem. Mater. 16: 4647-4656.
112. Kietzke T, Horhold H H, Neher D (2005) Efficient Polymer Solar Cells Based on M3EH – PPV. Chem. Mater. 17: 6532-6537.
113. Dittmer J J, Lazzaroni R, Leclere P, Moretti P, Granstrom M, Petritsch K, Marseglia E A, Friend R H, Bredas J L, Rost H, Holmes A B (2000) Crystal network formation in organic solar cells. Sol. Energy Mater. Sol. Cells 61: 53-61.
114. Eda G, Lin Y Y, Miller S, Chen C W, Su W F, Chhowalla M (2008) Transparent and conducting electrodes for organic electronics from reduced graphene oxide. Appl. Phys. Lett. 92: 233305.
115. Wang X, Zhi L J, Tsao N, Tomovic Ž, Li J L, Müllen K. (2008) Transparent carbon films as electrodes in organic solar cells. Angew. Chem.-Int. Edit. 47:2990–2992.
116. Becerril H A, Mao J, Liu Z, Stoltenberg R M, Bao Z, Chen Y (2008) Evaluation of solution-processed reduced graphene oxide films as transparent conductors. ACS Nano 2:463–470.
117. Sariciftci N S, Smilowitz L, Heeger A J, Wedl F (1992) Photoinduced electron transfer from a conducting polymer to buck minster fullerene. Science 258: 1474–1476.
118. Gao Y, Yip H -L, Chen K -S, O'Malley K M, Acton O, Sun Y, Ting G, Chen H, Jen A K Y (2011) Surface doping of conjugated polymers by graphene oxide and its application for organic electronic devices. Advanced Materials, 23: 1903–1908.
119. Ryu M S, Jang J (2011) Effect of solution processed graphene oxide/nickel oxide bi-layer on cell performance of bulk-hetero junction organic photovoltaic. Solar Energy Materials and Solar Cells, 95. 2893–2896.
120. Yun J -M, Yeo J -S, Kim J, Jeong H -G, Kim D -Y, Noh Y -J, Kim S -S, Ku B -Ch, Na S –I (2011) Solution-processable reduced graphene oxide as a novel alternative to PEDOT:PSS hole transport layers for highly efficient and stable polymer solar cells. Advanced Materials, 23: 4923–4928.

121. Schniepp H C, Li J -L, McAllister M J, Sai H, Herrera-Alonso M, Adamson D H, Prud'homme R K, Car R, Saville D A, Aksay I A (2006) Functionalized single graphene sheets derived from splitting graphite oxide. The Journal of Physical Chemistry B, 110: 8535–8539.
122. Hsu C L, Lin C T, Huang J H, Chu C W, Wei K H, Li L J (2012) Layer-by-Layer Graphene/ TCNQ Stacked Films as Conducting Anodes for Organic Solar Cells. ACS Nano 6 : 5031-5039.
123. Muhlbacher D, Scharber M, Morana M, Zhu Z, Waller D, Gaudiana R, Brabec C (2006) High photovoltaic performance of a low-bandgap polymer. Advanced Materials 18: 2884–2889.
124. Folgueras L D C, Nohara E L, Faez R, Rezende M C (2007) Dielec- tric microwave absorbing material processed by impregnation of carbon fiber fabric with polyaniline. Materials Research 10: 95–99.
125. Xu P, Han X J, Wang C, Zhou D H, Lv Z S, Wen A H, Wang X H, Zhang B (2008) Synthesis of Electromagnetic Functionalized Nickel/Polypyrrole Core/Shell Composites. J. Phys. Chem. B 112: 10443 – 10448.
126. Yan L G, Wang J B, Han X H, Ren Y, Liu Q F, Li F S (2010) Enhanced microwave absorption of Fe nanoflakes after coating with SiO_2 nanoshell. Nanotechnology 21: 095708.
127. Zhu W M, Wang L, Zhao R, Ren J W, Lu G Z, Wang Y Q (2011) Electromagnetic and microwave-absorbing properties of magnetic nickel ferrite nanocrystals. Nanoscale 3: 2862 – 2864.
128. Wang C, Han X J, Zhang X L, Hu S R, Zhang T, Wang J Y, Du Y C, Wang X H, Xu P (2010) Controlled Synthesis and Morphology-Dependent Electromagnetic Properties of Hierarchical Cobalt Assemblies. J. Phys. Chem. C 114: 14826 – 14830.
129. Novoselov K S, Geim A K, Morozov S V, Jiang D, Zhang Y, Dubonos S V, Grigorieva I V, Firsov A A (2004) Electric Field Effect in Atomically Thin Carbon Films. Science 306: 666 – 669.
130. Sun G B, Dong B X, Cao M H, Wei B Q, Hu C W (2011) Hierarchical Dendrite-Like Magnetic Materials of Fe_3O_4, γ-Fe_2O_3, and Fe with High Performance of Microwave Absorption. Chem. Mater. 23: 1587 – 1593.
131. Liu J W, Che R C, Chen H J, Zhang F, Xia F, Wu Q S, Wang M (2012) Microwave Absorption Enhancement of Multifunctional Composite Microspheres with Spinel Fe_3O_4 Cores and Anatase TiO_2 Shells. Small 8: 1214 – 1221.
132. Tong G X, Hu Q, Wu W H, Li W, Qian H S, Liang Y J (2012) Submicrometer-sized NiO octahedra: facile one-pot solid synthesis, formation mechanism, and chemical conversion into Ni octahedra with excellent microwave-absorbing properties. Mater. Chem. 22: 17494 – 17504.
133. Xu H L, Bi H, Yang R B (2012) Enhanced microwave absorption property of bowl-like Fe_3O_4 hollow spheres/reduced graphene oxide composites. J. Appl. Phys. 111: 07A522.
134. Sun X, He J P, Li G X, Tang J, Wang T, Guo Y X, Xue H R (2013) Laminated magnetic graphene with enhanced electromagnetic wave absorption properties. J. Mater. Chem. C 1: 765 – 777.
135. Ren Y L, Wu H Y, Lu M M, Chen Y J, Zhu C L, Gao P, Cao M S, Li C Y, Ouyang Q Y (2012) Quaternary Nanocomposites Consisting of Graphene, Fe_3O_4@Fe Core@Shell, and ZnO Nanoparticles: Synthesis and Excellent Electromagnetic Absorption Properties. ACS Appl. Mater. Interfaces 4: 6436 – 6442.
136. Wei T Y, Chen C H, Chang K H, Lu S Y, Hu C C (2009) Cobalt Oxide Aerogels of Ideal Supercapacitive Properties Prepared with an Epoxide Synthetic Route. Chem. Mater. 21: 3228 – 3233.
137. Chen K Y, Xiang C, Li L C, Qian H S, Xiao Q S, Xu F J (2012) A novel ternary composite: fabrication, performance and application of expanded graphite/polyaniline/ $CoFe_2O_4$ ferrite. Mater. Chem. 22: 6449 – 6455.
138. Singh K, Ohlan A, Pham V H, Balasubramaniyan R, Varshney S, Jang J, Hur S H, Choi W M, Kumar M, Dhawan S K, Kong B S, Chung J S (2013) Nanostructured graphene/

Fe_3O_4 incorporated polyaniline as a high performance shield against electromagnetic pollution. Nanoscale 5: 2411 – 2420.

139. Liu P B, Huang Y (2013) Synthesis of reduced graphene oxide-conducting polymers-Co_3O_4 composites and their excellent microwave absorption properties. RSC Adv. 3: 19033 – 19039.
140. Wang Y, Yang R, Shi Z, Zhang L, Shi D, Wang E, Zhang G(2011) Super- elastic graphene ripples for flexible strain sensors. ACS Nano 5: 3645 -3650.
141. Schedin F, Geim A K, Morozov S V, Hill E W, Blake P, Katsnelso M I, Novoselov K S (2007) Detection of individual gas molecules adsorbed on graphene. Nature Mater 6: 652-655.
142. Leenaerts O, Partoens B, Peeters F M (2008) Adsorption of H2O, NH3, CO, NO2 and NO on graphene: A first-principles study. Physical Review B 77: 125416-125421.
143. Fowler J D, Allen M J, Tung V C, Yang Y, Kaner R B, Weiller B H (2009) Practical chemical sensors from chemically derived grapheme. ACS Nano 3: 301-306.
144. Sundaram R S, Navarro C G, Balasubramaniam K, Burghard M, Kern K (2008) Electrochemical modification of grapheme. Advanced Materials 20: 3050-3053.
145. Hu H, Wang X, Xu C, Wang J, Wan L, Zhang M, Shang X (2012) Microwave-Assisted Synthesis of Graphene Nanosheets–Gold Nanocomposites with Enhancing Electrochemical Response. Fullerenes, Nanotubes and Carbon Nanostructures 20: 31-40.
146. Ghosh A, Late D J, Panchakarla L S, Govindaraj A, Rao C N R (2009) NO2 and humidity sensing characteristics of few-layergraphenes . Journal of Experimental Nanoscience. 4: 313-322.
147. Deng S, Tjoa V, Fan H M, Tan H R, Sayle D C, Olivo M, Mhaisalkar S, Wei J, Sow C H (2012) Reduced graphene oxide conjugated Cu2O nanowire mesocrystals for highperformance NO2 gas sensor. Journal of the American Chemical Society 134 : 4905-4917.
148. Lu C H, Yang H H, Zhu C L, Chen X, Chen G N (2009) A graphene platform for sensing biomolecules. Angewandte Chemie International Edition 48: 4785-4787.
149. Ponnamma D, Sadasivuni K K, Strankowski M, Guo Q, Thomas S (2013) Synergistic effect of multi walled carbon nanotubes and reduced graphene oxides in natural rubber for sensing application. Soft Matter 9: 10343-10353.
150. Jang H, Kim Y K, Kwon H M, Yeo W S, Kim D E, Min D H (1999) A grapheme based platform for the assay of dup lex-DNA unwinding by helicase. Angewandte Chemie International Edition 49: 5703-5707.
151. Bonnet G, Tyagi S, Libchaber A, Kramer F R (1999) Thermodynamic basis of the enhanced specificity of structured DNA probes. Proc. Natl. Acad. Sci. 96: 6171-6176.
152. Wang K, Tang Z, Yang C J, Kim Y, Fang X, Li W, Wu Y, Medley C D, Cao Z, Li J, Colon P, Lin H, Tan W (2009) Molecular engineering of DNA: Molecular beacons. Angewandte Chemie International Edition 48: 856-870.
153. Lu C H, Li J, Liu J J, Yang H H, Chen X, Chen G N (2010) Increasing the sensitivity and single-base mismatch selectivity of the molecular beacon using graphene oxide as the “nanoquencher”. Chemistry - A European Journal 16: 4889- 4894.
154. Li F, Huang Y, Yang Q, Zhong Z, Li D, Wang L, Song S, Fan C (2010) A grapheme enhanced molecular beacon for homogeneous DNA detection. Nanoscale 2: 1021-1026.
155. Lu C H, Zhu C L, Li J, Liu J J, Chen X, Yang H H (2010) Using graphene to protect DNA from cleavage during cellular delivery. Chemical Communications 46: 3116-3118.
156. Dong H, Gao W, Yan F, Ji H, Ju H (2010) Fluorescence resonance energy transfer between quantum dots and graphene oxide for sensing biomolecules. Analytical Chemistry 82: 5511-5517.
157. Balapanuru J, Yang J X, Xiao S, Bao Q, Jahan M, Polavarapu L, Wei J, Xu Q H, Loh K P (2010) A graphene oxide-organic dye ionic complex with DNA-sensing and optical-limiting properties. Angewandte Chemie International Edition 49: 6549-6553.
158. Wang Y, Li Z, Hu D, Lin C T, Li J, Lin Y (2010) Aptamer/ graphene oxide nanocomplex for in situ molecular probing in living cells. Journal of the American Chemical Society 132: 9274-9276.

159. Kodali V K, Scrimgeour J, Kim S, Hankinson J H, Carroll K M, De Heer W A, Berger C, Curtis J E (2011) Nonperturbative chemical modification of graphene for protein micropatterning. Langmuir 27: 863-865.
160. Dong X L, Cheng J S, Li J H, Wang Y (2010) Graphene as a novel matrix for the analysis of small molecules by MALDI-TOF MS. Analytical Chemistry 82 : 6208-6214 .
161. Tang L H, Wang Y, Liu Y, Li J (2011) DNA-directed self-assembly of graphene oxide with applications to ultrasensitive oligonucleotide assay. ACS Nano 5: 3817-3822 .
162. Prasad B B, Kumar D, Madhuri R, Tiwari M P (2011) Metal ion mediated imprinting for electrochemical enantioselective sensing of Lhistidine at trace level. Biosensors and Bioelectronics 28: 117-126.
163. Wang Z, Hu Y, Yang W, Zhou M, Hu X (2012) Facile One-Step Microwave-Assisted Route towards Ni Nanospheres/Reduced Graphene Oxide Hybrids for Non-Enzymatic Glucose Sensing. Sensors (Basel) 12: 4860- 4869.
164. Yang Y C, Dong S W, Shen T, Jian C X, Chang H J, Li Y, He F T, Zhou J X (2012) A Label-Free Amperometric Immunoassay for Thrombomodulin Using Graphene/Silver-Silver Oxide Nanoparticles as a Immobilization Matrix. Analytical Letters 45: 724-734.
165. Feng L, Gao G, Zhang C, Ma J, Cui D (2012) Electrochemical ascorbic acid/hydroquinone detection on graphene electrode and the electro-active site study. Journal of Experimental Nanoscience Version of record first published.
166. Lian H, Sun Z, Sun X, Liu B (2012) Graphene Doped Molecularly Imprinted Electrochemical Sensor for Uric Acid. Analytical Letters 45: 2717-2727.
167. Nayak T R, Andersen H, Makam V S, Khaw C, Bae S, Xu X, Ee P L R, Ahn J H, Hong B H, Pastorin G, Özyilmaz B(2011) Graphene for controlled and accelerated osteogenic differentiation of human mesenchymal stem cells. ACS Nano 5: 4670 – 4678.
168. Abdullahil K, Mithilesh Y, Sadasivuni K K, Mun S, Gao X, Kim J (2014) Synthesis and characterization of graphene/cellulose nanocomposite. Proc. SPIE 9060, Nanosensors, Biosensors, and Info-Tech Sensors and Systems 2014, 906008; doi:10.1117/12.2044964.
169. Ponnamma D, Thomas S (2013) Green Composites Green Methods to Synthesize and Recycle Materials- A promise to future, In Recent Advances in Rubber Recycling Edited by Yves Grohens, Kishor Kumar Sadasivuni and Abderrahim Boudenne, Apple Academic Press, ISBN: 9781926895277.
170. Jie W, Hao J (2014) Graphene-based hybrid structures combined with functional materials of ferroelectrics and Semiconductors, Nanoscale, 6: 6346-6362.
171. Lee W, Kahya O, Toh C T, zyilmaz B O, Ahn J H (2013) Flexible graphene–PZT ferroelectric nonvolatile memory, Nanotechnology, 24: 475202.
172. Ni G, Zheng Y, Bae S, YawTan C, Kahya O, Wu J, Hong B H, Yao K, Ozyilmaz B (2012) GrapheneFerroelectric Hybrid Structure for Flexible Transparent Electrodes, ACS Nano, 6: 3935–3942.
173. Yusuf M H, Nielsen B, Dawber M, Du X (2014) Extrinsic and Intrinsic Charge Trapping at the Graphene/Ferroelectric Interface, Nano Lett. 14: 5437 – 5444.
174. Rajapitamahuni A, Hoffman J, Ahn C H, Hong X (2013) Examining Graphene Field Effect Sensors for Ferroelectric Thin Film Studies, Nano Lett. 13: 4374 – 4379.

Electrical Properties of Graphene Polymer Nanocomposites

P. Noorunnisa Khanam, Deepalekshmi Ponnamma and M.A. AL-Madeed

Abstract Graphene, a monolayer of sp^2 hybridized carbon atoms arranged in a two dimensional lattice has attracted electronic industrial interest due to its exceptional electrical properties. One of the most promising applications of this material is in polymer nanocomposites in which the interface of graphene based materials and polymer chains merge to develop the most technologically promising devices. This chapter presents the electrical properties of such graphene based polymer nanocomposites and also discusses the effect of various factors on their electrical conductivity. Graphene enables the insulator to conductor transition at significantly lower loading by providing percolated pathways for electron transfer and making the polymers composite electrically conductive. The effect of processing conditions, dispersion, aggregation, modification and aspect ratio of graphene on the electrical conductivity of the graphene/polymer nanocomposites is conferred.

Keywords Conductivity · Percolation · Filler modification · Volume fraction · Fabrication

1 Introduction

Graphene, a two-dimensional, single-atom-thick structure of sp^2 bonded carbon atoms, has attracted tremendous research interest due to their excellent reinforcement, electrical properties, unique physical characteristics and high mechanical properties. Therefore, recent research has focused on developing high performance polymer nanocomposites, with the benefit of graphene nanotechnology, to

P.N. Khanam (✉) · M.A. AL-Madeed
Centre for Advanced Materials, Qatar University, Doha, Qatar
e-mail: noor.pathan@qu.edu.qa; pnkhanam_phd@yahoo.com

D. Ponnamma
School of Chemical Sciences, Mahatma Gandhi University, Kottayam 686560, Kerala, India

K.K. Sadasivuni et al. (eds.), *Graphene-Based Polymer Nanocomposites in Electronics*, Springer Series on Polymer and Composite Materials,
DOI 10.1007/978-3-319-13875-6_2

achieve novel composite materials for a wide range of industrial fields. Graphene dramatically improves the properties of polymer based composites at a very low loading and its most fascinating property is the very high surface conductivity leading to the formation of numerous electrically conductive polymer composites. Such conducting graphene nanocomposites have been widely applied in anti-static materials, electromagnetic interference (EMI) shielding, chemical sensor, bipolar plates for fuel cells etc. Other possible applications include radio-frequency interference shielding for electronic devices and electrostatic dissipation [1–3].

By using conventional processing methods, graphene composites can be easily fabricated into intricately shaped components with excellent preservation of the structure and properties. This is very important to make full use of the outstanding properties of graphene. Compared with carbon nanotubes (CNTs), graphene has a higher surface-to-volume ratio because of the inaccessibility of the CNT's inner tube surface to polymer molecules. This makes graphene potentially more favorable for improving the properties of polymer matrices, such as electrical properties. Therefore, graphene-based polymer composites have attracted both academic and industrial interest [3].

The present chapter gives an overview of the electrical properties of graphene based polymer nanocomposites. A brief description about the synthesis and characterization of graphene is also included in this chapter. Since the present book deals with the applications of graphene nanocomposites in various fields of flexible and wearable electronics, we think this chapter is of much significance as electrical conductivity is the basis for graphene's such applications. After giving an outline about the electrical properties of graphene polymer composites, the various factors affecting the conductivity such as filler aspect ratio, dispersion, modification of graphene surfaces etc. are also discussed here. The phenomenon of percolation threshold is also well pictured and finally this chapter ends with a few applications.

2 Synthesis and Characterization

2.1 Synthesis of Graphene

Graphite is available in large quantities as in the form of both natural and synthetic sources and is rather inexpensive [4]. The main graphite derivatives include EG, graphite oxide, graphene nanoplatelets (GNP), graphene oxide (GO), reduced graphene oxide (RGO), and graphene. Because the electronic, photonic, mechanical, and thermal properties of graphene depend on the number of layers [5] [although the monolayer (ML), bi-layer (BL), and tri-layer (TL) graphenes have practical significance] and its crystalline structure, the controlled synthesis of graphene with defined layers is rather significant. The mechanical peeling method by which graphene is first produced [6] is not used for an industrial scale of production. The GO and RGO derivatives are usually synthesized via solution-based oxidation and reduction by thermal and chemical methods, whereas graphene layers with superior electron transport characteristics are always synthesized using dry methods such as chemical vapor deposition (CVD) and surface segregation [6–13]. Although more than 95 % of graphene has been grown on Cu foil [5], this growth was not epitaxial, and thus complete growth over the entire substrate remains a major challenge. The surface of Ni(III) proved to be the best substrate for the epitaxial growth of structurally homogeneous graphene due to the small lattice mismatch of this surface with that of graphene and highly oriented pyrolytic graphite [14]. However, this method suffers from the disadvantage of carbon solubility in nickel, and thus achieving uniform thickness throughout the substrate is difficult. The simple method of surface segregation [15–17] was recently introduced to solve this problem and to epitaxially grow graphene over Ni film (~100 nm thick) [18]. Raman spectroscopy and scanning tunneling microscopy (STM) verified the homogeneity of the graphene layer over the entire Ni film [19–21].

2.2 Preparation of Graphene Polymer Composites

Various preparation methods employed to synthesize graphitic filler reinforced polymer nanocomposites include melt mixing, solution mixing and in situ polymerization. In addition to these three traditional polymer composite preparation methods, many other methods are also practiced for material fabrication. Since the common methods were the subjects for several reviews, a few other methods are introduced here. In a typical tape casting method [22], a mixture of natural flake graphite powders (NGP) and polyvinyl butyral (PVB) is stirred magnetically in ethanol, and after removing air bubbles by evacuation, this mixture is casted on a plastic film using a blade, as shown in Fig. 1a. The blade imparts a shearing force that orients the composite containing anisotropic graphite particles. This method was tested by varying the concentration of graphite powders from 10 to 95 wt% and varying the blade heights from 300 to 500 μm. A narrow blade gap is essential for producing a strong shearing force for better orientation, and the different testing conditions also affect thermal conductivity as well as the degrees of orientation [23].

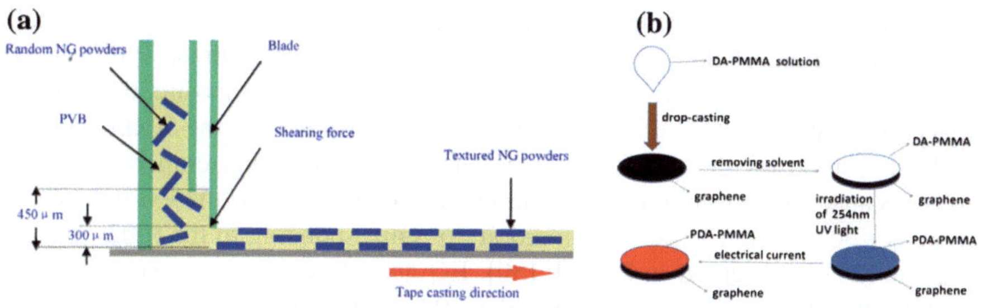

Fig. 1 Schematic of **a** tape-casting process and **b** electrochromic process [22]. Copyright 2012. Reproduced with permission from Elsevier Ltd. [24]. Copyright 2013. Reproduced with permission from American Chemical Society

In another preparation process, polydiacetylene (PDA)-Polymethylmethacrylate (PMMA)/graphene composites were developed that are capable of responding to electrical current with a color change. In the so-called electrochromic method (Fig. 1b), PDA acts as the electrochromic material and the graphene provides conductivity. The PMMA serves as the inert polymer matrix and improves the mechanical properties and colorimetric phenomenon. The blue-red phase transition is clearly visible in PDA-PMMA/graphene compared with that of PDA/graphene. The critical current for the color transition can be varied with the amount of graphene [24].

Mixing of functionalized GO with epoxy resin by sonication transfers the modified GO particles from the water to the epoxy. This is yet another fabrication method practiced. After decanting the water, heating the mixture forms a dark violet epoxy-modified GO composite, and the epoxy can be cured by adding hardener. The calculated volume fractions of functionalized GO in the final composites were 1.16 and 2.2 g/cm^3, respectively [25]. Copper oxide nanoparticle/graphene (CuO-GR) nanocomposites were prepared using GO synthesized by Hummers method. The copper-acetate-adsorbed GO acts as a precursor. The GO was washed with de-ionized water to remove the remaining metal ions and acid, and a copper(II) nitrate aqueous solution was added. Again, ammonium hydroxide was added under magnetic stirring, and the mixture was transferred to an autoclave at 100 °C. The black CuO-GR nanocomposites formed were washed with distilled water and ethanol [26].

2.3 Characterization of Polymer Nanocomposites

Many important characteristic techniques reveal the morphology, structure and superior properties of graphitic fillers and their polymer composites. Of the various characterization techniques, three important methods of analysis for the materials are explained in this section- X-ray diffraction spectroscopy (XRD), Fourier Transformation Infrared (FTIR) Spectroscopy and Raman spectroscopy. Figure 2 shows the XRD pattern for the graphite, GO and their mode of dispersion in polyaniline (PANI). Using the peak observed at 2θ/10.04°, the GO grain size was estimated as 5.1 nm. In the case of a polymer, the crystallized PANI gives a broad band extending from 15° to 34° [27]. For the PANI composite, the intensity of the peak at 10.04° is affected (Fig. 2a) [28]. Variation in the XRD spectrum is also observed for PDA composites prepared under different temperature conditions because of the influence of experimental conditions on the exfoliation rate of RGO fillers [28]. The peak centered at 26° corresponding to the (002) plane of graphite [29] is observed in Fig. 2b [30]. EG also shows the same peak, whereas for GO, the diffraction peak is shifted to 9.8° corresponding to a *d*-spacing of 0.9 nm [30]. The weak diffraction maximum of EG compared to graphite is attributed to its exfoliated nature. In GO, the *d*-spacing depends on the method of preparation and the number of layers of water trapped in the structure of the material

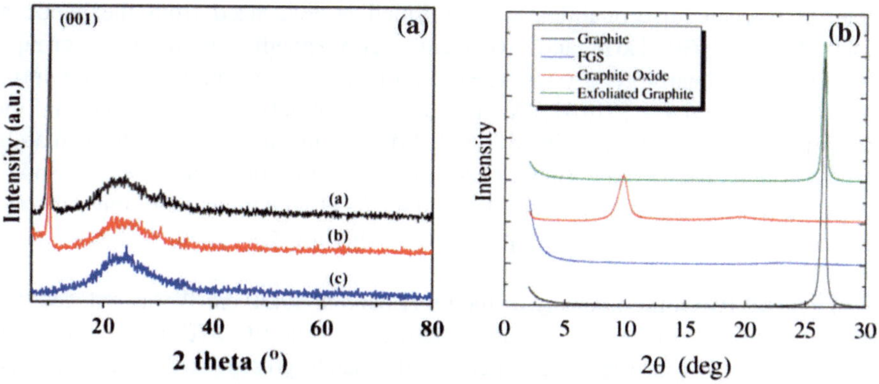

Fig. 2 **a** XRD patterns of the *a* GO film, *b* PANI/GO nanocomposite film, and *c* PANI film [28]. Copyright 2013. Reproduced with permission from Elsevier Ltd. **b** XRD Pattern of FGS, GO, and EG [30]. Copyright 2009. Reproduced with permission from John Wiley & Sons

[30]. Functionalised graphene sheets (FGS) show no characteristic peak indicating the loss of long-range stacking order in the material. Also in the graphene case, the Bragg peak is totally absent because pure graphene contains no stacks.

Another important characterization tool is FTIR, which confirms the chemical structure of nanocomposites and interactions among them. The FTIR spectra obtained for graphite, GO, magnetite–graphene (MG), and Pd/Fe_3O_4/graphene (PMG) nanocomposites are shown in Fig. 3 [31]. In GO, the presence of different

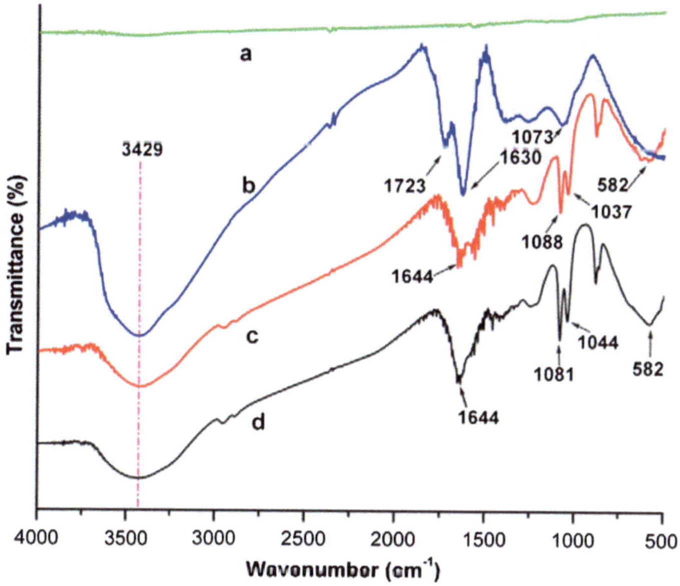

Fig. 3 FTIR spectra of **a** graphite, **b** GO, **c** MG and **d** PMG nanocomposite [31]. Copyright 2012. Reproduced with permission from Elsevier Ltd.

types of oxygen functionalities are confirmed as evidenced from the bands at 3,429, 1,723, 1,630, 1,618 and 1,073 cm^{-1} corresponding to the O–H groups, C = O carbonyl/carboxyl groups, C = C aromatic groups and C–O in the epoxide group, respectively [31, 32]. As a result GO interacts with metal ions and the additional peak at 582 cm^{-1} in MG and PMG composites implies the formation of Fe–O bond. The two additional peaks at 1,037 and 1,088 cm^{-1} for MG and at 1,044 cm^{-1} for PMG corresponds to distinct C–O stretching vibrations, involving the formation of metal nanoparticles by binding C–O with different metal ions [31].

In general RGO, the intensity of the O–H band at 3,430 cm^{-1} was reduced due to the de-oxygenation of the GO functionalities. For the RGO-(PMMA) nanocomposites obtained from the in situ method, the bands present at 3,420, 1,726 and 1,620 cm^{-1} are due to the presence of O–H, C = O and C = C groups, respectively [33]. When GO-PMMA is reduced in the case of R-(GO-PMMA), the intensity of the C = C bands increases, whereas that of the C = O band decreases.

The best technique for discriminating graphite and graphene is Raman spectroscopy, which is explained with the aid of Figs. 4, 5, and 6 [34–36]. This simple spectroscopy technique is capable of identifying the number of graphene layers, orientation and crystalline quality of the graphene layers [5, 37]. Figure 4a [34] shows the typical Raman signals of graphene recovered from dispersions of 3-glycidoxypropyl trimethoxysilane (GPTMS) and phenyl triethoxysilane (PhTES), and both are similar. In particular, graphene obtained from PhTES exhibits a G band at 1,577 cm^{-1}, a 2D band at 2,696 cm^{-1}, and a D peak at 1,346 cm^{-1}, whereas for the other graphene obtained from GPTMS, the G band is located at 1,574 cm^{-1}, the 2D band at 2,701 cm^{-1}, and the D band at 1,345 cm^{-1}. The 2D bands are de-convoluted in Fig. 4b [34], and the components of the 2,701 cm^{-1} peak resemble those of bilayer graphene [38]. The 2D peak of graphite consists of two components, and the main peak is upshifted to 2,713 cm^{-1}.

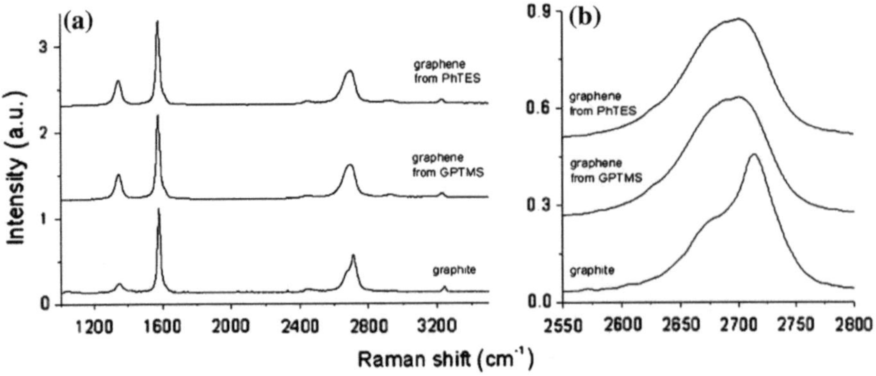

Fig. 4 Raman spectra of graphene obtained by sonication in PhTES and GPTMS from 5 wt% of the initial graphite compared with graphite (**a**) 2D peaks evaluation for this systems (**b**) [34]. Copyright 2009. Reproduced with permission from Springer

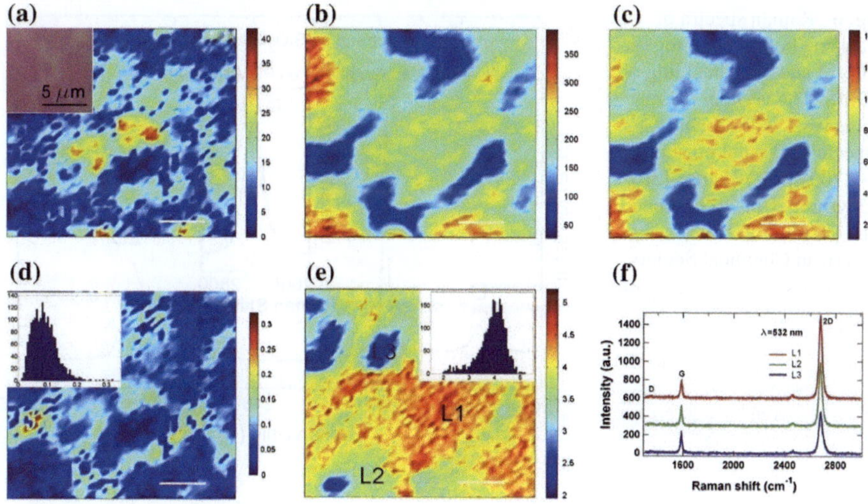

Fig. 5 Raman characterization of CVD graphene. All Raman maps for I_D (**a**), I_G (**b**), I_{2D} (**c**), I_D/I_G (**d**), and I_{2D}/I_G. In the histograms, *x*-axis indicates the ratios of I_D/I_G and I_{2D}/I_G and *y*-axis indicates counts. All of scale bars in maps are 2 μm, except (**a**) inset. Raman spectra shown in (**f**) measured from the marked location *L*1, *L*2 and *L*3 in (**e**), respectively. The D, G, and 2D bands are labeled in the spectra [35]. Copyright 2010. Reproduced with permission from Elsevier Ltd.

Spatially resolved Raman measurements on the CVD graphene films from a SiO_2/Si substrate were collected to probe the sample uniformity. A Raman map showing the intensity of the D, G and 2D bands measured in a 100 m^2 area of the CVD graphene sample is presented in Fig. 5a–c [35]. The insets of Fig. 5a, d and e [35] correspond to the optical microscope images of the scanned area and the histograms of I_D/I_G and I_{2D}/I_G. The data are fitted with Lorentzian formula, and the intensity of a given band is determined in terms of the amplitude value of the Lorentzian function fit. For fitting of the range, the D, G, and 2D bands were considered at 1,320–1,380, 1,560–1,620 and 2,640–2,720 cm^{-1} intervals.

From Fig. 5a–c [35], it is clear that the intensity maps of the G and 2D bands are correlated, whereas the D band is not. This occurs because the spatially non-uniform adhesion of transferred graphene causes the high-density locations of the G and 2D bands to be the same and also affects the intensity of the Raman spectra. It is noteworthy that the D band is always low, indicating the high quality of the graphene [5, 39]. The Raman map of I_D/I_G of the same scanned area is given in Fig. 5d [35]. In this figure, the $I_D/I_G < 0.1$ again confirms the low defect density in the graphene.

From Fig. 5 [35], for most of the mapped area, the value of $I_{2D}/I_G > 2$ indicates monolayer graphene. Additionally, as shown in Fig. 5f [35] from the corresponding marked spots in Fig. 5e [35], substantial variations from $I_{2D}/I_G > 2$ are observed at monolayer locations. This observation and notably large $I_{2D}/I_G > 5$ values are the reasons for spatially non-uniform adhesion between the graphene film and the

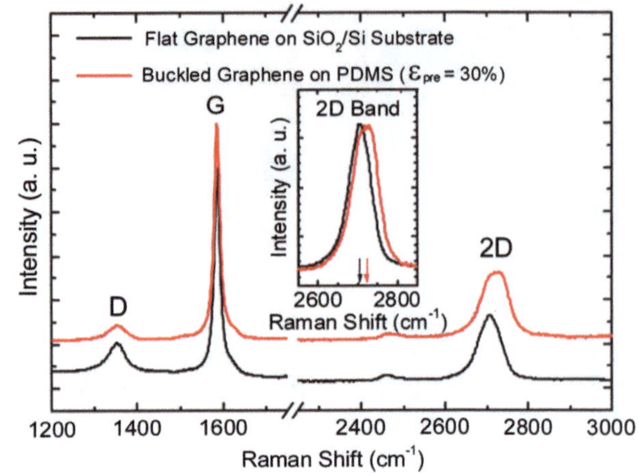

Fig. 6 Raman spectra of graphene before buckling on SiO_2/Si and after buckling on PDMS. The *vertical arrows* in the inset indicate the expected 2D band position (~2,700 cm^{-1}) change [36]. Copyright 2011. Reproduced with permission from American Chemical Society

underlying SiO_2 substrate [40]. The adhesion can significantly affect the G and 1D band but to different extents. Quantum Hall measurements also support the presence of monolayer graphene [41]. In short, Fig. 5 provides information on the high quality and uniform monolayers of CVD-grown graphene films.

The Raman spectra of graphene on a SiO_2/Si substrate (black curve) and buckled on PDMS (with a pre-strain of 30 %; red curve) are illustrated on Fig. 6 [36]. Due to the difference in the elastic moduli of graphene and PDMS, graphene ripples are spontaneously formed during pre-strain release. During buckling, compressive strain occurs in the graphene basal plane in addition to a geometry change. The spectra are obtained by subtracting the PDMS substrate Raman signal (red in figure). The D or 2D bands of graphene are quite sensitive to strain [42, 43] and the absence of any obvious variations in either the G and D peak positions or the I2D/IG ratios after many stretch-and-release cycles indicates that the defects of the graphene ribbons are independent of the buckling process. However, a blue shift of 15 cm^{-1} for the 2D peak is observed for graphene on PDMS compared with that on SiO_2/Si. This observation goes against the observation reported by Mohiuddin et al. [44] in which blue shifts were noted for both the G and 2D peaks for graphene formed on PMMA. Because the shift in the G band is affected only by doping, the shift in the 2D peak is the only parameter for estimating uniaxial strain in the Raman measurements. This uniaxial strain can be quantified as well.

3 Electrical Conductivity and Percolation Threshold in Graphene Polymer Nanocomposites

Superior electrical conductivity is the most important property of graphene. When graphene fills the insulating polymer matrix, conductive polymer composites result. Various polymers, including PMMA, PVA, PVC, PP, PE, PA12, PS etc. [45, 46]

have been used as matrices to prepare electrically conductive graphene/polymer composites. Such composite materials generally exhibit a non-linear increase of the electrical conductivity as a function of the filler concentration. The two parameters, electrical conductivity and percolation threshold are together associated with. At a certain filler loading fraction, which is known as the percolation threshold (p_c), the fillers form a network leading to a sudden rise in the electrical conductivity of the composites [45, 46]. Sometimes addition of a very low amount of conducting particles can make filler contact to form effective conducting paths and thus making the whole composite conductive.

A theoretical study by Xie et al. [47] predicted that graphene is more effective for conductivity improvement than competing nanofillers such as CNTs because of their large specific surface area. An outstanding electrically conductive graphene/polymer composite is expected to have lower percolation threshold and higher conductivity at a lower graphene loading, which can not only decrease the cost of filler but also preserve the processability of the composite. Ruoff et al. [48] synthesized graphene/PS composites and they observed a low percolation threshold at 0.1 vol% of graphene. The electrical conductivity variation in composites occurs in three stages, as illustrated in Fig. 7. Here the process is explained with a Graphene filled polymer. At first, the conductivity is quite low (Fig. 7a) due to a smaller number of additives, but large clusters gradually begin to form (Fig. 7b) with a slight increase in conductivity. At this stage, tunneling effects occur between neighboring graphene flakes, making it useful in Sensing materials.

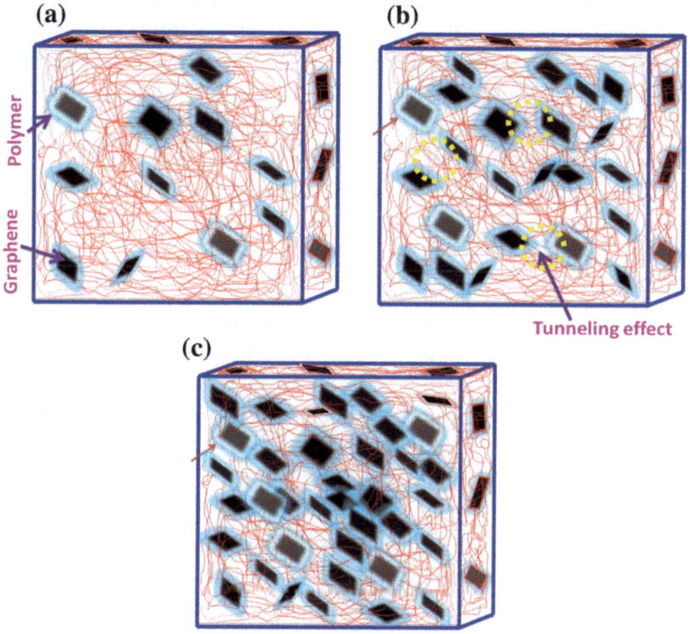

Fig. 7 Percolation process in conductive composites

As the graphene flakes increases, a complete conductive path (red) is formed by the contacting flakes (Fig. 7c) at the percolation, and further increase in the conducting particles enhances the number of conducting networks, as shown in Fig. 7c, until the conductivity levels off [49].

This explanation based on the way those nanoparticles form conducting network when dispersed in polymer matrix is called percolation theory. Various factors influence the electrical conductivity and the percolation threshold of the composites such as concentration of filler, aggregation of filler, processing methods, functionalization and aspect ratio of graphene sheets, inter-sheet junction, distribution in the matrix, wrinkles and folds etc. A more detailed study of these factors is done in the subsequent sections.

3.1 Effect of Graphene Concentration

In nanocomposites containing conducting fillers conduction takes place via tunneling between the thin polymer layers surrounding the filler particles, and this tunneling resistance is said to be the limiting factor in the composite conductivity. So in order to get a current flow in the composite, the direct contact between the filler is not necessary. However percolation of filler particles occurs and this enhances the electrical conductivity. The concentration of the filler for creating the percolation threshold varies from polymer to polymer. For instance Liang et al. reported a very low percolation threshold of 0.1 vol% for the solution-processable functionalized graphene filled epoxy composites compared to the percolation of 0.53 vol% for the neat graphene/epoxy nanocomposite [50]. For the graphene/ultra high molecular weight polyethylene composite, the percolation threshold was 0.070 vol% [51]. The graphene/polyethylene terephthalate (PET) nanocomposite [52] fabricated by melt compounding exhibited a percolation at 0.47 vol%. Kim et al. [53] reported lower percolation threshold of <0.5 vol% for TRGO while >2.7 vol% for graphite.

The synergy between MWCNTs and GNPs is successfully used to fabricate their polystyrene (PS) composites exhibiting a good response in terms of its DC conductivity value. Figure 8 illustrates the room temperature conductivity of PS/MWCNT/GNP nanocomposites prepared by in situ suspension polymerization of PS−GNP beads (50, 60 and 70 wt%) and MWCNT (0.1, 0.2 and 0.3 wt%). The increase in weight percent of both PS−GNP beads increased the conductivity of the composites, as shown in the figure [54].

The maximum electrical conductivity of $\sim 9.47 \times 10^{-3}$ S cm^{-1} is obtained in the case of PS–GNP (70 wt%) beads polymerized with 30 wt% PS-MWCNT containing 0.3 wt% MWCNT loading. Similar to the previous case, PS/MWCNT without PS–GNP beads was non-conducting with a 0.1 wt% loading of MWCNT. However, in situ polymerization of styrene−MWCNT in the presence of 50 wt% PS–GNP showed a conductivity value of $\sim 8.26 \times 10^{-7}$ S cm^{-1}. The MWCNT concentration in the in situ polymerized PS increases if PS−GNP is present and

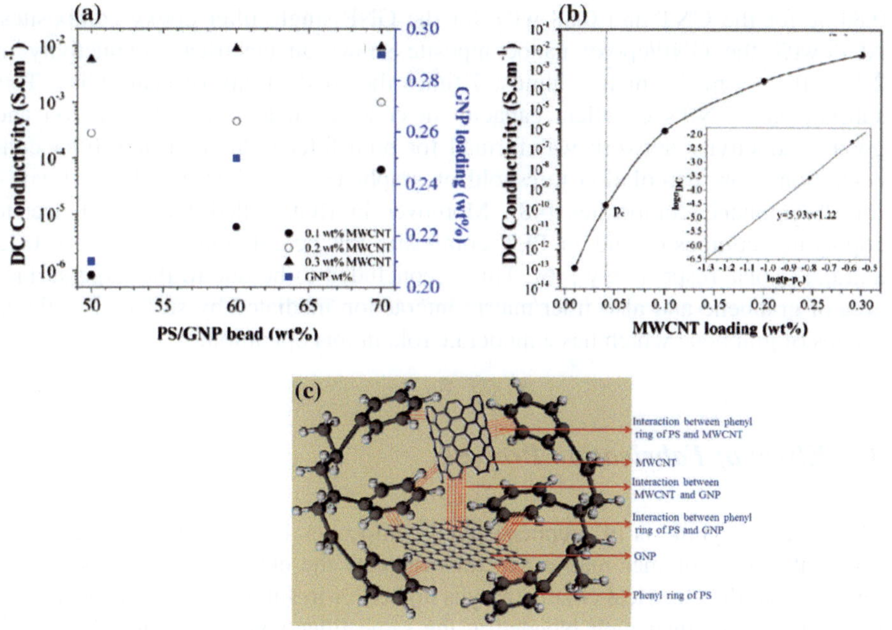

Fig. 8 DC conductivity of PS/MWCNT/GNP nanocomposites (**a**) at different wt% of PS–GNP beads and GNP loading with various MWCNT loadings (**b**) with MWCNT loading. (inset) log–log plot of σ_{DC} versus $(p–p_c)$ for the nanocomposites. *Straight line* in the inset is the model fit with the values $p_c = 0.041$ and $t = 5.93$. **c** Schematic representation shows π–π interactions between GNP, MWCNT, and PS in the composites [54]. Copyright 2013. Reproduced with permission from American Chemical Society

thus develops a continuous GNP–CNT–GNP conductive network path. These interactions that occur in the composite are schematically shown in Fig. 8c [54], where the π–π interaction between the phenyl rings of the PS and GNP sheets and the MWCNT is clear. For the excluded volume of PS–GNP beads in PS, the stick-like MWCNTs cannot enter, and thus the concentration of MWCNT will be high in the region of the in situ polymerized continuous PS phase adjoining the PS–GNP beads. These continuous GNP–CNT–GNP conductive network paths and π–π interactions with PS throughout the matrix cause the DC conductivity to enhance. In the case of PET/graphene composites electrical conductivity increases rapidly to 7.4×10^{-2} S m^{-1} from 2.0×10^{-13} S m^{-1} with a slight increase in graphene content from 0.47 to 12 vol%. Here the percolation of the graphene filled composites was achieved at a 2.4 % loading [55]. At this filler concentration PET/graphene exhibits higher increment in electrical conductivity compared to PET/graphite composites.

Yue et al. [56] prepared epoxy/GNP/CNT hybrid composites at different mixing ratios by dissolving the fillers in the epoxy resin using simultaneous magnetic stirring and sonication in warm bath. The electrical conductivity studies for the composites with different filler content reported a percolation threshold of about

0.84 % for the CNT and 0.88 wt% for the GNP single filler epoxy composites. At 4 wt% the GNP/epoxy nanocomposite shows an electrical conductivity of 2.1×10^{-5} S m^{-1} which is almost 7 times the conductivity of neat epoxy. This substantiates GNP's excellent capacity in conduction at low concentration and good conductivity network was formed for both filler systems at low filler concentration. Low percolation threshold of graphene is also observed at 0.1 vol% for PS/graphene composites [48]. Moreover in RGO filled PVC/vinyl acetate copolymer composite and in PET composite the percolation values were 0.15 and 0.47 vol% respectively [46]. This is concluded to be due to the large surface area of graphene and also filler/matrix interaction mediated by surface functional groups of graphene which has a moderate role in this operation.

3.2 Effect of Fabrication Process

The mode of synthesis of graphene filled composites influence the filler distribution within the polymer matrix and thus regulate the electrical conductivity. It is reported that the electrical conductivity is higher for in situ polymerized and solvent blended samples than melt blended at the same filler volume fraction. This indicates better dispersion from solvent based strategies [46]. However the conductivity observed for melt processed samples is attributed to the annealing happened during the process by which particles disorient and regain contacts between one another.

The static and electrical conductivity of the polyvinylidene difluoride (PVDF) composites containing two different fillers, solvothermally reduced graphene (SRG) and RGO were checked and a lower percolation threshold was observed for the SRG/PVDF composites [57, 58] compared to the graphene/PVDF composite prepared by direct blending of chemically/thermally reduced GO sheets with polymers [59]. In the composite [57] the SRG sheets were homogeneously dispersed, which is attributed to the low p_c value. Additionally, the SRGs remain stable in the PVDF solution for weeks, whereas without PVDF, they precipitate in dymethyl formamide (DMF) after 1 day. During the fabrication step of the solvothermal process, the GO sheets enclosed by the PVDF chains were reduced and did not fold easily or form aggregates. This behavior facilitated the formation of a conducting network and resulted in a low percolation threshold. The large aspect ratios of the SRGs make the percolation threshold even smaller and its value calculated mathematically yield a percolation value at 0.31 vol%. Equation 1 gives the mathematical formula used for calculating the percolation threshold.

$$\sigma(p) = \sigma_0(P - P_0)^t, \quad \text{for } P > P_c \tag{1}$$

where p_c is the percolation threshold, p is the filler content, and t is the critical exponent.

It is found that at 2.2 wt% RGO, the electrical conductivity of the RGO-filled PVDF increases enormously to 0.04 S m^{-1}, which is 14 orders of magnitude higher than that of neat PVDF. However at high concentrations of RGO, conductivity was notably low due to poor dispersion of the nanofiller in the polymer.

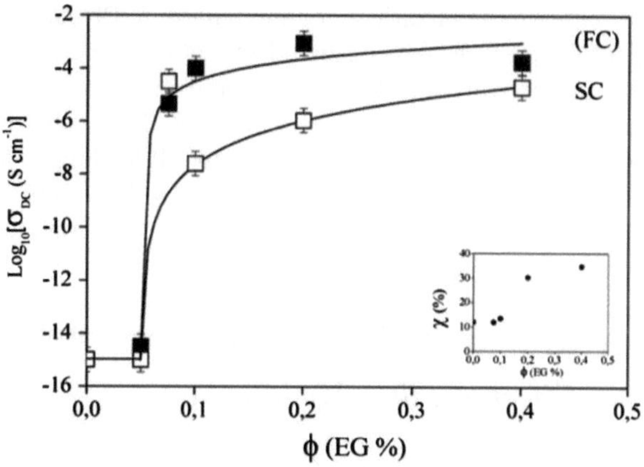

Fig. 9 Logarithm of the DC electrical conductivity versus EG concentration for the fast (FC) and slow (SC) cooled nanocomposites. *Continuous lines* are the predictions of percolation theory [27]. Copyright 2012. Reproduced with permission from John Wiley & Sons

Once percolation is achieved, the RGO-PVDF shows high conductivity, and this material finds many electrical applications. The electrical conductivities of the RGO-PVDF composite again increase with the amount of RGOs due to the above-mentioned conductivity and percolation effects. He et al. [57] reported a very low percolation threshold of 0.31 vol% for the chemically reduced GO filled PVDF prepared by direct blending, which is lower than that of graphene/PVDF composites due to the homogeneous graphene dispersion achieved in the former case.

The electrical conductivity of fast- and slow-cooled nanocomposites containing EG is shown in Fig. 9 [27]. The slow-cooled sample containing 0.075 wt% EG was excluded from the continuous line fit (according to Eq. 1). This sample with 0.075 wt% EG exhibited the highest conductivity of all the nanocomposites and therefore was not considered in the fitting. Additionally, the values calculated for the fast-cooled nanocomposites were well within the expectations of the percolation theory, whereas those for the slow-cooled case were much greater. The surface resistivity of a material is the electrical resistance to a leakage current along the surface of the insulator and the effect of GNS filler on the surface resistivity of unsaturated polyester resin (UPR) nanocomposites is reported. It is found that the surface resistivity decreases with increase in the concentration of GNS due to the platelet structure, the larger surface area of GNS and the larger free path available within the polymer matrix for the free electrons to propagate. The surface resistivity value varies between 0.99 × 1,015 and 0.14 × 1,015 Ω cm from 0.01 to 0.1 concentrations of GNS.

The PS/GNS composite prepared using in situ emulsion polymerization showed and electrical conductivity of ~2.9 × 10^{-2} S m^{-1} at 2.0 wt% of GNS [60]. The higher conductivity occurred here is attributed to the compatibility between the PS microspheres and GNS which is sufficient to obtain nanosized dispersions without any additional surface treatment. The electrical conductivity of WPU/FGS nanocomposite prepared by in situ polymerization method [61] was also increased up to 10^{15} fold compared to pristine WPU. This is due to the homogeneous dispersion of FGS particles in the WPU matrix and the formation of a conducting network throughout WPU cause an abrupt change in electrical conductivity. Here the percolation threshold was obtained at 2 wt% FGS loading. FGS can be used to improve the electrical conductivity of WPU as effectively as that of CNTs. The percolation threshold observed for PVDF were 2 wt% FGS and 5 wt% EG loading respectively and these nanocomposites were prepared by solution processing and compression molding method [30]. The higher aspect ratio of FGS compared to EG, makes better conductive network leading to a lower percolation threshold. Polycarbonate composites reinforced with graphite and FGS produced by melt compounding also results in high electrical conductivity with lower percolation threshold for FGS than graphite [62].

3.3 Effect of Filler Modification

There are two factors limiting the application of graphene based polymer composites, i.e. (1) poor dispersion of graphene in the given matrix due to their high specific surface area and strong intermolecular interactions between graphene sheets (2) at low filler content, graphene sheets are covered by polymer chains and thus difficult to achieve the percolation. These issues can be solved by the modification of graphene which impart the desired properties to the platelets. Generally covalent or noncovalent functionalization is employed to increase the effective dispersion of graphene sheets within the polymer. It is established that the chemical functionalization facilitates the dispersion, stabilize graphene and prevent agglomeration [63, 64]. The functional groups attached to graphene can be small molecules or sometimes large polymer chains. The chemical functionalization of graphene is particularly attractive as it can improve the solubility and processability as well as enhance the interactions with organic polymers [65]. Amination, esterification [65], isocyanate modification [66] and polymer wrapping were used in the literature for the functionalization. The electrochemical modification of graphene using ionic liquids has also been reported [67]. Other than the covalent [65, 66] and non covalent modifications [68], other methods such as reduction of GO in a stabilized medium [69], nucleophilic substitution to epoxy groups [70], and diazonium salt coupling [71] are also being practiced.

A homogeneous aqueous suspension of chemically modified graphene in water by means of strong base reduction of graphite oxide in a stabilization medium is reported by Park et al. [69]. Electrostatic stabilization used by Li et al. [72] enables the development of a facile approach to the large scale production of aqueous graphene dispersions. The different surface modifying agents like organic amines, alkyl lithium reagents, isocyanates, and diisocyanate compounds reduce the hydrophilic character of GO sheets by forming amide and carbamate ester bonds with the carboxyl and hydroxyl groups, respectively [66]. Graphite fluoride with alkyl lithium reagents were used for preparation of soluble graphene layers with covalent attachment of alkyl chains to the graphene layers [73].

In another work, the two dimensional GO was modified by isocyanation and dispersed in PS using compression molding and a low percolation threshold at 0.1 vol% was achieved [48]. This value was about 3 times lower than that obtained for other fillers. The reasons for this high electrical conductivity is attributed to the high aspect ratio of the graphene sheets and their homogenous dispersion in PS as that reported for epoxy/graphene composites [3]. Conductivity of these composites satisfied the approximately 0.15 vol% loading of the antistatic criterion (10^{-6} S m^{-1}) for the films. The value increased rapidly over a 0.4 vol% range and at 2.5 vol% loading, the electrical conductivity of composites showed ~0.1 to ~1 S m^{-1}. GNP also enhances the electrical conductivity as seen in its PS composite where 0.38 vol% GNP enhance the value to 5.77 S m^{-1} [46]. Eda et al. prepared PS/FGS composite thin films by [74] a solution blending method and the films exhibited a semi conducting and ambipolar behavior. These composites thin films were electrically conducting with a conductivity ranging from 1 to 24 S m^{-1}.

The electrical conductivity of Poly Aniline (PANI), PANI/GO and PANI/graphene prepared using the chemically modified graphene and PANI nano fibre by in situ polymerization of aniline monomer in the presence of GO under acidic conditions were reported to be 10.6, 23.12 and 168.7 S m^{-1} respectively. The conductivity of PANI/graphene composites was slightly lower than that of the PANI/GO composites, probably due to decrease in the degree of doping in PANI and a change in the morphology of the composite during the reduction, reoxidation and reprotanation processes [75].

Figure 10 [3] shows the plot of the electrical conductivity σ_c of chemically reduced graphene oxide (CRGO)/Polystyrene composites as a function of filler volume fraction φ. It can be seen that in situ CRGO/PS composites exhibit a typical percolation behavior and the introduction of CRGO to PS increase the conductivity to higher than 10 orders of magnitude. Percolation in the composites occurs when the filler concentration is near 0.1 vol%. At 1 vol% CRGO loading, the composite has a conductivity of $\sim$0.1 S m^{-1}. The right and left insets respectively show the plot of log σ_c against $\log(\varphi - \varphi_c)$ with φ_c the percolation threshold and the four probe setup for measurements with the computed distributions of the current density (contour lines) with directions and magnitude (arrows).

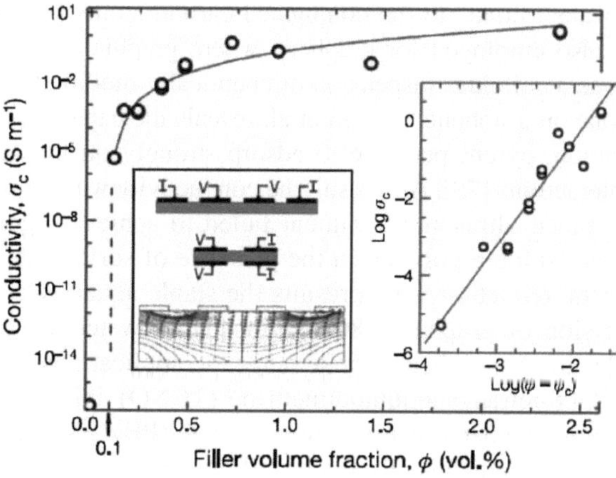

Fig. 10 Electrical conductivity of insitu CRGO/PS composites as a function of filler volume fraction [3]. Copyright 2012. Reproduced with permission from John Wiley & Sons

In another work by Chen et al. the electrical properties of both CRGO and thermally reduced graphene oxide (TRGO)/Poly dimethyl siloxane (PDMS) composites observed are much lower than CVD graphene/PDMS composites [76]. They directly synthesized three-dimensional graphene by template directed CVD, in which an interconnected flexible network of graphene exist as the fast transport channel of charge carriers for high electrical conductivity. Even with 0.5 wt% of graphene loading, the PDMS composite shows a very high electrical conductivity of 1,000 S m^{-1}, which is much higher than CRGO and TRGO/PDMS. This good conductivity with a low graphene loading is mainly due to the high electrical conductivity of CVDG and the interconnected 3D network structure in which all the graphene sheets are in direct contact with one another without breaks.

Xu et al. [77] modified graphene via the covalent attachment of a porphyrin ring on the GO surfaces. Thionyl chloride was used to activate the carboxylic acid group in the presence of porphyrin using DMF. Organic isocyanates were used for graphene modification by DMF solvent with methylene chloride and these isocyanate compound was attached to the hydroxyl and carboxyl groups of GO via the formation of carbamate and amide functionalities [66]. The lowest percolation occurred at 0.1 vol% in situ of CRGO reported by Stankovich et al. [68] is attributed to the excellent homogeneous dispersion with solution phase mixing of exfoliated phenyl isocyanate-treated GO sheets with PS.

Modification of graphene was also done by nucleophilic substitution in which aliphatic amine was used to modify GO [68] and the substitution occurs on the epoxy groups of GO and grafting was completed at room temperature [70]. For long chain aliphatic amines (octadecylamine), the reaction mixture was heated under reflux for 24 h. These modified GO derivatives were dispersed more easily in organic solvents. Bourlinos et al. [70] also used an alkaline solution of amino acids was used to modify GO. Nucleophilic attack of the $-NH_2$ end group on the epoxide groups of GO suggests a flat orientation of the amino acid molecules in the interlayer zone of GO. Dodecyl benzene sulfonate was used in reduction of GO with hydrazine and functionalized by a treatment with aryl diazonium salts [71]. Epitaxial graphene grown on SiC wafers were modified chemically with aryl groups via the formation of covalent bonds to the conjugated carbon atoms [78]. Electrochemical modification is also employed for graphene where graphite is electrochemically treated to produce a colloidal suspension of chemically modified graphene. The pyrene treatment done on graphene by Kim et al. reveals the capability of polycyclic aromatic hydrocarbon, pyrene/perylene to adsorb strongly on the graphene surface through π–π interactions [79]. As a result the conductivity was significantly improved.

Since ultrasonic treatment failed to achieve a good dispersion of graphene in water soluble polymer in the presence of surfactants, polymeric anions were used in the reduction which results the stable aqueous dispersion polymer-grafted dispersion of graphene [80]. High quality water soluble and organic solvent soluble graphene sheets for a range of applications were obtained from EG using 7,7,8,8-tetracyanoquinodimethane (TCNQ) anion as a stabilizer [81]. Water soluble pyrene derivative,1-pyrenebutyrate (PB) also acted as a stabilizer for the preparation of aqueous dispersions of graphene sheets [82].

3.4 Other Factors

In addition to the three factors discussed in the previous sub sections, several miscellaneous points also influence electrical conductivity of graphene and its derivative filled polymers. Zhang and co-workers [80] studied the effect of oxygen content on graphene sheets on the electrical property of graphene-PMMA nanocomposites. The percolation threshold increased with increasing oxygen content and the composite with lowest oxygen content (graphene-13.2) show a dramatic increment in electrical conductivity of over 12 orders of magnitude, from 3.33×10^{-14} S m^{-1} with 0.4 vol% of graphene to 2.38×10^{-2} S m^{-1} with 0.8 % of graphene. At 2.67 vol% the electrical conductivity reached up to 10 S m^{-1} indicating interconnected graphene network. Moreover the composite with lowest oxygen content exhibited much higher conductivity, in the percolation transition range than composites with higher content of oxygen (graphene-9.6 and graphene-5.0). The presence of oxygen-containing groups on graphene has been proved to disrupt its graphitic sp^2 network and decrease its intrinsic conductivity leading to a final conclusion that the higher the oxygen content, the lower the intrinsic conductivity.

Orientation of graphene platelets in polymer is another factor influencing composite conductivity. When graphene sponge containing randomly distributed graphene sheets are added to epoxy (insulator with conductivity around at 10^{-13} S cm^{-1}) [83] the magnitude of conductivity increased 12 times attributed to the compactly interconnected graphene network constructed within the polymer. Further improvement was also noticed upon graphene sponge treatment [84]. The filler surface area is yet another factor which causes the conductivity variation as noticed in the case of PVA/RGO nanocomposites [85]. Both large area (LRGO) and small area RGOs (SRGO) were used to fabricate the nanocomposites and found higher electrical conductivity for PVA/LRGO than that of SRGO at the same filler content. Percolation happened at 0.189 wt% of filler. The conductivity increased with specific surface area as well as with the reduction temperature. Conductivity improves when thermal reduction temperature exceeds critical temperature. Also high temperature is more efficient for reducing the GO sheets than low temperature. The electrical conductivity increased to the highest value at the reduction temperature 200 °C which indicating that the critical temperature for the thermal reduction in PVA/GO system is 200 °C. Swelling was also employed to understand the mechanism involved and found a decrease in conductivity after swelling due to the destroyed conductive networks. During swelling, water molecules interact between the RGO sheets and destroy the electrical networks and the conductivity is lower for PVA/LRGO than PVA/SRGO nanocomposites after swelling. This indicates easy destruction of the SRGO network after treatment [85].

Thermal reduction is an efficient way to produce the nanocomposite with low percolation threshold and high electrical conductivity and also it's suitable for preparing the thermoset or thermoplastic polymer/RGO nanocomposites through reduction of GO sheets at appropriate temperature. The long reduction time improves the electrical conductivity of nanocomposites [85]. It is reported that thermally reduced

GO has higher electrical conductivity than chemically reduced GO due to the absence of oxygenated functional groups in TRGO [45, 46]. This effect of thermal and chemical reduction of GO on the electrical properties of graphene/PU composites was studied by Kim et al. [53]. The electrical properties of graphene and graphite in terms of conductivity or sheet resistance can be modulated by controlling the number of graphene layers [86]. For this purpose, the sheet resistance values were checked for graphene and graphite films containing different numbers of layers. The sheet resistance was reduced by a factor of approximately 25 as the number of layers increased from 24 to 850 layers. Additionally, the current versus back-gate characteristics showed much stronger modulation in 24-layer graphene compared with a negligible change in the thick graphite (850 layers) due to the stronger screening effect as the number of layers increases. The low sheet resistance and field-effect response of graphite are advantageous for applications in conductive films or electrodes. In contrast, the superior transconductance level of the 24-layer graphene is appropriate for the active channels of field-effect transistors (FETs).

On the basis of the above results, we can conclude that many factors including filler aspect ratio, surface area, concentration, dispersion state, mode of synthesis and contact resistance are key factors affecting the electrical properties of graphene/polymer composites.

4 Applications

Based on this chapter, it is clear that the reinforcement of graphene and its polymer nanocomposites have shown very promising results in enhancing the electrical conductivity at low percolation threshold. The discovery of graphene as nanofiller has opened a new dimension for the production of light weight, low cost, and high performance composite materials for a range of applications. The electrically conductive polymer/graphene nanocomposites have been widely used for making various sensors, memory and energy storage, antistatic coatings, EMI etc. and have potential applications in the biomedical fields, such as ultraminiaturized low cost sensors for the analysis of blood and urine. Moreover the polymer/graphene flexible electrode has some commercial applications in LEDs, transparent conducting coatings for solar cells and displays.

5 Conclusion

The development of a nanolevel dispersion of graphene particles in a polymer matrix has opened a new and interesting area in materials science in recent years. Its unique properties make it suitable to improve the electrical properties of polymer composites. This chapter summarizes the electrical properties of various polymer/graphene composites and the different factors affecting electrical conductivity. We discussed the percolation threshold based on filler volume fraction, processing methods, aspect ratio, surface area, orientation etc. In order to improve conductivity, the dispersion of graphene in polymer matrices and the graphene–polymer interaction needs to be improved, which are achieved by the surface modification of graphene. Finally a few electronic applications of these high-performance graphene composite materials is mentioned.

References

1. Huang X, Yin Z, Wu S, Qi X, He Q, Zhang Q, Yan Q, Boey F, Zhang H (2011) Graphene-Based Materials: Synthesis, Characterization, Properties, and Applications. Small 7: 1876–1902.
2. Dong L X, Chen Q (2010) Properties, synthesis, and characterization of grapheme. Frontiers of Materials Science in China 4: 45–51.
3. Du J, Cheng H M (2012) The Fabrication, Properties, and Uses of Graphene/Polymer Composites. Macromolecular Chemistry and Physics 213: 1060–1077.
4. Cai W W, Piner R D, Stadermann F J, Park S, Shaibat M A, Ishii Y, Yang D X, Velamakanni A, An S J, Stoller M, An J H, Chen D M, Ruoff R S S (2008) Synthesis and solid-state NMR structural characterization of C-labeled graphite oxide. Science 321: 1815–1817.
5. Li X S, Cai W W, An J H, Kim S, Nah J, Yang D X, Piner R, Velamakanni A, Jung I, Tutuc E, Banerjee S K, Colombo L, Ruoff R S (2009) Large-area synthesis of high-quality and uniform graphene films on copper foils. Science 324: 1312–1314.
6. Novoselov K S, Geim A K, Morozov S V, Jiang D, Zhang Y, Dubonos S V, Grigorieva I V, Firsov A A (2004) Electric field effect in atomically thin carbon films. Science 306: 666–669.
7. Novoselov K S, Geim A K, Morozov S V, Jiang D, Katsnelson M I, Grigorieva I V, Dubonos S V, Firsov A A (2005) Two-dimensional gas of massless Dirac fermions in grapheme. Nature 438: 197–200.
8. Zhang Y, Tan Y W, Stormer H L, Kim P (2005) Experimental observation of the quantum Hall effect and Berry's phase in grapheme. Nature 438: 201–204.
9. Geim A K, Novoselov K S (2007) The rise of graphene. Nature Mater 6: 183–191.
10. Dreyer D R, Ruoff R S, Bielawski C W (2010) From conception to realization: an historial account of graphene and some perspectives for its future. Agnewandle Chemie International Edition 49: 9336–9345.
11. Boehm H P, Setton R, Stumpp E (1986) Nomenclature and terminology of graphite intercalation compounds. Carbon 24: 241–245.
12. Novoselov K S, Jiang D, Schedin F, Booth T, Khotkevich V V, Morozov S V, Geim A K (2005) Two-dimensional atomic crystals. Procedings of the National Academy of Sciences 102: 10451–10453.
13. Park S, Ruoff R S (2009) Chemical methods for the production of graphenes. Nature Nanotechnology 4: 217–224.
14. Fujita D, Yoshihara K(1994) Surface precipitation process of epitaxially grown graphite (0001) layers on carbon-doped nickel(III) surface. J Vac Sci Technol A 1994, 12, 2134–2139.
15. Gao J H, Fujita D, Xu M S (2010) Unique synthesis of few-layer graphene films on carbon-doped $Pt_{83}Rh_{17}$ surfaces. ACS Nano 4: 1026–1032.
16. Xu M S, Fujita D, Chen H Z, Hanagata N(2011) Formation of monolayer and few-layer hexagonal boron nitride nanosheets via surface segregation. Nanoscale 3: 2854–2858.
17. Fujita D (2011) Nanoscale synthesis and characterization of graphene based objects. Science and Technology of Advanced Materials 12: 044611 (1–10).
18. Xu M S, Fujita D, Sagisaka K, Watanable E, Hanagata N (2011) Production of extended single-layer graphene. ACS Nano 5: 1522–1528.

19. Xu M S, Endres R G, Tsukamoto S, Kitamura M, Ishida S, Arakawa Y (2005) Conformation and local environment dependent conductance of DNA molecules. Small 1: 1168–1172.
20. Xu M S, Tsukamoto S, Ishida S, Kitamura M, Arakawa Y, Endres R G, Shimoda M (2005) Conductance of single thiolated poly(GC)-poly(GC) DNA molecules. Applied Physics Letters 87: 083902 (1–3).
21. Merino P, Svec M, Pinardi A L, Otero G, Martin-Gago J A (2011) Strain-driven moiré superstructures of epitaxial graphene on transition metal surfaces. ACS Nano 5: 5627–5634.
22. Zhou S X, Zhu Y, Du H D, Li B H, Kang F Y (2012) Preparation of oriented graphite/polymer composite sheets with high thermal conductivities by tape casting. New Carbon Mater 27: 241–249.
23. Zhou S, Chiang S, Xu J, Du H, Li B, Xu C, Kang F (2012) Modeling the in-plane thermal conductivity of a graphite/polymer composite sheet with a very high content of natural flake graphite. Carbon 50: 5052–5061.
24. Zhang W, Xu H, Chen Y, Cheng S, Fan L (2013) Polydiacetylene-polymethylmethacrylate/graphene composites as one-shot, visually observable, and semiquantative electrical current sensing materials. ACS Applied Materials & Interfaces 5: 4603–4606.
25. Gudarzi M M, Sharif F (2012) Enhancement of dispersion and bonding of graphene-polymer through wet transfer of functionalized graphene oxide. Express Polymer Letters 6: 1017–1031.
26. Yu H Y, Xu M Q, Yu S H, Zhao G C (2013) A novel non-enzymatic glucose sensor based on CuO - graphene nanocomposites. International Journal of Electrochemical Science 8: 8050–8057.
27. Paszkiewicz S, Szymczyk A, Spitalsky S, Soccio M, Mosnacek J, Ezquerra T A, Roslaniec Z (2012) Electrical conductivity of poly(ethylene terephthalate)/expanded graphite nanocomposites prepared by in situ polymerization. Journal of Polymer Science Part B: Polymer Physics 50: 1645–1652.
28. Wei H, Zhu J, Wu S, Wei S, Guo Z (2013) Electrochromic polyaniline/graphite oxide nanocomposites with endured electrochemical energy storage. Polymer 54: 1820–1831.
29. Chen G L, Shau S M, Juang T Y, Lee R H, Chen C P, Suen S Y, Jeng R J (2011) Single-layered graphene oxide nanosheet/polyaniline hybrids fabricated through direct molecular exfoliation. Langmuir 27: 14563–14569.
30. Ansari S, Emmanuel P G (2009) Functionalized graphene sheet-poly(vinylidene fluoride) conductive nanocomposites. Journal of Polymer Science Part B: Polymer Physics 47: 888–897.
31. Chandra S, Bag S, Das P, Bhattacharya D, Pramanik P (2012) Fabrication of magnetically separable palladium–graphene nanocomposite with unique catalytic property of hydrogenation. Chemical Physics Letters 519: 59–63.
32. Matusinovic Z, Rogosic M, Sipusic J (2009) Synthesis and characterization of poly(styrene-co-methyl methacrylate)/layered double hydroxide nanocomposites via in situ polymerization. Polymer Degradation and Stability 94: 95–101.
33. Aldosari M, Othman A, Alsharaeh E (2013) Synthesis and characterization of the in situ bulk polymerization of PMMA containing graphene sheets using microwave irradiation. Molecules 18: 3152–3167 (3).
34. Nuvoli D, Alzari V, Sanna R, Scognamillo S, Piccinini M, Peponi L, Kenny J M, Mariani A (2012) The production of concentrated dispersions of few-layer graphene by the direct exfoliation of graphite in organosilanes. Nanoscale Research Letters 7: 674 (1–7).
35. Wu W, Liu Z, Jauregui L A, Yu Q, Pillai R, Cao H, Bao J, Chen Y P, Pei S S (2010) Wafer-scale synthesis of graphene by chemical vapor deposition and its application in gas sensing. Sensors and Actuators B 150: 296–300.
36. Wang Y, Yang R, Shi Z, Zhang L, Shi D, Wang E, Zhang G (2011) Super-elastic graphene ripples for flexible strain sensors. ACS Nano 5: 3645–3650.
37. Sojoudi H, Graham S (2013) Transfer-free selective area synthesis of graphene using solid-state self-segregation of carbon in Cu/Ni bilayers. ECS Journal of Solid State Science and Technology 2: M17–M21.

38. Kim K S, Zhao Y, Jang H, Lee S Y, Kim J M, Kim K S, Ahn J H, Kim P, Choi J Y, Hong B H (2009) Large-scale pattern growth of graphene films for stretchable transparent electrodes. Nature 457: 706–710.
39. Reina A, Jia X T, Ho J, Nezich D, Son H B, Bulovic V, Dresselhaus M S, Kong J (2009) Large area, few-layer graphene films on arbitrary substrates by chemical vapor deposition. Nano Letters 9: 30–35.
40. Berciaud S, Ryu S, Brus L E, Heinz T F (2009) Probing the intrinsic properties of exfoliated graphene: Raman spectroscopy of free-standing monolayers. Nano Letters 9: 346–352.
41. Cao H, Yu Q, Jauregui L A, Tian J, Wu W, Liu Z, Jalilian R, Benjamin D K, Jiang Z, Bao J, Pei S S, Chen Y P (2010) Electronic transport in chemical vapor deposited graphene synthesized on Cu: quantum Hall effect and weak localization. Applied Physics Letters 96: 122106 (1–3).
42. Ferrari A C, Meyer J C, Scardaci V, Casiraghi C, Lazzeri M, Mauri F S, Piscanec F S, Jiang D, Novoselov K S, Roth S, Geim A K (2006) Raman spectrum of graphene and graphene layers. Physical Review Letters 97: 187401 (1–4).
43. Peng Xu, James Loomis, and Balaji Panchapakesana. load transfer and mechanical properties of chemically derived single layer graphene reinforcements in polymer composites. Nanotechnology. 2012 December 21; 23(50): 505713
44. Mohiuddin T M G, Lombardo A, Nair R R, Bonetti A, Savini G, Jalil R, Bonini N, Basko D M, Galiotis C, Marzari N, Novoselov K S, Geim A K, Ferrari A C (2009) Uniaxial strain in graphene by Raman spectroscopy: G peak splitting, Grüneisen parameters, and sample orientation. Physical Review B 79: 205433.
45. Galpaya D, Wang M, Liu M, Motta N, Waclawik E, Yan C (2012) Recent advances in fabrication and characterization of graphene-polymer nanocomposites. Graphene 1: 30–49
46. Kuilla T, Bhadrab S, Yao D, Kim N H, Bose S, Lee J H (2010) Recent advances in graphene based polymer composites. Progress in Polymer Science 35: 1350–1375
47. Xie S H, Liu Y Y, Li J Y (2008) Comparison of the effective conductivity between composites reinforced by graphene nanosheets and carbon nanotubes. Applied Physics Letters 92 : 243121 (1–3).
48. Stankovich S, Dikin D A, Dommett G H B, Kohlhaas K M, Zimney E J, Stach E A, Piner R D, Nguyen S T, Ruoff R S (2006) Graphene-based composite materials. Nature 442: 282–286.
49. Alamusi N H, Hisao F, Satoshi A, Yaolu L, Jinhua L (2011) Piezoresistive strain sensors made from carbon nanotubes based polymer nanocomposites. Sensor 11: 10691–10723.
50. Liang J J, Wang Y, Huang Y, Ma Y, Liu Z, Cai J, Zhang C, Gao H, Chen Y (2009) Electromagnetic interference shielding of graphene/epoxy composites. Carbon 47: 922–925.
51. Pang H, Chen T, Zhang G, Zeng B, Li Z M (2010) An electrically conducting polymer/graphene composite with a very low percolation threshold. Materials Letters 64: 2226–2229.
52. Zhang H B, Zheng W G, Yan Q, Yang Y, Wang J W, Lu Z H, Ji G Y, Yu Z Z (2010) Electrically conductive polyethylene terephthalate/graphene nanocomposites prepared by melt compounding. Polymer 51: 1191–1196.
53. Kim H, Y. Miura Y, Macosko C W (2010) Graphene/Poly- urethane Nanocomposites for Improved Gas Barrier and Electrical Conductivity. Chemistry of Materials 22: 3441–3450.
54. Maiti S, Shrivastava N K, Suin S, Khatua B B (2013) Polystyrene/MWCNT/graphite nanoplate nanocomposites: efficient electromagnetic interference shielding material through graphite nanoplate–MWCNT–graphite nanoplate networking. ACS Applied Materials & Interfaces 5: 4712–4724.
55. Zhang H B, Zheng W G, Yan Q, Yang Y, Wang J W, Lu Z H, Ji G Y, Yu Z Z (2010) Electrically conductive polyethylene terephthalate/graphene nanocomposites prepared by melt compounding. Polymer 51: 1191–1196.
56. Liang Y, Gholamerza P, Seyed A Monemian, Ica M Zioczowe (2014) Epoxy composites with carbon nanotubes and graphene nanoplatetletes-dispersion and synergy effects. Carbon 78: 268–278.
57. He L, Tjong S C (2013) Low percolation threshold of graphene/polymer composites prepared by solvothermal reduction of graphene oxide in the polymer solution. Nanoscale Research Letters 8: 132 (1–7).

58. Nan C W, Shen Y, Ma J (2010) Physical properties of composites near percolation. Annual Review of Materials Research 40: 131–151.
59. Cui L L, Lu X F, Chao D M, Liu H T, Li Y X, Wang C (2011) Graphene-based composite materials with high dielectric permittivity via an in situ reduction method. physica status solidi (a) 208: 459–461.
60. Hu H, Wang X, Wang J, Wan L, Liu F, Zheng H, Chen R, Xu C (2010) Preparation and properties of graphene nanosheets-polystyrene nanocomposites via in situ emulsion polymerization. Chemical Physics Letters 484 : 247–253.
61. Lee Y R, Raghu A V, Jeong H M, Kim B K (2009) Properties of waterborne polyurethane/functionalized graphene sheet nanocomposites prepared by an in situ method. Macromolecular Chemistry and Physics 210 : 1247–1254.
62. Kim H, Macosko C W (2009) Processing–property relationships of polycarbonate/ graphene nanocomposites. Polymer 50:3797– 3809.
63. Geng Y, Wang S J, Kim J K (2009) Preparation of graphite nanoplatelets and graphene sheets. Journal of Colloid and Interface Science 336 : 592–598.
64. Wei T, Luo G, Fan Z, Zheng C, Yan J, Yao C, Li W, Zhang C (2009) Preparation of graphene nanosheet/polymer composites using in situ reduction-extractive dispersion. Carbon 47:2296–2299.
65. Niyogi S, Bekyarova E, Itkis M E, McWilliams J L, Hamon M A, Haddon R C (2006) Solution properties of graphite and graphene. Journal of American Chemical Society 128:7720–7721.
66. Stankovich S, Piner R D, Nguyen S T, Ruoff R S (2006) Synthesis and exfoliation of isocyanate-treated graphene oxide nanoplatelets. Carbon 44:3342–3347.
67. Liu N, Luo F, Wu H, Liu Y, Zhang C, Chen J (2008) One step ionic-liquid assisted electrochemical synthesis of ionic-liquid-functionalized graphene sheets directly from graphene. Advanced Functional Materials 18:1518–1525.
68. Stankovich S, Piner R D, Chen X, Wu N, Nguyen S T, Ruoff R S (2006) Stableaqueous dispersions of graphitic nanoplatelets via the reductionof exfoliated graphite oxide in the presence of poly(sodium 4-styrenesulfonate). Journal of Materials Chemistry 16:155–158.
69. Park S, An J, Piner R D, Jung I, Yang D, Velamakanni A, Nguyen S T, Ruoff R S (2008) Aqueous suspension and characterization of chemically modified grapheme sheets. Chemistry of Materials 20:6592–6594.
70. Bourlinos A B, Gournis D, Petridis D, Szabo T, Szeri A, Dekany I (2003) Graphite oxide: chemical reduction to graphite and surface modificationwith primary aliphatic amines and amino acids. Chemistry of Materials 19: 6050–6055.
71. Lomeda J R, Doyle C D, Kosynkin D V, Hwang W F, Tour J M (2008) Diazonium functionalization of surfactant-wrapped chemically converted graphene sheets. Journal of American Chemical Society 130:16201–16206.
72. Li D, Muller M B, Gilje S, Kaner R B, Wallac G G (2007) Processable aqueous dispersions of graphene nanosheets. Nature Nanotechnology 3:101–105.
73. Worsley K A, Ramesh P, Mandal S K, Niyogi S, Itkis M E, Haddon R C (2007) Soluble graphene derived from graphite fluoride. Chemical Physics Letters 445: 51–56.
74. Eda G, Chhowalla M (2009) Graphene-based composite thin films for electronics. Nano Letters 9: 814–818.
75. Bhadra S, Khstagir D, Singh A K, Lee J H (2009) Progress in preparation, processing and applications of polyaniline. Progress in Polymer Science 34:783–810.
76. Chen Z, Ren W, Gao L, Liu B, Pei S, Cheng H M (2011) Three-dimensional flexible and conductive interconnected graphene networks grown by chemical vapour deposition. Nature Materials 10: 424–428.
77. Xu Y, Liu Z, Zhang X,Wang Y, Tian J, Huang Y (2009) A graphene hybrid material covalently functionalized with porphyrin: synthesis and optical limiting property. Advanced Mater 21: 1275–1279.

78. Berger C, Song Z, Li T, Li X, Ogbazghi AY, Feng R (2004) Ultrathin epitaxial graphite: 2D electron gas properties and a route toward graphene-based nanoelectronics. Journal of Physics and Chemistry B 108:19912–19916.
79. Kim SC, Lee H, Jeong HM (2010) Effect of pyrene treatment on the properties of graphene/epoxy nanocomposites. Macromolecular Research 18(11): 1125–1128.
80. Zhang H B, Zheng W G, Yan Q, Yang Y, Wang J W, Lu Z H, Ji G Y, Yu Z Z (2010) Electrically conductive polyethylene terephthalate/graphene nanocomposites prepared by melt compounding. Polymer 51:1191–1196.
81. Hao R, Qian W, Zhang L, Hou Y (2008) Aqueous dispersions of TCNQ-anion stabilized graphene sheets. Chem Commun 48: 6576–6578.
82. Xu Y, Bai H, Lu G, Li C, Shi G (2008) Flexible graphene films via the filtration of water-soluble noncovalent functionalized graphene sheets. Journal of American Chemical Society 130:5856–5857.
83. Wang T X, Liang G Z, Yuan L, Gu A (2014) Unique hybridized graphene and its high dielectric constant composites with enhanced frequency stability, low dielectric loss and percolation threshold. Carbon 77: 920–932.
84. Li Y, Samad Y A, Polychronopoulou K, Alhassan S M, Liao K (2014) Highly Electrically Conductive Nanocomposites Based on PolymerInfused Graphene Sponges. Scientific Reports 4:4652 (1–6).
85. Zhou T N, Qi X D, Fu Q (2013) The preparation of the poly(vinyl alcohol)/graphene nanocomposites with low percolation threshold and high electrical conductivity by using the large-area reduced graphene oxide sheets. Express Polymer Letters 7: 747–755.
86. Park J U, Nam S W, Lee M S, Lieber C M (2012) Synthesis of monolithic graphene–graphite integrated electronics. Nature Materials 11: 120–125.

Graphene/Polymer Nanocomposites with High Dielectric Performance: Interface Engineering

Dongrui Wang, Feng You and Guo-Hua Hu

Abstract High permittivity polymer-based composites are highly desired due to their potential applications as high energy density capacitors and inherent advantages of easy processing, flexibility and light weight. Graphene, a two-dimensional nanomaterial with a layer thickness of one atom, has showed many outstanding properties and aroused tremendous research enthusiasm. Its large aspect ratio, high surface area and high electric conductivity make it an ideal filler for fabricating polymer-based percolative nanocomposites with high dielectric performance. This chapter reviews progresses in graphene-filled polymer nanocomposites with high permittivity that have been made over the past few years. The basic theory of percolation and the interface effect of graphene/polymer on the dielectric properties of nanocomposites are discussed.

Keywords Graphene · Nanocomposite · Dielectric permittivity · Percolation · Energy storage

1 Introduction

Dielectric materials have long been investigated and widely applied in power and electronic industries. They exhibit polarization and conducting phenomena under an applied electric field through charge migration. Due to their insulating

D. Wang
Department of Polymer Science and Engineering, School of Chemistry and Biological Engineering, University of Science and Technology Beijing, 100083 Beijing, China

F. You
School of Material and Engineering, Wuhan Institute of Technology, 430073 Wuhan, China

G.-H. Hu (✉)
Laboratory of Reactions and Process Engineering (CNRS UMR 7274), Université de Lorraine - CNRS, ENSIC, 1 rue Grandville, BP 20451, 54001 Nancy, France
e-mail: guo-hua.hu@univ-lorraine.fr

K.K. Sadasivuni et al. (eds.), *Graphene-Based Polymer Nanocomposites in Electronics*, Springer Series on Polymer and Composite Materials,
DOI 10.1007/978-3-319-13875-6_3

nature, charge in dielectric materials is restricted in the dielectric molecular or local space, which results in local migration under the electric field and generates induced dipole moment. Therefore the dielectric properties of a material, including dielectric constant, dielectric loss, and dielectric breakdown strength, are closely related to polarization. The complex permittivity of a material can be described as:

$$\varepsilon^* = \varepsilon' - j\varepsilon'' \tag{1}$$

where ε' and ε'' are the real and imaginary parts of the complex permittivity and $j = \sqrt{-1}$. The real part of the permittivity is given by:

$$\varepsilon' = \varepsilon_0 \varepsilon_r \tag{2}$$

where ε_0 is the permittivity of vacuum (8.85×10^{-12} F/m) and ε_r is the relative permittivity or dielectric constant of the material. Conventionally, the dielectric constant can also be expressed as "*k*" for dielectrics in engineering applications. The magnitude of dielectric constant can reflect the storage capability of electrical energy from the applied electric field. A parallel plate capacitor with area *A* and thickness *d* shows a capacitance given by:

$$C = \varepsilon_0 \varepsilon_r A/d \tag{3}$$

Meanwhile, the imaginary part of the complex permittivity ε'' is called the dielectric loss, which means some of the energy is dissipated due to the charge migration and molecular vibration under an applied electric field. For an energy storage device, we would like to minimize the dielectric loss for a given dielectric constant. Therefore, for a given dielectric material, a quantity named loss factor (or loss tangent) is often used to describe the dielectric loss:

$$\tan \delta = \varepsilon''/\varepsilon' \tag{4}$$

The polarization of dielectric materials may vary with external conditions, such as temperature and electric field frequency. Consequently, the dielectric properties of materials may vary upon changes in these conditions. Different applications may require different values of dielectric constant. High dielectric constant (high-*k*) and high dielectric strength are essential for capacitors with a high energy storage density, while a low dielectric constant (low-*k*) is needed for insulative packing material in micro-electronics industry to decrease the circuit capacitance.

Nowadays, demands have become more and more stringent for dielectric materials. Traditional polymeric or ceramic dielectric materials can hardly meet all the requirements for dielectric properties for specific applications. For instance, the dielectric constant and energy storage density of a biaxially oriented polypropylene (BOPP) dielectric film in capacitor are only 2 and 1 J/cm^3, respectively. If the dielectric constant of the BOPP film can be increased by a factor of 100 while its dielectric loss and strength are retained, its energy storage density can be increased to 100 J/cm^3, a value very close to that of lithium cell or electrochemical supercapacitors. However, this goal is almost impossible to achieve by optimizing the chemical structure of PP alone. An efficient method is to incorporate functional fillers to obtain polymer-based composites with high-*k* materials [1, 2]. Many research groups have reported that the dielectric constant of polymeric matrices can be efficiently enhanced through incorporating high-*k* ceramic nanoparticles such as barium titanate ($BaTiO_3$) whose *k* is 1,700 at room temperature [3, 4]. However, the high-*k* of the resulting nanocomposites can only be achieved when the volume fraction of ceramic fillers exceeds 50 %. Such high loadings of inorganic fillers severely suppress the merits of flexibility, low density, and low cost of polymers and deteriorate the final performance of composites [5].

Recently, fabrication of nanocomposites with high-*k* through incorporating conductive nanoparticles (such as metal nanoparticles, carbon nanotubes, etc.) into polymer matrices has aroused intensive interests. Generally, the dielectric constant of a polymer matrix can be enhanced by a factor of 10 by adding only less than 10 vol% of conducting nano-fillers. For example, the dielectric constant of polyvinylidenefluoride (PVDF) can be increased by 400 and 500 times with the addition of 17 vol% of Ni nanoparticles [6] and 8 vol% of multi-walled carbon nanotubes [7], respectively. The dramatic increase in dielectric constant is related to the formation of a conducting percolation network of the conductive fillers in the polymer matrix. The percolation threshold in nanocomposites is closely dependent of the shape and dimension of nano-fillers, which decrease with increasing aspect ratio of nano-fillers [8]. Therefore, one can expect to obtain dielectric nanocomposites with high-*k* by incorporating a lower fraction of nano-fillers with a larger aspect ratio.

Since its first experimental finding by Geim's group in 2004 [9], graphene, a monolayer of graphite, has attracted the most intense interest in recent years. Owing to its unique structure in which sp^2 hybridized carbon atoms packed into a two-dimensional honey-comb lattice, graphene exhibits many superior fundamental physical properties such as high mobility of charge carriers (200,000 $cm^2\ V^{-1}\ s^{-1}$), high thermal conductivity (5,000 $W\ m^{-1}K^{-1}$), excellent Young's modulus (~1,100 GPa) and fracture strength (125 GPa) [10, 11]. Because of these properties, graphene is considered as a very promising material for a wide range of applications, including advanced field effect transistors [12, 13], sensors [14, 15], transparent conductive electrodes [16, 17], energy storage devices [18, 19], high-performance polymer composites [20–22], and so on. Strictly speaking, the term graphene only refers to the monolayer graphite flake without any structural defects. However, in this chapter, this term will be generalized to a class of graphene-like substances, including the strict sense of graphene nanosheets, chemically modified graphene (CMG), graphene oxide (GO), etc. Recently, interest among chemists and researchers from other disciplines has increasingly shifted from graphene as an individual component to graphene-based hybrid materials in which graphene nanosheets are functionalized with other matters such as organic functional groups, metal oxide nanoparticles, and polymers [23, 24]. Graphene-based hybrid materials have opened up more possibilities in practical applications ranging from polymer electronics, energy materials, to biomaterials.

Owing to the large aspect ratio and superior conducting nature of graphene, nanocomposites with high-*k* can be realized by filling the polymer matrix with graphene nanosheets at relatively low loading. Compared to zero-dimensional metallic conducting particles or one-dimensional carbon nanotubes, two-dimensional graphene nanosheets can form a conducting network in the polymer matrix through contact and overlap with each other. Consequently, the percolation threshold in the resulting nanocomposites is lower than that in metal nanoparticles or carbon nanotube-filled nanocomposites. This means that high-*k* can be achieved by using smaller amounts of fillers while the merits of polymers, including flexibility, easy-processing, and low-cost, can be retained to a great extent. Actually, research on the dielectric properties of graphene-filled nanocomposited started to appear in the literature in 2009 and has aroused many interests over the past five years. The dielectric properties of this type of nanocomposite can be well-tuned and optimized by controlling the state of dispersion of graphene in polymer matrices and the interfacial interactions between graphene and polymer.

2 Basic Theory of Percolative Dielectrics

Significant enhancement in permittivity of polymer matrices can be realized by incorporating only a small amount of graphene. The main reason is that two-dimensional graphene nanosheets dispersed in polymer matrices act as large amounts of micro-capacitors, which improve the electric energy storage capability of the nanocomposites. As graphene nanosheets are being incorporated into polymer matrices, they tend to interconnect with each other to form a conducting network. When the graphene loading reaches a critical value, the conducting network is built up and the nanocomposite is transformed from an insulator to a conductor. This transition is called "percolation transition" and the critical fraction of fillers is known as "percolation threshold".

As the volume fraction of conductive fillers approaches the percolation threshold, the conductivity of the nanocomposites can be described as follows [8],

$$\sigma \propto \sigma_1(\varphi - \varphi_c)^t, \quad \text{when } \varphi > \varphi_c \tag{5}$$

$$\sigma \propto \sigma_2(\varphi_c - \varphi)^{-s}, \quad \text{when } \varphi < \varphi_c \tag{6}$$

where φ is the volume fraction of conducting filler, φ_c is the percolation threshold, σ_1 and σ_2 are the electrical conductivity of the conductive filler and insulative matrix, respectively, t and s are the dimensional parameters. Generally, the values of t and s for two-dimensional materials are in the range of 1.6–2, while those for three-dimensional materials are in the range of 1–1.3 [25]. In addition, the percolation threshold is closely related to the shape of the fillers. When the fillers are randomly dispersed in the matrix, the percolation threshold decreases with increasing aspect ratio of the fillers [8]. This is why two-dimensional graphene nanosheets with an ultrahigh aspect ratio have attracted so much attention to be utilized as fillers for high conductive and high-*k* nanocomposites.

For the insulator-conductor composites discussed above, when the percolation threshold is approached from a lower filler loading, the dielectric constant k of the composite also undergoes a dramatic transition as follows,

$$k \propto k_{\mathrm{m}}(\varphi - \varphi_{\mathrm{c}})^{-s} \tag{7}$$

where k and k_{m} are the dielectric constants of the composite and the polymer matrix, respectively. This equation suggests that a large dielectric constant can be achieved in composites when the fraction of fillers is very close to the percolation threshold. Moreover, the dielectric constant of the polymer matrix plays an important role in the dielectric properties of resulting composites. Choosing a polymer with a higher dielectric constant as the matrix is more efficient to prepare composites with a higher energy storage capacity.

In such percolative composite systems, the conductivity and dielectric permittivity show strong dependency on the frequency of the applied electric field. According to the percolation theory, the relationships between the conductivity (or the permittivity) and the frequency can be expressed by the following exponential equation when the content of fillers approaches the percolation threshold:

$$\sigma(\omega, \varphi_{\mathrm{c}}) \propto \omega^{u} \tag{8}$$

$$k(\omega, \varphi_{\mathrm{c}}) \propto \omega^{-v} \tag{9}$$

where ω is the angular frequency, u and v are the critical parameters. Theoretically, there is a relationship for the four parameters (u, v, t, s) [25]:

$$u = t/(t+s),\ v = s/(t+s),\ u + v = 1 \tag{10}$$

The dielectric permittivity of such percolative nanocomposites also shows strong dependency on temperature. The dielectric spectroscopy, i.e. the dielectric constant and loss in wide frequency and temperature ranges, can provide information about dielectric relaxations, including the relaxation of polymer chain segments and filler/matrix interfaces. The temperature dependence of the relaxation frequency obeys the Arrhenius equation:

$$f = f_0 \exp\left(\frac{\Delta E_{\mathrm{a}}}{k_{\mathrm{B}} T}\right) \tag{11}$$

where ΔE_{a} is the activation energy of a dielectric relaxation process in the composites, f_0 is a characteristic frequency, k_{B} is the Boltzmann constant, and T is the absolute temperature. The temperature dependence of the dielectric constant or loss is often used to investigate the relaxation mechanisms in nanocomposites. The movement of dipoles in nanocomposites becomes easier as the temperature increases, which will lead to two different results. On the one hand, more dipoles will be induced and higher orientation degree of the dipoles can be reached in the applied electric field. On the other hand, the disorientation of dipoles will also become easier because of the high kinetic energy at a high temperature [26]. These two competitive processes complicate the temperature dependence of percolative composites' dielectric properties.

3 Graphene-Filled Nanocomposites with High Permittivity

Basically, fabrication methods for graphene-filled nanodielectrics are the same with those of conventional composites. Different compounding methods, including solution mixing, melt blending, and in situ intercalation polymerization, can be utilized according to the nature of the polymer matrix and that of the filler. Each method has its advantages and disadvantages. Many review articles have summarized preparation strategies for the graphene/polymer nanocomposites [21, 22]. Therefore, it will not be discussed in detail here.

3.1 Dielectric Properties of Graphene/Polymer Nanocomposites Through Simple Compositing

For the graphene-filled percolative nanocomposites, as described above, the selection of polymer matrix is very important for the final dielectric properties. A polymer matrix with a relatively high dielectric constant is more efficient to fabricate nanocomposites with higher permittivity. PVDF is a ferroelectric polymer with a relative dielectric constant of 9–11, which is the highest among common polymers. Therefore, many researchers have used PVDF as the polymer matrix to fabricate graphene-filled nanocomposites with high dielectric performance. However, the final dielectric properties of resultant nanocomposites vary greatly with the fabrication procedures.

He et al. [27] reported on graphene-filled nanocomposites with high-*k* by using PVDF as the matrix. Exfoliated graphite nanoplatelets (xGnPs) were mixed with PVDF through a solution mixing process. As shown in Fig. 1, the percolation threshold was low (1.0 vol%) and the dielectric constant was as high as 200 near the percolation threshold at 1 kHz. It should be mentioned that the dielectric loss increased to 0.48 near the percolation threshold, which is relatively high and is not suitable for energy storage.

The literature reported on the use of GO as the starting material to mix with PVDF, followed by in situ chemical reduction to fabricate graphene/PVDF nanocomposites with high-*k*. Li et al. [28] realized in situ reduction of GO in PVDF matrix by using hydrazine as a reduction agent. The resulting nanocomposite showed a percolation threshold of 2.5 wt% of reduced graphene oxide (rGO). Shang et al. [29] used the same reduction method to prepare a series of rGO/PVDF nanocomposites with a percolation threshold of 1.3 vol%. Their rGO/PVDF nanocomposites showed a high-*k* of 63 at 100 Hz near the percolation threshold. Cui et al. [30] reduced GO in PVDF solution by using phenylhydrazine. They found that the dielectric constant of the resulting nanocomposite at $\varphi_c = 4.1$ vol% was as high as 2,080 at 1 kHz. Fan et al. [31] also prepared rGO/PVDF nanocomposites through solution mixing and in situ phenylhydrazine reduction method. The resulting nanocomposites showed a very low percolation threshold of

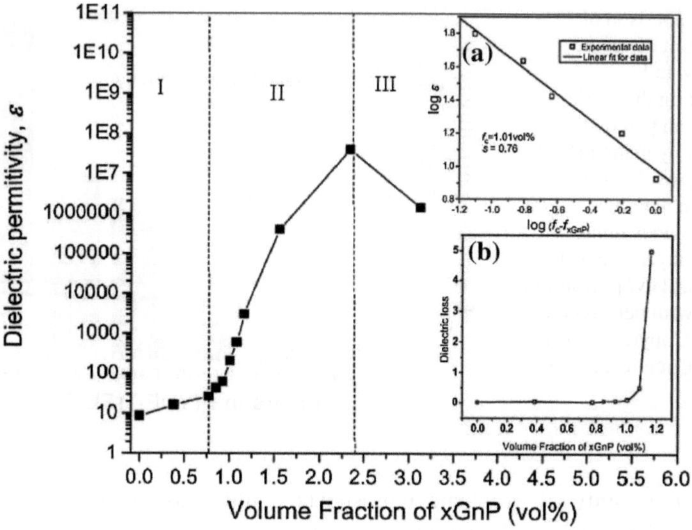

Fig. 1 Permittivity of PVDF/xGnPs nanocomposites with various xGnP loadings, measured at 1 kHz and room temperature. Reprinted with permission from Ref. [27]. Copyright 2009 John Wiley & Sons, Inc

0.18 vol%. Near the threshold, the dielectric constant of the rGO/PVDF nanocomposites reached 340 at 100 Hz.

Besides chemical reduction, GO can also be thermally reduced, which avoids the use of toxic reduction agents. The as-prepared graphene is called thermally reduced graphene oxide (TRGO). Tang et al. [32] developed a simple thermal in situ reduction method of GO in PVDF matrix to prepare TRGO/PVDF nanocomposites and the percolation threshold was 1.60 vol%. Yu et al. [33] fabricated TRGO/PVDF nanocomposites through solution mixing. The percolation threshold of this nanocomposite was 4.5 wt% and the dielectric constant near the threshold was about 40 at 100 Hz. Recently, Almadhoun et al. [34] carefully examined the effect of reduction methods of GO on the dielectric performance of graphene/PVDF nanocomposites. They prepared two types of rGO through hydrothermal reduction (HT-rGO) and hydrazine reduction (HZ-rGO) and incorporated those two different types of fillers into poly(vinylidene fluoride-trifluoroethylene-chlorofluoroethylene) (P(VDF–TrFE–CFE)). The results, as shown in Fig. 2, revealed that nanocomposites with a higher dielectric constant can be achieved by using the HZ-rGO as the filler near the percolation threshold (ca. 1 vol%).

The following conclusion can be drawn from the literature results described above: the percolation threshold and dielectric constant of the graphene/PVDF nanocomposites depend very much on the nature of graphene nanosheets and the fabrication processes of nanocomposites. Compared with TRGO, rGO-filled nanocomposites generally show a lower percolation threshold, which should be attributed to the fact that they have larger aspect ratio than TRGO. Furthermore, the higher

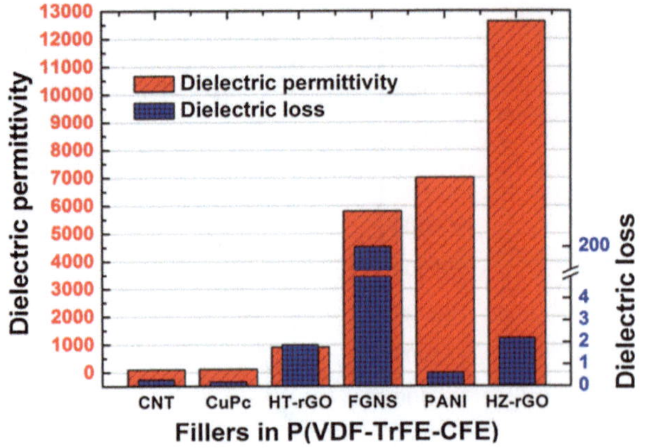

Fig. 2 Comparison of the maximum dielectric permittivity and corresponding dielectric loss in typical percolative composites by using P(VDF–TrFE–CFE) as the matrix. *CNT* carbon nanotube, *CuPc* copper phthalocyanine, *FGNS* few-layer graphene nanosheets, *PANI* polyaniline. Reprinted with permission from [34]. Copyright 2014 American Chemical Society

of the conductivity of graphene nanosheets could lead to the higher dielectric constant of the corresponding nanocomposite.

In addition to PVDF, there is a rich literature on the incorporation of graphene nanosheets into other polymer matrices to prepare high-*k* nanocomposites. Dimiev et al. [35] compared the effects of graphene and carbon nanotubes on the dielectric properties of a silicon rubber. Their results revealed that graphene nanosheet filled silicon rubber exhibited a higher dielectric constant and a lower dielectric loss than carbon nanotube filled one. Das et al. [36] obtained similar results and found that graphene also had a greater effect on improving the microwave absorbance of thermoplastic polyurethanes than carbon nanotubes. Our group prepared rGO-filled PP nanocomposites by in situ reducing GO in PP emulsion. The percolation threshold of the resulting nanocomposite was as low as 0.03 vol% [37]. Shevchenko et al. [38] used an in situ polymerization method to fabricate rGO/PP nanocomposites and obtained a percolation threshold of 0.25 vol%. Srivastava et al. [39] also prepared graphene-filled polystyrene (PS) nanocomposites by an in situ polymerization process, and found that the dielectric constant of the nanocomposites near the percolation threshold ($\varphi_c = 1.10$ vol%) was 12 at 100 Hz. Panwar et al. [40] fabricated graphene-filled acrylonitrile-styrene copolymer (SAN) nanocomposites by melt blending. Their results revealed that the percolation threshold was 2.71 vol% and the corresponding dielectric constant near the threshold reached 468 at 100 kHz. Romasanta et al. [41] investigated the effect of graphene loadings on the dielectric constant of polydimethylsiloxane (PDMS). They found a percolation threshold of 2.0 wt% and a dielectric constant of 23 at 10 Hz near the percolation threshold. Bhadra et al. [42] fabricated rGO by using a hydrothermal reduction technique and found that the percolation threshold of rGO-filled polyvinylalcohol (PVA) nanocomposites was 0.41 vol%. The dielectric constant of the resulting nanocomposites near the percolation threshold reached 1,000 at 1 kHz.

Though the incorporation of graphene nanosheets can significantly enhance the dielectric constant of the polymer matrix, the dielectric loss of the nanocomposites also increases to a relatively high degree due to the increase in leakage current induced by the conducting fillers. As a substitute, GO can not only enhance the dielectric constant but also maintain the dielectric loss of resultant nanocomposites at a low level. For instance, Wang et al. [43] have reported that the dielectric constant of PDMS can be increased to 8 by blending with 5 phr of GO while the dielectric loss is only 0.02.

3.2 *Effects of Interfacial Functionalization on Dielectric Properties of Graphene/Polymer Nanocomposites*

It has long been known that aggregation of nano-fillers must be efficiently suppressed to obtain nanocomposites with superior performance. For graphene-filled polymers, interactions between graphene and polymer matrices are generally weak. Thus graphene nanosheets tend to aggregate to decrease the interfacial energy, resulting in a decrease in the aspect ratio and deterioration of the final performance of the nanocomposites. Furthermore, incorporation of conducting graphene nanosheets also gives rise to a large increase in dielectric loss, which must be avoided for insulating materials. Therefore, interfaces between graphene nanosheets and polymer matrices should be well-designed and tailored to improve the dispersion of graphene and the final properties of nanocomposites. To date, several efficient interfacial functionalization approaches have been developed to fabricate graphene/polymer nanocomposites with high dielectric performance.

Chu et al. [44] prepared porous graphene nanosheets through etching graphene in sodium hydroxide solution during a hydrothermal process. Thus the graphene changed from a conductor to a semiconductor because its conjugating structure was destroyed. The structure of the resultant graphene nanosheets is shown as Fig. 3. They further prepared "semiconductor-conductor-semiconductor" sandwich hybrid nano-fillers by using the porous graphene and unmodified one and incorporated this type of filler into PVDF matrix. Their results revealed that the hybrid nano-fillers could be well dispersed in the PVDF matrix because of the strong interaction between the porous graphene and the PVDF. In addition, the leakage current could be effectively suppressed due to the low electrical conductivity of the outer layer of the hybrid nano-filler. Consequently, the dielectric constant of resulting nanocomposites was increased to 350 at 1 kHz while the dielectric loss was only 0.52 near the percolation threshold ($\varphi_c = 3.0$ vol%).

More works have concentrated on improving the filler-matrix interactions through tailoring the surface properties of graphene nanosheets to develop nanocomposites with superior dielectric performance. For example, Wang et al. [45] made (1-hexadecyl) triphenylphosphonium bromide (HTPB)-functionalized graphene through non-covalent absorption of HTPB on graphene surfaces. Owing to good dispersion of the HTPB-functionalized graphene in PVDF matrix, the nanocomposites showed a low percolation threshold of 0.66 wt%. They also found that the dielectric constant of the nanocomposite with 0.86 wt% of functionalized

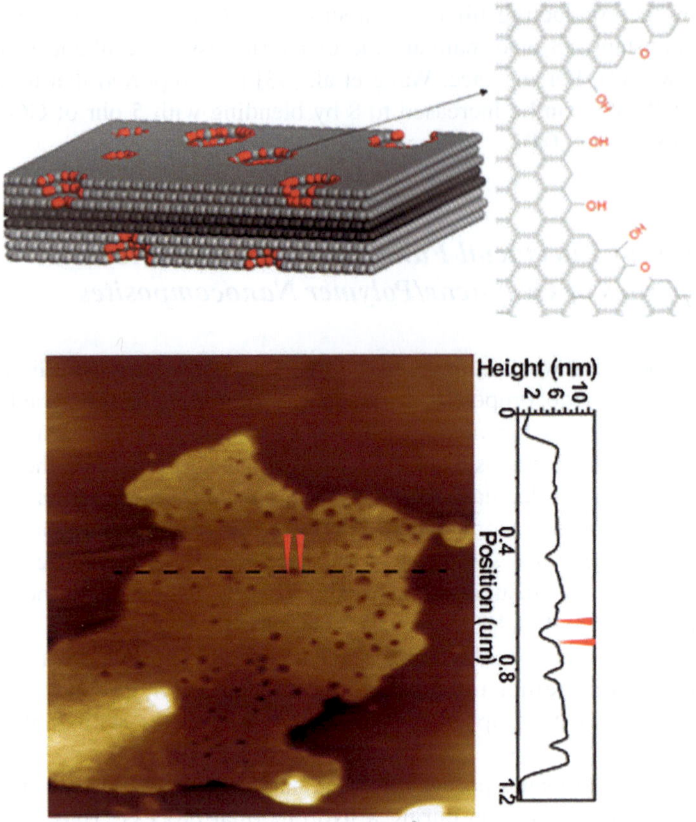

Fig. 3 Schematic structure and the AFM image of porous "semiconductor-conductor-semiconductor" sandwich graphene nanosheets. Adapted with permission from [44]. Copyright 2013 Elsevier

graphene was 3 times higher than that of the neat PVDF, while the dielectric loss was as low as 0.07.

Besides the non-covalent functionalization manner, the surface properties of graphene nanosheets can also be modified through the covalent bonds. For example, our group investigated the effect of PVA modified rGO (rGO-PVA) on the dielectric properties of PVDF matrix [46]. The results are shown in Fig. 4. Although the percolation threshold was increased from 0.61 vol% for rGO filled PVDF nanocomposites to 2.24 vol% for rGO-PVA filled nanocomposites, the dielectric loss of the latter near the percolation threshold was only 0.5, which was lower than that of the former. Two hypotheses were made to explain this interesting phenomenon. One is strong hydrogen bonding between rGO-PVA and PVDF matrix which could lead to more homogenous dispersion of nano-fillers. The other is that the PVA layer on rGO surfaces could suppress charge transportation between rGO nanosheets, which would effectively decrease dielectric loss.

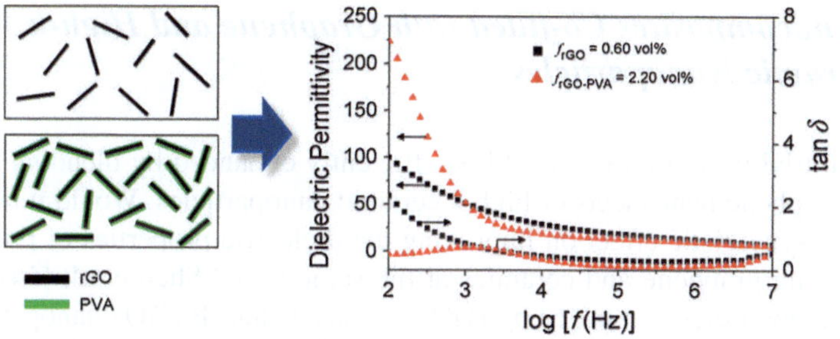

Fig. 4 Schematic structure of rGO-PVA and comparison in dielectric properties of rGO/PVDF and rGO-PVA/PVDF near the percolation threshold. Reprinted with permission from Ref. [46]. Copyright 2012 American Chemical Society

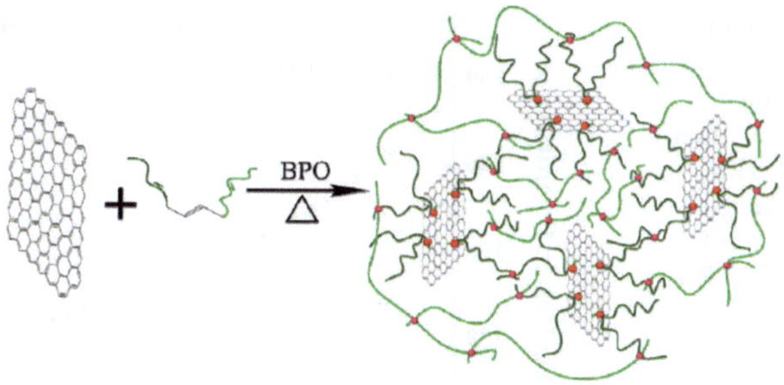

Fig. 5 Schematic description of the reaction in P(VDF–TrFE–DB)/GNS crosslinked nanocomposites. Reprinted with permission from Ref. [47]. Copyright 2013 American Chemical Society

Recently, Wen et al. [47] functionalized graphene with PVDF derivatives (poly(vinylidenefluoride-trifluorethylene with double bonds) and obtained P(VDF–TrFE–DB)) covalently grafted on surfaces. Figure 5 depicts the preparation procedure. The covalent grafting can improve the dispersion of graphene in PVDF matrix, leading to a decrease in dielectric loss. The percolation threshold of the nanocomposites was about 4 wt% and the dielectric constant was 74 at 1 kHz near the percolation while the dielectric loss was as low as 0.08.

There is still a rich literature on the preparation of graphene/polymer nanocomposites with high dielectric performance by using surface functionalized graphene to fill polymer matrices [48–53]. The basic idea is to anchor insulative small molecules or polymer chains onto graphene surfaces by covalent or non-covalent bonding. Consequently, the functionalized graphene can be well dispersed in polymer matrices due to improved interfacial interactions. In addition, the current leakage can be effectively suppressed. The fact that the surfaces of graphene are coated with insulating materials may result in a decrease in dielectric loss of resulting nanocomposites.

3.3 Nanocomposites Co-filled with Graphene and High-k Ceramic Nanoparticles

The permittivity of polymers could be efficiently enhanced by blending with conductive graphene nanosheets or high-*k* ceramic nanoparticles. Would it be possible to reach a synergistic effect on improving the dielectric properties of polymers by incorporating graphene and ceramics at the same time? Shen et al. [54] prepared hybrid nanoparticles composed by rGO nanosheets and $BaTiO_3$ nanoparticles and filled PVDF with them. The results revealed that the rGO nanosheets improved the state of dispersion of $BaTiO_3$ nanoparticles in PVDF, and the $BaTiO_3$ prevented rGO from forming a conducting network. Thus the resulting nanocomposites showed a high dielectric constant and a low dielectric loss. Furthermore,

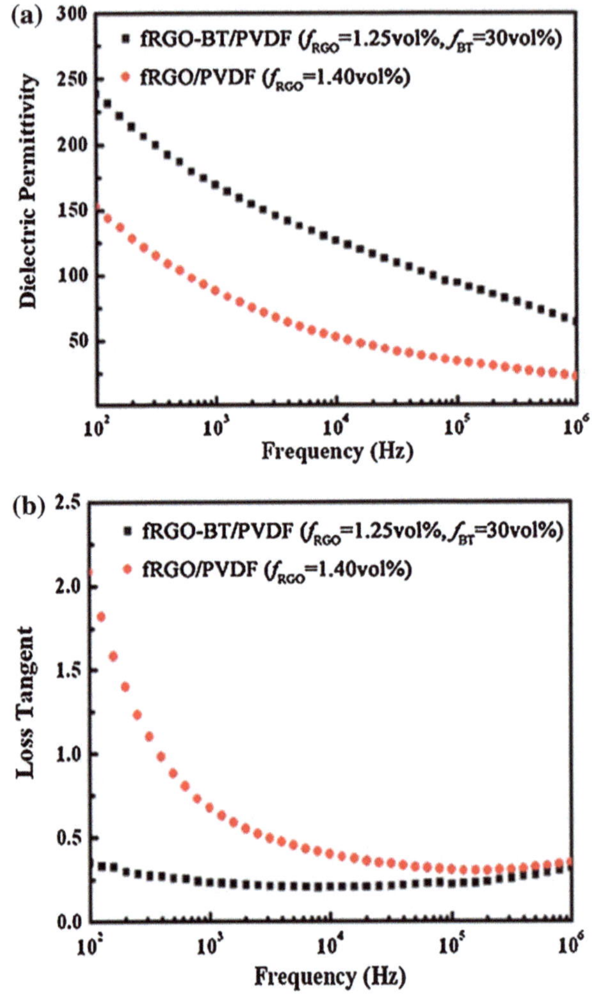

Fig. 6 Dielectric permittivity **a** and loss tangent **b** of RGO/PVDF nanocomposites near the percolation threshold, with or without the addition of $BaTiO_3$ nanoparticles. Adapted with permission from Ref. [55]. Copyright 2013 RSC Publishing

the dielectric strength was as high as 150 kV/mm, which would be suitable for practical applications. Our group used functionalized rGO and $BaTiO_3$ as co-fillers for PVDF and the resulting nanocomposites exhibited high permittivity and low dielectric loss, as shown in Fig. 6 [55].

Wu et al. [56] investigated the effect of hybrid nanoparticles, TiO_2 nanorod modified graphene, on the dielectric properties of PS. They found that the dielectric constant of nanocomposites with 10.9 vol% of hybrid fillers reached 1,741 at 100 Hz, which was 643 times that of the neat PS (2.7), and the dielectric loss was only 0.39. The permittivity of the nanocomposites was further tuned by controlling the amount of TiO_2 nanorods on the graphene nanosheets. The high dielectric performance could be attributed to anchoring of TiO_2 nanorods onto the graphene surfaces, which prevented conducting graphene nanosheets from direct contact and suppressed the leakage current. Similarly, Wang et al. [57] fabricated $BaTiO_3$ nanoparticle modified graphene and then used the hybrid filler to enhance the permittivity of PDMS. The results confirmed that the introduction of $BaTiO_3$ nano-fibers efficiently prevented dielectric loss from increasing by the conducting graphene. By carefully tuning the ratio of $BaTiO_3$ nano-fibers to graphene nanosheets, PDMS-based nanocomposites with high permittivity and low dielectric loss were obtained. There are still some examples on the enhancement of dielectric properties of polymers by co-filling with graphene and high-*k* ceramics [58, 59]. Improved dielectric properties, including high-*k*, high breakdown strength, and low loss may be realized at the same time through optimizing the multi-phase structures of the nanocomposites.

4 Summary and Outlook

According to the percolation theory, a high dielectric permittivity can be achieved by introducing conductive fillers into an insulative polymer matrix. Over the past five years, great progresses have been made in fabricating graphene-filled polymer nanocomposites with high dielectric properties. Compared with 0D nanospheres and 1D nanotubes, the graphene nanosheets with 2D structures can create more interfacial area, thus can trap more charges at the interfaces. The works in this field during the past few years have demonstrated the great potential of graphene/polymer nanocomposites. To date, research interest in this field has been shifted from simply blending graphene nanosheets with polymers to tuning the hierarchical structures of the nanocomposites to afford advanced nanodielectrics with desired dielectric properties. By appropriate structural design and good control of the preparation process, researchers have already successfully fabricated graphene/polymer nanocomposites with permittivity of several hundreds and dielectric loss of smaller than 0.1, which are very close to the requirements of dielectrics for high energy density storage.

Despite the great achievement in fabricating high-*k* graphene/polymer nanocomposites, there are still some severe challenges to conquer. One is the dielectric breakdown strength of the resultant nanocomposites. Although it is rarely mentioned in the literature, the fact is that the dielectric breakdown strength of polymer matrix may undergo a dramatical decrease by blending with conductive fillers due to the increase of conductance. The lowered breakdown strength offsets the advantage of an increased dielectric constant. Such nanocomposites may not be used as energy storage materials. This limitation may be partially addressed by tailoring the electronic transport properties of graphene. Recently it is reported that the fluorinated graphene shows a high breakdown strength of 10 MV/cm, which is about ten times that of graphene [60]. In view of this, one can expect that graphene/polymer nanocomposites with an acceptable breakdown strength can be obtained through optimizing the structure of graphene nanosheets.

Another challenge is how to fabricate high quality graphene/polymer nanocomposites with superior dielectric performance. Many works on graphene/polymer nanocomposites have been reported and have claimed that monolayer graphene is obtained and well-dispersed in polymer matrices. The fact is that most of them are graphene nanosheets containing several graphene layers, but not monolayer graphene. This is a reason that the properties of reported nanocomposites are often far below theoretical predictions. Thus it is very important and critical to develop specific preparation methods of monolayer graphene. In addition, fabrication approach is also crucial to the ultimate properties of graphene/polymer nanocomposites. Avoiding aggregation of monolayer graphene during the dispersion process is a huge challenge.

Furthermore, the dielectric properties of the nanocomposites strongly depend on the chemical structures of graphene and the fabrication methods. However, it is still very challenging to develop theoretical tools to predict dielectric properties and to design the structure of graphne/polymer nanocomposites that will provide prescribed properties.

In summary, graphene/polymer nanocomposites with high dielectric performance are very promising for energy storage and electromagnetic shielding applications. Nevertheless big efforts are still needed to develop relationships among fabrication processes, hierarchical structures, and dielectric properties of graphene/polymer nanocomposites.

Acknowledgment The authors gratefully acknowledge the financial support from NSFC (51103011), Beijing Natural Science Foundation (2132029, 2142023), and Beijing Higher Education Young Elite Teacher Project (YETP0392).

References

1. Dang Z-M, Yuan J-K, Zha J-W, Zhou T, Li S-T, Hu G-H (2012) Fundamentals, processes and applications of high-permittivity polymer matrix composites. Prog Mater Sci 57 (4):660–723.
2. Wang Q, Zhu L (2011) Polymer nanocomposites for electrical energy storage. J Polym Sci B: Polym Phys 49 (20):1421–1429.
3. Kim P, Simon C, Hotchkiss PJ, Joshua N, Kippenlen B, Marder SR, Perry JW (2007) Phosphonic acid-modified barium titanate polymer nanocomposites with high permittivity and dielectric strength. Adv Mater 19:1001–1005.

4. Kim P, Doss NM, Tillotson JP, Hotchkiss PJ, Pan MJ, Marder SR, Li JY, Calame JP, Perry JW (2009) High Energy Density Nanocomposites Based on Surface-Modified BaTiO3 and a Ferroelectric Polymer. ACS Nano 3 (9):2581–2592.
5. Barber P, Balasubramanian S, Anguchamy Y, Gong S, Wibowo A, Gao H, Ploehn H, Loye H (2009) Polymer Composite and Nanocomposite Dielectric Materials for Pulse Power Energy Storage. Materials 2:1697–1733.
6. Dang ZM, Lin YH, Nan CW (2003) Novel ferroelectric polymer composites with high dielectric constants. Adv Mater 15:1625–1629.
7. Dang ZM, Wang L, Yin Y, Zhang Q, Lei QQ (2007) Giant dielectric permittivities in functionalized carbon-nanotube/electroactive-polymer nanocomposites. Adv Mater 19 (6):852–857.
8. Nan CW, Shen Y, Ma J (2010) Physical Properties of Composites Near Percolation. Annu Rev Mater Res 40:131–151.
9. Novoselov KS, Geim AK, Morozov SV, Jiang D, Zhang Y, Dubonos SV, Grigorieva IV, Firsov AA (2004) Electric field effect in atomically thin carbon films. Science 306 (5296):666–669.
10. Zhu YW, Murali S, Cai WW, Li XS, Suk JW, Potts JR, Ruoff RS (2010) Graphene and Graphene Oxide: Synthesis, Properties, and Applications. Adv Mater 22 (35):3906–3924.
11. Park S, Ruoff RS (2009) Chemical methods for the production of graphenes. Nat Nanotechnol 4 (4):217–224.
12. Novoselov KS, Jiang D, Schedin F, Booth TJ, Khotkevich VV, Morozov SV, Geim AK (2005) Two-dimensional atomic crystals. Proc Nat Acad Sci 102 (30):10451–10453.
13. Gilje S, Han S, Wang M, Wang KL, Kaner RB (2007) A chemical route to graphene for device applications. Nano Lett 7 (11):3394–3398.
14. Schedin F, Geim AK, Morozov SV, Hill EW, Blake P, Katsnelson MI, Novoselov KS (2007) Detection of individual gas molecules adsorbed on graphene. Nat Mater 6 (9):652–655.
15. Robinson JT, Perkins FK, Snow ES, Wei ZQ, Sheehan PE (2008) Reduced Graphene Oxide Molecular Sensors. Nano Lett 8 (10):3137–3140.
16. Eda G, Fanchini G, Chhowalla M (2008) Large-area ultrathin films of reduced graphene oxide as a transparent and flexible electronic material. Nat Nanotechnol 3 (5):270–274.
17. Li XL, Zhang GY, Bai XD, Sun XM, Wang XR, Wang E, Dai HJ (2008) Highly conducting graphene sheets and Langmuir-Blodgett films. Nat Nanotechnol 3 (9):538–542.
18. Stoller MD, Park SJ, Zhu YW, An JH, Ruoff RS (2008) Graphene-Based Ultracapacitors. Nano Lett 8 (10):3498–3502.
19. Ruoff RS, Zhu YW, Murali S, Stoller MD, Ganesh KJ, Cai WW, Ferreira PJ, Pirkle A, Wallace RM, Cychosz KA, Thommes M, Su D, Stach EA (2011) Carbon-Based Supercapacitors Produced by Activation of Graphene. Science 332 (6037):1537–1541.
20. Stankovich S, Dikin DA, Dommett GHB, Kohlhaas KM, Zimney EJ, Stach EA, Piner RD, Nguyen ST, Ruoff RS (2006) Graphene-based composite materials. Nature 442 (7100):282–286.
21. Kuilla T, Bhadra S, Yao D, Kim NH, Bose S, Lee JH (2010) Recent advances in graphene based polymer composites. Prog Polym Sci 35:1350–1375.
22. Kim H, Abdala AA, Macosko CW (2010) Graphene/Polymer Nanocomposites. Macromolecules 43 (16):6515–6530.
23. Kim F, Cote LJ, Huang JX (2010) Graphene Oxide: Surface Activity and Two-Dimensional Assembly. Adv Mater 22 (17):1954–1958.
24. Kim J, Cote LJ, Huang JX (2012) Two dimensional soft material: New faces of grahpene oxide. Acc Chem Res 45 (8):1356–1364.
25. Nan C-W (1993) Physics of inhomogeneous inorganic materials. Prog Mater Sci 37 (1):1–116.
26. Li B, Zhong W-H (2011) Review on polymer/graphite nanoplatelet nanocomposites. J Mater Sci 46 (17):5595–5614.
27. He F, Lau S, Chan HL, Fan JT (2009) High Dielectric Permittivity and Low Percolation Threshold in Nanocomposites Based on Poly(vinylidene fluoride) and Exfoliated Graphite Nanoplates. Adv Mater 21 (6):710–715.

28. Li YC, Tjong SC, Li RKY (2010) Electrical conductivity and dielectric response of poly(vinylidene fluoride)-graphite nanoplatelet composites. Synth Met 160 (17–18):1912–1919.
29. Shang J, Zhang Y, Yu L, Shen B, Lv F, Chu PK (2012) Fabrication and dielectric properties of oriented polyvinylidene fluoride nanocomposites incorporated with graphene nanosheets. Mater Chem Phys 134 (2–3):867–874.
30. Cui L, Lu X, Chao D, Liu H, Li Y, Wang C (2011) Graphene-based composite materials with high dielectric permittivity via an in situ reduction method. Physica Status Solidi A 208 (2):459–461.
31. Fan P, Wang L, Yang J, Chen F, Zhong M (2012) Graphene/poly(vinylidene fluoride) composites with high dielectric constant and low percolation threshold. Nanotechnology 23 (36):365702.
32. Tang H, Ehlert GJ, Lin Y, Sodano HA (2012) Highly efficient synthesis of graphene nanocomposites. Nano Lett 12 (1):84–90.
33. Yu JH, Huang XY, Wu C, Jiang PK (2011) Permittivity, Thermal Conductivity and Thermal Stability of Poly(vinylidene fluoride)/Graphene Nanocomposites. IEEE Transactions on Dielectrics and Electrical Insulation 18 (2):478–484.
34. Almadhoun MN, Hedhili MN, Odeh IN, Xavier P, Bhansali US, Alshareef HN (2014) Influence of Stacking Morphology and Edge Nitrogen Doping on the Dielectric Performance of Graphene-Polymer Nanocomposites. Chem Mater 26 (9):2856–2861.
35. Dimiev A, Lu W, Zeller K, Crowgey B, Kempel LC, Tour JM (2011) Low-loss, high-permittivity composites made from graphene nanoribbons. ACS Appl Mater Interfaces 3 (12):4657–4661.
36. Das CK, Bhattacharya P, Kalra SS (2012) Graphene and MWCNT: Potential Candidate for Microwave Absorbing Materials. J Mater Sci Res 1 (2):126–132.
37. Wang D, Zhang X, Zha J-W, Zhao J, Dang Z-M, Hu G-H (2013) Dielectric properties of reduced graphene oxide/polypropylene composites with ultralow percolation threshold. Polymer 54 (7):1916–1922.
38. Shevchenko VG, Polschikov SV, Nedorezova PM, Klyamkina AN, Shchegolikhin AN, Aladyshev AM, Muradyan VE (2012) In situ polymerized poly(propylene)/graphene nanoplatelets nanocomposites: Dielectric and microwave properties. Polymer 53 (23):5330–5335.
39. Srivastava NK, Mehra RM (2008) Study of structural, electrical, and dielectric properties of polystyrene/foliated graphite nanocomposite developed viain situpolymerization. J Appl Polym Sci 109 (6):3991–3999.
40. Panwar V, Kang B, Park J-O, Park S, Mehra RM (2009) Study of dielectric properties of styrene-acrylonitrile graphite sheets composites in low and high frequency region. Eur Polym J 45 (6):1777–1784.
41. Romasanta LJ, Hernandez M, Lopez-Manchado MA, Verdejo R (2011) Functionalised graphene sheets as effective high dielectric constant fillers. Nanoscale Research Letters 6:508.
42. Bhadra D, Sannigrahi J, Chaudhuri BK, Sakata H (2012) Enhancement of the transport and dielectric properties of graphite oxide nanoplatelets-polyvinyl alcohol composite showing low percolation threshold. Polym Compos 33 (3):436–442.
43. Wang Z, Nelson JK, Hillborg H, Zhao S, Schadler LS (2012) Graphene oxide filled nanocomposite with novel electrical and dielectric properties. Adv Mater 24 (23):3134–3137.
44. Chu L, Xue Q, Sun J, Xia F, Xing W, Xia D, Dong M (2013) Porous graphene sandwich/poly(vinylidene fluoride) composites with high dielectric properties. Compos Sci Technol 86:70–75.
45. Wang J, Wu J, Xu W, Zhang Q, Fu Q (2014) Preparation of poly(vinylidene fluoride) films with excellent electric property, improved dielectric property and dominant polar crystalline forms by adding a quaternary phosphorus salt functionalized graphene. Compos Sci Technol 91:1–7.
46. Wang D, Bao Y, Zha J-W, Zhao J, Dang Z-M, Hu G-H (2012) Improved Dielectric Properties of Nanocomposites Based on Poly(vinylidene fluoride) and Poly(vinyl alcohol)-Functionalized Graphene. ACS Appl Mater Interfaces 4 (11):6273–6279.

47. Wen F, Xu Z, Xia W, Wei X, Zhang Z (2013) High dielectric permittivity and low dielectric loss nanocomposites based on poly(VDF–TrFE–CTFE) and graphene nanosheets. J Adv Dielectrics 3 (2):1350010.
48. Kim JY, Lee WH, Suk JW, Potts JR, Chou H, Kholmanov IN, Piner RD, Lee J, Akinwande D, Ruoff RS (2013) Chlorination of reduced graphene oxide enhances the dielectric constant of reduced graphene oxide/polymer composites. Adv Mater 25 (16):2308–2313.
49. Jeong HK, Jin MH, An KH, Lee YH (2009) Structural Stability and Variable Dielectric Constant in Poly Sodium 4-Styrensulfonate Intercalated Graphite Oxide. J Phys Chem C 113 (30):13060–13064.
50. Li M, Huang X, Wu C, Xu H, Jiang P, Tanaka T (2012) Fabrication of two-dimensional hybrid sheets by decorating insulating on reduced graphene oxide for polymer nanocomposites with low dielectric loss and high dielectric constant. J Mater Chem 22:23477–23484.
51. Tong W, Zhang Y, Yu L, Luan X, An Q, Zhang Q, Lv F, Chu PK, Shen B, Zhang Z (2014) Novel Method for the Fabrication of Flexible Film with Oriented Arrays of Graphene in Poly(vinylidene fluoride-co-hexafluoropropylene) with Low Dielectric Loss. J Phys Chem C 118 (20):10567–10573.
52. Tian M, Zhang J, Zhang L, Liu S, Zan X, Nishi T, Ning N (2014) Graphene encapsulated rubber latex composites with high dielectric constant, low dielectric loss and low percolation threshold. J Colloid Interface Sci 430:249–256.
53. Sharmila BTK, Nair AB, Abraham BT, Beegum PMS, Thachil ET (2014) Microwave exfoliated reduced graphene oxide epoxy nanocomposites for high performance applications. Polymer 55 (16):3614–3627.
54. Shen Y, Guan Y, Hu Y, Lei Y, Song Y, Lin Y, Nan C-W (2013) Dielectric behavior of graphene/BaTiO3/polyvinylidene fluoride nanocomposite under high electric field. Appl Phys Lett 103 (7):072906.
55. Wang D, Zhou T, Zha J-W, Zhao J, Shi C-Y, Dang Z-M (2013) Functionalized graphene-BaTiO3/ferroelectric polymer nanodielectric composites with high permittivity, low dielectric loss, and low percolation threshold. J Mater Chem A 1 (20):6162–6168.
56. Wu C, Huang XY, Xie LY, Yu JH, Jiang PK (2011) Morphology-controllable graphene-TiO2 nanorod hybrid nanostructures for polymer composites with high dielectric performance. J Mater Chem 21 (44):17729–17736.
57. Wang Z, Nelson JK, Miao J, Linhardt RJ, Schadler LS (2012) Effect of High Aspect Ratio Filler on Dielectric Properties of Polymer Composites: A Study on Barium Titanate Fibers and Graphene Platelets. IEEE Transactions on Dielectrics and Electrical Insulation 19 (3):960–967.
58. Wang T, Liang G, Yuan L, Gu A (2014) Unique hybridized graphene and its high dielectric constant composites with enhanced frequency stability, low dielectric loss and percolation threshold. Carbon 77:920–932.
59. Zhu Z, Sun X, Xue H, Guo H, Fan X, Pan X, He J (2014) Graphene-carbonyl iron cross-linked composites with excellent electromagnetic wave absorption properties. J Mater Chem C 2 (32):6582–6591.
60. Ho K-I, Huang C-H, Liao J-H, Zhang W, Li L-J, Lai C-S, Su C-Y (2014) Fluorinated Graphene as High Performance Dielectric Materials and the Applications for Graphene Nanoelectronics. Scientific Reports 4: doi: 10.1038/srep05893.

Multi Functional and Smart Graphene Filled Polymers as Piezoelectrics and Actuators

Kishor Kumar Sadasivuni, Abdullahil Kafy, Lingdong Zhai, Hyun-U Ko, Seong Cheol Mun and Jaehwan Kim

Abstract Graphite and its derivative materials are widely used in fabricating energy harvesters and are known as materials of this generation. The excellent applications of these materials in technology come from their superior electronic properties. Piezoelectric, Actuator and other tactile materials based on graphene have come up with substantially improved properties. The present chapter deals with these aspects of graphene filled polymer nanocomposites where a thorough investigation of the design and properties of them is carried out. Effect of homogeneous distribution of graphene within the matrix, interfacial interaction and functionalization of fillers are discussed to bring dynamic control to nanoscale actuators and piezoelectrics. In addition to explaining the fundamental requirements to make the best piezoelectric and actuator materials, the existing confronts to guide future progress is also undertaken in this study.

Keywords Graphene · Electronics · Elastomers · Piezoelectric · Actuator

1 Introduction

As part of the technological evolution, miniaturization has attained tremendous importance and many flexible and light weight conducting materials form the basics of wearable electronics. The two dimensional graphene is a zero band gap semi-metallic conductor where the valence and conduction bands intersect at a single Dirac point and various composites are widely used in electrical and electronic applications. Our basic aim here is to review the major applicability and use of such graphene filled polymers as piezoelectric and actuators. The piezoelectricity is not an intrinsic property of a particular material phase and it is often engineered

K.K. Sadasivuni · A. Kafy · L. Zhai · H.-U. Ko · S.C. Mun · J. Kim (✉)
Center for EAPap Actuator, Department of Mechanical Engineering, Inha University, 253 Yonghyun-Dong, Nam-Ku, Incheon 402-751, South Korea
e-mail: jaehwan@inha.ac.kr

K.K. Sadasivuni et al. (eds.), *Graphene-Based Polymer Nanocomposites in Electronics*, Springer Series on Polymer and Composite Materials,
DOI 10.1007/978-3-319-13875-6_4

into intrinsically non-piezoelectric and mechanically flexible graphene [1]. Functionally graded graphene reinforced composite plates used as good piezoelectrics in electronics are reported [2–5]. These types of piezoelectrics are frequently used for dynamical control of material deformation and are useful to fabricate pressure sensors [6], acoustic transducers [7] and high voltage generators [8].

In general the vibration to electrical conversion follows electromagnetic, electrostatic and piezoelectric transductions mechanisms [9, 10] of which the latter is of great importance due to its large power density, easy application (piezoelectric devices can be fabricated on both the macro and the micro scale due to well established production techniques), and the fact that the output voltage is obtained directly from the piezoelectric material itself, which removes the requirement of an external voltage input [10]. The energy conversion in piezoelectric materials is based on variations in the dipolar moment when a strain is applied, and therefore the formation of a potential difference which can be used to power devices [11]. Zhu et al. [12] investigated the design and analysis of a broadband piezoelectric energy harvester that used a simply supported piezoelectric beam compressed by dynamic loading. They have used parametrically excited cantilever beams and the method of multiple scales in their study to obtain analytical expressions for the tip deflection, output voltage and harvested power near the first principal parametric resonance.

On the other hand the actuators undergo shape or mechanical change in response to an appropriate external stimulus [13]. Actuators based on polymers especially electroactive polymers (EAPs) give larger displacements at less force unlike their conventional ceramic counterparts and possess light weight, good processability and low cost [14]. The lower power requirement of EAPs makes it applicable in wireless sensor networks for monitoring in which the devices work using the vibrational energy available in their environment, thus avoiding the maintenance cost of external power sources [11]. Actuators find use in industrial automation tasks, such as robotic assembly of small, delicate objects, medical equipment during surgery [15–20], artificial muscles and energy harvesting materials [11]. Within the pure polymer/polymer composite actuators, an 'energy transfer' unit absorbs the light energy and transfers to a 'molecular switch' unit which undergoes the mechanical change [21]. The molecular switch is the polymer network itself whereas the energy transfer unit can either be a functional group as in the case of pure polymers or a filler material as in composites [22]. Many polymer composite optical actuators with improved mechanical and thermal properties and tunable actuation have published so far [23–35]. The optical actuator have additional advantages like wireless actuation and remote control and also useful when stimulus other than electricity or heat is preferred especially in biomedical field [20].

By incorporating graphene nanomaterials into polymer performance of actuators are developed that convert external stimuli such as thermal, light, electrical or chemical energy to mechanical energy. Reversible mechanical actuators based on graphene and polymers have suggested applications in robotics, sensors, mechanical instruments, microscopy tips, switches, and memory chips [36, 37]. Recently, electromechanical resonators composed of single and multilayer graphene sheets were reported [38]. Another report illustrates a novel macroscopic graphene-based actuator depending on variation in humidity and/or temperature [39] where the aromatic sp^2 carbon networks of graphene extend structural compatibility and a stable interface. Such developed bilayer paper actuator is anticipated to have applications in gas sensors as well as in micro and nanoelectromechanical systems (NEMS) [40–42]. The derivative materials of graphene including graphene oxide, reduced graphene oxide, graphene ribbons and functionalized graphene are used to fabricate photomechanical actuators [43]. For instance the sulfonated graphene/TPU and isocyanate-graphene/TPU composites possess photomechanical properties [43]. The highly efficient actuators similar to human muscles are applied in micro robotics and artificial muscles [44, 45].

Before going to the details of piezoelectrics and actuators, a brief discussion about the synthesis and characterization of graphene and its polymer nanocomposites is first done.

2 Synthesis and Characterization

2.1 Graphene and Its Derivatives

The micromechanical exfoliation of graphene from graphite using the simple adhesive tape is well known [46]. Direct growth on suitable substrate and chemical reaction are also employed. Depending on the preparation method, the graphene flakes differ considerably in size (from nanometers to several tens of micrometers) and thickness. A liquid phase preparation enables highly processable graphene and its easy dispersion in suitable matrix. Simple sonication of graphite in organic solvent by applying voltage produces high quality graphene with low complexity and high production [47, 48]. The principle of liquid-phase exfoliation can also be used to exfoliate graphite oxide as well, by which hydrophilic and water soluble GOs are formed. The oxidation and micromechanical cleavage of graphite also generates GO and its further thermal or chemical reduction, the graphene [49]. The growth of graphene on a surface occurs by chemical vapour deposition (CVD) and epitaxial growth (simply heating and cooling graphene on SiC crystal) [50, 51]. In CVD gas mixture of H_2, CH_4 and Ar decompose on a Ni film at about 1,000 °C [52]. By Nickel patterning and using polymer support on the top of graphene, the shape of graphene can be controlled [50, 53]. By this method graphene layers of 30-in. thickness can be grown. CVD also allows doping of graphene in order to decrease the resistance and offers graphene with high optical and electrical performance, more or less unlimited size, high controllability, low complexity and large production.

In the epitaxial method, single- or bi-layer graphene forms on the Si face and few-layer graphene on the ‘C’ face of the crystal [51]. Moreover the growth of graphene depends on temperature, heating rate and pressure and if the conditions are high, nanotubes grow instead of graphene. Thus the grown graphene is not perfectly homogeneous and of good quality except grown on a perfect single crystal. This method is controllable and less complex, but the limited size of graphene layers and low production are considered to be some demerits.

2.2 Polymer Nanocomposites

It is really important to disperse the graphene in the given matrix polymer to synthesize composites with improved properties. Depending on the polarity, molecular weight, hydrophobicity and reactive groups of the polymer, graphene and solvent [50], there are three general ways for fabricating graphene filled polymer composites. In the in situ *intercalative polymerization*, graphene or modified graphene is first swollen within the liquid monomer by adding an initiator which initiates the polymerization either by heat or radiation [54, 55]. A large number of polymer nanocomposites such as polystyrene (PS)/graphene have been prepared by this method [54, 55]. In the *solution intercalation*, both the polymer or pre-polymer and graphene or modified graphene layers are allowed to swell in the suitable solvent system (water, acetone, chloroform, tetrahydrofuran (THF), dimethyl formamide (DMF), toluene etc.) and thereafter mixing. Polymer adsorbs on to the delaminated graphene sheets and finally the solvent is evaporated [56] to obtain the nanocomposites such as polyethylene-grated maleic anhydride (PE-g-MA)/graphite [57], PS/graphene [58], polypropylene (PP)/graphene [59], plyvinylalcohol (PVA)/graphene, TRGO/TPU [43] etc. *Melt intercalation* allows mixing the filler and polymer in molten state. Usually thermoplastic polymers are mixed at elevated temperatures using conventional methods like extrusion and injection molding [58–60]. A wide range of polymer nanocomposites such as PP/expanded graphite (EG) [59], High dense polyethylene (HDPE)/EG [61] etc. have been prepared by this method. In a non liquid-phase, graphene is dispersed by vacuum filtration, simple drop-casting, spin coating etc. of which the spin coating allows homogeneous and large area graphene dispersion.

3 Piezoelectric and Actuating Properties

3.1 Mechanism Involved in Piezoelectrics

Piezoelectric materials or piezoelectrics produce electricity upon mechanical stress (squeezing or stretching). Upon pressure, a small voltage is produced from the changing charge created by the moving electrons and changes their shape in some way. Crystalline materials like quartz come under this category and there are several gadgets such as phonographs, microphones, loudspeakers, signal transducers and barbecue lighters work on this principle [62]. The mechanism

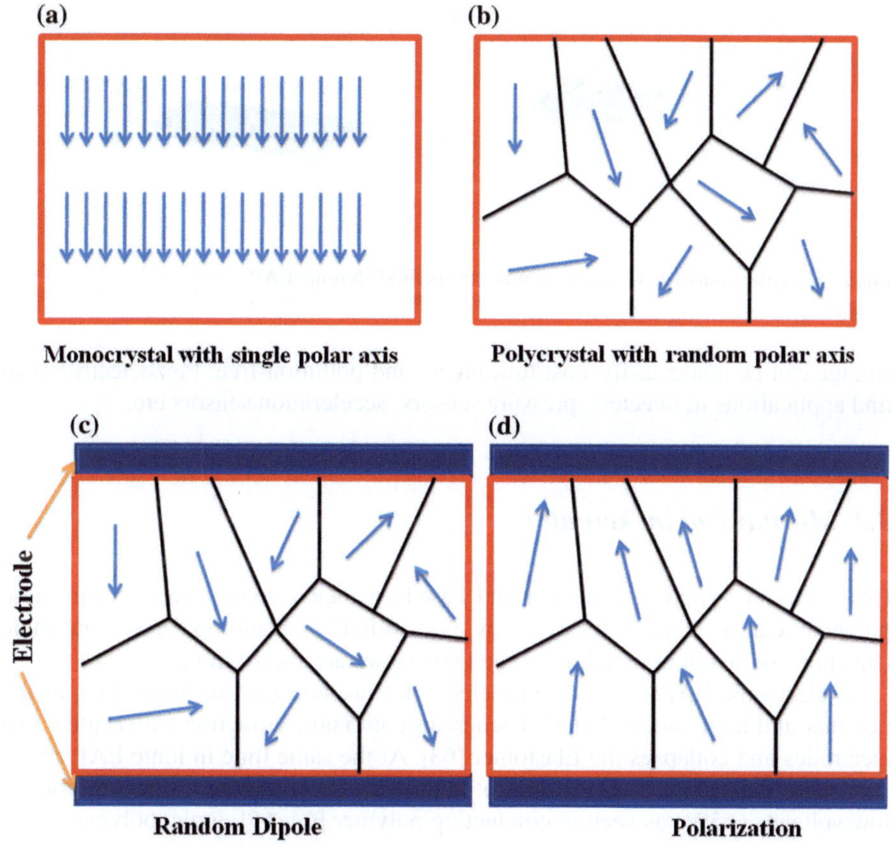

Fig. 1 Dipoles representation in **a** monocrystal and **b** polycrystal, **c** and **d** polarization of ceramic material to generate piezoelectric effect

involved in piezoelectricity is explained by considering the polarization of individual crystal molecules. It is assumed that the polar axis passes through the center of both charges on the molecule, and in a Monocrystal (Fig. 1a), the polar axes of all the dipoles lie in one direction whereas in a polycrystal (Fig. 1b), different regions within the material have different polar axes. A polycrystal is heated by applying strong electric field by which the molecules move more freely and the electric field forces all of the dipoles in the crystal to line up in nearly the same direction and thus make it piezoelectric (Fig. 1c, d).

A piezoelectric crystal can be made into various shapes at different frequencies and can achieve different vibration modes and thus realize small, cost effective, and high performance products capable of working in the kHz-MHz range. The material distorts upon voltage and the electricity generated by deformation is high-voltage and low current characterizing a transducer. Since walking generates pressure, and is applied on a piezoelectric ceramic electrical energy production and

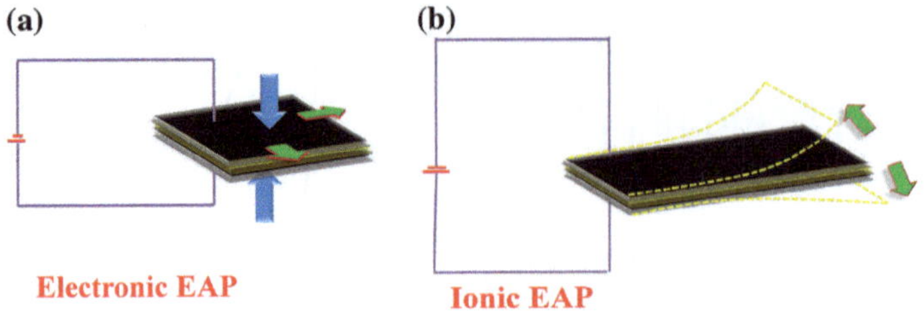

Fig. 2 Classification of EAP actuators **a** electronic EAP. **b** Ionic EAP

storage can be made easily, cost low, green, and pollution-free. Piezoelectrics also find applications in tweeters, pressure sensors, acceleration sensors etc.

3.2 Mechanism in Actuator

EAPs are very good at actuation and based on their actuation mechanism these polymers can be electronic (driven by electric field or coulomb forces) and ionic (functions with a liquid medium). The two classes are shown in Fig. 2.

In electronic EAPs, two flexible electrodes are arranged on both sides of the polymer and high voltage (>1 kV) generates coulombic attraction between the two electrodes and collapses the Elastomer [63]. At the same time in ionic EAP actuators, large deformation occurs due to the movement of charge by flexible ions at low voltages (<5 V) as seen in conducting polymer [64, 65], ionic–polymer metal composite [66, 67], and CNT based variations [68]. For a dielectric elastomer, the electrodes designed comply with the elastomer deformations and thus gives rise to coulomb forces between opposite charges, generating a pressure, known as the Maxwell stress. This stress causes the electrodes to move closer and thus squeezes the polymer and it elongates in the directions perpendicular to the applied force. The dielectric elastomer is basically a compliant capacitor and its applications include artificial muscles, robotics, micropumps, moulding in aviation and automotive microactuators.

3.3 Piezoelectric Properties

The superior properties of graphene imparts piezoelectric to its polymer nanocomposites. In a typical rGO–PVDF nanocomposite film, the output voltage gives rise to peaks in each frequency with rGO content (0.05 and 0.2 wt%, 30 Hz) [69] and compared with the pure PVDF, the signals possess higher amplitudes, good responsiveness and traceability to the excitation and so useful in strain sensor applications. It is also found that small rGO contents do not have major influence

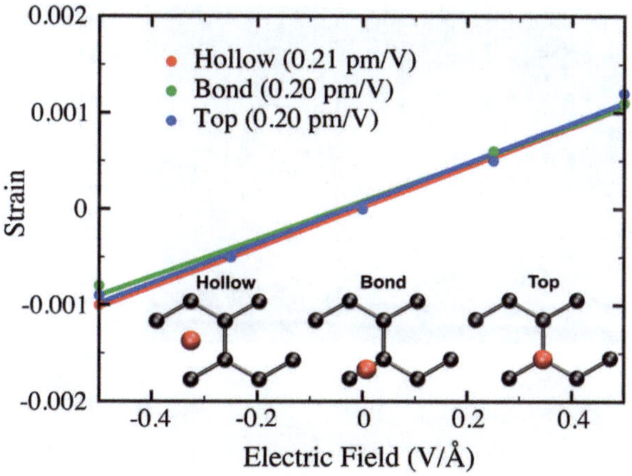

Fig. 3 Equibiaxial in-plane strain as a function of applied electric field for different positions (*hollow*, *bond* and *top site*) of Li on graphene for C8Li. Unit cells for each of these positions are given in the inset [70]. Copyright 2012. Reproduced with permission from American Chemical Society

on the film stiffness and the more the b-crystalline phase is present, the better is the piezoelectric performance of the material. The output voltage of nanocomposite films increased with the addition of rGO as well.

Sometimes atoms of potassium and lithium are present on the surface of graphite and graphene [70–72] and may diffuse away from the configurations. The dependence of such atom positions on the piezoelectric response is investigated and found an absence of centrosymmetry regardless of the adatom location. Figure 3 illustrates how equibiaxial in-plane strain varies under applied electric fields perpendicular to the surface for the Li adatom positions at the hollow, bond, or top site in a sample of C8Li. The results indicate that the piezoelectric response does not strongly depend on adatom location.

ZnO nanorods were fabricated on the single layer graphene coated PET polymer substrate and its piezoelectric response was studied. Using AFM in contact mode the bending of well aligned ZnO nanorods was investigated as shown in Fig. 4. Bending resulted in an average output potential of 80–100 mV (Fig. 4b) and upon releasing the nanorods the output potential turned to be zero. From the directly recorded topography image of the bending distance in Fig. 4c the mechanism behind the generation of the piezoelectric potential can be understood in terms of the potential created by the relative displacement of the Zn^{2+} cations with respect to the O^{2-}. However, these ionic charges cannot freely move and cannot recombine without releasing the strain. The symmetric ZnO nanorods are in contact with the graphene single layer coated PET polymer substrate as well. It is clear from Fig. 4d that at the time of applied external force, the ZnO nanorods were deformed and a piezoelectric potential was produced along the top and bottom electrodes due to short lived presence of charges. This allows the electrons to flow along the external force at the point of equilibrium. But the piezoelectric nanogenerator gains its original position when the applied force is cut-off thus diminishing the piezoelectric potential and making the reverse movement of stored electrons. Here the AFM tip acts as the applied external force on the nanorods and a Schottky barrier develops between the ZnO nanorods and the electrode.

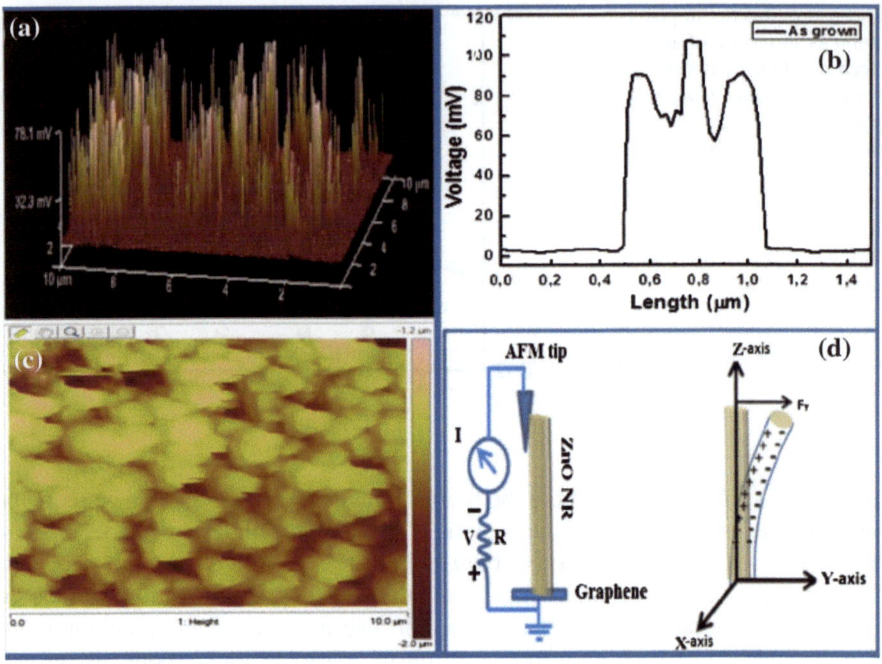

Fig. 4 Piezoelectric power generation from as-grown ZnO nanorods fabricated on single layer graphene coated PET polymer substrate. **a** Three-dimensional plot of the output voltage. **b** Typical AFM tip scanning the surface of ZnO nanorods in micrometers. **c** AFM opography image. **d** Mechanism of the electrical pulse generation by ZnO nanorods [73]. Copyright 2014. Reproduced with permission from John Wiley & Sons

The idea of conserving electricity from human body movement was first demonstrated by Wang and co-workers through the conversion of bio-mechanical energy from muscle movement to power wireless devices and charge mobile phones. They have made self-powered ZnO-based NGs with nanowire diameter 100–800 nm and lengths ~100–500 μm. The SWG was attached to the joint position of the index finger as shown in Fig. 5a and repeated bending of the finger produce a cycled strain in the nanowire. As a result the nanowire deforms, creates a piezoelectric potential within it and drives the flow of external electrons thus producing electric power output as shown in Fig. 5b. The I–V characteristic shows Schottky behavior and the side with Shottky contact is identified as the positive side.

The forward and reverse connections are respectively made by connecting the positive and negative sides of the single wire generator (SWG) with the positive and negative probe of the measuring instrument and vice versa. The magnitude of the signal differs with different connecting methods due to the contribution from the bias current of the measurement system, which is usually a few picoamperes. But to ensure true signal generated by SWG, the switching polarity has to be satisfied. Figure 5c, d respectively represent the open-circuit voltage and short-circuit

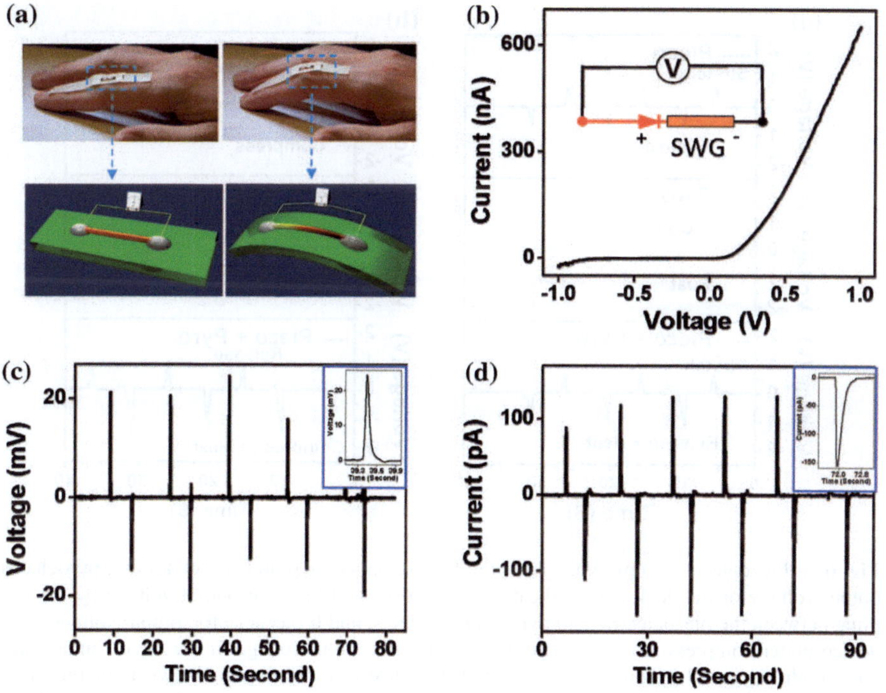

Fig. 5 Energy harvesting from an oscillating human index finger using an SWG. **a** An SWG attached to a human index finger, which drove the SWG to bend and produce power output. **b** I–V characteristic of the SWG. The *inset* illustrates the connection configuration of the SWG in reference to the measurement system. Open-circuit voltage (**c**) and short-circuit current power output (**d**) from the SWG when the finger was periodically bent back and forth [74]. Copyright 2009. Reproduced with permission from American Chemical Society

current output from the SWG when the index finger oscillates at a relatively slow rate. The periodic motion of the finger was recorded and the average straining rate of the SWG is found to be (4–8) $\times$ 10^{-3} s^{-1} at a maximum nanowire strain of ~0.2 %. From a single SWG device voltage output up to 25 mV and current output more than 150 pA are obtained. The variation in bending speed, finger oscillation and irregular finger movement can cause fluctuation in the output voltage and the current.

Figure 6 shows both the piezoelectric and pyroelectric effect evidenced for a Hybrid Stretchable NanoGenerator (HSNG) device. Here stretching and releasing enhanced the piezoelectric output voltage up to 0.7 V whereas the thermal heating and cooling condition increased the pyroelectric output voltage up to 0.4 V. This substantiates the possibility of integrating the device to get a total output voltage of up to 1.1 V by simultaneously applying the strain (stretch-release mode) and thermal gradient (heat and cool mode) (Fig. 6a). The output voltage can again be enhanced to 1.4 V by applying a different mode of strain like compress-release (output voltage ≈ 1.0 V) and heat and cool (output voltage ≈ 0.4 V) as illustrated

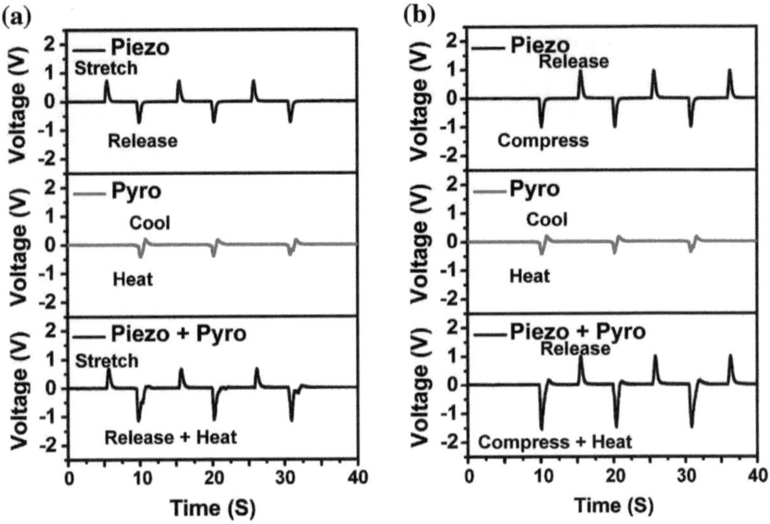

Fig. 6 **a** Piezoelectric output voltage of the HSNG under stretchrelease condition, pyroelectric output voltage of the device under heatcool condition, and the total output voltage by the coupling between the piezoelectric and pyroelectric effects, and **b** piezoelectric output voltage of the device under compress-release condition, pyroelectric output voltage of the device under heat-cool condition, and the total output voltage of the HSNG by the coupling between the piezoelectric and pyroelectric effects [75]. Copyright 2014. Reproduced with permission from John Wiley & Sons

in Fig. 6b. This stands out as the first demonstration of simultaneous harvesting the mechanical and thermal energies from a single cell of highly stretchable NG. Another work has also demonstrated a piezoelectric power generation device which includes a piezoelectric ceramics in the insole and the external integrated circuit outside of the shoes and the research explained an easy to replace, low cost and easy-to-production piezoelectric energy harvesting device useful to charge the battery for continuous and intermittent charging.

3.4 Polymer/Graphene Composite Actuators

Various polymers reinforced with graphene and its derivative nanofillers have proved their capability in actuation by different modes. Figure 7a represents a GNP/elastomer photo-thermal actuator with the composite expressed as sample [15] and the pre-strained and heated samples as sample [16] and sample [17]. The GNP/polydimethylsiloxane (PDMS) possesses random polymeric chain arrangement/entanglement and the addition of weight to its free-end stretches the composite to a more ordered arrangement by reducing the entropy. Upon heating the composite by IR illumination, GNPs absorb light and optical energy is efficiently

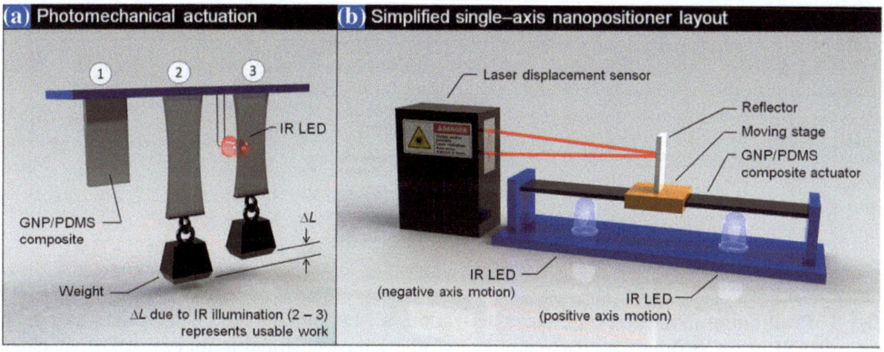

Fig. 7 **a** Photo-thermal actuation schematic. Three thin GNP/PDMS composite strips (samples *1*, *2* and *3*) are mounted with their upper ends fixed to a rigid plate. Illumination via an IR LED results in energy transduction to the polymeric chains, causing a contraction in the actuator (and thus usable work). **b** Simplified single-axis nanopositioner layout. A laser displacement sensor is used to measure stage position. Independently controlled diodes on either side of the stage allow for differential positive or negative axis stage motion [23]. Copyright 2013. Reproduced with permission from Nature

transduced into thermal energy through phonons in the sp^2 graphene sheets [76]. The high macroscopic thermal conductivity of graphene [77] (300 W/m/K) and good dispersion of GNPs within the PDMS result in heat percolation and polymer chain contraction. When the temperature is increased the spring constant also increases [78]. The constant weight applied to the composite contracts it and lifts the weight, resulting in usable work done through the system by so called "rubbery elasticity". The large amplitudes of photo-thermal actuation is the result of the strong covalent sp^2 bonding in GNPs resulting in efficient heat transfer by lattice vibrations and accomplished through heat transduction from the GNP lattice to the PDMS [78, 79].

A simplified model of a single-axis photo-thermal nanopositioning system using the thin GNP/PDMS composite strip held at a fixed pre-strain is presented in Fig. 7b. The nanopositioning stage is mounted in the composite center, and IR light-emitting diodes (LEDs) on either side via both positive and negative stage actuation. The working of the device is explained as follows. When the left LED is illuminated the GNP/PDMS composite around the light spot gets heated and the polymer chain in that region contracts and subsequently transfers the effect to the right. Similarly the LED on the right side causes the stage to translate right. The dynamic modulation of IR intensity balances a force on either side of the stage and maintains the desired displacement. The intensity of the LED is continuously tuned by a proportional-integral-derivative control loop which monitors stage position and compensates the thermal drift even when the stage is at a set position. A high speed laser displacement sensor measures the stage displacement and in short this highly dynamic nanopositioning system works by controlling polymeric chain extension/contraction through thermal energy modulation.

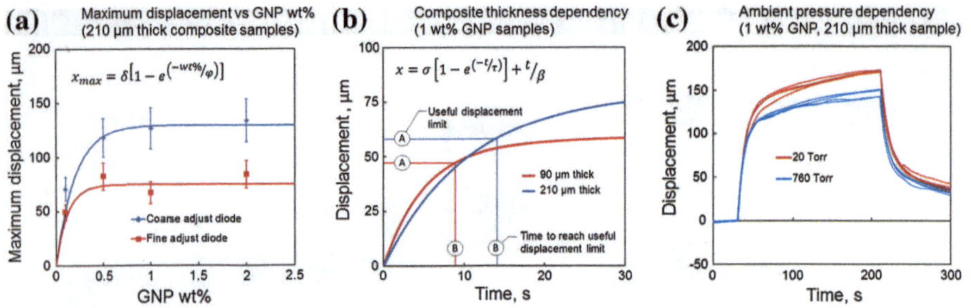

Fig. 8 Positioning diode response characterization. **a** Maximum stage displacement as a function of GNP wt% loading for coarse adjust and fine adjust positioning diodes. **b** Displacement kinetics as a function of composite actuator thickness for a 1 wt% GNP sample. Useful displacement limit and time are also indicated. **c** Displacement versus time curves shown for a 1 wt% GNP sample at ambient pressure (760 Torr) and at medium vacuum (20 Torr) [23]. Copyright 2013. Reproduced with permission from Nature

The maximum displacement for a coarse and fine adjust diode is plotted as a function of GNP wt% as in Fig. 8a. Since the steady-state equilibrium positions are reached only after long wait times (180 s), these nanopositioning devices are not practically useful. Consequently the selection of a tripwire value for slope of the displacement versus time curve was set to determine useful displacement versus maximum displacement and attained a balance between quick actuation times, maximum possible displacement and allowed ample additional displacement control. This fact is considered and a displacement of 47.2 mm is achieved at 8.5 s for the sample as derived from the Fig. 8b. Whereas for a thicker sample, a displacement of 58 mm or 60 % of maximum is obtained at 14 s. Moreover the actuation speed was calculated by dividing the usable displacement by the time required to reach those values and found the values as 5.5 mm/s for thin sample and 4.1 mm/s for the thick. In order to improve the actuation speed, ultra thin composite samples must be used, but this requires additional preparation steps. However integration of thinner films onto MEMS substrates with on-chip positioning stages improves photo-thermal actuation velocity. The ambient pressure dependency of the actuator was also checked as given in the displacement versus time curves of Fig. 8c. The analysis done on a 1 wt% 210 mm thick composite sample at atmospheric pressure and in vacuum indicates a drop in composite heat loss as pressure decreases due to natural convection. At vacuum pressure displacement was 30 % higher than that at atmospheric pressure and this behavior depends on the fluid density and thus the heat transfer coefficient.

In another work the photomechanical actuation of TRGO/TPU composites at different prestrains and filler loadings were checked [43] and found better actuation. For instance the TRGO/TPU at 1 wt% loading gives a photomechanical force of 0.383 N with a stress of 1,532 kPa at ~120 % prestrain and it can lift a weight of 39.08 g to a height of 10.8 mm. This was really high when compared to the GNP/PDMS at 2 wt% (50 kPa) and sulfonated graphene/TPU composites at

(0.21 N) at 1 wt% [80] filler loadings. The actuation also depends on the intensity of IR light as the same TRGO/TPU at 1 wt% gave a photomechanical stress of 1,357 kPa (0.475 N) at 166 % prestrain at lower light intensity. Similarly at 2 wt% TRGO/TPU can lift 42.85 g to a height of 25 mm under IR irradiation. This is attributed to the fact that upon IR absorption the composites restore the sp^2 carbon network in TRGO and causes in a homogeneous distribution of TRGO in TPU. There is a strong absorption of IR light in TRGO due to the resonant induction by edge oxygen motion of mobile electrons localized in the vicinity of the oxygen and it acts as energy transfer unit. Whereas neat TPU is transparent to IR light and hence shows negligible photomechanical response.

By varying the voltage (1–4 kV) and frequency (1–5 Hz) the performance of an FLG-driven actuator was monitored using a laser displacement sensor [25]. When the actuating voltage was increased at 1 Hz, the displacement increased with a maximum reaching at 946 lm. This effect was attributed to the variation of the compression force directly as a function of applied voltage [81]. Moreover the displacement was slightly increasing during 1–2 Hz and thereafter decreasing at 2–5 Hz (916–674 lm) due to the decrease in molecular orientation or relaxation time with frequency increase [82, 83]. At 4 kV the transparency of the FLG-driven actuator was analysed with the help of a digital single lens reflex camera [25]. At "ON" voltage, the actuator bends in the meniscus shape and the manuscript image becomes out of focus and blurred due to the altered focal length triggered by the actuator, which result from the thick center than that of the boundary [83]. This confirms the possibility to alter the focal length by varying the voltage input at the FLG-driven actuator and ensures its applicability in opto-electromechanical devices such as lenses, sensors and touch screens. However the authors couldn't report the lifetime and durability of such devices.

Actuators find excellent applications in fiber and textile devices. In this regard flexible G/GO fibers having a tensile strength of 100 MPa are fabricated and the oxygen-rich functional groups of GOs allow fast and reversible expansion/contraction of GO layers through the adsorption and desorption of water vapor [84, 85]. Thus the moisture-sensitive asymmetric G/GO fiber actuator shows a rapid bending to the G side when (Fig. 9c) exposed to moist air having 80 % relative humidity (Fig. 9d), and returns to the initial state at ambient conditions (Fig. 9c). This is a reversible process occurring per 88 s and thus the fiber behaves like shape memory material.

Figure 9b indicates the influence of humidity on the G/GO fiber actuation and observed a fiber deformation to the GO side due to the contraction of GO layers caused by loss of adsorbed water. The bending degree of the fiber is plotted against the relative humidity in Fig. 9e. As the R_H increases from 25 to 80 % the G/GO fiber rapidly bends from 0 to 140°. At lower than 25 % RH the fiber moves in the opposite direction indicated by a bending angle of −30° at 10 % RH. More facile deformation is observed towards the G side than the GO side, because of the tensile strength from the moisture-insensitive G layers. Also the increased reduction of GO led to improved actuation response. The humidity is altered at 10–80 % and the response is found to be stable even after 1,000 cycles (Fig. 9f). Using the

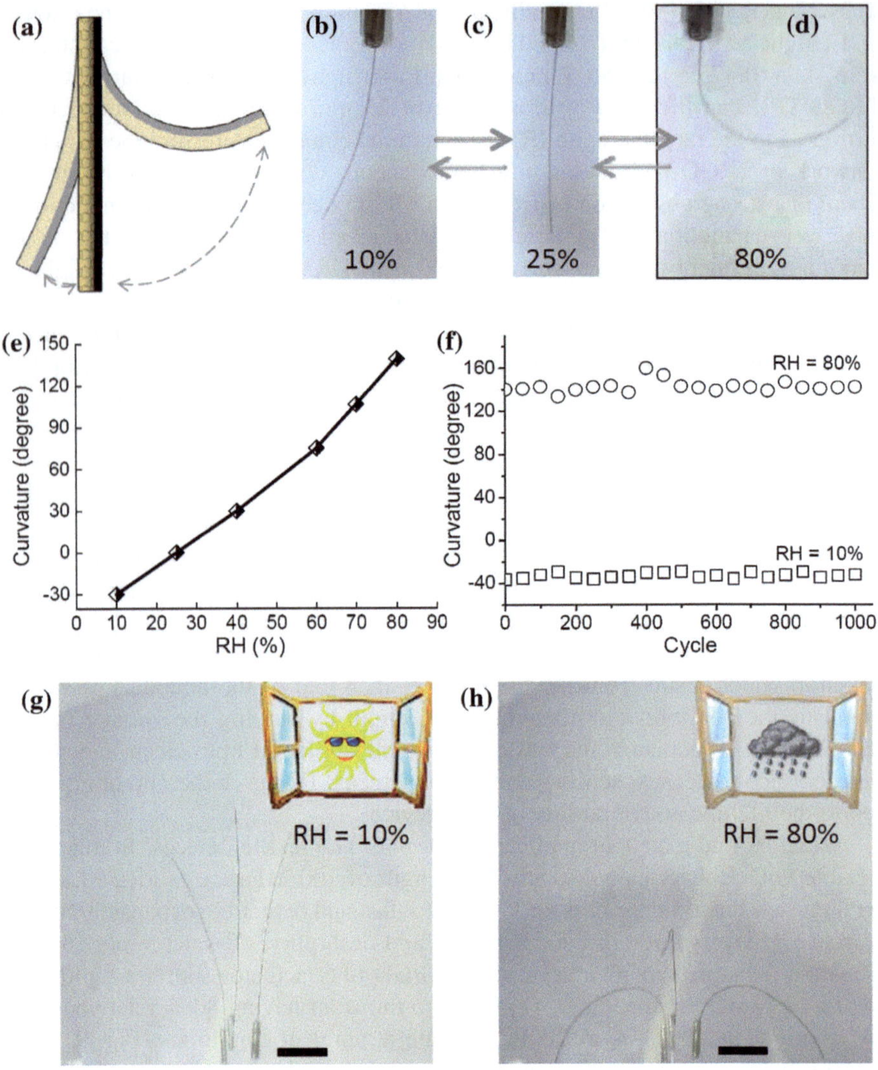

Fig. 9 **a** Representation of the possible bending of a G/GO fiber exposed to different relative humidities. **b–d** Photographs of a G/GO fiber (2 cm in length) under different relative humidities. **e** Plot of the curvature of a G/GO fiber versus RH. **f** Durability test for a G/GO fiber undergoing repeated changes in relative humidity. **g–h** Photographs of three-wire moisture tentacles made of the G/GO fibers on a sunny and a rainy day, respectively; scale bars: 5 mm [24]. Copyright 2013. Reproduced with permission from John Wiley & Sons

sensitive G/GO fibers a three-wire moisture tentacle was made (Fig. 9g, h) and its performance on a sunny and a rainy day was monitored respectively to be standing and lowering their heads. This substantiates its use in devices like moisture-triggered claw. Laser irradiation causes spot reduction in the fiber and thus the

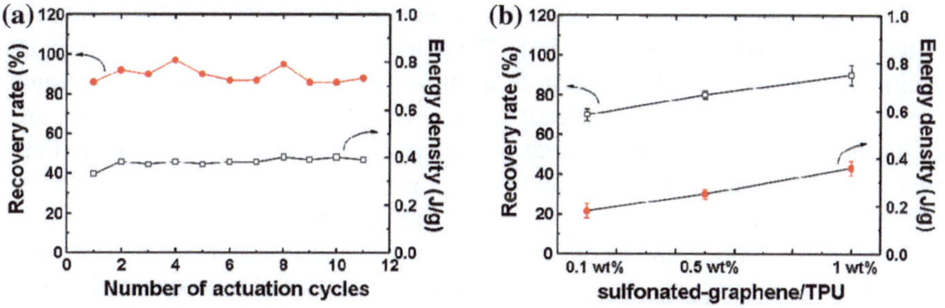

Fig. 10 IR-triggered actuation performance for the sulfonated-graphene/TPU nanocomposites. Recovery rate (*left axis*) and energy densities (*right axis*) for the IR-induced actuation behavior of composites **a** at 1 wt% filler loading for 11 cycles, **b** at 0.1, 0.5 and 1 wt% loading. The weights lifted by all the samples are about 21.6 g [26]. Copyright 2009. Reproduced with permission from American Chemical Society

moisture-insensitive G region can be extended in G/GO. In this case the G region deforms like a hinge to allow the reversible folding of the fiber and such region-specific G/GO fibers can be applied for complex deformation-predefined actuators. The fibers alternate its response upon moisture exposure and removal and exhibit good durability at different relative humidities as well.

In another example of IR-induced actuation behavior, the sulfonated-graphene/TPU nanocomposites were analysed at different filler loadings. At 1 wt% loading (Fig. 10a) the energy densities and recovery rate sustain stabilization after 10 cycles without significant attenuation indicating the repeatable response to IR light. The actuation performance was studied at different concentrations and found its dependence (Fig. 10b) as the energy densities and shape recovery rate increased with the increasing contents of sulfonated-graphene. For all the samples the energy densities and the recovery rate fluctuated between 0.33–0.40 J/g and 88–95 %, respectively. The increased performance at higher concentration is attributed to the higher thermal energy generated in the composites at higher loading [86, 87]. But this phenomenon was noticed only up to 1 wt% and the reasons behind this need to be further understood.

Bi et al. demonstrated the actuation of a G/GO bilayer paper as a function of applied voltage. The actuator was made by assembling the bilayer papers (G thickness 3.61 mm and GO 1.62, 2.69, 3.77, and 4.85 mm) in U-shaped beams and the free ends curled upward with respect to the applied voltage. The beam actually rolled towards GO face at low voltages and thereafter rolled with higher curvature [27]. As the voltage varied the displacement changed linearly and finally the beam was unrolled and returned to its original shape when the power was cut off. The displacement of the beam became negligible at a voltage called threshold voltage and it decreased slightly with the thickness of the GO layers (for the thinner the displacement can be up to 9 mm which is 14 times higher). While checking the durability and stability of the G/GO actuator for a month, the displacement was observed to decrease by only 5, 7 and 3 % over 100 measurement cycles respectively at 10, 15 and 20 V indicating sufficient stability, good consistency and good performance [86]. The actuation of the bilayer paper was also checked as a function of temperature using a thermal infrared imager to monitor the temperature changes of the beam surface. With applied voltage the temperature of the beam rises gradually leading to greater displacement and the actuation of beams may be attributed to the internal heat generated by the current (Joule heat) [87, 88]. The thermal reduction of the GO paper under the working voltages is not a concern and in addition, when the actuators were inserted into a drying oven with controlled temperature, the beam started to curl as expected. For U-shaped beams, the GO layer showed the lowest electrical conductivity while the graphene layer showed a resistance of several thousand ohms due to their completely different surface chemistry. These layers could generate a large amount of heat when an electric current passes through them, causing a rise in local temperature. As a result a significant thermal stress prevails at the G/GO interface and the rise in temperature of the GO layer through heat transfer generates a temperature gradient across the interface [89].

A site-specific, hinge-like bending was observed in a hydrogel actuator in response to the nIR-laser irradiation (Fig. 11a). Regardless of the surface exposed to the beam, the hydrogels always bent maintaining the porous surface on the concave side. The bending rates and angles increased with increasing laser intensity and rGO concentration in the bending hydrogels (Fig. 11b, c). The gels unbent when the nIR light is removed, recovering ~74–84 % within 10 s. The durability was ensured by following the nature of the hydrogels during 100 repeated cycles of bending and unbending. Whereas in the case of composite hydrogels synthesized with isotropic networks similar motions were not observed even after 1 min

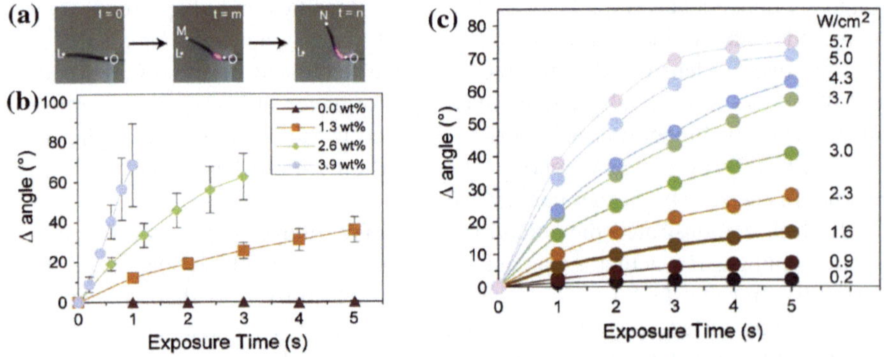

Fig. 11 Light-controlled bending of ELP-rGO composite hydrogels. **a** Images of site-specific nIR-induced bending of an anisotropic hydrogel. Measurement of bending angle, θ, as a function of exposure time, t. $\theta(0) = \angle LOL = 0$. $\theta(m) = \angle LOM$. $\theta(n) = \angle LON$. **b** Change in bending angle in response to nIR laser exposure for hydrogels with varying rGO concentrations. wt% correspond to the mass of V50GB-rGO hybrids divided by the total mass of V50GB-rGO hybrids, V50CK1, and 4-arm PEG cross-linker. **c** Change in bending angle of 2.6 wt% gels in response to nIR laser exposure of different power densities [28]. Copyright 2013. Reproduced with permission from American Chemical Society

of nIR irradiation, other than small shape changes. This is consistent with the slow kinetics observed with previous macroscale photothermally driven hydrogels [90, 91] and thus porosity in the given hydrogels creates photomechanical responses that do not require extended heating times.

The performance of a GO-Nafion and pure Nafion membrane actuators was illustrated in Fig. 12. It is clear that one end of the membrane is fixed in the

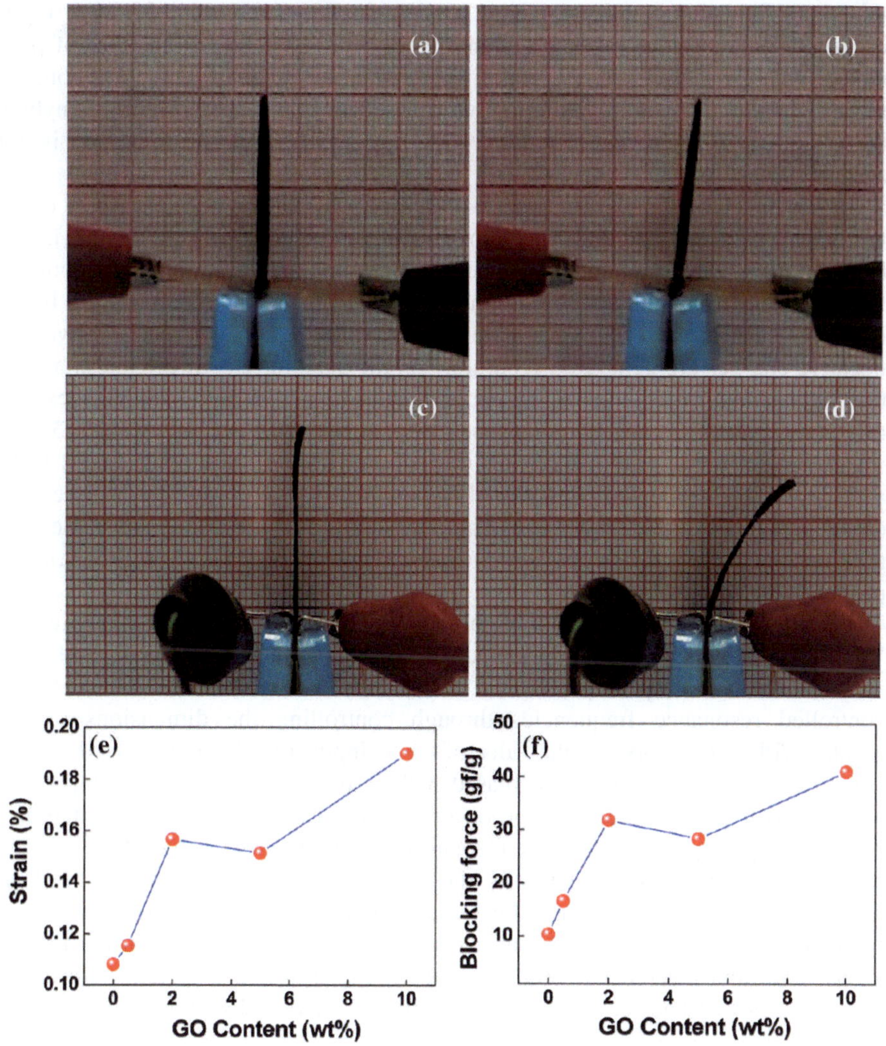

Fig. 12 Displacement photo of an IPMC of pure Nafion membrane and Nafion hybrid membrane (**a** and **c**) before test, **b** and **d** driven under a 3 V electrical field. Dependence of **e** the generated strain (ε), **f** the blocking force on the GO content of the membranes [92]. Copyright 2012. Reproduced with permission from American Chemical Society

apparatus whereas the other end is free to deflect (Fig. 12a–d). At 3 V the 10 wt% GO-Nafion actuator exhibits a perceptibly larger displacement (Fig. 12d) than that of the pure Nafion (Fig. 12b). The generated strain (ε) was calculated according to the tip displacement and a larger strain showed a higher deflection of membrane and a better displacement behavior (Fig. 8e). With GO content the strains increased and all hybrid membranes exhibit excellent displacement compared to pure Nafion membrane. The sample containing 10 wt% GO exhibited the largest displacement due to its higher conductivity. It is also found that the composite membranes have higher blocked force (Fig. 8f) with highest performance for membrane containing 10 wt% GO (four times that of pure nafion). This was a great result as the CNT/Nafion actuator at 1 wt% concentration showed a blocking force of two times than that of pure Nafion. Anyhow the effect of size and dispersibility of GO on the actuation performance is not addressed here.

Using conducting polymers, GO actuators are made where the GOs covalently attach on the polymer to improve its electrochemical properties [93]. The bending displacement response of the air-operable conducting polymer actuator for different electrolyte concentrations at 3 V and 1 Hz showed enhanced bending displacement with the electrolyte concentration. Upon voltage, positive and negative charges are generated separately on the actuator electrodes and to maintain charge neutrality, ions in electrolytes move towards opposite charges and this is the reason for the bending displacement. At different voltages (2, 3, and 4 V) bending displacement increases (maximum strain of 0.78, 1.40, and 2.18 % at 2, 3, and 4 V, respectively) with voltage as more ions from the electrolyte penetrate into the actuator backbone. The neat sample of conducting polymer did not show actuation due to its very low conductivity and poor electrochemical properties.

Finally actuation behavior of graphene/PDA composite is shown in Fig. 13. When the frequency of the applied ac (0.58 A/mm^2 density) reached 17 Hz, the actuator showed a tip displacement of 4.8 mm. The authors have established controlled resonance frequencies through controlling the dimensions of the graphene/PDA actuators and the value of the ac input (Fig. 13b). At a fixed dimension, the resonance frequency increased with ac input and at a fixed ac input, the actuator with larger dimensions produced resonance at lower frequency and at a fixed ac frequency, it can generate resonance below the lower value of the ac. The resonance identifies the tendency of the material to absorb energy and oscillate when the frequency of the applied external stimulus matches its intrinsic resonant frequency of vibration. This intrinsic resonant frequency depends on the nature, shape, size, and mechanical properties of the material and a system with larger size has a lower natural frequency as indicated in Fig. 13b for graphene/PDA actuators.

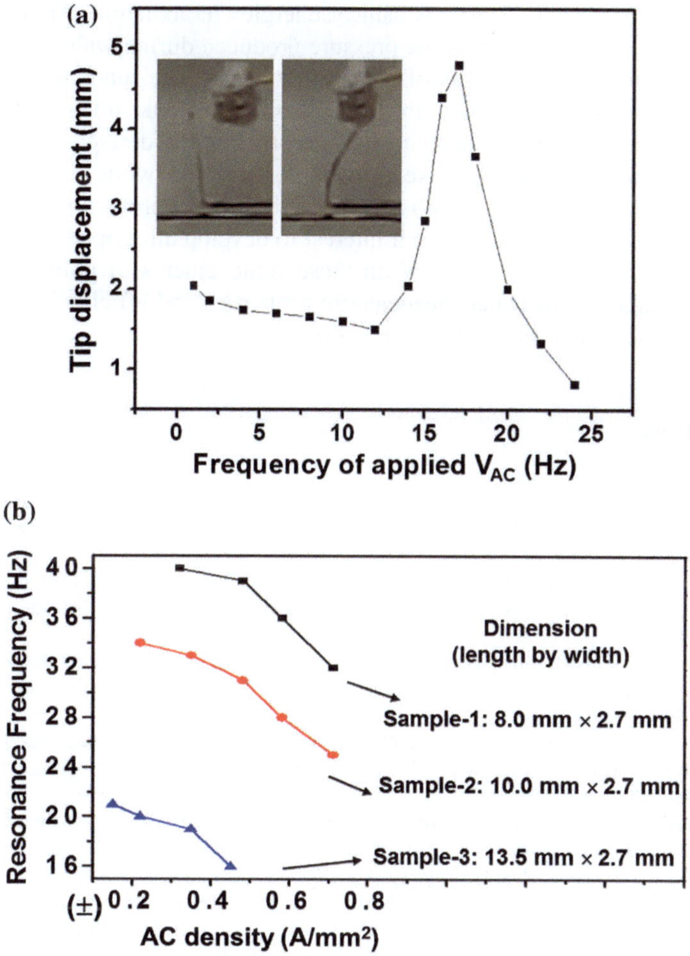

Fig. 13 **a** Dependence of frequency of applied ac (0.57 A/mm^2 density) on the tip displacement for the graphene/PDA actuators (9.0 mm × 1.2 mm). *Inset* represents the movement of the tip (b) ac density dependence of the resonance frequency for graphene/PDA actuator with three different dimensions [29]. Copyright 2010. Reproduced with permission from American Chemical Society

4 Conclusions and Perspectives

Integrating graphene into smart systems that spontaneously respond to external stimuli is important to develop advanced smart devices such as actuator, piezoelectric switches and artificial muscles. In this featured article, we have systematically reviewed the recent progress in piezoelectric and stimulus-responsive graphene actuation systems that are able to convert electrical, chemical, photonic, thermal and other types of energies to mechanical energies in actuation. Application like harvesting energy by converting the pressure produced during walking on an inexpensive piezoelectric material is of great significance. The fundamental principle of piezoelectric film producing power is also studied, the transfer relationship between piezoelectric patch and the storage capacitor is derived. These kind of systems have many advantages like wearing comfortable, while occupying space, easy replacement, low cost, and it is green environmental protection. In the future, it is of great theoretical and practical interest to develop all-graphene based actuators working using other stimuli. With these achievements of polymer graphene contains actuation with better piezoelectric properties and efficiency of system is envisaged to have wide applications in various fields.

References

1. Novoselov K S, Geim A K,Morozov S V, Jiang D, Zhang Y, Dubonos S V, Grigorieva I V, Firsov A A. Electric Field Effect in Atomically Thin Carbon Films. Science. 2004; 306: 666–669.
2. Rafiee M, Yang J, Kitipornchai S. Thermal bifurcation buckling of carbon nanotube reinforced composite beams. Computers & Mathematics with Applications. 2013; 66: 1147–1160.
3. Balamurugan V, Narayanan S. A piezoelectric higher-order plate element for the analysis of multilayer smart composite laminates. Smart Materials and Structures. 20007; 16: 2026–2039.
4. Narayanan S, Balamurugan V. Finite element modeling of piezolaminated smart structures for active vibration control with distributed sensors and actuators. Journal of Soundand Vibration. 2003; 262: 529– 562.
5. Narayanan S, Balamurgan V. Active control of FGM plates using distributed piezoelectric sensors and actuators. ICTAM04: Proc. 21st Int. Congr. of Theoretical and Applied Mechanics (Warszawa, Poland, Aug. 2004) CDROM.
6. Morten B, Decicco G, Prudenziati M. Resonant Pressure Sensor Based on Piezoelectric Properties of Ferroelectric Thick-Films. Sensors and Actuators A: Physical. 1992; 31: 153–158.
7. Jaffe H, Berlincourt D A. Piezoelectric Transducer Materials. Proc. IEEE. 1965; 53: 1372–1386.
8. Wang Z L, Song J H. Piezoelectric Nanogenerators Based on Zinc Oxide Nanowire Arrays. Science. 2006; 312: 242–246.
9. Anton S R, Sodano H A. A review of power harvesting using piezoelectric materials (2003–2006). Smart Materials and Structures. 2007; 16: R1.
10. Williams C B, Yates R B. Analysis of a micro-electric generator for microsystems. Sensors Actuators. 1996; 52: 8 – 11.
11. Anton S R and Sodano H A 2007 A review of power harvesting using piezoelectric materials (2003–2006) Smart Mater. Struct. 16 R1.
12. Zhu Y, Zu J, Su W. Broadband energy harvesting through a piezoelectric beam subjected to dynamic compressive loading. Smart Materials and Structures. 2013; 22: 045007.
13. Ahir S V, Squires A M, Tajbakhsh A R, Terentjev E M. Infrared actuation in aligned polymer-nanotube composites Physical Review B. 2006; 73: 085420.
14. Bao Z,Miao F, Chen Z, Zhang H, Jang W Y, Dames C, Lau C N. Controlled Ripple Texturing of Suspended Graphene and Ultrathin Graphite Membranes. Nature Nanotechnology. 2009; 4: 562–566.

15. Lee M H, Nicholls H R. Tactile sensing for mechatronics—a state of the art survey. Mechatronics. 1999; 9: 1-31.
16. Voges U. Laparoscopic technique—which developments are possible? Urologe A. 1996; 3: 208 -214 (in German).
17. Eltaib M E H, Hewit J R. Tactile sensing technology for minimal access surgery—a review. Mechatronics. 2003; 13: 1163-1177.
18. Gray B L, Fearing R S, A surface micromachined microtactile sensor array. Proceedings of the IEEE International Conference on Robotics and Automation. 1996; vol. 1: Minneapolis, MN, USA, April 22–28, 1996, pp. 1–6.
19. Takashima K, Yoshinaka K, Okazaki T, Ikeuchi K. An endoscopic tactile sensor for low invasive surgery. Sensors and Actuator A. 2005; 119: 372 – 383.
20. Lu S, Panchapakesan B. Photomechanical responses of carbon nanotube/polymer actuators. Nanotechnology 2007;18: 305502.
21. Lendlein A, Jiang H,Junger O, Langer R.Light-induced shape-memory polymers. Nature2005;434: 879-882.
22. Jiang H,Kelch S,Lendlein A.Polymers Move in Response to Light.Adv. Mater.2006;18: 1471-1475.
23. Loomis J, Fan X, Khosravi F, Xu P, Fletcher M, Cohn R W, Panchapakesan B. Graphene/ elastomer composite-based photo-thermal nanopositioners. Scientific Reports 2013; 3. 1900; DOI: 10.1038/srep01900.
24. Cheng H, Liu J, Zhao Y, Hu C, Zhang Z, Chen N, Jiang L, Qu L,Graphene Fibers with Predetermined Deformation as Moisture- Triggered Actuators and Robots Angew. Chem. Int. Ed.2013; 52: 10482 –10486.
25. Hwang T, Kwon H Y, Oh J S, Hong J P, Hong S C, Lee Y, Choi H R, Kim K J, Bhuiya M H, Nam J D. Transparent actuator made with few layer graphene electrode and dielectric elastomer, for variable focus lens. Applied Physics Letters. 2013; 103: 023106.
26. Liang J, Xu Y, Huang Y, Zhang L, Wang Y, Ma Y, Li F, Guo T, Chen Y. Infrared-Triggered Actuators from Graphene-Based Nanocomposites, The Journal of Physical Chemistry C. 2009; 113: 9921–9927.
27. Bi H, Yin K, Xie X, Zhou Y, Wan S, Banhartb A F, Sun L. Microscopic bimetallic actuator based on a bilayer of graphene and graphene oxide. Nanoscale. 2013; 5: 9123 – 9128.
28. Wang E, Desai M S, Lee S W. Light-Controlled Graphene-Elastin Composite Hydrogel Actuators. Nano Letters. 2013; 13: 2826–2830.
29. Liang J. J, Huang L, Li N, Huang Y, Wu Y. P, Fang S. L, Oh J, Kozlov M, Ma Y. F, Li F. F, Baughman R, Chen Y. S. Electromechanical Actuator with Controllable Motion, Fast Response Rate, and High-Frequency Resonance Based on Graphene and Polydiacetylene. ACS nano, 2012; 6: 4508–4519.
30. Ahir S. V, Huang Y, Terentjev E. M. Polymers with aligned carbon nanotubes: Active composite materials. Polymer 2008; 49: 3841-3854.
31. Levitsky I. A,Kanelos P. T, Woodbury D. S, Euler W. B.Photoactuation from a Carbon Nanotube–Nafion Bilayer Composite.J. Phys. Chem. B2006;110: 9421-9425.
32. Ahir S. V, Terentjev E. M. Fast Relaxation of Carbon Nanotubes in Polymer Composite Actuators Phys. Rev. Lett. 2006; 96: 133902.
33. Ahir S. V,Terentjev E. M.Photomechanical actuation in polymer–nanotube compositesNature Mater.2005;4: 491-495.
34. Yang L, Setyowati K, Li A, Gong S, Chen J. Reversible Infrared Actuation of Carbon Nanotube–Liquid Crystalline Elastomer Nanocomposites. Adv. Mater. 2008; 20: 2271-2275.
35. Koerner H, Price G, Pearce N. A, Alexander M, Vaia R. A. Remotely actuated polymer nanocomposites—stress-recovery of carbon-nanotube-filled thermoplastic elastomers. Nature Mater. 2004; 3: 115-120.
36. Osada Y, Okuzaki H, Hori H. A polymer gel with electrically driven motility. Nature, 1992; 355: 242 - 244.

37. Sidorenko A, Krupenkin T, Taylor A, Fratzl P, Aizenberg J.Reversible switching of hydrogel-actuated nanostructures into complex micopatterns. Science. 2007; 315: 487 - 490.
38. Bunch J S, Zande A M Van Der, Verbridge S S, Frank I W, Tanenbaum D M, Parpia J M, Craighead H G, McEuen P L.Electromechanical Resonators from Graphene Sheets. Science. 2007; 315: 490 - 493.
39. Ikuno T, Honda S I, Yasuda T, Oura K, Katayama M, Lee J G, Mori H. Thermally driven nanomechanical deflection of hybrid nanowires. Applied Physics Letters. 2005; 87: 213104.
40. Craighead H G. Nanoelectromechanical Systems. Science. 2000; 290: 1532–1535.
41. Fennimore A M, Yuzvinsky T D, Han W Q, Fuhrer M S., Cumings J, Zettl A. Rotational Actuators Based on Carbon Nanotubes. Nature. 2003; 424: 408–410.
42. Park S, An J, Suk J W, Ruoff R S. Graphene-Based Actuators. Small. 2010; 6: 210–212.
43. Muralidharan M N, Ansari S. Thermally reduced graphene oxide/thermoplastic polyurethane nanocomposites as photomechanical actuators. Advanced Materials Letters. 2013; 4: 927-932.
44. Gerratt A. P, Bergbreiter S. Incorporating compliant elastomers for jumping locomotion in microrobots. Smart Mater. Struct. 2013; 22: 014010.
45. Brochu P, Stoyanov H, Niu X, Pei Q. All-silicone prestrain-locked interpenetrating polymer network elastomers: free-standing silicone artificial muscles with improved performance and robustness. Smart Mater. Struct. 2013; 22: 055022 DOI:10.1088/0964-1726/22/5/055022.
46. Novoselov K. S, Geim A. K, Morozov S. V, Jiang D, Zhang Y, Dubonos S. V, Grigorieva I. V, Firsov A. A. Electric Field Effect in Atomically Thin Carbon Films. Science 2004; 306: 666-669.
47. Lotya M, King P. J, Khan U, De S, Coleman J. N. High-Concentration, Surfactant-Stabilized Graphene Dispersions. ACS Nano 2010; 4: 3155-3162.
48. Su C. Y, Lu A. Y, Xu Y, Chen F. R, Khlobystov A. N, L L. J. High-Quality Thin Graphene Films from Fast Electrochemical Exfoliation. ACS Nano 2011; 5: 2332-2339.
49. Tkachev S. V, Buslaeva E. Y, Gubin S. P. Graphene: A novel carbon nanomaterial. Inorganic Materials 2011; 47: 1-10.
50. Forbeaux I,Themlin J. M, Debever J. M.Heteroepitaxial graphite on 6H–SiC(0001): Interface formation through conduction-band electronic structure. Phys. Rev. B 1998; 58: 16396-16406.
51. Cambaz Z. G,Yushin G, Osswald S, Mochalin V, Gogotsi Y. Noncatalytic synthesis of carbon nanotubes, graphene and graphite on SiC. Carbon 2008; 46: 841-849.
52. Kim K. S, Zhao Y, Jang H, Lee S. Y, Kim J. M, Kim K. S, Ahn J. H, Kim P, Choi J. Y, Hong B. H. Large-scale pattern growth of graphene films for stretchable transparent electrodes. Nature 2009; 457: 706-710.
53. Xuesong L, Cai W. W, An J, Kim S. Y, Nah J, Yang D. X, Piner R, Velamakanni A, Jung I, Tutuc E, Banerjee S. K, Colombl L, Ruoff R. S. Large-Area Synthesis of High-Quality and Uniform Graphene Films on Copper Foils. Science 2009; 324: 1312-1314.
54. Zheng W, Lu X, Wong S.C. Electrical and mechanical properties of expanded graphite-reinforced high-density polyethylene. J. Appl. Polym. Sci. 2004; 91: 2781–2788.
55. Hu H.T, Wang J.C, Wan L, Liu F.M, Zheng H, Chen R, Xu C. H. Preparation and properties of graphene nanosheets – polystyrene nanocomposites via insitu emulsion polymerization. Chem. Phy. Letts.2010; 484: 247–253.
56. Lee W.D, Im S.S. Thermomechanical properties and crystallization behavior of layered double hydroxide/poly(ethylene terephthalate) nanocomposites prepared by in-situ polymerization. J. Polym. Sci. Pt. B Polym. Phys. 2007; 45: 28–40.
57. Hussain F, Hojjati M, Okamoto M, Gorga R.E. Review article: polymer–matrix nanocomposites, processing, manufacturing, and application: an overview. J. Compos. Mater. 2006; 40: 1511–1575.
58. Wanga W.P, Pana C.Y. Preparation and characterization of polystyrene/graphite composite prepared by cationic grafting polymerization. Polymer 2004; 45: 3987–3995.
59. Kalaitzidou K, Fukushima H, Drzal L.T. A new compounding method for exfoliated graphite-polypropylene nanocomposites with enhanced flexural properties and lower percolation threshold. Compos. Sci. Technol. 2007; 67: 2045–2051.

60. Kim S.K, Kim N.H, Lee J.H. Effects of the addition of multiwalled carbon nanotubes on the positive temperature coefficient characteristics of carbon-black-filled high density polyethylene nanocomposites. Scripta. Mater. 2006; 55: 1119–1122.
61. Kim, S.; Do, I.; Drzal, L.T. Thermal stability and dynamic mechanical behavior of exfoliated graphite nanoplatelets-LLDPE nanocomposites. Polym. Compos. 2009, 31, 755–761.
62. Bai S, Xu Q, Gu L, Ma F, Qin Y, Wang Z L, (2012) Single crystalline lead zirconate titanate (PZT) nano/micro-wire based self-powered UV sensor Nano Energy, DOI: 10.1016/j.nanoen.2012.09.001.
63. R. Pelrine, R. Kornbluh, Q. B. Pei and J. Joseph, High-speed electrically actuated elastomers with strain greater than 100 %. Science, 2000; 287: 836–839.
64. Kaneto K, Kaneko M, Min Y, MacDiarmid A. G."Artificial muscle": Electromechanical actuators using polyaniline films. Synth. Met. 1995; 71: 2211–2212.
65. Baughman R. H. Conducting polymer artificial muscles Synth. Met., 1996; 78: 339–353.
66. Smela E. Conjugated Polymer Actuators for Biomedical Applications. Adv. Mater., 2003; 15: 481–494.151 Asaka K, Oguro K, Nishimura Y, Mizuhata M, Takenaka H. Bending of Polyelectrolyte Membrane–Platinum Composites by Electric Stimuli I. Response Characteristics to Various Waveforms. Polym. J., 1995; 27: 436–440.
67. Shahinpoor M. Ionic polymer–conductor composites as biomimetic sensors, robotic actuators and artificial muscles—a review. Electrochim. Acta, 2003; 48: 2343–2353.
68. Baughman R. H, Cui C, Zakhidov A. A, Iqbal Z, Barisci J. N, Spinks G. M, Wallace G. G, Mazzoldi A, De Rossi D, Rinzler A. G, Jaschinski O, Roth S, Kertesz M. Carbon Nanotube Actuators. Science, 1999; 284: 1340–1344.
69. Alamusi, Xue J. M, Wu L. K, Hu N, Qiu J. H, Chang C, Atobe S, Fukunaga H, Watanabe T, Liu Y. L, Ning H. M, Li J. H, Li Y, Zhao Y. H. Evaluation of piezoelectric property of reduced graphene oxide (rGO)–poly(vinylidene fluoride) nanocomposites. Nanoscale, 2012; 4: 7250-7255.
70. Ong M. T, Reed E. J, Engineered Piezoelectricity in Graphene, ACS Nano, 2012: 1387–1394.
71. Osterlund L, Chakarov D. V, Kasemo B. Potassium Adsorption on Graphite(0001). Surf. Sci. 1999; 420: 174–189.
72. Virojanadara C,Watcharinyanon S,Zakharov A. A, Johansson L. I. Epitaxial Graphene on 6*H*-SiC and LiIntercalation. Phys. Rev. B 2010; 82: 205402.
73. Hussain M, Abbasi M. A, Ibupoto Z. H, Nur O, Willander M. The improved piezoelectric properties of ZnO nanorods with oxygen plasma treatment on the single layer graphene coated polymer substrate, Phys. Status Solidi A, 2014; 211: 455–459.
74. Yang R, Qin Y, Li C, Zhu G, Wang Z. L. Complex Crystal Structures Formed by the Self-Assembly of Ditethered Nanospheres. NanoLett. 2009; 9: 1201-1205.
75. Lee J. H, Lee K. Y, Gupta M. K, Kim T. Y, Lee D. Y, Oh J, Ryu C. K, Yoo W. J, Kang C. Y, Yoon S. J, Yoo J. B, Kim S. W. Highly Stretchable Piezoelectric-Pyroelectric Hybrid Nanogenerator. Adv. Mater. 2014; 26: 765–769.
76. Balandin A A. Thermal properties of graphene and nanostructured carbon materials. Nature Materials. 2011; 10, 569–581.
77. Wu H, Drzal L T. Graphene nanoplatelet paper as a light-weight composite with excellent electrical and thermal conductivity and good gas barrier properties. Carbon. 2012; 50: 1135–1145.
78. Balandin A A, Ghosh S, Bao W, Calizo I, Teweldebrhan D, Miao F, Lau C N. Superior thermal conductivity of single-layer graphene. Nano Letters. 2008; 8: 902–907.
79. Choi S U S, Zhang Z G, Yu W, Lockwood F E, Grulke E A. Anomalous thermal conductivity enhancement in nanotube suspensions. Applied Physics Letters. 2001; 79:2252–2254.
80. Loomis J, King B, Panchapakesan B. Layer dependent mechanical responses of graphene composites to near-infrared light. Applied Physics Letter. 2012; 100: 073108.
81. Pelrine R, Kornbluh R, Pei Q, Joseph J. High-Speed Electrically Actuated Elastomers with Strain Greater Than 100 %.Science.2000;287: 836 – 839.
82. Nam J D, Hwang S D, Choi H R, Lee J H, Kim K J, Heo S. Electrostrictive polymer nanocomposites exhibiting tunable electrical properties. Smart Materials and Structures. 2005; 14: 87.

83. Son S I, Pugal D, Hwang T, Choi H R, Koo J C, Lee Y, Kim K, Nam J D. Electromechanically driven variable-focus lens based on transparent dielectric elastomer. Applied Optics. 2012; 51: 2987 – 2996.
84. Nair R R, Wu H A, Jayaram P N, Grigorieva I V, Geim A K. Unimpeded Permeation of Water through Helium-Leak–Tight Graphene-Based Membranes. Science. 2012; 335: 442 – 444.
85. Jian Z, Christine M A, Jia D X, Ayyalusamy R, Thomas T, Nicholas A K. Pseudonegative Thermal Expansion and the State of Water in Graphene Oxide Layered Assemblies. ACS Nano.2012; 6: 8357 – 8365.
86. Lee S K, Jang H Y, Jang S, Choi E, Hong B H, Lee J,Park S, Ahn J H. All Graphene-Based Thin Film Transistors on Flexible Plastic Substrates. Nano Letters. 2012; 12: 3472 – 3476.
87. Jeong H Y, Kim J Y, Kim J W, Hwang J O, Kim J E, Lee J Y, Yoon T H, Cho B J, Kim S O, Ruoff R S,Choi S Y. Graphene Oxide Thin Films for Flexible Nonvolatile Memory Applications. Nano Letters. 2010; 10: 4381 – 4386.
88. He Q Y, Sudibya H G, Yin Z Y, Wu S X, Li H, Boey F, Huang W, Chen P, Zhang H. Centimeter-Long and Large-Scale Micropatterns of Reduced Graphene Oxide Films: Fabrication and Sensing Applications. ACS Nano. 2010; 4: 3201 – 3208.
89. Timoshenko S. Analysis of bi-metal thermostats. J. Opt. Soc. Am., 1925; 11: 233-255.
90. Lo C.W, Zhu D, Jiang H. An infrared-light responsive graphene-oxide incorporated poly (N-isopropylacrylamide) hydrogel nanocomposite. Soft Matter. 2011; 7: 5604-5609.
91. Zhu C. H, Lu Y, Peng J, Chen J.F, Yu S.H. Photothermally Sensitive Poly (N-isopropylacrylamide)/Graphene Oxide Nanocomposite Hydrogels as Remote Light-Controlled Liquid Microvalves. Adv. Funct.Mater. 2012; 22: 4017–4022.
92. Lian Y, Liu Y, JiangT, Shu J, Lian H, Cao M. Enhanced Electromechanical Performance of Graphite Oxide-Nafion Nanocomposite Actuator. Journal of Physical Chemistry C. 2010; 114: 9659–9663.
93. Ramasamy M. S, Mahapatra S. S, Yoo H. J, Kim Y. A, Cho J. W. Soluble conducting polymer-functionalized graphene oxide for air-operable actuator, fabrication. J. Mater. Chem. A, 2014; 2:4788-4794.

Graphene Polymer Nanocomposites for Fuel Cells

Jinghan Zhu, Fei Liu, Nasir Mahmood and Yanglong Hou

Abstract Fuel cells have long been considered as highly efficient devices for energy conversion which transform chemical energy directly into electricity, without any venomous pollutants emitted into ambient environment. In recent years, graphene-polymer nanocomposites have attracted intense interest as functional components in fuel cells. This chapter focuses on the potential applications of graphene-polymer composites in fuel cells. Recent advancement in the synthesis of graphene, polymer, graphene-polymer composites will be presented. Then, latest explorations of graphene-polymer composites applied as membranes, anode and cathode materials will be summarized. Furthermore, the vital roles of graphene-polymer to support noble metal catalysts will be illustrated. Finally, prospects of graphene-polymer composites for fuel cells will be outlined for further development.

Keywords Graphene · Polymer · Nanocomposites · Fuel cells · Membrane · Anode · Cathode · Electrocatalysts

Abbreviation

BMI	1-butyl-3-methylimidazolium
CC	Carbon cloth
CCG	Chemically converted graphene
CDP	β-Cyclodextrin polymer
CV	Cyclic voltammetry

J. Zhu · F. Liu · N. Mahmood · Y. Hou (✉)
Department of Materials Science and Engineering,
College of Engineering, Peking University, 100871 Beijing, China
e-mail: hou@pku.edu.cn

K.K. Sadasivuni et al. (eds.), *Graphene-Based Polymer Nanocomposites in Electronics*, Springer Series on Polymer and Composite Materials,
DOI 10.1007/978-3-319-13875-6_5

CVD	Chemical vapor deposition
DGO	Modified graphene oxide
DMFC	Direct methanol fuel cells
DFAFC	Direct formic acid fuel cells
ECSA	Electrochemical active surface area
EG	Expanded graphite
ERGO	Electrochemically reducing graphene oxide
GC	Glass carbon
GCN	Graphitic carbon nitride
GF	Graphite felt
GM-silica	Graphene-based silica
GNP	Graphite nanoplatelet
GNS	Ggraphene nanosheets
GO	Graphene oxide
K-L	Koutecky-Levich
LbL	Layer by layer
LSV	Linear sweep voltammetry
MFC	Microbial fuel cell
MNF	Mesonaphthobifluorene
NG	Nitrogen-doped graphene
NPs	Nanoparticles
O-EDOT	Oxidized Ethylenedioxythiophene
ORR	Oxygen reduction reaction
PAA	Polyallylamine
PANI	Polyaniline
PB	Platinum black
PBI	Polybenzimidazole
PBS	Phosphate buffer solution
PDDA	Poly(diallyldimethylammonium chloride)
PEDOT	Poly(3,4-ethylenedioxythiophene)
PEEK	Polyether ether ketone
PEM	Proton exchange membranes
PEMFC	Proton exchange membrane fuel cells
PMF	Phenol–melamine–formaldehyde
PMAA	Poly(methacrylic acid sodium salt)
PSS	Poly(sodium 4-styrenesulfonate)
PPy	Polypyrrole
PSF	Polysulfone
PVA	Polyvinyl alcohol
PVP	Polyvinylpyrrolidone
RGO	Reduced graphene oxide
SGO	Sulfonated graphene oxide
SPES	Sulfonated poly(ether sulfone)
SPI	Sulfonated polyimide
SPEEK	Sulfonated polyether ether ketone

STA	Silicotungstic acid
TCNQ	7,7,8,8-Tetracyanoquinodimethane
3D	Three-dimensional

1 Introduction

Rapid consumption and over-exploration of conventional fossil fuels such as coals, petroleum and natural gases have consequently given rise to heavy natural resource depletion. During the utilization process of fossil fuels, green-house gases and even some poisonous byproducts are inevitably generated, harassing the environmental balance and aggravating the climate change. Many scientists are devoted to fulfilling the crucial task of developing clean and sustainable energy for commercial applications, in advance of the complete exhaustiveness of fossil fuels. Fuel cells have long been considered as highly efficient devices for energy conversion which transform chemical energy directly into electricity, without any venomous pollutants emitted into ambient environment [1–4]. Since the primitive invention of a fuel cell prototype over 160 years ago, fuel cells have attracted a lot of research interests [5]. Various kinds of fuels such as hydrogen, methanol, formic acid and even waste water have been explored to continuously power the fuel cells. However, several obstacles hindered the commercialization of fuel cells, including high cost, counterproductive power output and inefficacious stability [6–9].

Graphene, as a unique one-atom-thick layer composed of sp^2-hybridized carbon atoms aligned in a honeycomb lattice, possesses many unique properties such as extraordinary electrical conductivity, greatly optimized specific surface area (up to 2,600 m^2/g), mechanical strength and resilience and chemical stability. Great research efforts have confirmed that graphene could play a vital role in many application fields [10–14]. Polymers are frequently blended with graphene materials to form functional nanocomposites, and to achieve the processibility and surface modification. The multifunctional properties endowed graphene-polymer composites copious potentials in the advance of various energy conversion and storage systems, such as fuel cells [15–17], supercapacitors [18–20], lithium ion batteries [21–24], solar cells [11, 15, 17, 25].

Graphene-polymer nanocomposites which possess excellent electrical conductivity, exceptional mechanical properties, abundant surface functionality, and outstanding electrochemical properties have been investigated as promising candidates to build the ideal fuel cells. In this chapter, we aim to outline recent progress in the preparation and potential applications of graphene-polymer composites in fuel cells. The chapter fist discusses the synthesis of graphene and graphene-polymer nanocomposites. It will then summarize the latest explorations of graphene-polymer composites applied as membranes, anode, cathode materials. Toward the end, the vital roles of graphene-polymer to support noble metal NPs are illustrated. Further, this chapter will lead towards possible solution to the challenges associated with graphene-polymer nanocomposites and their applications in the field of fuel cell.

2 Synthesis of Graphene-Polymer Nanocomposites

2.1 Synthesis of Graphene

Since the first time that graphene was successfully produced via mechanical cleavage in 2004, many synthesis methods have been developed to pursue high quality graphene nanosheets. Currently, there are several dominant methods including chemical vapor deposition (CVD) on metal substrates, epitaxial growth on single-crystal SiC, oxidation-exfoliation-reduction of graphite powder, direct exfoliation of graphite via sonication, and organic coupling reactions [12]. In terms of these methods, the oxidation-exfoliation-reduction synthesis provided great opportunities for synthesis of functional hydrophilic graphene nanosheets in large scale at low cost. Several oxidation protocols such as Hofmann (1937), Hummers and Offeman (1958) and Tour (2010) methods have been developed towards Graphene Oxide (GO) [26]. The as-obtained GO can be well dispersed as single-layer nanosheets in water and other solvents, under the assistance of stirring or sonication. Given that GO layers possess abundant functional groups containing hydroxyl, epoxy and carboxylic groups, further modification with organic or inorganic components can be easily achieved [27, 28].

While severe oxidation process can greatly enrich the surface functionality of GO, it brings about plenty of structural defects and attenuation in some properties such as electrical conductivity. Thus several strategies were adopted to reduce the oxygen contents of GO, among them thermal annealing and use of reducing agents are mostly used [29]. Figure 1 represents the various characteristics of Reduced graphene oxide (RGO) (synthesized via post-reduction treatment of GO) over GO [30]. Thermal annealing and reducing agents are commonly adopted to achieve RGO nanosheets [26, 30, 31]. Research analysis revealed that RGO showed an obvious decline in oxygen content and revitalized a graphitic structure. These features, to some extent, contributed to the enhanced electrical conductivity and structure stability of RGO.

Apart from the above mentioned oxidation-reduction synthesis methods to produce GO-derived RGO nanosheets, it is also applicable in other liquid-phase synthetic adopting heat treatment or microwave irradiation to expand the graphite precursor. Our group has developed a facile liquid synthesis to prepare high quality graphene well dispersed in aqueous suspensions. In the study, TCNQ anions were employed as a stabilizer and expanded graphite (EG) as the starting material [32]. Figure 2 displayed the synthesis process. Transmission electron microscope (TEM) images demonstrated that the products were mainly composed of isolated graphene layers, as shown in Fig. 3. We also generated monolayer and bilayer graphene via a simple solvothermal-assisted exfoliation with the assistance of acetonitrile, a highly polar organic solvent [33], as shown in Fig. 4. This method showed advantages in maintaining the defect-free state and intrinsic graphene without any stabilizer or modifier. Moreover, it shed light on the large scale synthesis of high quality graphene.

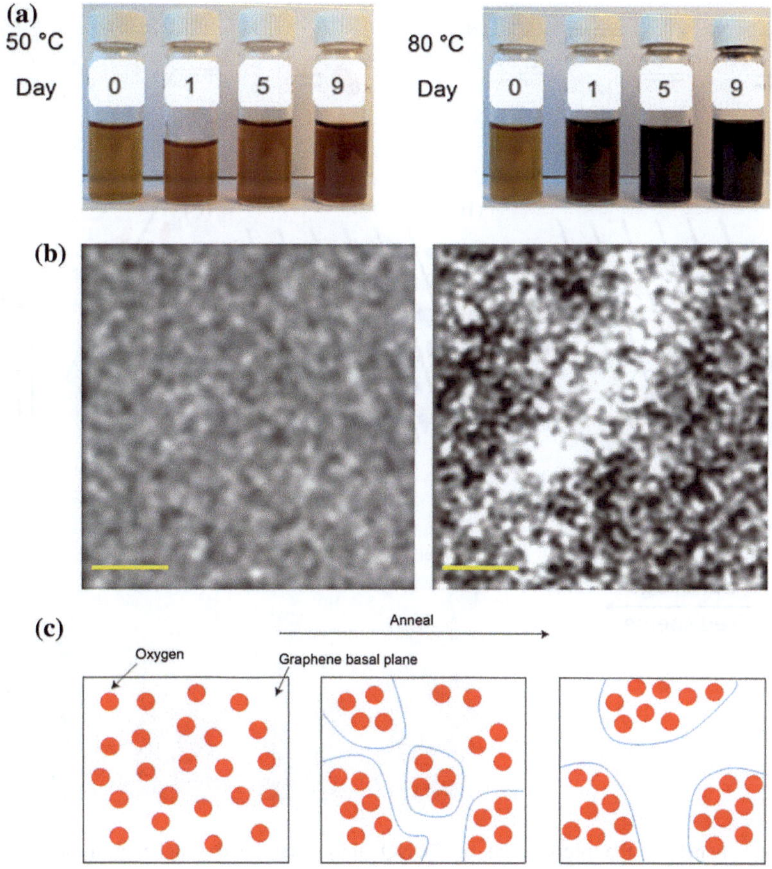

Fig. 1 Conversion of GO to RGO via reduction: **a** pictures of GO suspensions aged at 50 and 80 °C at different time intervals. **b** AES oxygen mapping of as-synthesized and annealed GO films. The *white* spots indicate oxygen-rich regions and the *black* spot indicate oxygen-poor regions or, in other words, carbon-rich regions. *Scale bar*, 2 mm. **c** Schematic depicting the proposed phase-separation process in as-synthesized GO structures. Synthetic protocols of GO structures lead to a mixed sp^2–sp^3 phase that has the potential to separate into two distinct oxidizedand graphene phases through diffusion of oxygen atoms on the graphene basal plane under the influence of an external stimulus. a.u., arbitrary units [30]. Copyright 2013. Reproduced with permission from nature

2.2 Synthesis of Graphene-Polymer Nanocomposites

Generally, graphene-polymer composites are fabricated by blending graphene and polymers in liquid solution. According to the reacting condition and parameters, blending methods are classified into three types: solvent mixing, melt blending and in situ polymerization [34, 35]. A solution mixing synthesis is feasible for soluble and malleable polymers, e.g., Polyvinyl alcohol (PVA) [36, 37], Nafion [38],

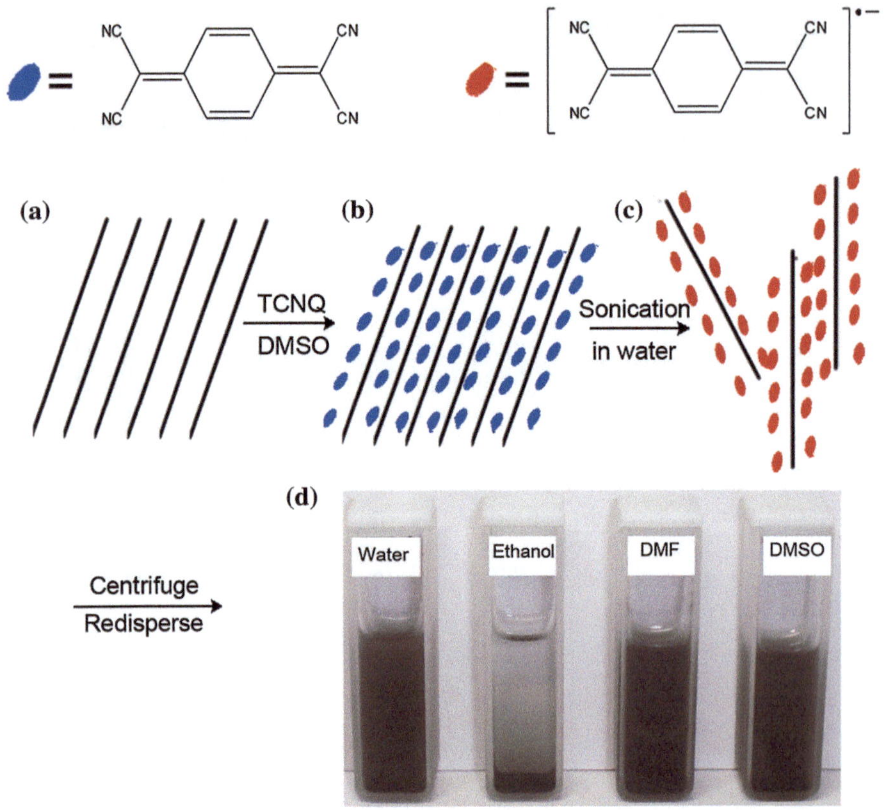

Fig. 2 Scheme of the preparation of TCNQ anion stabilized graphene: **a** pretreated EG; **b** TCNQ insertion into the interlayer of EG; **c** TCNQ anion stabilized graphene; **d** photograph of TCNQ anion stabilized graphene dispersed in different solvents (water, ethanol, DMF and DMSO) [32]. Copyright 2008. Reproduced with permission from Royal Society of Chemistry

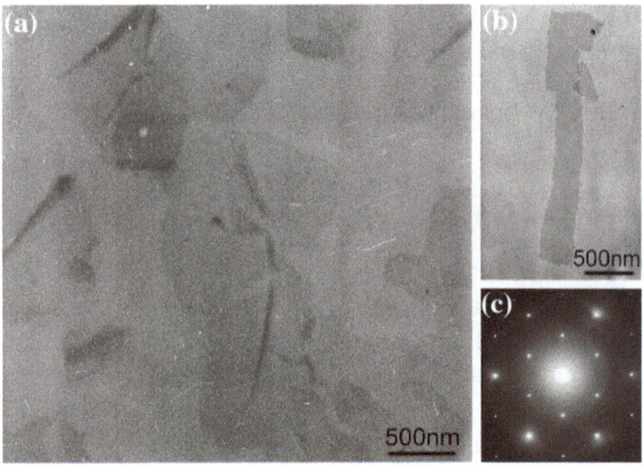

Fig. 3 TEM images of **a** several isolated graphene and **b** GNR; **c** the SAED pattern of graphene [32]. Copyright 2008. Reproduced with permission from Royal Society of Chemistry

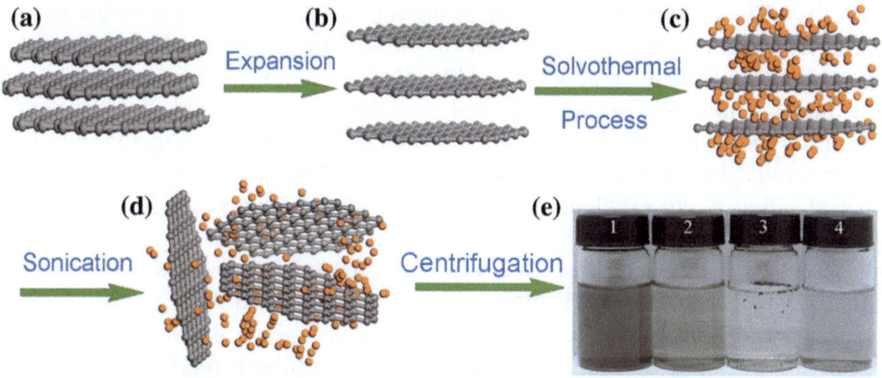

Fig. 4 Scheme of the preparation of graphene via solvothermal process in acetonitrile: **a** pristine expandable graphite; **b** EG; **c** acetonitrile molecules insertion into the interlayer of EG; **d** exfoliated graphene dispersed in acetonitrile; **e** photograph of four samples obtained under different conditions [33]. Copyright 2009. Reproduced with permission from Royal Society of Chemistry

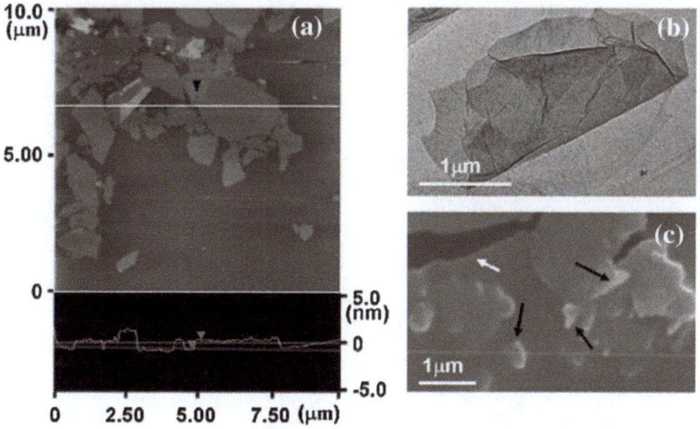

Fig. 5 Characterization of individual graphene sheets and dispersion state of GO in the polymer matrix. **a** Typical tapping-mode AFM image of graphene sheets deposited on mica substrate from an aqueous dispersion; the height difference between the *arrows* is ~0.8 nm, indicating the typical height of an individual GO sheet. **b** Typical TEM image of an individual graphene sheet in water solution. **c** SEM cross-sectional image of graphene/PVA nanocomposites with GO concentration of 0.7 wt%. The *black arrows* indicate GO sheets, and the *white arrow* points to a fracture [41]. Copyright 2009. Reproduced with permission from John Wiley & Sons

polyurethane [39, 40]. Usually, graphene and polymer are evenly amalgamated in designated solvents, and after the homogenous distribution of graphene in polymer matrix through stirring solvent was evaporated by simple heating. Chen et al. prepared PVA-GO nanocomposites via a facile aqueous solution blending method and testified the molecule-level dispersion, as shown in Fig. 5. Composites with only 0.7 wt% GO exhibited a 76 % increase in tensile strength, compared with pristine

PVA [41]. The disadvantage of solution mixing is creation of voids by the solvent during evaporation process that introduces the structural defects which brings poor thermal and electrical conductivity [42]. Besides simple mixing method, melt blending has also been developed, and considered more feasible for the simple reason of being free of solvents. In this synthesis, graphene was mixed with melt polymers via shear force to generate composites. Despite the facile manipulation, melt blending is not efficient enough to achieve well dispersion of high content graphene, due to the heavy viscosities of melt polymers. Thus the aggregated graphene exist in composite that inversely effect the final properties and limits the applications. In situ polymerization of monomers premixed with graphene or GO suspensions can also be used to produce highly uniform composites. The degree of polymerization is of vital importance on the properties of final products, which can be tailored by reaction parameters.

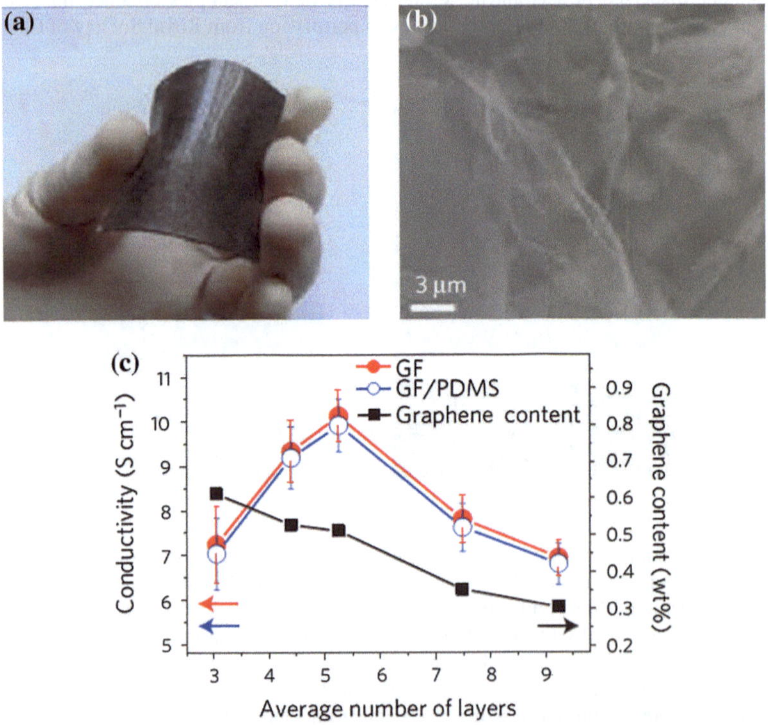

Fig. 6 Morphology, fracture surface, electrical conductivity and mechanical properties of GF/PDMS composites. **a** Photograph of a bent GF/PDMS composite, showing its good flexibility. **b** SEM image of the fracture surface of the composite. **c** Electrical conductivity of GFs and GF/PDMS composites as a function of the number of graphene layers (corresponding to different weight fractions of GFs in the composites). The average conductivity of GFs and GF/PDMS composites was calculated over five specimens for each sample and error bars for standard deviation are shown [49]. Copyright 2011. Reproduced with permission from Nature

Recently, 3D graphene monolithic templates have been developed to support polymers. Graphene frameworks with continuous and interconnected channels were constructed using several methods like vacuum filtration of graphene suspension, CVP on nickel foams, hydrothermal synthesis, thermal reduction and lyophilization [43–46]. 3D graphene provided a superior platform to form high quality composites [47, 48]. For instance, Cheng et al. successfully loaded poly(dimethyl siloxane) into 3D graphene foams by infiltration, as shown in Fig. 6. The composites possessed extraordinary electrical conductivity and flexibility, determined by the porous network structure [49].

3 Graphene-Polymer Nanocomposites Directly Applied in Fuel Cells

3.1 Fuel Cells

Fuel cell is a device of energy storage and conversion, which transforms the chemical energy of fuels into electrical energy, under the assistance of effective catalysts. Various kinds of fuels such as hydrogen, methanol, formic acid and even waste water have been explored to continuously power the fuel cells. Typically, Proton Exchange Membrane Fuel Cells (PEMFCs), Direct Methanol Fuel Cells (DMFCs), and Direct Formic Acid Fuel Cells (DFAFCs) have been considered promising candidates with high efficiency and high energy density. Generally, a classical PEMFC is amalgamation of three main units, including cathode, anode and electrolyte membrane [4] (Fig. 7), which takes use of the hydrogen fuel. At the anode, H_2 molecules are oxidized into H^+ cations, while at the cathode, O_2 molecules are reduced into O^{2-}, which further react with H^+ and form H_2O

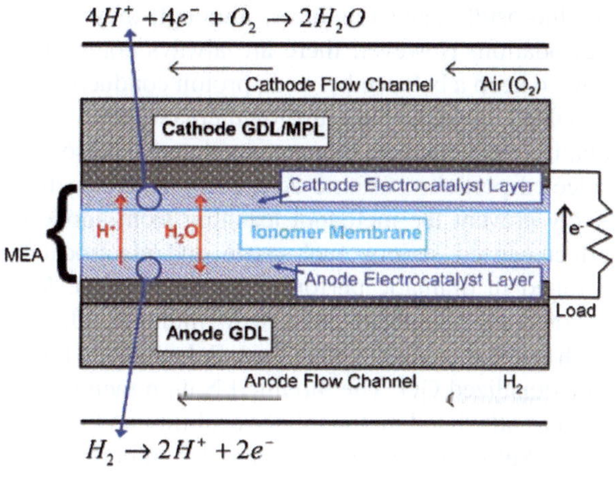

Fig. 7 Schematic cross section of a typical PEM fuel cell. PEM fuel cells are composed of anode, cathode and membrane. At the anode, H_2 molecules are oxidized into H^+ cations, while at the cathode, O_2 molecules are reduced into O^{2-}, which further react with H^+ and form H_2O molecules [4]. Copyright 2007. Reproduced with permission from American Chemical Society

molecules. This process is clean since the only chemical product is water. In other words, fuel cells are promising alternatives for traditional energy forms such as fossil fuels, which caused heavy air pollution and global climate change.

The components of PEMFC are exposed to a harsh environment consisting of strong oxidizing atmosphere, liquid media, strongly acidic conditions, high electrochemical potentials, reactive intermediate reaction products and high electric current. Thus, many scientists have made great efforts to design a well-developed fabrication to lower the negative effects. As shown in Fig. 7, the anode and cathode each have an electrocatalyst layer and a gas diffusion layer to increase the electron diffusion rate and energy transfer efficiency. Fuel cells are stacked together to obtain a considerable voltage, with each cell providing 0.6–0.7 V. Parallel to the bipolar electrodes are the GDLs, which contain two macroporous substrate layer and a microporous layer, in order to direct gases to the catalyst layer, providing an interconnected network of paths for liquid water to flow fluently from outwards the electrode.

3.2 Membranes in Fuel Cells

In fuel cells, especially PEMFC, proton-conducting membranes account for one of the most important part. There are usually several evaluation parameters to evaluate the quality of PEM, mainly including the following: (1) excellent mechanical, chemical and thermal properties; (2) fast proton transfer and good electrical conductivity; (3) low methanol and gas permeability; (4) geometry stability without swelling [8].

Primarily, various kinds of polymers were adopted and fabricated into membrane-like products for fuel cells. It is revealed that the properties of proton conducting polymer mainly depend on the degree of sulfonation and the nano-phase separation. Commonly used electrolyte membranes are made of perfluorosulfonated ionomers, although they possess high cost and conductivity degradation. However, there are always contradicting problems to solve such as how to keep a balance between proton conductivity and geometry stability [50].

DMFC, which make use of methanol fuel, have special requirement for high quality membrane to prevent methanol crossover effect. However, traditionally selected perfluorosulfonic acid membranes, such as Nafion, are unable to effectively prevent the methanol leaching from anode to cathode unit. This crossover effect caused serious fuel waste and electrocatalyst poisoning, and eventually resulted in dramatic performance degradation. Many research efforts have been devoted to combine Nafion with multiple additives, including silica, polymers such as Polypyrrole (PPy) and Polyaniline (PANI), polycarbon nanotubes and functionalized GO. The modified Nafion membranes exhibited supreme properties such as alleviated methanol permeability and maintenance of good proton conductivity. Among them, GO is unique due to profound functional groups, high thermal and electrical conductivity as well as large surface area.

GO with luxuriant oxygen containing functional groups, high surface area, robust strength and remarkable resilience has long been considered to work as nanofillers so as to enhance the proton transport, ionic conductivity, chemical stability and mechanical properties of PEMs. Besides, Nafion can be incorporated with functionalized graphene in order to tailor the ionic structure and physical properties. Choi et al. [51] have integrated Nafion with sulfonated GO, transport properties of which were superior to those of pristine Nafion and GO-Nafion nanocomposites. The confined bound water resulted to enhanced properties including high proton conductivity and low methanol crossover, for the purpose of high performance fuel cells. While most studies utilized a similar mixing approach to combine GO with the base polymer followed with the simple casting to obtain membrane, Lin et al. [52] reported the fabrication of parallel-oriented GO paper via vacuum filtration and subsequent adhesion of laminated Nafion 115 membrane via transfer printing and hot-pressing, as shown in Fig. 8. The as-obtained dual-layer laminate membrane took advantage of 2D GO paper as efficient methanol barrier under high methanol feed concentrations. The thickness of GO paper in the composite membrane was calculated to be 1.0 mm thick, according to scanning electron microscopy (SEM) analysis. Compared with GO-PVA composite membrane made by simple casting method, the dual layer GO-Nafion nanocomposites possessed 70 % lower methanol penetration at a cost of 22 % decrease in proton conductivity, revealing great potential as membranes for DMFCs. Layer by layer (LbL) deposition technique is widely adopted to fabricate GO/PDDA onto Nafion membranes to suppress the methanol crossover and enhance the strength [53]. To decrease the methanol crossover effect of PEM membranes for direct methanol fuel cells, LbL synthesis is adopted to fabricated PDDA-GO onto the surface of

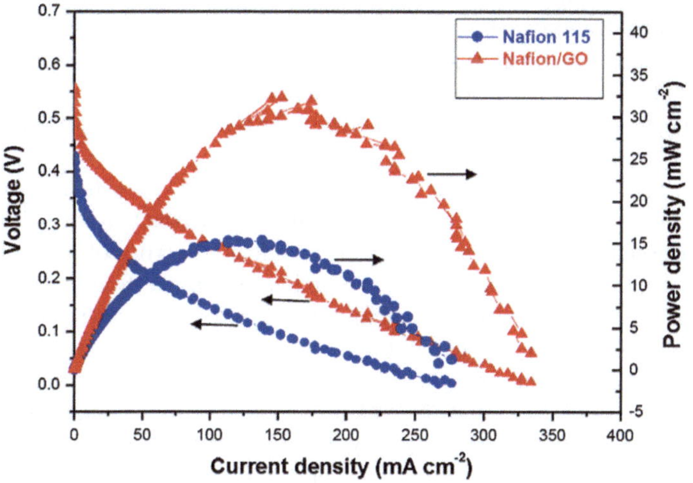

Fig. 8 Comparison of DMFC performances at 8 M methanol feed concentration by using Nafion 115 membrane and GO-laminated Nafion membrane [52]. Copyright 2013. Reproduced with permission from Elsevier

Nafion membranes. It is revealed that the bilayers formed a uniform and thin film on Nafion®, and thus induced high methanol-blocking properties. The methanol diffusion efficiency was oppressed to 67 % that of the pristine membrane [53].

Graphene/polymer nanocomposites have been explored as remarkable membranes for fuel cells integrating both high proton and electron conductivity and anti-methanol crossover properties. It is still challenging to blend pristine graphene with polymer molecules and obtain atomically well-dispersed nanohybrid because of its novel nature due to existence of aromatic structure. GO nanosheets, decorated with plentitude of functional groups such as hydroxyl and carboxyl ones and were speculated as effective intermediate to disperse polymer electrolytes. In general, GO is advantageous in chemical compatibility with polymers over hydrophobic graphene. Besides, GO surface modification is easy to manipulate with ionic or cationic functional groups. Following thermal treatment or reducing process convert GO into RGO.

For instance, Ansari prepared GO-Nafion nanocomposites initially by suspension solution casting and then chemically reduced the membrane via hydrazine to obtain G-Nafion nanocomposites. This two-step approach is critical to achieve well-oriented structure where GO platelets were aligned parallel to the membrane surface. While simple mixing of hydrophobic graphene with Nafion via solution casting results in random distribution of graphene in matrix. GO are well accustomed to the polymer matrix than graphene, as confirmed by XRD analysis. The G-Nafion nanocomposites showed dramatic increase in electrical conductivity compared with pure Nafion, due to the highly conductive reduced graphene [54].

Limitation of Nafion, sulfonated perfluoropolymers and flemion membranes include limited operation temperature (0–80 °C) and high cost. And Nafion is of little function under low humidity environment [55]. The decrease in proton conductivity above 80 °C hinders the use of Nafion for large scale applications. Alternatives to Nafion membrane aromatic polymers, such as PEEK, PBI, and PSF, have received significant attention to meet the operating requirements for fuel cell applications [56] GO nanosheets have also been investigated as contributors to escalate the proton conductivity and methanol crossover resistance of these binary and ternary systems.

Mishra reported that the addition of Sulfonated polyether ether ketone (SPEEK) into Nafion will greatly improve the proton conductivity. The utility of two dimensional GO could enhance the mechanical properties of the membrane to a great extent [57]. Sulfonated Graphene Oxide (SGO) and Sulfonated Poly(ether sulfone)(SPES) were formed into membrane nanocomposites via sonication and solution casting, which possessed improved proton conductivity, ion exchange capacity and mechanical strength because of the strong interaction between the large surface area of SGO and the SPES. Nanocomposites containing 5 % SGO was revealed with the best proton conductivity and large methanol crossover resistance, as an ideal candidate for fuel cells and other high-temperature applications [58].

A set of SPES decorated with MNF moiety were synthesized polycondensation, Friedel-Crafts reaction, and sulfonation reaction, which held graphene-like structure. The SPES membrane showed excellent proton conductivity, as high as that of Nafion 211®, which is suitable for fuel cell application [55].

Size effect of GO integrated into Sulfonated Polyimide (SPI) membrane on performance enhancement of fuel cell membranes were revealed. GO nanosheets ranging from 60 nm to 1 μm were synthesized via modified Hummers method by tuning graphite size, oxidation time, centrifugation and sonication time, which were hybridized into SPI membrane to achieve well-defined microstructures and well-connected proton transport channels owing to strong hydrogen bonding interaction. The properties mainly include proton conductivity, methanol permeability and mechanical property were affected by size of GO nanosheets. It is worth to note that the SPI–0.5 %-GO composite membrane with the smallest size GO showed the best and outstanding fuel cell performance compared to that of pure SPI and other SPI–GO composite membranes [50].

SPEEK-SGO nanocomposites as membrane electrolytes were prepared for fuel cells. The proton transport channels within the polymer were interconnected by SGO. The uniform distribution of GO within the polymer structure were confirmed by SEM, and functionalization between SGO and SPEEK was revealed by FTIR analysis. The composite membrane showed excellent proton conductivity (0.055 S cm^{-1})and fuel cell performance measured at 80 °C and 30 % RH, compared with that of recast SPEEK [56].

Ye et al. [59] reported synthesis of graphene-PVA nanocomposites by a simple blending method. Graphene formed uniformly well-connected ionic channels in the membrane and lead to enhanced ionic conductivity. 0.7 wt% addition of graphene resulted in 126 % enhancement in ionic conductivity and 55 % reduction in methanol permeability, as shown in Fig. 9. Strong mechanical properties were observed since graphene possess robust strength within PVA matrix because of strong adhesion among matrix and graphene. The PVA/graphene membranes were investigated in alkaline fuel cells, which showed 148 % increase of maximum power density in the fuel cell performance. It is believe that the addition of graphene into fuel cell membranes could modify the transport kinetics.

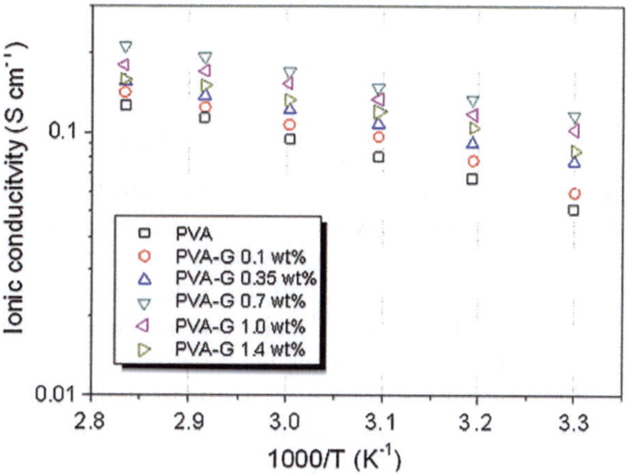

Fig. 9 Ionic conductivities of PVA and PVA/graphene composite membranes plotted as function of inverse temperature [59]. Copyright 2013. Reproduced with permission from Elsevier

PEM used in specific environment such as under high temperature and anhydrous media was investigated. It is reported that the introduction of functionalized graphene oxide could large enhance the ion conductivity, thermal and mechanical properties and wide applicability under harsh conditions. Polydopamine-modified graphene oxide (DGO) sheets were incorporated into SPEEK matrix to obtain anhydrous PEM. The DGO bore $-NH_2$ and –NH–groups conjugated with the SPEEK acidic polymer to form acid–base-paired nanocomposite membranes. This interaction could act as long-range and low-energy barrier pathways for proton hopping and imparted increased proton transfer. The interconnected DGO nanofillers through electrostatic interaction could effectively adjusted the thermal and mechanical behaviors of membrane, with only 2.5 % DGO raising the Young's modulus and tensile strength up to 896.1 and 57.5 MPa. By using the DGO-SPEEK membranes, the cell worked at up to 120 °C in anhydrous conditions yielded a 47 % enhancement in maximum current density and a 38 % increase in power density [60].

3.3 Cathode

At the cathode of fuel cells, electrocatalysts for ORR are core materials. There are two common pathways for reduction of O_2: a direct four-electron transfer route, where O_2 is reduced to H_2O or OH^- in an acidic or alkaline media; a two-electron transfer pathway, where O_2 is partly reduced to H_2O_2 or OOH^- in an acidic or alkaline media. Empirically, cyclic voltammetry (CV) analysis is usually adopted to examine the ORR pathways. Typically, a featured cathodic peak can be obviously identified in the presence of O_2, while no current shift is observed near the same potential range under N_2 protection.

LSV measurements provide efficient characterization results about the catalytic activity. The kinetic parameters including electron transfer number (n) were analyzed on the basis of K-L equations:

$$\frac{1}{J} = \frac{1}{J_K} + \frac{1}{B\omega^{1/2}} \tag{1}$$

$$B = 0.2nFC_O(D_O)^{2/3}(\upsilon)^{-1/6} \tag{2}$$

in which J is the measured current density, J_K is the kinetic-limiting current density, ω is the electrode rotating rate, n is the electron transfer number, F is the Faraday constant ($F = 96{,}485$ C mol^{-1}), C_O is the bulk concentration of O_2 ($C_O = 1.2 \times 10^{-6}$ mol cm^{-3}), D_O is the diffusion coefficient of O_2 in 0.1 M KOH ($D_O = 1.9 \times 10^{-5}$ cm^2 s^{-1}), and υ is the kinetic viscosity ($\upsilon = 0.01$ cm^2 s^{-1}). The constant 0.2 is adopted when the rotation rate is expressed in rpm.

Electrocatalysts are key material to build high quality fuel cell cathode and anode. However, dominant high cost noble metal based catalysts, together with tremendous oxygen crossover from anode to cathode are major obstacles for the commercialization of fuel cells. Polymer-graphene nanocomposites have been explored as both cathode and anode electrocatalysts, owing to their low cost, rich resources, excellent electron conductivity and remarkable electrochemical properties. In this section, graphene-polymer nanocomposites and derived nitrogen-doped graphene applied as ORR electrocatalysts are summarized.

4 Graphene-Polymer Nanocomposites as ORR Catalysts

Polymer-functionalized graphene, especially nitrogen-containing polymer, have been extensively investigated as promising metal-free electrocatalysts towards ORR at the cathode in fuel cells and lithium air batteries. These metal free nanomaterials provided efficacious alternatives to supplant traditional Pt-based catalysts badly interrupted with high cost, scarcity, and poor stability.

According to quantum mechanics calculation, the electron-accepting nitrogen atoms imparted a high positive charge on adjacent carbon atoms via intramolecular charge transfer. The charge delocalization could induce readily attraction of O_2 springing from anode, and convert the O_2 chemisorption from the common end-on to a side-on mode, which effectively facilitate the ORR process [61, 62]. Many research efforts have been made to pursue optimized carbon-based electrocatalysts with appropriate and tunable nitrogen-carbon atomic interactions.

Wang et al. reported synthesis of PDDA-decorated graphene nanohybrids via reduction of PDDA-GO mixture with $NaBH_4$. PDDA acting as an electron acceptor to impart electrocatalytic activity for ORR. PDDA-G showed a superior onset potential of 0.15 V (vs. SCE) compared with pristine graphene electrode (0.25 V), and exhibited better fuel selectivity, anti-CO poisoning ability and good stability compared to commercial Pt/C catalysts [63].

Sun immobilized graphitic carbon nitride (GCN) onto chemically converted graphene (CCG) substrates by polymerizing melamine molecules adsorbed on graphene under high temperature and obtained GCN-CCG nanocomposites, as shown in Fig. 10. The addition of graphene largely increased the poor electric conductivity of GCN and endowed the nanocomposites with large specific surface area, which consequently induced better electrocatalytic activity and stability for ORR with Pt/CCG nanocomposites. GCN-CCG is considered to be a low cost, versatile metal free electrocatalysts for fuel cells [64].

O-EDOT functionalized two dimensional graphene nanocomposites were synthesized via a simple, one-step route combining reduction of GO to graphene and oxidation of EDOT to O-EDOT. The polymer induced functional groups residing on planar graphene assisted to tailor the physical and chemical properties of graphene. Compact interaction between the polymer and graphene contributed to superior electrocatalytic activity for the reduction of oxygen and I^{3-} to I^- compared with pristine graphene. In this case, the nanocomposites can be used as an efficient cathode electrocatalyst for alkaline anion exchange membrane fuel cells and dye-sensitized solar cells [65].

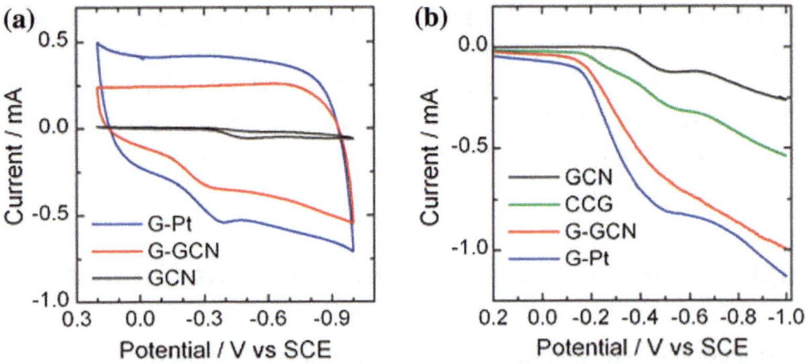

Fig. 10 **a** CVs of GCN, G-GCN and G-Pt electrodes at a scan rate of 100 mV sand **b** RDE voltammograms of GCN, CCG, G-GCN and G-Pt electrodes at a scan rate of 10 mV $s^{-1}$and a rotation rate of 1,500 rpm in O_2-saturated 0.1 mol L^{-1}KOH solution [64]. Copyright 2010. Reproduced with permission from Royal Society of Chemistry

Carbon nitride, especially g-C_3N_4, is a type of well-developed synthetic polymer, which can be synthesized via polymerization of cyanamide, dicyandiamide or melamine. Featured with high N content, stable chemical structure and thermal stability, g-C_3N_4 have long been considered as promising metal-free ORR electrocatalyts [66–68].

Lyth et al. first examined ORR electrocatalytic properties of pristine g-C_3N_4 in acidic medium, which exhibited higher onset potential (0.69 V vs. NHE) than carbon black (0.45 V). However, the current density of g-C_3N_4 was too low, probably because of limited surface area, which greatly hindered potential applications of g-C_3N_4 in fuel cells. Further investigation confirmed that blending carbon black support with g-C_3N_4 could induce a considerable increase in current density as well as onset potential to 0.76 V [69].

Many other studies proposed combination of g-C_3N_4 with conductive carbon nanomaterials to improve the electron transfer rate and electrochemical properties [70]. Among the common used carbon substrates, 2D graphene nanosheets with excellent electron accumulation ability and high surface area have been revealed to be an optimal choice to synthesize g-C_3N_4-carbon based nanocomposites [66, 71, 72]. Since g-C_3N_4 nanosheets had analogous planar organization with graphene, they attached firmly with each other in the LbL mode, providing a superior transport channel for electrons.

Sun et al. adopted a high temperature polymerization of melamine in the presence of graphene to decorate g-C_3N_4 onto graphene nanosheets, and obtained g-C_3N_4/graphene nanocomposites. The addition of graphene greatly enhanced electrocatalytic activity and CO tolerance ability of g-C_3N_4 for ORR, which approached that of 23 wt% Pt/graphene catalysts. They provided a proposition that the catalytic mechanism of on g-C_3N_4/graphene resembled that of other well developed metal-free ORR catalysts such as nitrogen doped carbon nanotubes [64].

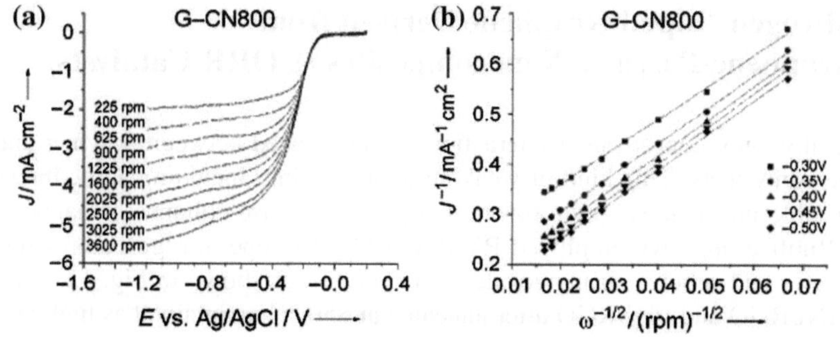

Fig. 11 **a** RDE linear sweep voltammograms of G-CN800 in O_2-saturated 0.1 M KOH with various rotation rates at a scan rate of 5 mVs^{-1}. **b** Koutecky–Levich plots of G–CN800 derived from RDE voltammograms in **a** at different electrode potentials [73]. Copyright 2011. Reproduced with permission from John Wiley & Sons

Yang et al. developed a novel synthesis method to produce graphene-carbon nitride nanosheets, as shown in Fig. 11. They employed GM-silica nanosheets as hard templates, ethylenediamine and carbon tetrachloride as CN precursors. Compared with pristine carbon nitride nanosheets without graphene, the obtained G-CN nanosheets exhibited high activity and selectivity for 4e ORR pathway, as well as long term durability. The excellent electrocatalytic performance were attributed to high nitrogen content, large surface areas, thin thicknesses and enhanced electrical conductivity. Besides, they revealed that the amount of pyridinic N had an enormous effect on the catalytic activity [73].

In terms of nitrogen-carbon catalytic system, major ORR active sites depended on the type and amount of nitrogen, which created net positive charge density on the abut carbon atoms, especially those at quaternary sites at the graphene edges, according to theoretical simulation results [71, 72, 74, 75].

MFC is a typical energy conversion device to convert chemical energy stored in organic wastes at places like municipal sludge and submarine sediment into electricity power. Similar to other types of fuel cells, cathode is of vital importance for improving the electricity generation of MFCs [76]. Graphene-polymer conjugated nanocomposites were thoroughly investigated as cathode electrode for rapid power output of MFC, due to their remarkable electron transfer efficiency and fast oxygen reduction rate.

PANI-GNS modified cathodes were fabricated via in situ polymerization of aniline in well-dispersed GNS solutions. The PANI-GNS electrodes participated as the cathode of MFCs to improve the electricity generation capacity and ORR efficiency. The properties of modified cathode can be readily tuned by the mass ratio of aniline and GNS in the precursor solution and optimized performance was achieved at the ratio of 9:1. The addition of highly conductive GNS assisted PANI-GNS electrode to outperform the PANI electrode with the highest output voltage of 640 mV and maximum power density of 99 mW/m^2, much higher than the blank MFC [76].

5 Nitrogen-Doped Graphene Derived from Graphene-Polymer Nanocomposites as ORR Catalysts

Generally, only non-covalent interaction can be created between polymer and graphene nanosheets. This kind of weak attraction mainly including static interaction is unstable under harsh environment and prone to cause collapse of nanocomposites. Ruoff et al. have employed PANI and PPy to dope nitrogen atoms into GO substrates and achieve strong covalent interactions, as shown in Fig. 12. N content of PANI/RGO and PPy/RGO after annealed at 850 °C maintained as high as 10 %, due to high nitrogen percentage in these two polymers. By ranging the annealing temperature from 550 to 1,000 °C, nitrogen state and total amount in the final products were readily tuned, resulting in variations in electrocatalytic properties. In this sense, the generated N-graphene with controlled stoichiometry not only provided a series of highly active catalysts, but also act as a suitable platform to investigate the intrinsic relation between N dopant species and catalytic activity [77].

Up to date, many advanced synthesis protocols have been explored to synthesize NG nanosheets, such as CVD [78, 79], solverthermal method [80], thermal annealing of nitrogen sources and graphene oxide [81, 82]. However, it remained an enigma to clearly distinguish the effect of nitrogen states on the electrochemical

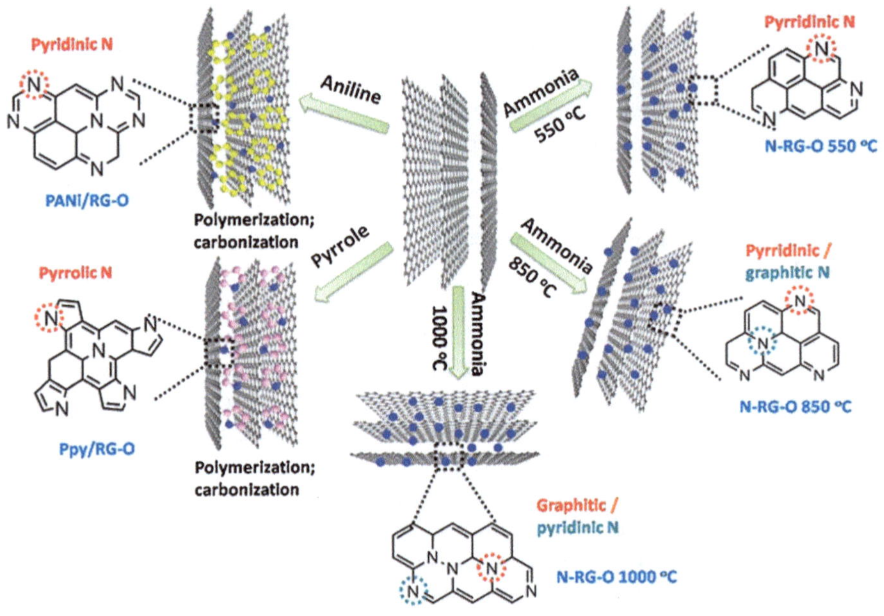

Fig. 12 Schematic diagram for preparation of N doped graphene with different N states. N-RG-O 550, 850, and 1,000 °C are prepared by annealing of G-O powder at temperatures of 550, 850 and 1,000 °C under a NH_3 precursor. PANi/RG-O and Ppy/RG-O are prepared by annealing of PANi/G-O and Ppy/G-O composites at 850 °C [77]. Copyright 2012. Reproduced with permission from Royal Society of Chemistry

properties. Our group have synthesized NG nanosheets with superior electrocatalytic performance towards ORR [62], as shown in Fig. 13. By systematically investigating the roles of various nitrogen states in annealing derived NG, it is revealed that the enhanced catalytic activity towards ORR are mainly due to the pyridinic N or graphitic N, while some other factors (e.g., the graphitic degree) might to some extent influence the results. Noteworthy, amino N with considerable stability, has been proved to be critical in escalating electrocatalytic performance of NG. Furthermore, the roles of pyridinic, pyrrolic and graphitic nitrogen states in ORR were extensively studied via a simple annealing method via NG with tunable nitrogen dopants, as shown in Fig. 14. Similarly, we demonstrated that phosphorus dopants also contribute greatly to the increase in electrocatalytic performance in phosphorus doped graphene [83].

Sun et al. have synthesized nitrogen-doped graphene via carbonization of a mixture of PMF pre-polymer and GO under the assistance of F127 soft template. The obtained G-PMF nanocomposites were composed of sandwich structures,

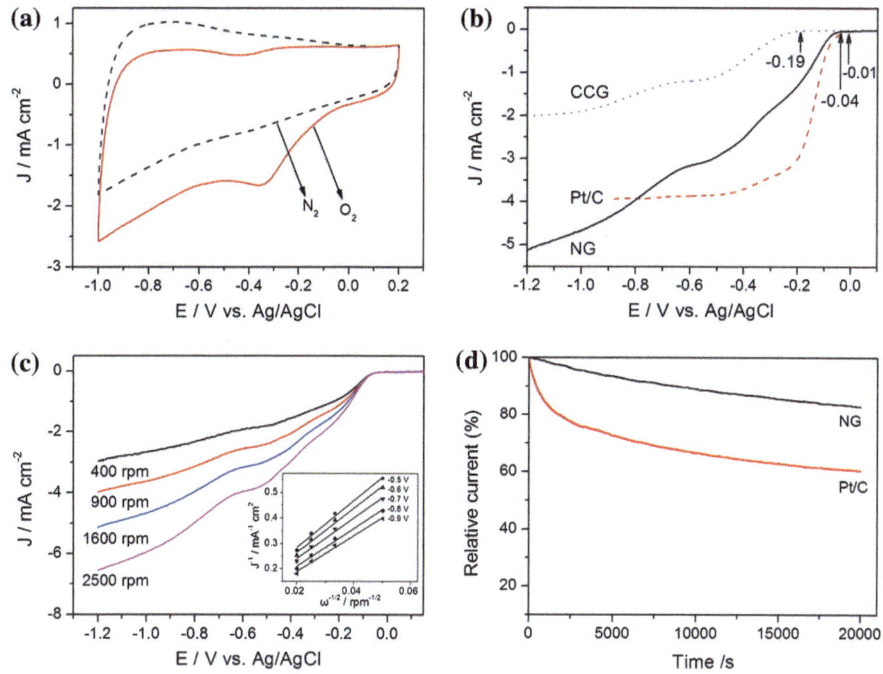

Fig. 13 **a** CV curves of nitrogen-doped graphene in a N_2- and O_2-saturated 0.1 M KOH solution; **b** LSV curves of CCG, Pt/C and nitrogen-doped graphene in an O_2-saturated 0.1 M KOH solution at a rotation speed of 1,600 rpm; **c** RDE curves of nitrogen-doped graphene in an O_2-saturated 0.1 M KOH solution with various rotation speeds at a scanning rate of 10 mV s^{-1}. Inset shows the K-L plots of J^{-1} versus $\omega^{-1/2}$ at different electrode potentials derived from RDE measurements; **d** current-time response of nitrogen-doped graphene and Pt/C at −0.28 V in an O_2-saturated 0.1 M KOH at a rotation speed of 1,600 rpm [62]. Copyright 2013. Reproduced with permission from Elsevier

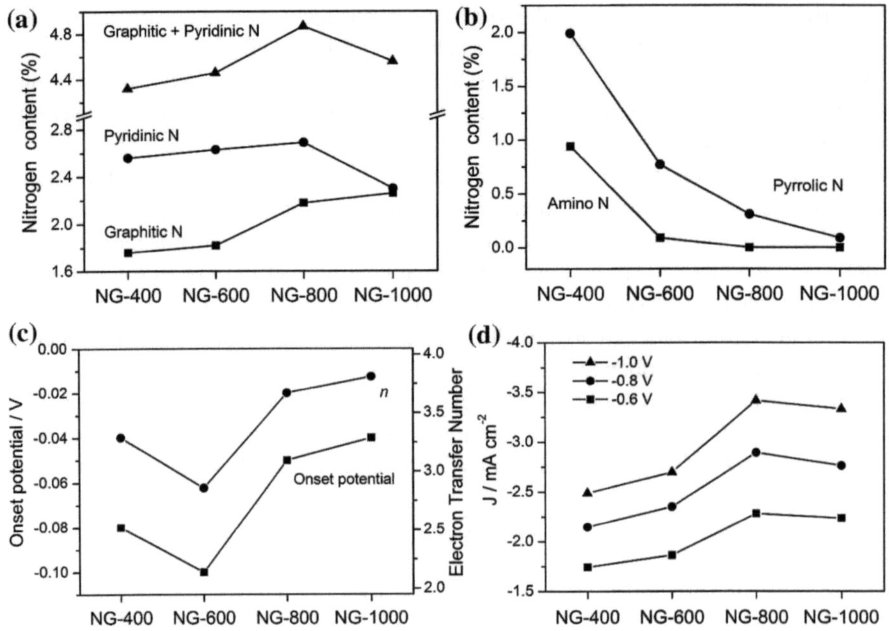

Fig. 14 The content of **a** pyridinic N, graphitic N, **b** pyrrolic N and amino N in nitrogen-doped graphene; the change of **c** onset potential, electron transfer number and **d** current density at different potentials in nitrogen-doped graphene [62]. Copyright 2013. Reproduced with permission from Elsevier

in which a graphene layer was covered by two PMF layers. G-PMFs possessed large specific surface areas up to 630 m^2/g, with good electrocatalytic activity and robust stability for ORR. Furthermore, a Zn–air battery equipped with a G-PMF anode showed similar capacity to that of the Zn–air cell with a commercial Pt/C anode [84].

5.1 Anode

At the anode, continuous inflows of fuels are pumped across the electrocatalysts to convert chemical energy of fuels into electricity. Various types of fuels such as methanol, ethanol, and formic acid have been thoroughly explored as energy resources for fuel cells. Dominant anode electrocatalysts for the oxidation of these fuels are Pt and Pd based nanomaterials. MFC, environmentally benign and sustainable energy devices that employ organic wastes as fuel sources, have aroused great research efforts. Since MFCs depend on microbial metabolism rather than electrocatalysts to trigger the energy conversion, graphene-polymer nanocomposites have found a good position acting as ideal supports for microbials. Due to their outstanding abilities to provide rapid extracellular electron transfer and to keep microbials from aggregation, graphene-polymer substrates have been considered promising anode materials in MFCs.

Yong et al. [85] utilized 3D monolithic graphene/PANI anode for MFC for the first time. Since the 3D graphene fabricated via CVD showed hydrophobicity which hindered the bacteria loading, conducting PANI molecules were in situ polymerized and decorated onto the graphene surface to achieve good biocompatibility. Due to the large specific surface area and the capability to form three dimensional structure with bacterial biofilm, the 3D nanocomposites owned high bacteria loading. Meanwhile, the interconnected highly conductive networks were responsible for the fast extracellular electron transfer efficiency. As expected, the 3D graphene/PANI anode functioned MFC yielded higher power output (up to 768 mW/m^2) than carbon cloth anode (158 mW/m^2), as shown in Fig. 15. Besides, another noteworthy merit is that the performance of 3D structures can be easily boosted by simply raising the thickness. Thus, the 3D graphene/PANI nanocomposites showed promising potential in high-power large scale MFCs.

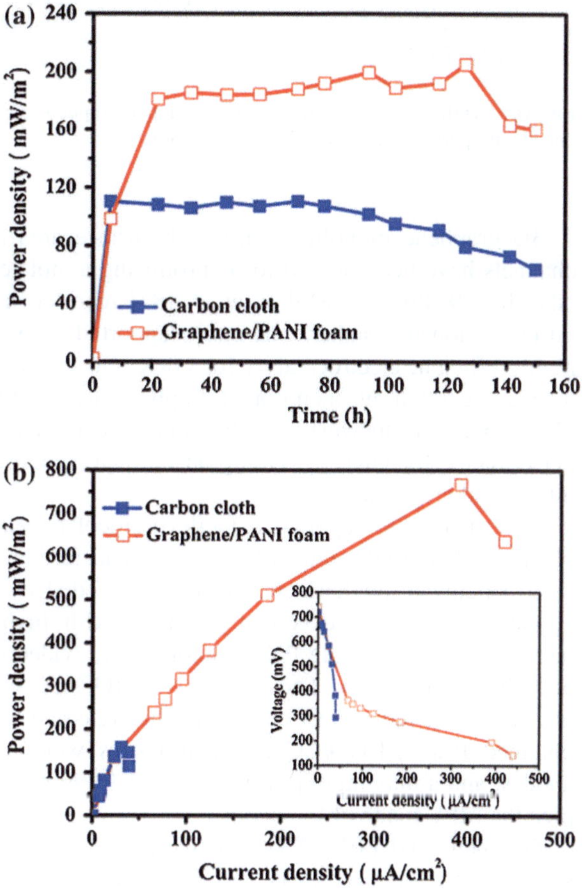

Fig. 15 **a** Time courses of the power density output ofthe MFCs equipped with a carbon cloth anode or a graphene/PANI foam anode. **b** Polarization curves of the twotypes of MFCs. The inset shows the I-V relation [85]. Copyright 2012. Reproduced with permission from American Chemical Society

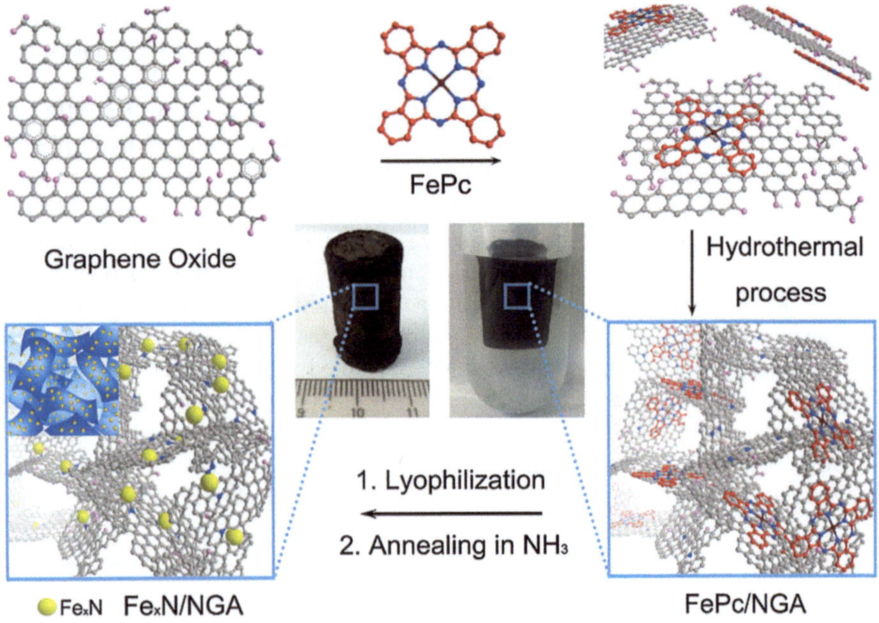

Fig. 16 Synthesis of iron nitride-nitrogen doped graphene aerogel [48]. Copyright 2014. Reproduced with permission from John Wiley & Sons

3D graphene monoliths with high surface area and sufficient interconnected channels have been proved to be promising templates to support metal nanocrystals [46, 49, 86, 87]. Mullen et al. [87] reported hydrothermal synthesis of 3D nitrogen doped graphene aerogel supported Fe_3O_4 nanoparticles which were highly efficient electrocatalyts for ORR. Our group have fabricated a novel iron nitride decorated nitrogen doped graphene aerogel [48], as shown in Fig. 16. The 3D graphene framework greatly improved the surface area and electrocatalytic performance for ORR, revealing an intimate synergistic interaction between iron nitride and graphene.

Hou et al. [88] reported a facile protocol by electrochemically reduced graphene oxide (ERGO) onto CC substrate and further coated PANI nanofibers. The electrochemical synthesis was easily proceeded by cyclic voltammetry in PBS. Graphene served as excellent substrate with high electrical conductivity and large surface area, while PANI nanofibers provided efficient redox properties and good biocompatibility. Thus the PANI/ERGO/CC nanocomposites owned several advantages including fast charge transfer rate and high bacterial loading. The as-obtained PANI/ERGO/CC nanocomposites worked well as microbial fuel cell anodes, and a maximum power output of 1,390 mW/m^2 was generated, which tripled the MFC made of CC anode.

Graphene and PEDOT conductive polymer were hybridized through galvanostatic electropolymerization and G/PEDOT nanocomposites were examined as anode for a MFC. The negatively charged bacteria and positively charged PEDOT formed a dense biofilm. Compared with three similar anodes such as carbon paper, graphene modified carbon paper, and PEDOT-modified carbon paper, the G/PEDOT anode possessed lower resistance, good stability, high voltage output and larger active surface area [89].

PPy/GO nanocomposites were immobilized onto the graphite felt (GF) substrates by a facile electropolymerization method. The PPy/GO-modified anode was cycled in MFCs and their power performance was tested. PPy/GO nanocomposites were produced to integrate good stability and high conductivity, and to overcome the low stability of PPy polymer as well as poor adhesion of GO used alone as anode in MFCs. Compared with PPy-GF, ERGO-GF and GF anodes, the PPy/GO-GF anode generated the maximum cell potential (0.33 V), highest short-circuit currents (7.53 A m^{-2}), lowest ohmic resistance (1.30 Ω) and charge transfer resistance (2.6 Ω). SEM analysis revealed that interfacial surface area and improved adhesion of bacteria on the anode material might attribute to the enhanced performance of PPy/GO-GF anode [90].

Khilari et al. reported a proton-exchange polymer membrane separator composed of GO, PVA and STA. The nanocomposite membranes were investigated as cathode materials in single-chambered MFC. The power density and Columbic efficiency were monitored to evaluate the performance of PVA-STA-GO membrane made sMFC. GO concentration had an observable influence on the proton conductivity and mechanical properties, as well as power output and Columbic efficiency of sMFC. The addition of GO increased proton conductivity due to wealthy ion-exchange groups (–COOH), and the interconnected ion transport networks. Appropriate amount of GO additives will largely enhance the tensile strength to 58.25 MPa at 0.9wt% GO loadings, because GO fillers firmly contacted with the polymer matrix and generated interconnected structure. In sMFC measurement, the 0.5 wt% GO-modified PVA-STA nanocomposite generated a maximum power density of 1.9 W/m^3, utmost voltage of 0.21 V, and weakened oxygen permeability, compared with a Nafion 117 membrane. Meanwhile, GO content should not be higher than 0.9 % which caused blocking effect in properties and performance [37].

6 Graphene-Polymer Nanocomposites Supported Noble Metal NPs as Electrocatalysts

Noble metalbased nanomaterials, especially Pt and Pd, are the dominant catalysts for both ORR and fuel oxidation reaction in acid media. Theoretical results confirmed that, Pt owned unique electronic structure and d-band center which is benefit for the fast growth of ORR. Pd, with approaching catalytic activity and lower price compared with Pt, have also been considered as promising catalyst for both cathode and anode of fuel cells. Meanwhile, Pd possessed better durability than Pt in presence of methanol, indicating its potential of enhanced cycling performance.

Both Pt and Pd are expensive catalysts, which raised the cost of fuel cells. To lower the cost and further enhance the properties, PtM (M = Fe [91, 92], Co [93, 94], Ni [95–97], Au [98, 99], Bi [100], Cu [94]) and PdM (M = Co [101], Bi [100, 102]) alloy electrocatalysts have been developed. Besides the development of noble metal catalysts, many research efforts were devoted to the advance of non-precious catalysts. Plenty of transition metal nanomaterials such as transition metal oxide and chalcogenide (Fe_3O_4 [86, 87], Co_3O_4 [103, 104], Co_3S_4 [105], Mn_3O_4 [106, 107]), and M/N/C structure [108–111] have been investigated as electrocatalysts in alkaline media.

Recently, Hou's group have reported a facile synthesis of FePc supported on NG nanocomposites, denoted as FePc/NG [112]. Compared with pristine FePc, mechanically mixed FePc-NG and Pt/C, the FePc/NG hybrid exhibited remarkable catalytic activity and stability towards ORR, as shown in Fig. 17. The enhanced performance is attributed to the tight contact between aromatic planes of graphene and FePc.

CCG evolved from GO, with large surface area and abundant functional groups, have been extensively investigated to support noble metal NPs. CCG/

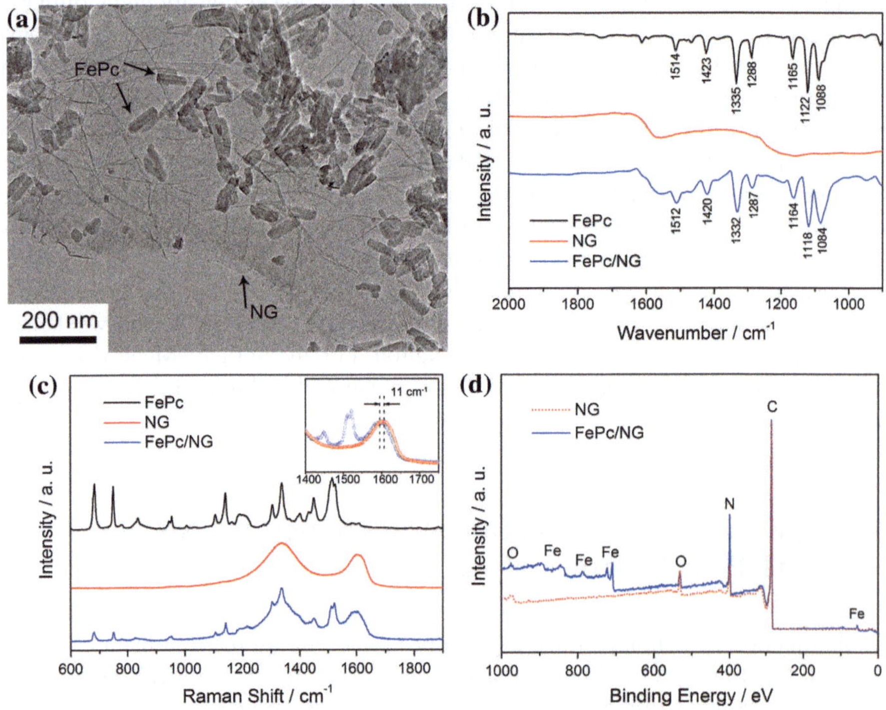

Fig. 17 **a** TEM image of FePc/nitrogen-doped graphene composite; **b** FTIR and **c** Raman spectra of FePc, nitrogen-doped graphene and FePc/nitrogen-doped graphene composite; **d** XPS survey spectra of nitrogen-doped graphene and FePc/nitrogen-doped graphene composite [112]. Copyright 2012. Reproduced with permission from American Chemical Society

noble metal nanocomposites represent a new class of robust fuel cell electrocatalysts. Generally, chemical synthesis via reduction and decomposition of noble metal salts in presence of CCG nanosheets were adopted. Noble metal NPs could selectively anchor onto the two dimensional graphene layer, driven by strong electrostatic attraction and low interface free energy. However, in most cases, noble metal nanostructures tended to reside randomly on the graphene substrate rather than evenly distributed. Meanwhile, it is difficult to control the size and shape of noble metal NPs and prevent particle aggregations, just like the challenges we faced in controlling the uniformity of graphene surface functional groups and defects.

However, directly deposited noble metal NPs (such as Pt, Pd, and Au) and alloys on GO substrates were not homogeneously distributed and the size distribution was too wide to produce excellent cathode and anode. Many efforts have been made to develop facile and effective protocols to improve the even distribution and uniformity of size, in order to guarantee the overall controlled performance of fuel cells. Many researchers adopted polymer functionalized graphene as substrates to deposit noble metal nanostructures because of several advantages. First, the negatively charged GO conjugated with positively charged polymers such as PAMAM, PDDA, PEI expose positive surface, which is suitable for anchoring of noble metal NPs through reduction of negative metal salts. Second, the noble metal NPs could be easily settled in polymer pore on graphene planar sheets [113, 114], and resulted in small size and narrow size distribution. Third, the polymer could partly reduce the GO through electrostatic reaction, and further improve the electrical conductivity, chemical stability and mechanical properties. Fourth, polymer molecules could effectively constrain the embedded metal NPs and prevents the occurrence of aggregation, thus maintaining the high stability.

Polymers have greatly relieved the stress and provided a useful tool to improve the quality of CCG/noble metal electrocatalysts for fuel cells [115]. The hydrophilic polymer additives behaved well in several aspects, such as saving graphene from stacking into precipitates, acting as intermediate linker between graphene and noble metal NPs, as well as raising electrical transfer rates. Furthermore, narrow size range and even distribution of noble metal NPs were achieved, since some charged polymers, comparable to surfactants, closely contact with NPs and stabilize the surface morphology. For instance, Guo et al. reported that could effectively induce the formation of small Pt NPs (2.6 nm) dispersed on graphene.

Widely used polymers are classified into two major types, polyelectrolyte and conducting polymer [116]. Both types have been investigated to functionalize graphene and to manipulate the growth of metal NPs.

6.1 Graphene-Polyelectrocatalytic Nanocomposites

Polyelectrocatalyte has been illustrated to show great potential in producing polymer/graphene substrates for noble metal catalysts, since it can induce large amount of surface charge density and fortify the electronic attraction with metal salt precursors [117, 118]. As an example, PDDA is a widely used cationic polyelectrolyte and could act as capping ligand to stabilize metal nanocrystals. As revealed via XPS and PA-FTIR analysis, PDDA could adhere onto carbon nanotubes and graphene through π–π interaction and electrostatic attraction [119]. It is confirmed that this unique PDDA-assisting synthesis guaranteed small sized Pt NPs on graphene [120–122]. Qiu et al. reported that large amount of available amine moieties on PDDA provided plenty of positively charged sites, which readily captured negatively charged $[PtCl_6]^{2-}$ precursor and converted into tiny Pt NPsupon reduction. In this study, Pt loading amount were ranged from 30 to 78 wt% by adjusting the mass ratio of PDDA-GO to H_2PtCl_6 precursor. It is noteworthy that the average size of Pt NPs remained almost the same (4.6 nm) at varying concentrations, as shown in Fig. 18. This synthesis method successfully realized size control of Pt NPs on graphene under high loading density. Small size and uniform distribution of Pt NPs endowed the electrocatalysts with large electrochemical active surface area (ECSA) up to 141.6 m^2/g, which conveyed higher utility and electrocatalytic activity of Pt atoms, compared with previously reported Pt-based electrocatalysts. In consequence, Pt-PDDA/graphene nanocomposites showed greatly enhanced catalytic activity and anti-CO poisoning ability for methanol oxidation, and moreover also worked well for ORR [123].

Besides, Pt/PDDA-graphene nanocomposites were synthesized for catalysis of methanol oxidation as anode materials for fuel cells. It is confirmed that Pt/PDDA-graphene anode electrode possessed larger ECSA of 66 m^2/g than the pristine Pt/graphene electrode of 55 m^2/g. The role of PDDA is to assist the formation of small Pt NPs (1.8 nm), and to provide a bridge between both negatively charged GO and Pt precursor. Raman and SEM analysis confirmed the mono dispersed distribution of Pt/PDDA-graphene anode materials. The unique sandwich structure also endowed excellent stability and durability against CO gas poisoning [124].

Huang el al. prepared a uniform aqueous solution containing PDDA, Poly(sodium 4-styrenesulfonate) (PSS) and RGO. Then they employed a layer-by-layer assembly method to fabricate [117] multilayer films. The film was further adopted as novel substrates to electrodeposit Pt NPs. Cyclic voltammetry and chronoamperometric analysis confirmed the high electrocatalytic activity of Pt/{PDDA-GN/PSS-GN}n multilayer films for methanol oxidation [117].

Au/PDDA/graphene nanocomposites were prepared via similar method, with highly uniform 4.1 nm Au NPs. PDDA acted as both a reducing agent and stabilizer. The obtained hybrids showed high electrocatalytic performance in ascorbic acid media [125]. Zhang et al. confirmed that PDDA acted not only as a stabilizing but also reducing agent in GO aqueous suspensions, as XPS results confirmed obvious decline in oxygen content. Moreover, PDDA/graphene nanocomposites loaded with Pt or Pd NPs were demonstrated to be promising electrocatalysts for formic acid oxidation [126].

Similar synthesis methods were applied to deposit Pt(Pd)-based bimetallic or trimetallic alloys. He et al. have deposited PtPd alloys with varying compositions (Pd_4Pt_1, Pd_3Pt_1, Pd_2Pt_1, and Pd_1Pt_1) under the reduction of ethylene glycol.

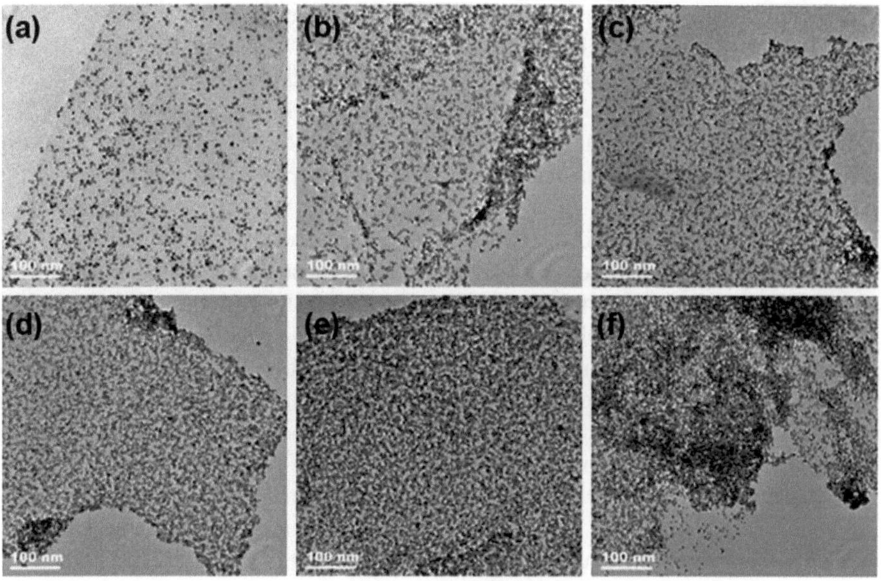

Fig. 18 TEM images of Pt/graphene nanocomposites synthesized from precursors with different mass ratios of PDDA-GO to H_2PtCl_6: **a** 1:1, **b** 1:1.5, **c** 1:2.5, **d** 1:3.5, **e** 1:6.0, and **f** 1:8.5 [123]. Copyright 2011. Reproduced with permission from American Chemical Society

These novel PtPd-based catalysts all demonstrated excellent ORR durability in the presence of PDDA, confirming the critical role of PDDA in embedding and stabilizing PtPd NPs [127]. PtAu NPs with an average diameter of 3.2 nm were successfully blended onto PDDA-modified graphene via reduction of by $NaBH_4$. PDDA formed suitable templates for preparation of PtAu alloy NPs, and further guaranteed the even distribution of PtAu NPs on graphene. After degradation test, PtAu/graphene showed a current density of 0.120 A mg^{-1}, much larger than that of PtAu/CB (0.091 A mg^{-1}) and commercial Pt/C (0.004 A mg^{-1}). This result revealed that PDDA/G substrates greatly contributed to the catalytic stability in direct formic acid fuel cells [98].

6.2 Graphene-Conducting Polymer Nanocomposites

Numerous investigations have confirmed that conducting polymers are promising reinforcers to endow graphene with low resistance, high electrical conductivity, reversible doping/de-doping process, and excellent electrochemical stability. In this sense, graphene-conducting polymer nanocomposites are also promising support for noble metal electrocatalysts. Mayavan et al. demonstrated that Pt NPs on poly(sodium styrene sulfonate)-functionalized graphene showed better catalytic activity and durability for methanol oxidation, compared with Pt NPs directly supported on pristine graphene [128]. Huang et al. have deposited Pt NPs onto PANI-modified GNP using an aqueous solution blending approach. The as obtained Pt/PANI/GNP composite exhibited remarkable electrocatalytic activity, nearly two times that of Pt/GNP catalysts. Uniform size (4–5 nm) of Pt NPs and the noncovalent interaction between PANI and GNP are the main reasons for the outstanding performance [129]. Zhao et al. reported synthesis of Pd/PPy-graphene nanocomposites. In this method, PPy-graphene nanosheets were first fabricated via reduction and in situ polymerization process. Then the as-obtained carbon supports were employed to grow uniformly distributed Pd NPs. Compared with Pd/C and Pd/graphene, the Pd/PPy-graphene showed remarkable electrocatalytic activity and durability for methanol oxidation. This enhancement in performance might be that π-bonded graphene surface strongly conjugated with the pyrrole ring, which endowed the nanocomposites with high electron transfer and conductivity. Meanwhile, smaller size and better dispersion of Pd contributed to higher utilization of Pd atoms, increased the anti-CO poisoning ability. The forward peak current density of the three catalysts were in the order of Pd/PPy–graphene > Pd/graphene > Pd/Vulcan, indicating that Pd/PPy-graphene is a more promising catalyst for methanol oxidation [130]. Zong et al. decorated the Pt NPs onto PPy-graphene nanosupport through in situ synthetic method. Interaction between graphene and PPy interface attribute to the modification of the π–π conjugation and the electron density structure of the nanocomposites, and further enhanced the physical and chemical properties. The Pt/PPy-graphene electrocatalysts showed enhanced activity towards methanol oxidation reaction and stronger anti-CO_{ads} oxidation of PPy molecules [131].

PAIn polymer was electropolymerized on graphene decorated GC electrode and further electrodeposited Pt NPs to obtain Pt/PAIn/G/GC electrode for methanol oxidation in alkaline medium. PAIn could modify the surface of GE and provide large amount of anchoring sites for Pt NPs to achieve homogeneous distribution. After 300 cycles of stability test, the PAIn/G/GC electrode showed only 33 % current degradation, superior than that of PAIn/GC electrode (56 %). Since the structure of 5-Aminoindole(AIn) is similar to both aniline and pyrrole, this conducting polymer also played a part in the surface functionality of graphene. The unique properties of GE such as large surface area, excellent electrical conductivity and fast electron transfer rate attributed to the enhanced electrochemical activity for methanol electrooxidation in alkaline medium [132].

ERGO nanosheets were blended with PEDOT, and then act as high quality supports to deposit Pt NPs. As electrocatalysts for ethanol oxidation, the prepared Pt-PEDOT/ERGO nanocomposites exhibited remarkable peak current density (up to 390 A g^{-1}), long-term stability and robust anti-poisoning ability, compared with commercial JM Pt/C (20 %). Since PEDOT exhibited remarkably high electrical conductivity and excellent stability, among various conducting polymers, the combination of PEDOT and graphene greatly buttressed ionic transport and cyclic stability. These features together with well dispersion of Pt NPs resulted in the high ECSA of Pt (47.1 m^2/g). Thus, PEDOT/ERGO was a promising support for noble metal catalysts [133].

Moreover, the catalytic activity of Pt/PAIn/GE/GC electrode was compared with Pt/GC and Pt/PAIn/GC electrode. The electronic interaction between Pt and polymer matrixes induced well dispersed Pt NPs onto the PAIn/GE substrate and assisted the fast electron transfer between Pt and polymer and consequently resulted in enhanced methanol oxidation catalytic activity [132]. Shi et al. utilized graphene to catalyze the electrochemical polymerization of pyrogallol to achieve poly(pyrogallol) (PPG)/graphene composite. The composite rich in oxygen-containing functional groups served as a supreme platform to control the well dispersion of Pt NPs. According to cyclic voltammetry and chronoamperometry techniques, the order of anodic peak current density and anti-CO poisoning stability of three catalysts was as follows: Pt/PPG/RGO>Pt/PPG > Pt [134].

Guo etl al. developed a wet-chemical synthesis of well-controlled three-dimensional Pt-on-Pd bimetallic nanodendrites decorated on PVP-functionalized graphene substrates. This novel method involved three major steps. PVP-graphene was at first reduced by hydrazine and then decorated with Pd nanocrystals. Porous Pt-on-Pd nanodendrites well dispersed on PVP-G were finally obtained via decomposition of Pt precursor using ascorbic acid as reducing agent. The electrochemical results indicate that the Pt-on-Pd nanodendrites-PVP-graphene nanocomposites exhibited much better electrocatalytic activity for methanol oxidation reaction than PB and commercial E-TEK Pt/C catalysts, as shown in Fig. 19 [135].

Zhang et al. reported a versatile and environmentally benign aqueous solution synthesis to produce Au-PAA/graphene nanocomposites. PAA was used to reduce and stabilize Au NPs. The hybrid film exhibited excellent electrocatalytic activity towards reduction of both O_2 and H_2O_2, indicating promising potential application in fuel cells [136]. Nitrogen and other heteroatom doped graphene nanosheets were revealed to possess high catalytic activity towards ORR due to their two dimensionality and high conductivity. Blending graphene nanosheets with nitrogen-containing polymers to form G/P nanocomposites will shed light on developing advanced catalysts of more distinct physical and chemical properties.

It is reported that in methanol oxidation reaction, the best substrate for Pt colloidal NPs is sulfonated graphene rather than other carbon derivatives such as carbon black and graphene. And PVA exhibit much higher activity as the anionic conductive polymer rather than Nafion [137].

PBI-modified graphene nanosheets were produced via PBI assisted graphite exfoliation via sonication process, without excessive oxidative sites on graphene, which acted as well defined substrates to homogeneously deposit Pt NPs to obtain Pt-PBI-G electrocatalysts. The surface area of prepared electrocatalysts maintained almost the same even after 2,000 cycles of cyclic voltammogram tests, which confirmed the exhibited excellent electrochemical durability, superior than the conventional Pt/C catalyst. The presence of PBI molecules over the surface of graphene were of vital importance for the homogeneous distribution of Pt nanoparticles on the 2D substrates and prevented aggregation of particles during the catalytic process [138].

Pd and PdCo alloy NPs supported on polypropylenimine dendrimere grafted graphene (Pd and PdCo/PPI-g-G) were prepared, which exhibited excellent

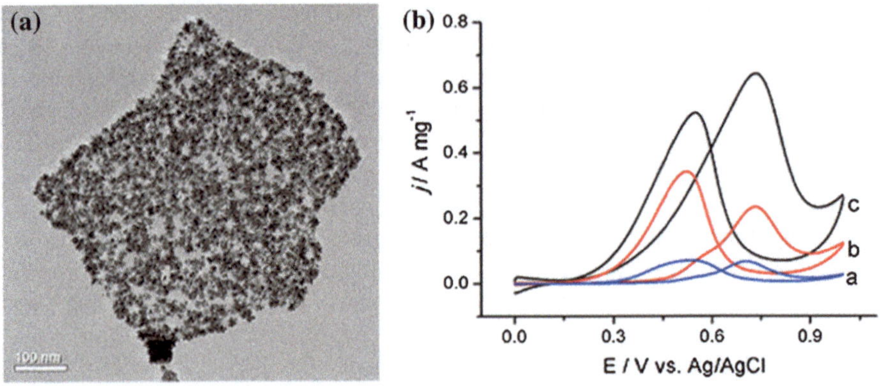

Fig. 19 **a** TEM of three-dimensional (3D) Pt-on-Pd bimetallic nanodendrites supportedon graphene nanosheets. **b** Cyclic voltammogram of PB (*trace a*), E-TEK catalyst (*trace b*), and TP-BNGN (*trace c*) [135]. Copyright 2009. Reproduced with permission from American Chemical Society

performances for formic acid electro-oxidation. This enhancement in catalytic activity may be attributed to well dispersed metallic NPs with a lower particle size achieved with the PPI-G supported materials [139]. β-cyclodextrin polymer (CDP)-functionalized carbon nanomaterials (including fullerenes, single-walled carbon nanotubes, multi-walled carbon nanotubes, and graphene sheets) were produced via a one-step synthesis. Noncovalent interaction between carbon and CDP polymer endowed these nanocomposites good solubility and stability in water. Further conjugation of PATP onto CDP produced PATP-CDP-MWCNTs, which acted as ideal support for heterogeneous distribution of noble metal NPs such as Au and Pt for catalysis of methanol oxidation in direct methanol fuel cells [140].

Ionic liquid, a highly conductive type of polymer, can be also integrated into the graphene system to form graphene-polymer nanocomposites. GO nanosheets functionalized by 1-butyl-3-methylimidazolium (BMI), denoted as BMI-GONs, were prepared by a simple electrochemical method. Pt NPs were easily assembled onto the ordered sandwich structure and formed novel BMI-GONs/Pt nanohybrid, which exhibited excellent catalytic activity towards methanol oxidation with current density of 41.8 mA mg^2 and activity retention of 78.9 % after 1,000 cycles at 50 mV s^{-1}. The in-between BMI ionic liquid molecules provided high electron transfer and enhanced the close interaction between GONs and Pt. The anti-CO poisoning ability (the ratio of I_f/I_b equals 1.97) of the nanohybrids was improved in this manner [141].

7 Conclusions

Graphene-polymer nanocomposites which possessed excellent electrical conductivity, exceptional mechanical properties, abundant surface functionality, and outstanding electrochemical properties have been investigated as promising candidates to build the ideal fuel cells. GO nanosheets synthesized via liquid phase synthesis are feasible and easy to manipulate in various solvents, prone to form molecule interactions and well dispersions in polymer. Commonly used synthesis methods of graphene-polymer nanocomposites included solution mixing, melt blending, and in situ polymerization. Since the uniformity and quality of graphene-polymer nanocomposites are responsible for their mechanical and electrochemical properties, selecting the supreme synthesis protocols becomes extremely important, which results in homogeneous dispersion and strong adhesion among graphene and polymeric matric. Integrating graphene with sulfonated polymers such as Nafion can largely enhance the proton conductivity and prohibit the methanol crossover effect, providing robust and reliable long-term performance of fuel cells, especially PEMFC and DMFC. Graphene-polymer nanohybrids decorated with microbial have been applied as MFC anode, taking advantage of both their electrochemical properties and biocompatibility. Nitrogen containing polymer and graphene can form covalent interaction between N–C atoms, generating ideal N–C active sites for ORR activity. Carbon nitride and other polymers with high nitrogen content have been investigated as apropos sources to dope nitrogen into graphene. Moreover, graphene-polymer nanocomposites have been extensively probed as feasible substrates to support noble metal NPs such as Pt, Pd, Au and their alloys. The composite substrate with tunable functional groups and superior electrochemical properties not only provide effective platforms to guarantee uniformity and small size of grown metal NPs, but also buttressed rapid electron transfer and longevity of both cathode and anode of fuel cells. Therefore, large-scale synthesis of high-quality graphene at low cost is required. Commonly used techniques, which take use of either exfoliation or deposition principles, require precise control of reaction parameters. Moreover, yet the productivity of well-developed graphene nanosheets cannot be guaranteed. In addition pristine graphene nanosheets have been proved to be highly conductive and well-dispersed structures, the CCG nanosheets, which are commonly used as substrates for the fabrication of graphene/polymer nanocomposites, have comparatively low conductivities owing to the structural defects. Thus, the majority of graphene/polymer nanocomposites are salvaged with low electric current in use. Finally, the stability characteristics for long-term use of these nanocomposites membranes have not been fully understood or evaluated. Although some recent samples reported investigations on this topic, more focused efforts are urgently required.

However, there are still some knots related to the graphene-polymer nanocomposites are waiting to be unraveled as following: (1) synthesis methods of graphene-polymer nanocomposites are still limited, difficult to provide enough choices to meet the numerous requirements for structures and properties; (2) it is challenging to assure the ideal molecule and electron structure of graphene-polymer composites which determined the electrocatalytic activity and chemical stability; (3) practical applications of graphene-nanocomposites in commercialized fuel cell still keep in stasis, demanding standard evaluation system for membranes, anode and cathode materials. Great research efforts are essential and imperative to advance the application of graphene-polymer nanocomposites in fuel cells. This chapter is dedicated to provide a summary of current progress, to inspire appropriate solutions to break the obstacles, and to pave the way for commercialization of graphene-polymer nanocomposites and fuel cells. The better interaction between graphene and polymer to build continues and faster conductive path is highly desired along with stable performance for long cyclic life of fuel cell.

References

1. Demirdöven N, Deutch J (2004) Hybrid Cars Now, Fuel Cell Cars Later. Science 305 (5686):974-976. doi:10.1126/science.1093965
2. Steele BCH (2001) Material science and engineering: The enabling technology for the commercialisation of fuel cell systems. Journal of Materials Science 36 (5):1053-1068. doi:10.1023/a:1004853019349
3. Winter M, Brodd RJ (2004) What Are Batteries, Fuel Cells, and Supercapacitors? Chemical Reviews 104 (10):4245-4270. doi:10.1021/cr020730k
4. Borup R, Meyers J, Pivovar B, Kim YS, Mukundan R, Garland N, Myers D, Wilson M, Garzon F, Wood D, Zelenay P, More K, Stroh K, Zawodzinski T, Boncella J, McGrath JE, Inaba M, Miyatake K, Hori M, Ota K, Ogumi Z, Miyata S, Nishikata A, Siroma Z, Uchimoto Y, Yasuda K, Kimijima K-i, Iwashita N (2007) Scientific Aspects of Polymer Electrolyte Fuel Cell Durability and Degradation. Chemical Reviews 107 (10):3904-3951. doi:10.1021/cr050182l
5. Grove WR (1839) On voltaic series and the combination of gases by platinum. Philosophical Magazine Series 3 14 (86):127-130. doi:10.1080/14786443908649684
6. Markovic NMPNR, Ross PN (2000) New Electrocatalysts for Fuel Cells from Model Surfaces to Commercial Catalysts. CATTECH 4 (2):110-126. doi:10.1023/a:1011963731898
7. Service RF (2006) New Polymer May Rev Up the Output of Fuel Cells Used to Power Cars. Science 312 (5770):35. doi:10.1126/science.312.5770.35a
8. Brandon NP, Skinner S, Steele BCH (2003) Recent Advances in Materials for Fuel Cells. Annual Review of Materials Research 33 (1):183-213. doi:10.1146/annurev.matsci.33.022802.094122
9. Zhao F, Slade RCT, Varcoe JR (2009) Techniques for the study and development of microbial fuel cells: an electrochemical perspective. Chemical Society Reviews 38 (7):1926-1939. doi:10.1039/b819866g
10. Novoselov KS, Falko VI, Colombo L, Gellert PR, Schwab MG, Kim K (2012) A roadmap for graphene. Nature 490 (7419):192-200
11. Luo B, Liu S, Zhi L (2012) Chemical Approaches toward Graphene-Based Nanomaterials and their Applications in Energy-Related Areas. Small 8 (5):630-646. doi:10.1002/smll.201101396
12. Sun Y, Wu Q, Shi G (2011) Graphene based new energy materials. Energy & Environmental Science 4 (4):1113-1132. doi:10.1039/c0ee00683a
13. Zhu Y, Murali S, Cai W, Li X, Suk JW, Potts JR, Ruoff RS (2010) Graphene and Graphene Oxide: Synthesis, Properties, and Applications. Advanced Materials 22 (35):3906-3924. doi:10.1002/adma.201001068
14. Allen MJ, Tung VC, Kaner RB (2009) Honeycomb Carbon: A Review of Graphene. Chemical Reviews 110 (1):132-145. doi:10.1021/cr900070d
15. Wang L, Lu X, Lei S, Song Y (2014) Graphene-based polyaniline nanocomposites: preparation, properties and applications. Journal of Materials Chemistry A 2 (13):4491-4509. doi:10.1039/c3ta13462h
16. Sun Y, Shi G (2013) Graphene/polymer composites for energy applications. Journal of Polymer Science Part B: Polymer Physics 51 (4):231-253. doi:10.1002/polb.23226

17. Xie J, Guo CX, Li C (2014) Construction of One-Dimensional Nanostructures on Graphene for Efficient Energy Conversion and Storage. Energy & Environmental Science. doi:10.1039/c4ee00531g
18. Nanjundan AK, Baek J-B (2014) Electrochemical supercapacitors from conducting polyaniline-graphene platforms. Chemical Communications. doi:10.1039/c4cc01049c
19. Ashok Kumar N, Baek J-B (2014) Electrochemical supercapacitors from conducting polyaniline-graphene platforms. Chemical Communications. doi:10.1039/c4cc01049c
20. Wang Y, Xia Y (2013) Recent Progress in Supercapacitors: From Materials Design to System Construction. Advanced Materials 25 (37):5336-5342. doi:10.1002/adma.201301932
21. Osiak M, Geaney H, Armstrong E, O'Dwyer C (2014) Structuring materials for lithium-ion batteries: advancements in nanomaterial structure, composition, and defined assembly on cell performance. Journal of Materials Chemistry A. doi:10.1039/c4ta00534a
22. NOORDEN RV (2014) A Better Battery. Nature 507:26-28
23. Xin S, Guo Y-G, Wan L-J (2012) Nanocarbon Networks for Advanced Rechargeable Lithium Batteries. Accounts of Chemical Research 45 (10):1759-1769. doi:10.1021/ar300094m
24. Palacin MR (2009) Recent advances in rechargeable battery materials: a chemist's perspective. Chemical Society Reviews 38 (9):2565-2575. doi:10.1039/b820555h
25. Li L, Wu Z, Yuan S, Zhang X (2014) Advances and Challenges for Flexible Energy Storage and Conversion Devices and Systems. Energy & Environmental Science. doi:10.1039/c4ee00318g
26. Chua CK, Pumera M (2014) Chemical reduction of graphene oxide: a synthetic chemistry viewpoint. Chemical Society Reviews. doi:10.1039/c3cs60303b
27. Eda G, Fanchini G, Chhowalla M (2008) Large-area ultrathin films of reduced graphene oxide as a transparent and flexible electronic material. Nat Nano 3 (5):270-274. doi:http://www.nature.com/nnano/journal/v3/n5/suppinfo/nnano.2008.83_S1.html
28. Huang H, Song Z, Wei N, Shi L, Mao Y, Ying Y, Sun L, Xu Z, Peng X (2013) Ultrafast viscous water flow through nanostrand-channelled graphene oxide membranes. Nat Commun 4. doi:10.1038/ncomms3979
29. Zhang C, Mahmood N, Yin H, Hou Y (2014) Graphene-Based Nanomaterials for Energy Conversion and Storage. In: Handbook of Carbon Nano Materials, vol 6. World Scientific Series, pp 51-82. doi:10.1142/9789814566704_0006
30. Kumar PV, Bardhan NM, Tongay S, Wu J, Belcher AM, Grossman JC (2013) Scalable enhancement of graphene oxide properties by thermally driven phase transformation. Nat Chem advance online publication. doi:10.1038/nchem.1820 http://www.nature.com/nchem/journal/vaop/ncurrent/abs/nchem.1820.html#supplementary-information
31. Dikin DA, Stankovich S, Zimney EJ, Piner RD, Dommett GHB, Evmenenko G, Nguyen ST, Ruoff RS (2007) Preparation and characterization of graphene oxide paper. Nature 448 (7152):457-460. doi:http://www.nature.com/nature/journal/v448/n7152/suppinfo/nature06016_S1.html
32. Hao R, Qian W, Zhang L, Hou Y (2008) Aqueous dispersions of TCNQ-anion-stabilized graphene sheets. Chemical Communications 0 (48):6576-6578. doi:10.1039/b816971c
33. Qian W, Hao R, Hou Y, Tian Y, Shen C, Gao H, Liang X (2009) Solvothermal-assisted exfoliation process to produce graphene with high yield and high quality. Nano Res 2 (9):706-712. doi:10.1007/s12274-009-9074-z
34. Mahmood N, Islam M, Hameed A, Saeed S (2013) Polyamide 6/Multiwalled Carbon Nanotubes Nanocomposites with Modified Morphology and Thermal Properties. Polymers 5 (4):1380-1391
35. Mahmood N, Islam M, Hameed A, Saeed S, Khan AN (2013) Polyamide-6-based composites reinforced with pristine or functionalized multi-walled carbon nanotubes produced using melt extrusion technique. Journal of Composite Materials:0021998313484779
36. Bao C, Guo Y, Song L, Hu Y (2011) Poly(vinyl alcohol) nanocomposites based on graphene and graphite oxide: a comparative investigation of property and mechanism. Journal of Materials Chemistry 21 (36):13942-13950. doi:10.1039/c1jm11662b

37. Khilari S, Pandit S, Ghangrekar MM, Pradhan D, Das D (2013) Graphene Oxide-Impregnated PVA–STA Composite Polymer Electrolyte Membrane Separator for Power Generation in a Single-Chambered Microbial Fuel Cell. Industrial & Engineering Chemistry Research 52 (33):11597-11606. doi:10.1021/ie4016045
38. Enotiadis A, Angjeli K, Baldino N, Nicotera I, Gournis D (2012) Graphene-Based Nafion Nanocomposite Membranes: Enhanced Proton Transport and Water Retention by Novel Organo-functionalized Graphene Oxide Nanosheets. Small 8 (21):3338-3349. doi:10.1002/smll.201200609
39. Kim H, Miura Y, Macosko CW (2010) Graphene/Polyurethane Nanocomposites for Improved Gas Barrier and Electrical Conductivity. Chemistry of Materials 22 (11):3441-3450. doi:10.1021/cm100477v
40. Sun X, Sun H, Li H, Peng H (2013) Developing Polymer Composite Materials: Carbon Nanotubes or Graphene? Advanced Materials 25 (37):5153-5176. doi:10.1002/adma.201301926
41. Liang J, Huang Y, Zhang L, Wang Y, Ma Y, Guo T, Chen Y (2009) Molecular-Level Dispersion of Graphene into Poly(vinyl alcohol) and Effective Reinforcement of their Nanocomposites. Advanced Functional Materials 19 (14):2297-2302. doi:10.1002/adfm.200801776
42. Hameed A, Islam M, ahmad I, Mahmood N, Saeed S, Javed H (2014) Thermal and mechanical properties of carbon nanotube/epoxy nanocomposites reinforced with pristine and functionalized multiwalled carbon nanotubes. Polymer Composites:n/a-n/a. doi:10.1002/pc.23097
43. Chabot V, Higgins DC, Yu A, Xiao X, Chen Z, Zhang J (2014) A Review of Graphene and Graphene Oxide Sponge: Material Synthesis and Applications towards Energy and the Environment. Energy & Environmental Science. doi:10.1039/c3ee43385d
44. Cao X, Yin Z, Zhang H (2014) Three-Dimensional Graphene Materials: Preparation, Structures and Application in Supercapacitors. Energy & Environmental Science. doi:10.1039/c4ee00050a
45. Nardecchia S, Carriazo D, Ferrer ML, Gutierrez MC, del Monte F (2013) Three dimensional macroporous architectures and aerogels built of carbon nanotubes and/or graphene: synthesis and applications. Chemical Society Reviews 42 (2):794-830. doi:10.1039/c2cs35353a
46. Jiang H, Lee PS, Li C (2013) 3D carbon based nanostructures for advanced supercapacitors. Energy & Environmental Science 6 (1):41-53. doi:10.1039/c2ee23284g
47. Qiu L, Liu D, Wang Y, Cheng C, Zhou K, Ding J, Truong V-T, Li D (2014) Mechanically Robust, Electrically Conductive and Stimuli-Responsive Binary Network Hydrogels Enabled by Superelastic Graphene Aerogels. Advanced Materials 26 (20):3333-3337. doi:10.1002/adma.201305359
48. Yin H, Zhang C, Liu F, Hou Y (2014) Hybrid of Iron Nitride and Nitrogen-Doped Graphene Aerogel as Synergistic Catalyst for Oxygen Reduction Reaction. Advanced Functional Materials24(20): 2930-2937. doi:10.1002/adfm.201303902
49. Chen Z, Ren W, Gao L, Liu B, Pei S, Cheng H-M (2011) Three-dimensional flexible and conductive interconnected graphene networks grown by chemical vapour deposition. Nat Mater 10 (6):424-428. doi:http://www.nature.com/nmat/journal/v10/n6/abs/nmat3001.html#supplementary-information
50. He Y, Tong C, Geng L, Liu L, Lü C (2014) Enhanced performance of the sulfonated polyimide proton exchange membranes by graphene oxide: Size effect of graphene oxide. Journal of Membrane Science 458 (0):36-46. doi:http://dx.doi.org/10.1016/j.memsci.2014.01.017
51. Choi BG, Hong J, Park YC, Jung DH, Hong WH, Hammond PT, Park H (2011) Innovative Polymer Nanocomposite Electrolytes: Nanoscale Manipulation of Ion Channels by Functionalized Graphenes. ACS Nano 5 (6):5167-5174. doi:10.1021/nn2013113
52. Lin CW, Lu YS (2013) Highly ordered graphene oxide paper laminated with a Nafion membrane for direct methanol fuel cells. Journal of Power Sources 237 (0):187-194. doi:http://dx.doi.org/10.1016/j.jpowsour.2013.03.005

53. Yuan T, Pu L, Huang Q, Zhang H, Li X, Yang H (2014) An effective methanol-blocking membrane modified with graphene oxide nanosheets for passive direct methanol fuel cells. Electrochimica Acta 117 (0):393-397. doi:http://dx.doi.org/10.1016/j.electacta.2013.11.063
54. Ansari S, Kelarakis A, Estevez L, Giannelis EP (2010) Oriented Arrays of Graphene in a Polymer Matrix by in situ Reduction of Graphite Oxide Nanosheets. Small 6 (2):205-209. doi:10.1002/smll.200900765
55. Lim Y-D, Seo D-W, Lee S-H, Choi S-Y, Lee S-Y, Jin L, Tan F, Kim W-G (2014) The Sulfonated poly(ether sulfone ketone) ionomers containing partial graphene of mesonaphthobifluorene for PEMFC. Electronic Materials Letters 10 (1):205-207. doi:10.1007/s13391-013-7001-7
56. Kumar R, Mamlouk M, Scott K (2014) Sulfonated polyether ether ketone - sulfonated graphene oxide composite membranes for polymer electrolyte fuel cells. RSC Advances 4 (2):617-623. doi:10.1039/c3ra42390e
57. Mishra AK, Kim NH, Jung D, Lee JH (2014) Enhanced mechanical properties and proton conductivity of Nafion–SPEEK–GO composite membranes for fuel cell applications. Journal of Membrane Science 458 (0):128-135. doi:http://dx.doi.org/10.1016/j.memsci.2014.01.073
58. Gahlot S, Sharma PP, Kulshrestha V, Jha PK (2014) SGO/SPES-Based Highly Conducting Polymer Electrolyte Membranes for Fuel Cell Application. ACS Applied Materials & Interfaces 6 (8):5595-5601. doi:10.1021/am5000504
59. Ye Y-S, Cheng M-Y, Xie X-L, Rick J, Huang Y-J, Chang F-C, Hwang B-J (2013) Alkali doped polyvinyl alcohol/graphene electrolyte for direct methanol alkaline fuel cells. Journal of Power Sources 239 (0):424-432. doi:http://dx.doi.org/10.1016/j.jpowsour.2013.03.021
60. He Y, Wang J, Zhang H, Zhang T, Zhang B, Cao S, Liu J (2014) Polydopamine-modified graphene oxide nanocomposite membrane for proton exchange membrane fuel cell under anhydrous conditions. Journal of Materials Chemistry A. doi:10.1039/c3ta15301k
61. Gong K, Du F, Xia Z, Durstock M, Dai L (2009) Nitrogen-Doped Carbon Nanotube Arrays with High Electrocatalytic Activity for Oxygen Reduction. Science 323 (5915):760-764. doi:10.1126/science.1168049
62. Zhang C, Hao R, Liao H, Hou Y (2013) Synthesis of amino-functionalized graphene as metal-free catalyst and exploration of the roles of various nitrogen states in oxygen reduction reaction. Nano Energy 2 (1):88-97. doi:http://dx.doi.org/10.1016/j.nanoen.2012.07.021
63. Wang S, Yu D, Dai L, Chang DW, Baek J-B (2011) Polyelectrolyte-Functionalized Graphene as Metal-Free Electrocatalysts for Oxygen Reduction. ACS Nano 5 (8):6202-6209. doi:10.1021/nn200879h
64. Sun Y, Li C, Xu Y, Bai H, Yao Z, Shi G (2010) Chemically converted graphene as substrate for immobilizing and enhancing the activity of a polymeric catalyst. Chemical Communications 46 (26):4740-4742. doi:10.1039/c001635g
65. Unni SM, Bhange SN, Anothumakkool B, Kurungot S (2013) Redox-Mediated Synthesis of Functionalised Graphene: A Strategy towards 2D Multifunctional Electrocatalysts for Energy Conversion Applications. ChemPlusChem 78 (10):1296-1303. doi:10.1002/cplu.201300153
66. Zheng Y, Jiao Y, Jaroniec M, Jin Y, Qiao SZ (2012) Nanostructured Metal-Free Electrochemical Catalysts for Highly Efficient Oxygen Reduction. Small 8 (23):3550-3566. doi:10.1002/smll.201200861
67. Thomas A, Fischer A, Goettmann F, Antonietti M, Muller J-O, Schlogl R, Carlsson JM (2008) Graphitic carbon nitride materials: variation of structure and morphology and their use as metal-free catalysts. Journal of Materials Chemistry 18 (41):4893-4908. doi:10.1039/b800274f
68. Wang Y, Wang X, Antonietti M (2012) Polymeric Graphitic Carbon Nitride as a Heterogeneous Organocatalyst: From Photochemistry to Multipurpose Catalysis to Sustainable Chemistry. Angewandte Chemie International Edition 51 (1):68-89. doi:10.1002/anie.201101182
69. Lyth SM, Nabae Y, Moriya S, Kuroki S, Kakimoto M-a, Ozaki J-i, Miyata S (2009) Carbon Nitride as a Nonprecious Catalyst for Electrochemical Oxygen Reduction. The Journal of Physical Chemistry C 113 (47):20148-20151. doi:10.1021/jp907928j

70. Zheng Y, Liu J, Liang J, Jaroniec M, Qiao SZ (2012) Graphitic carbon nitride materials: controllable synthesis and applications in fuel cells and photocatalysis. Energy & Environmental Science 5 (5):6717-6731. doi:10.1039/c2ee03479d
71. Ritter KA, Lyding JW (2009) The influence of edge structure on the electronic properties of graphene quantum dots and nanoribbons. Nat Mater 8 (3):235-242. doi:http://www.nature.com/nmat/journal/v8/n3/suppinfo/nmat2378_S1.html
72. Deng D, Yu L, Pan X, Wang S, Chen X, Hu P, Sun L, Bao X (2011) Size effect of graphene on electrocatalytic activation of oxygen. Chemical Communications 47 (36):10016-10018. doi:10.1039/c1cc13033a
73. Yang S, Feng X, Wang X, Müllen K (2011) Graphene-Based Carbon Nitride Nanosheets as Efficient Metal-Free Electrocatalysts for Oxygen Reduction Reactions. Angewandte Chemie International Edition 50 (23):5339-5343. doi:10.1002/anie.201100170
74. Wang X, Dai H (2010) Etching and narrowing of graphene from the edges. Nat Chem 2 (8):661-665. doi:http://www.nature.com/nchem/journal/v2/n8/abs/nchem.719.html#supplementary-information
75. Li X, Wang X, Zhang L, Lee S, Dai H (2008) Chemically Derived, Ultrasmooth Graphene Nanoribbon Semiconductors. Science 319 (5867):1229-1232
76. Ren Y, Pan D, Li X, Fu F, Zhao Y, Wang X (2013) Effect of polyaniline-graphene nanosheets modified cathode on the performance of sediment microbial fuel cell. Journal of Chemical Technology & Biotechnology 88 (10):1946-1950. doi:10.1002/jctb.4146
77. Lai L, Potts JR, Zhan D, Wang L, Poh CK, Tang C, Gong H, Shen Z, Lin J, Ruoff RS (2012) Exploration of the active center structure of nitrogen-doped graphene-based catalysts for oxygen reduction reaction. Energy & Environmental Science 5 (7):7936-7942. doi:10.1039/c2ee21802j
78. Qu L, Liu Y, Baek J-B, Dai L (2010) Nitrogen-Doped Graphene as Efficient Metal-Free Electrocatalyst for Oxygen Reduction in Fuel Cells. ACS Nano 4 (3):1321-1326. doi:10.1021/nn901850u
79. Kong X-K, Chen C-L, Chen Q-W (2014) Doped graphene for metal-free catalysis. Chemical Society Reviews. doi:10.1039/c3cs60401b
80. Qian W, Cui X, Hao R, Hou Y, Zhang Z (2011) Facile Preparation of Nitrogen-Doped Few-Layer Graphene via Supercritical Reaction. ACS Applied Materials & Interfaces 3 (7):2259-2264. doi:10.1021/am200479d
81. Shao Y, Zhang S, Engelhard MH, Li G, Shao G, Wang Y, Liu J, Aksay IA, Lin Y (2010) Nitrogen-doped graphene and its electrochemical applications. Journal of Materials Chemistry 20 (35):7491-7496. doi:10.1039/c0jm00782j
82. Sheng Z-H, Shao L, Chen J-J, Bao W-J, Wang F-B, Xia X-H (2011) Catalyst-Free Synthesis of Nitrogen-Doped Graphene via Thermal Annealing Graphite Oxide with Melamine and Its Excellent Electrocatalysis. ACS Nano 5 (6):4350-4358. doi:10.1021/nn103584t
83. Wang K, Wang J, Fan J, Lotya M, O'Neill A, Fox D, Feng Y, Zhang X, Jiang B, Zhao Q, Zhang H, Coleman JN, Zhang L, Blau WJ (2013) Ultrafast Saturable Absorption of Two-Dimensional MoS2 Nanosheets. ACS Nano. doi:10.1021/nn403886t
84. Sun Y, Li C, Shi G (2012) Nanoporous nitrogen doped carbon modified graphene as electrocatalyst for oxygen reduction reaction. Journal of Materials Chemistry 22 (25):12810-12816. doi:10.1039/c2jm31525d
85. Yong Y-C, Dong X-C, Chan-Park MB, Song H, Chen P (2012) Macroporous and Monolithic Anode Based on Polyaniline Hybridized Three-Dimensional Graphene for High-Performance Microbial Fuel Cells. ACS Nano 6 (3):2394-2400. doi:10.1021/nn204656d
86. Wei W, Yang S, Zhou H, Lieberwirth I, Feng X, Müllen K (2013) 3D Graphene Foams Cross-linked with Pre-encapsulated Fe3O4 Nanospheres for Enhanced Lithium Storage. Advanced Materials 25 (21):2909-2914. doi:10.1002/adma.201300445
87. Wu Z-S, Yang S, Sun Y, Parvez K, Feng X, Müllen K (2012) 3D Nitrogen-Doped Graphene Aerogel-Supported Fe3O4 Nanoparticles as Efficient Electrocatalysts for the Oxygen Reduction Reaction. Journal of the American Chemical Society 134 (22):9082-9085. doi:10.1021/ja3030565

88. Hou J, Liu Z, Zhang P (2013) A new method for fabrication of graphene/polyaniline nanocomplex modified microbial fuel cell anodes. Journal of Power Sources 224 (0):139-144. doi:http://dx.doi.org/10.1016/j.jpowsour.2012.09.091
89. Wang Y, Zhao C-e, Sun D, Zhang J-R, Zhu J-J (2013) A Graphene/Poly(3,4-ethylenedioxythiophene) Hybrid as an Anode for High-Performance Microbial Fuel Cells. ChemPlusChem 78 (8):823-829. doi:10.1002/cplu.201300102
90. Lv Z, Chen Y, Wei H, Li F, Hu Y, Wei C, Feng C (2013) One-step electrosynthesis of polypyrrole/graphene oxide composites for microbial fuel cell application. Electrochimica Acta 111 (0):366-373. doi:http://dx.doi.org/10.1016/j.electacta.2013.08.022
91. Zhang J, Yang H, Yang K, Fang J, Zou S, Luo Z, Wang H, Bae I-T, Jung DY (2010) Monodisperse Pt3Fe Nanocubes: Synthesis, Characterization, Self-Assembly, and Electrocatalytic Activity. Advanced Functional Materials 20 (21):3727-3733. doi:10.1002/adfm.201000679
92. Wang Y, Sun Z, Kumbhar A, Luo Z, Wang C, Zhang J, Porter N, Xu C, Sun K, Martens B, Fang J (2013) Is CO adequate to facilitate the formation of Pt3 M (M = Fe, Ni and Co) nanocubes? Chemical Communications 49 (38):3955-3957. doi:10.1039/c3cc41424h
93. Zhang J, Fang J (2009) A General Strategy for Preparation of Pt 3d-Transition Metal (Co, Fe, Ni) Nanocubes. Journal of the American Chemical Society 131 (51):18543-18547. doi:10.1021/ja908245r
94. Yu Y, Yang W, Sun X, Zhu W, Li XZ, Sellmyer DJ, Sun S (2014) Monodisperse MPt (M = Fe, Co, Ni, Cu, Zn) Nanoparticles Prepared from a Facile Oleylamine Reduction of Metal Salts. Nano Letters. doi:10.1021/nl500776e
95. Wu Y, Wang D, Niu Z, Chen P, Zhou G, Li Y (2012) A Strategy for Designing a Concave Pt–Ni Alloy through Controllable Chemical Etching. Angewandte Chemie International Edition 51 (50):12524-12528. doi:10.1002/anie.201207491
96. Zhang J, Yang H, Fang J, Zou S (2010) Synthesis and Oxygen Reduction Activity of Shape-Controlled Pt3Ni Nanopolyhedra. Nano Letters 10 (2):638-644. doi:10.1021/nl903717z
97. Stamenkovic VR, Fowler B, Mun BS, Wang G, Ross PN, Lucas CA, Marković NM (2007) Improved Oxygen Reduction Activity on Pt3Ni(111) via Increased Surface Site Availability. Science 315 (5811):493-497. doi:10.1126/science.1135941
98. Zhang S, Shao Y, Liao H-g, Liu J, Aksay IA, Yin G, Lin Y (2011) Graphene Decorated with PtAu Alloy Nanoparticles: Facile Synthesis and Promising Application for Formic Acid Oxidation. Chemistry of Materials 23 (5):1079-1081. doi:10.1021/cm101568z
99. Venkateswara Rao C, Cabrera CR, Ishikawa Y (2011) Graphene-Supported Pt–Au Alloy Nanoparticles: A Highly Efficient Anode for Direct Formic Acid Fuel Cells. The Journal of Physical Chemistry C 115 (44):21963-21970. doi:10.1021/jp202561n
100. Liao H, Zhu J, Hou Y (2014) Synthesis and electrocatalytic properties of PtBi nanoplatelets and PdBi nanowires. Nanoscale 6 (2):1049-1055. doi:10.1039/c3nr05590f
101. Ho SFF, Garcia AM, Guo S, He K, Su D, Liu S, Metin O, Sun S (2014) A Facile Route to Monodisperse MPd (M = Co or Cu) Alloy Nanoparticles and Their Catalysis for Electrooxidation of Formic Acid. Nanoscale. doi:10.1039/c4nr01107d
102. Liao H, Hou Y (2013) Liquid-Phase Templateless Synthesis of Pt-on-Pd0.85Bi0.15 Nanowires and PtPdBi Porous Nanoparticles with Superior Electrocatalytic Activity. Chemistry of Materials 25 (3):457-465. doi:10.1021/cm3037179
103. Zhang G, Li C, Liu J, Zhou L, Liu R, Han X, Huang H, Hu H, Liu Y, Kang Z (2014) One-step conversion from metal-organic frameworks to Co3O4@N-doped carbon nanocomposites towards highly efficient oxygen reduction catalysts. Journal of Materials Chemistry A. doi:10.1039/c4ta00677a
104. Sa YJ, Kwon K, Cheon JY, Kleitz F, Joo SH (2013) Ordered mesoporous Co3O4 spinels as stable, bifunctional, noble metal-free oxygen electrocatalysts. Journal of Materials Chemistry A 1 (34):9992-10001. doi:10.1039/c3ta11917c
105. Mahmood N, Zhang C, Jiang J, Liu F, Hou Y (2013) Multifunctional Co3S4/Graphene Composites for Lithium Ion Batteries and Oxygen Reduction Reaction. Chemistry – A European Journal 19 (16):5183-5190. doi:10.1002/chem.201204549

106. Li Q, Xu P, Zhang B, Tsai H, Wang J, Wang H-L, Wu G (2013) One-step synthesis of Mn3O4/reduced graphene oxide nanocomposites for oxygen reduction in nonaqueous Li-O2 batteries. Chemical Communications 49 (92):10838-10840. doi:10.1039/c3cc46441e
107. Tan Y, Xu C, Chen G, Fang X, Zheng N, Xie Q (2012) Facile Synthesis of Manganese-Oxide-Containing Mesoporous Nitrogen-Doped Carbon for Efficient Oxygen Reduction. Advanced Functional Materials 22 (21):4584-4591. doi:10.1002/adfm.201201244
108. Ferrandon M, Kropf AJ, Myers DJ, Artyushkova K, Kramm U, Bogdanoff P, Wu G, Johnston CM, Zelenay P (2012) Multitechnique Characterization of a Polyaniline–Iron–Carbon Oxygen Reduction Catalyst. The Journal of Physical Chemistry C 116 (30):16001-16013. doi:10.1021/jp302396g
109. Wu G, More KL, Johnston CM, Zelenay P (2011) High-Performance Electrocatalysts for Oxygen Reduction Derived from Polyaniline, Iron, and Cobalt. Science 332 (6028):443-447. doi:10.1126/science.1200832
110. Jaouen F, Proietti E, Lefevre M, Chenitz R, Dodelet J-P, Wu G, Chung HT, Johnston CM, Zelenay P (2011) Recent advances in non-precious metal catalysis for oxygen-reduction reaction in polymer electrolyte fuel cells. Energy & Environmental Science 4 (1):114-130. doi:10.1039/c0ee00011f
111. Bashyam R, Zelenay P (2006) A class of non-precious metal composite catalysts for fuel cells. Nature 443 (7107):63-66. doi:http://www.nature.com/nature/journal/v443/n7107/suppinfo/nature05118_S1.html
112. Zhang C, Hao R, Yin H, Liu F, Hou Y (2012) Iron phthalocyanine and nitrogen-doped graphene composite as a novel non-precious catalyst for the oxygen reduction reaction. Nanoscale 4 (23):7326-7329. doi:10.1039/c2nr32612d
113. Wang Y-J, Wilkinson DP, Zhang J (2011) Noncarbon Support Materials for Polymer Electrolyte Membrane Fuel Cell Electrocatalysts. Chemical Reviews 111 (12):7625-7651. doi:10.1021/cr100060r
114. Chen Z, Higgins D, Yu A, Zhang L, Zhang J (2011) A review on non-precious metal electrocatalysts for PEM fuel cells. Energy & Environmental Science 4 (9):3167-3192. doi:10.1039/c0ee00558d
115. Guo S, Wen D, Zhai Y, Dong S, Wang E (2010) Platinum Nanoparticle Ensemble-on-Graphene Hybrid Nanosheet: One-Pot, Rapid Synthesis, and Used as New Electrode Material for Electrochemical Sensing. ACS Nano 4 (7):3959-3968. doi:10.1021/nn100852h
116. Huang H, Wang X (2014) Recent progress on carbon-based support materials for electrocatalysts of direct methanol fuel cells. Journal of Materials Chemistry A. doi:10.1039/c3ta14754a
117. Huang X, Li Z, Zhang X, He X, Lin S (2013) Preparation of Pt/{PDDA-GN/PSS-GN}n multilayer films and their electrocatalytic activity regarding methanol oxidation. Journal of Colloid and Interface Science 393 (0):300-305. doi:http://dx.doi.org/10.1016/j.jcis.2012.10.058
118. Cheng Y, Jiang SP (2013) Highly effective and CO-tolerant PtRu electrocatalysts supported on poly(ethyleneimine) functionalized carbon nanotubes for direct methanol fuel cells. Electrochimica Acta 99 (0):124-132. doi:http://dx.doi.org/10.1016/j.electacta.2013.03.081
119. Yang D-Q, Rochette J-F, Sacher E (2005) Spectroscopic Evidence for π–π Interaction between Poly(diallyl dimethylammonium) Chloride and Multiwalled Carbon Nanotubes. The Journal of Physical Chemistry B 109 (10):4481-4484. doi:10.1021/jp044511+
120. Shao Y, Zhang S, Wang C, Nie Z, Liu J, Wang Y, Lin Y (2010) Highly durable graphene nanoplatelets supported Pt nanocatalysts for oxygen reduction. Journal of Power Sources 195 (15):4600-4605. doi:http://dx.doi.org/10.1016/j.jpowsour.2010.02.044
121. Chen H, Wang Y, Dong S (2007) An Effective Hydrothermal Route for the Synthesis of Multiple PDDA-Protected Noble-Metal Nanostructures. Inorganic Chemistry 46 (25):10587-10593. doi:10.1021/ic7009572
122. Jiang SP, Liu Z, Tang HL, Pan M (2006) Synthesis and characterization of PDDA-stabilized Pt nanoparticles for direct methanol fuel cells. Electrochimica Acta 51 (26):5721-5730. doi:http://dx.doi.org/10.1016/j.electacta.2006.03.006

123. Qiu J-D, Wang G-C, Liang R-P, Xia X-H, Yu H-W (2011) Controllable Deposition of Platinum Nanoparticles on Graphene As an Electrocatalyst for Direct Methanol Fuel Cells. The Journal of Physical Chemistry C 115 (31):15639-15645. doi:10.1021/jp200580u
124. Park JY, Kim S (2013) Preparation and electroactivity of polymer-functionalized graphene oxide-supported platinum nanoparticles catalysts. International Journal of Hydrogen Energy 38 (14):6275-6282. doi:http://dx.doi.org/10.1016/j.ijhydene.2012.12.059
125. Le Z-G, Liu Z, Qian Y, Wang C (2012) A facile and efficient approach to decoration of graphene nanosheets with gold nanoparticles. Applied Surface Science 258 (14):5348-5353. doi:http://dx.doi.org/10.1016/j.apsusc.2012.01.169
126. Zhang S, Shao Y, Liao H, Engelhard MH, Yin G, Lin Y (2011) Polyelectrolyte-Induced Reduction of Exfoliated Graphite Oxide: A Facile Route to Synthesis of Soluble Graphene Nanosheets. ACS Nano 5 (3):1785-1791. doi:10.1021/nn102467s
127. He W, Jiang H, Zhou Y, Yang S, Xue X, Zou Z, Zhang X, Akins DL, Yang H (2012) An efficient reduction route for the production of Pd–Pt nanoparticles anchored on graphene nanosheets for use as durable oxygen reduction electrocatalysts. Carbon 50 (1):265-274. doi:http://dx.doi.org/10.1016/j.carbon.2011.08.044
128. Mayavan S, Jang H-S, Lee M-J, Choi SH, Choi S-M (2013) Enhancing the catalytic activity of Pt nanoparticles using poly sodium styrene sulfonate stabilized graphene supports for methanol oxidation. Journal of Materials Chemistry A 1 (10):3489-3494. doi:10.1039/c2ta00619g
129. Zhang X, Xia G, Huang C, Wang Y (2013) Preparation and characterization of Pt nanoparticles supported on modified graphite nanoplatelet using solution blending method. International Journal of Hydrogen Energy 38 (21):8909-8913. doi:http://dx.doi.org/10.1016/j.ijhydene.2013.05.038
130. Zhao Y, Zhan L, Tian J, Nie S, Ning Z (2011) Enhanced electrocatalytic oxidation of methanol on Pd/polypyrrole–graphene in alkaline medium. Electrochimica Acta 56 (5):1967-1972. doi:http://dx.doi.org/10.1016/j.electacta.2010.12.005
131. Zong J, Jin Q, Huang C (2013) Effect of wetted graphene on the performance of Pt/PPy-graphene electrocatalyst for methanol electrooxidation in acid medium. Journal of Solid State Electrochemistry 17 (5):1339-1348. doi:10.1007/s10008-012-1993-z
132. Yue R, Zhang Q, Wang C, Du Y, Yang P, Xu J (2013) Graphene–poly(5-aminoindole) composite film as Pt catalyst support for methanol electrooxidation in alkaline medium. Electrochimica Acta 107 (0):292-300. doi:http://dx.doi.org/10.1016/j.electacta.2013.06.021
133. Jiang F, Yao Z, Yue R, Du Y, Xu J, Yang P, Wang C (2012) Electrochemical fabrication of long-term stable Pt-loaded PEDOT/graphene composites for ethanol electrooxidation. International Journal of Hydrogen Energy 37 (19):14085-14093. doi:http://dx.doi.org/10.1016/j.ijhydene.2012.04.084
134. Shi Q, Mu S (2012) Preparation of Pt/poly(pyrogallol)/graphene electrode and its electrocatalytic activity for methanol oxidation. Journal of Power Sources 203 (0):48-56. doi:http://dx.doi.org/10.1016/j.jpowsour.2011.11.077
135. Guo S, Dong S, Wang E (2009) Three-Dimensional Pt-on-Pd Bimetallic Nanodendrites Supported on Graphene Nanosheet: Facile Synthesis and Used as an Advanced Nanoelectrocatalyst for Methanol Oxidation. ACS Nano 4 (1):547-555. doi:10.1021/nn9014483
136. Zhang Q, Ren Q, Miao Y, Yuan J, Wang K, Li F, Han D, Niu L (2012) One-step synthesis of graphene/polyallylamine–Au nanocomposites and their electrocatalysis toward oxygen reduction. Talanta 89 (0):391-395. doi:http://dx.doi.org/10.1016/j.talanta.2011.12.049
137. Yang J-M, Wang S-A, Sun C-L, Ger M-D (2014) Synthesis of size-selected Pt nanoparticles supported on sulfonated graphene with polyvinyl alcohol for methanol oxidation in alkaline solutions. Journal of Power Sources 254 (0):298-305. doi:http://dx.doi.org/10.1016/j.jpowsour.2013.12.120
138. Fujigaya T, Hirata S, Nakashima N (2014) A highly durable fuel cell electrocatalyst based on polybenzimidazole-coated stacked graphene. Journal of Materials Chemistry A 2 (11):3888-3893. doi:10.1039/c3ta14469k

139. Hosseini H, Mahyari M, Bagheri A, Shaabani A (2014) Pd and PdCo alloy nanoparticles supported on polypropylenimine dendrimer-grafted graphene: A highly efficient anodic catalyst for direct formic acid fuel cells. Journal of Power Sources 247 (0):70-77. doi:http://dx.doi.org/10.1016/j.jpowsour.2013.08.061
140. Zhang W, Chen M, Gong X, Diao G (2013) Universal water-soluble cyclodextrin polymer–carbon nanomaterials with supramolecular recognition. Carbon 61 (0):154-163. doi:http://dx.doi.org/10.1016/j.carbon.2013.04.079
141. Ming H, Li X, Wei Y, Bu L, Kang Z, Zheng J (2013) Facile synthesis of ionic-liquid functionalized graphite oxide nanosheets for a highly efficient fuel cell. RSC Advances 3 (11):3655-3660. doi:10.1039/c3ra23343j

Graphene Nanocomposites in Optoelectronics

Yuye Zhang, Zhixin Zhou, Jianhai Wang, Songqin Liu and Yuanjian Zhang

Abstract As a kind of emerging two-dimensional (2D) materials, graphene has attracted worldwide attentions both in fundamental studies and practical applications across many fields such as physics, chemistry, materials and electronics. Here, we will survey the recent advances in optoelectronics properties of graphene and graphene nanocomposites, as well as their potential applications. Moreover, the chemistry, the preparation techniques, and the structure–property relationships of the graphene nanocomposites would be highlighted.

Keywords Graphene · Chemical functionalization · Optoelectronics · Structure-property relationships

1 Introduction

As the most significant component of diverse compounds and the fundamental element of the building blocks for organisms, the chemistry of carbon has charmed researchers for generations. The electron surrounding carbon atom enables the formation of C–C covalent bond, thereby developing thousands of molecules. There are several allotropes of carbon, for instance, diamond, graphite and amorphous carbon. Different bonding between carbon atoms brings huge variation of the allotropes. For example, diamond is the hardest material found naturally on earth while graphite is soft enough to act as a lubricant for machines running under high temperature.

Y. Zhang · Z. Zhou · J. Wang · S. Liu · Y. Zhang (✉)
School of Chemistry and Chemical Engineering, Southeast University,
Nanjing 211189, China
e-mail: Yuanjian.Zhang@seu.edu.cn

K.K. Sadasivuni et al. (eds.), *Graphene-Based Polymer Nanocomposites in Electronics*, Springer Series on Polymer and Composite Materials,
DOI 10.1007/978-3-319-13875-6_6

Fullerene and carbon nanotube, the carbon allotropes which were found in 1985 [1] and 1991 [2] respectively, exhibit interesting properties due to the zero-dimensional (0D) or one-dimensional (1D) structures. The 2D carbon allotrope was not identified until 2004, when Andre Geim and Konstantin Novoselov produced the single-layer graphene by mechanical exfoliation [3]. Before that, strictly 2D crystals were considered unstable from both theoretical and experimental perspective [4–6]. And researchers had made some efforts to prepare thin films of graphite, but the result was not satisfied. Therefore, the discovery of graphene and its unique properties attract tremendous attention in the following years.

The distinguished features of graphene make it promising in many fields. For instance, owing to the excellent electrical properties, graphene is expected to be faster and more energy efficient than silicon when applied in transistors. This is pretty welcomed in the post silicon era. However, the actual application of graphene in electronics is hampered by the lacking of a band gap, although some methods were developed for the opening of a band gap [7–9]. In this sense, there is still a long way for graphene in replacing silicon in electronics.

Nevertheless, graphene has emerged in applications ranging from utilizing its optical properties and the electrical properties so far. For instance, the atomic thickness of graphene makes it a practically transparent material. Together with the outstanding electrical properties, graphene is deemed a promising candidate for optoelectronics, such as transparent conductors. Superior to the conventional transparent conductors such as indium tin oxide (ITO), graphene is flexible, inexpensive and more abundant, thus supposed to be the next generation of transparent conductor (Scheme 1a). In addition, due to facile structural manipulation, graphene nanoribbon or some graphene derivatives are photoactive and themselves can be used as photoluminescent probes in bioanalysis or major light absorption components in optoelectronic devices such as light-emitting diodes (LEDs) and transistors (Scheme 1b). Moreover, graphene is a kind of π-conjugated material. It may be used as an electron accepter, and capable of quenching other fluorescent

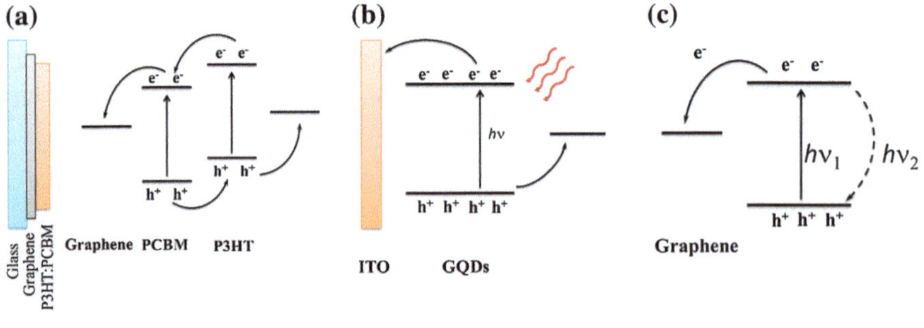

Scheme 1 General principle and charge-transfer pathway for graphene in optoelectronics. **a** Graphene as a transparent charge collector for poly(3-hexylthiophene) (*P3HT*)/[6,6]—phenyl C_{61}-butyric acid methyl ester (*PCBM*) donor/acceptor system, **b** graphene quantum dots (*GQDs*) as photoactive component to generate free electrons and holes, and **c** graphene as an electron acceptor in resonance energy transfer process

materials upon a proper resonance energy transfer, which is quite promising for optical devices in visualized biosensing (Scheme 1c). Besides, but not less important, based on the same proof-of-the-concept, graphene-based nanocomposites also show some successful applications in the sister field of optoelectronics such as photoelectric conversion and photocatalysis.

In recent years, researchers have seen remarkable electronic and optical properties of graphene, and many excellent studies were carried out on graphene, ranging from the methods for graphene growth, modification, and transfer, to the possible photonics and optoelectronics applications [10, 11]. Notably, many investigations were devoted to revealing the structure–property relationships of the graphene-based nanocomposites, and a wide variety of sophisticated methods has been developed to prepare them in a chemical way. Therefore, in this chapter, the fundamental optical and electric properties of graphene and its derivatives, the relevant emerging applications and the structure–property relationships of graphene nanocomposites, and the chemistry that was involved for preparing graphene-based materials would be discussed.

2 Transparent Conductors with Graphene and Graphene Nanocomposites

Transparent conductors are widely used in solar cells, displays, LEDs and etc. High optical transparency and electrical conductivity are the two most significant features of transparent conductors. Nowadays, the industry standards of transparent conductors are ITO and stannic oxide (SnO_2) due to the satisfied electrical and optical properties as well as the ease of fabrication. However, these materials suffered from some shortcomings such as the conductivity losses upon bending [12], which hamper the application in flexible devices. In addition, the resource of indium is depletable and the production cost of ITO increases continuously. These drawbacks push forward the search of alternatives to ITO.

As an allotrope of carbon, carbon nanotube was investigated as transparent conductor for various applications [13–15]. Besides carbon nanotube, graphene is also considered a potential candidate of transparent conductors. High optical transparency and carrier mobility of graphene meet the basic requirements of transparent conductors. It is worth noting that graphene is stretchable, which enable the fabrication of flexible devices. This is rather competitive in contrast with conventional transparent conductors. In addition, other merits such as thermal stability and abundant sources also make it a research hotspot of the next generation transparent conductors.

The high optical transparency of graphene in visible spectrum can be attributed to its atomical thickness nature. From atomic force microscopy (AFM) studies, single layer graphene (SLG) sheet possesses a height of 0.34–1.2 nm [16]. Meanwhile, the transmittance of SLG (97.1 %) is observed by UV-visible spectra. For bilayer graphene, it shows a transmittance of 94.3 % at the same wavelength of 550 nm [17].

2.1 Liquid Exfoliated Graphene from Graphite

Direct liquid-phase exfoliation of graphite is a feasible approach to produce graphene of high-quality in large scale. Due to the unique structure of graphene, only a few solvents, e.g. *N*-methyl-2-pyrrolidone (NMP) and N,*N*-dimethylacetamide (DMA), can be applied to disperse graphene [18–20]. It was revealed that the optimal solvents should have surface tension of ca. 40–50 mJ m^{-2} which matches the surface energy of graphite [18]. Although the exfoliation efficiency is relatively low (~0.01 mg mL^{-1}), graphene prepared from this method retains the original electronic properties owning to the absence of defects.

Blake et al. [21] reported that liquid-phase exfoliated graphene could be used to fabricate transparent electrodes for liquid crystal devices which showed excellent performance such as a high contrast ratio. They prepared graphene suspension by sonicating natural graphite in dimethyl formamide (DMF) and then centrifuging to remove thick flakes. The remaining suspension consisted of high concentration of SLG [up to 50 %, see transmission electron microscope (TEM) and electron diffraction pattern in Fig. 1a]. After depositing onto a glass substrate by spray-coating and annealed subsequently, a thin film of 4–5 layer of graphene was prepared. The scanning electron microscopy (SEM) image in Fig. 1b illustrated that the as-prepared graphene film consisted of overlapping individual graphene and few-layer graphene (FLG) flakes. The transparency of these graphitic layers was

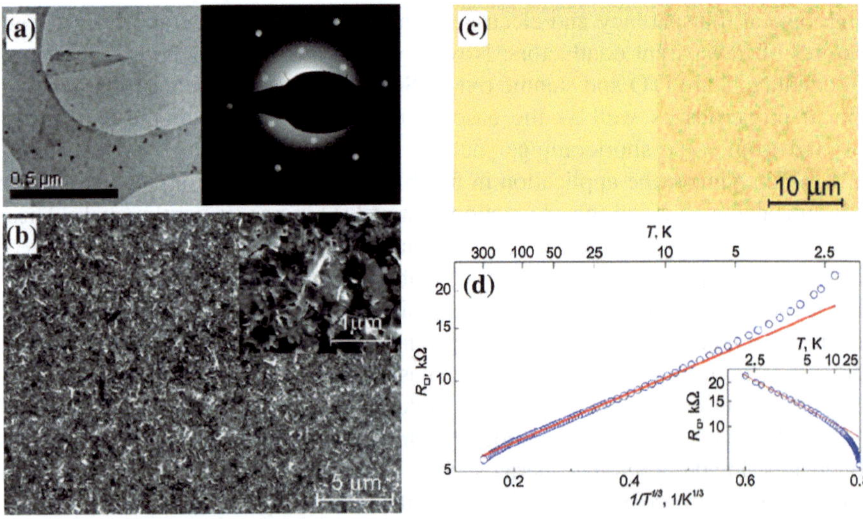

Fig. 1 **a** TEM image and electron diffraction pattern of a graphene flake obtained by the chemical exfoliation. **b** SEM of a thin graphitic film on the glass substrate obtained by spray-coating and annealing. *Inset* the enlarged image. **c** Light transmission through an original glass slide (*left*) and the one covered with the graphitic film (*right*). **d** Temperature dependence of the film's sheet resistance. *Inset* the same data but for the low temperature interval. Blake et al. [21] Copyright 2008. Reproduced with permission from American Chemical Society

approximately 90 % (Fig. 1c). The four-probe resistance was further measured as a function of temperature (Fig. 1d). It was found that the room temperature sheet resistance was of the order of 5 kΩ. Considering the high transparency, the film was already acceptable for some optoelectronic devices, such as liquid crystals.

Except for direct exfoliation of graphite in organic solvents, another method for the preparation of graphene is to disperse graphite in water-surfactant solutions [22]. Since the surface energy of water is too high for the exfoliation, surfactants were utilized to adjust the surface energy so as to disperse graphene. For instance, with the aid of sodium dodecylbenzenesulfonate (SDBS), graphene can be suspended in water stably by Coulomb repulsion, which enabled the fabrication of conductive films by vacuum filtration [22].

As the preliminary study of graphene as transparent conductors, the performance of films prepared from liquid exfoliated graphene remains to be improved. The sheet resistance of films is still too high, and most of the solvents used during the process are not environment friendly. However, these pioneering works paved the potential applications of graphene as transparent conductors in many devices.

2.2 *Graphene Oxide (GO)*

The mass production is essential to the application of graphene in transparent conductor field. However, for the liquid-phase exfoliation strategy, the concentration of graphene dispersion is very low (~0.01 mg mL^{-1}) even after days of sonication, thus not practical for large scale applications. For this, a feasible alternative approach is to synthesize GO (see proposed molecular structure in Fig. 2a) [23] from graphite and then reduce it to restore the electrical conductivity subsequently. The oxidation can be easily reached by several methods including Brodie method [24], Staudenmaier method [25] and Hummers method [26]. Among them, the Hummers method is most widely used to oxidize graphite to graphite oxide in a large scale by using concentrated sulfuric acids and highly oxidizing agents [26]. After that, graphite oxide could be exfoliated to obtain single layer or few layered GO sheets. Li et al. [27] proposed that due to the existing of large amount of oxygen-containing groups, GO could be further dispersed in water solution easily through electrostatic stabilization (Fig. 2b). Notably, such dispersion did not need any polymeric or surfactant stabilizers. Furthermore, water is a non-toxic solvent with lower boiling point, much greener than commonly used organic solvents for dispersing pristine graphene. Thus, it offered a low-cost solution processing technique to prepare graphene materials in large scale. As a result, it was demonstrated that a 10-μm-thick chemically converted graphene paper that exhibited a shiny metallic lustre could be prepared by vacuum filtration of chemically converted graphene dispersion through an alumina membrane [27].

The water-soluble GO enable the further processing which require sample in liquid phase, for instance, spin coating and dip coating. For instance, Wang et al. [28] reported that exfoliated GO was used to fabricate ultrathin graphene films

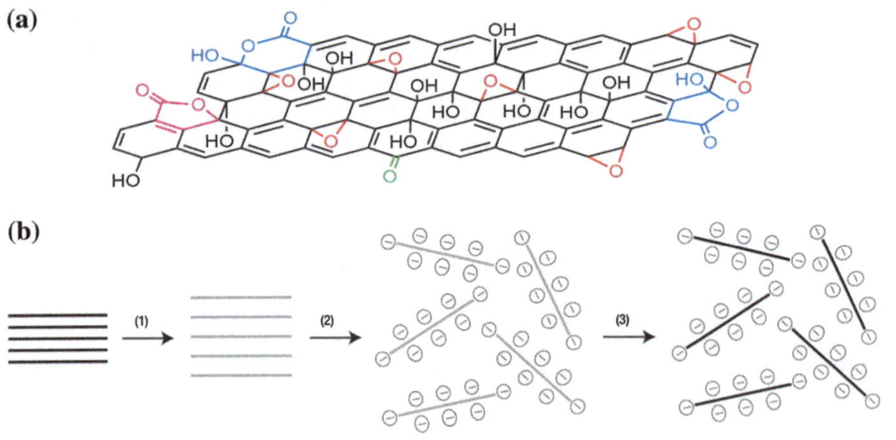

Fig. 2 **a** The proposed molecular structural of GO. **b** Scheme of the chemical route to the synthesis of aqueous graphene dispersions from GO. *1* Oxidation of graphite to graphite oxide; *2* exfoliation of graphite oxide in water by sonication to obtain GO colloids that are stabilized by electrostatic repulsion; *3* controlled conversion of GO to reduced graphene oxide (rGO) by hydrazine reduction. Gao et al. [23] Copyright 2009. Reproduced with permission from Nature. Li et al. [27] Copyright 2008. Reproduced with permission from Nature

by dip coating and a subsequent thermal reduction. The as-obtained films showed a high conductivity (550 S/cm) and transparency (70 % between the wavelengths of 1,000–3,000 nm). In another interesting work, Langmuir-Blodgett (LB) technique was exampled to assemble GO single layers [29]. It was found that GO single layers could float on a water surface because of the electrostatic repulsion. Thus, the SLG films without overlapping during compression were successfully prepared on hydrophilic substrates.

Spin coating is also widely used to fabricate uniform thin films with controllable thicknesses [30, 31]. GO films on quartz by spin coating were reported by Becerril et al. [32]. The sheet resistance and optical transparency were measured to evaluate the efficiencies of different reduction treatments, including chemical and thermal reductive treatments. Their results showed that in comparison with chemical reduction using hydrazine a thermal graphitization procedure at high temperature was the most effective procedure to produce films with sheet resistances as low as 10^2–10^3 $\Omega/\square$ (ohms per square) with 80 % transmittance at 550 nm (Fig. 3).

rGO was also reported as the precursor of flexible graphene films [33]. For instance, during the reduction process, pyrenebutyric acid and NaOH were added to stabilize the rGO through the strong π–π interactions between 1-pyrenebutyrate (PB^-) and basal plane of graphene sheets. The resulting PB^--rGO can be stably dispersed in water and the suspension was used to prepare flexible PB^--rGO films with layered structures by filtration.

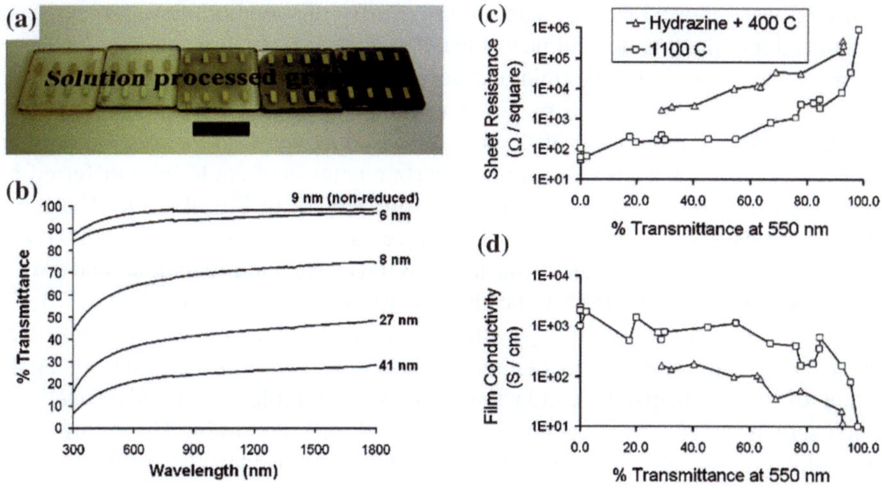

Fig. 3 Optical and electrical characterization of spin-coated GO films on quartz. **a** Photograph of an unreduced (*leftmost*) and a series of high-temperature reduced GO films of increasing thickness. *Black scale bar* is 1 cm. **b** Optical transmittance spectra of the films in (**a**) with the film thickness indicated. **c** Comparison of the dependence of sheet resistance versus optical transmittance to 550 nm light for films reduced using chemical reduction or high temperature graphitization. **d** Film conductivity versus % transmittance for films shown in (**c**). Becerril et al. [32] Copyright 2008. Reproduced with permission from American Chemical Society

2.3 Chemical Vapor Deposition (CVD) Synthesized Graphene

CVD is another promising method for synthesizing large area defect-free graphene [34–36], which is preferable in the fabrication of transparent conductor. This method is usually carried out in a tubular furnace, selected substrate is exposed to carbon-containing volatile precursors (e.g. CH_4), which decompose and deposit on the substrate surface at high temperature condition.

Graphene synthesized by CVD was firstly deposited on nickel substrates [37]. CH_4 mixed with H_2 and Ar in particular proportion was used as the reaction gas and react on the substrate surface at 1,000 °C. After the substrate was cooled down to room temperature rapidly, two methods can be applied to transfer the graphene films to other stretchable substrates, either by etching the nickel substrate or using the polydimethylsiloxane (PDMS) stamp. The resulting graphene films showed a low sheet resistance of 280 Ω/□ at ~80 % optical transparency.

Li et al. reported that copper foil was another fine substrate for the preparation of graphene by CVD [35]. Unlike the nickel substrate, graphene grown on copper foils were predominantly of single-layer. This might be due to the low solubility of carbon atoms in copper, which appeared to make the growth process self-limiting. An improved transfer process of graphene on Cu foils using polymethyl-methacrylate

(PMMA) was reported by the same group [38]. By this method, fewer cracks were introduced into the graphene films to show better devices performance.

Since the Cu foils are flexible and easy to roll up in limited space, preparing graphene of ultra large scale by CVD in tubular furnace is possible. Bae et al. demonstrated that monolayer 30-inch graphene films were synthesized onto rolled copper foils in an 8-inch-wide tubular quartz reactor, and could be transferred by roll-to-roll method (see detailed synthesis procedures in Fig. 4) [39]. The monolayer graphene films showed sheet resistances as low as ~125 Ω/□ with 97.4 % optical transmittance. Further using layer-by-layer stacking technique and *p*-type doping with nitric acid (HNO_3), the sheet resistances of the as-prepared four-layer film would decrease to ~30 Ω/□ with optical transparency of ~90 %. These data had been already superior to commercial ITO. The proof-of-the-concept application of such transparent conductive film was exampled as a fully functional touch-screen panel device capable of withstanding high strain. The future commercialization of large scale transparent electrodes replacing ITO by this method still needs detailed investigation.

2.4 Graphene Based Hybrid Materials

As an emerging material with excellent properties, graphene can be hybrid with plenty of other materials for additional interesting performance [40–44]. The first example of a transparent and electrically conductive ceramic composite of graphene-based sheets was reported in 2007 by Watcharotone et al. [40]. Transparent and electrically conductive composite silica films were fabricated on glass and hydrophilic SiO_x/silicon substrates by incorporation of individual GO sheets into silica sols followed by spin-coating, chemical reduction, and thermal curing.

Since carbon nanotubes is another promising candidate for transparent conductors, nanocomposite comprised of carbon nanotubes and chemically converted graphene was prepared and deposited onto a variety of substrates by spin-coating [41]. A resistance as low as 240 Ω/□ was observed at 86 % transmittance from the resulting films after doping. Yan et al. [45] reported that carbon nanotubes could be used to reinforce graphene through both π–π stacking domains and covalent bonding. The synthesized graphene sheets showed improved sheet resistance at higher transparency as compared to general polycrystalline CVD graphene. This made it a promising candidate for flexible all-carbon transparent electrode. Electroluminescent structures were also fabricated from transparent electrodes based on graphene nanoplatelets and carbon nanotubes by spray-coating [46]. Compared to ITO based structures, the graphene based structures exhibited higher mechanical endurance in periodical mechanical stress test and better performance during accelerated aging process.

Khrapach et al. [42] reported that ferric chloride could improve the electrical conductivity of FLG. This can be reached by intercalating FLG with ferric chloride. Transparent conductors from this material showed a sheet resistance

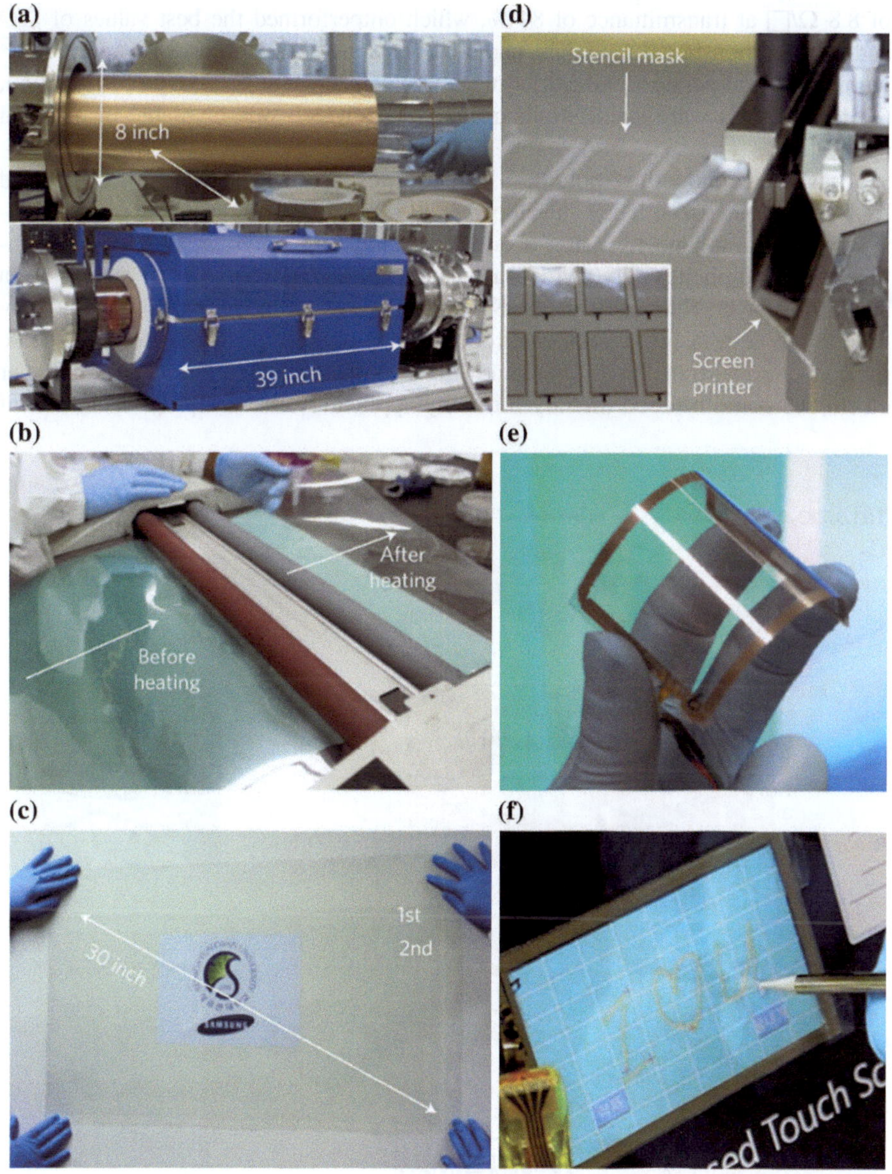

Fig. 4 Photographs of the roll-based production of graphene films. **a** Copper foil wrapping around a 7.5-in. quartz tube to be inserted into an 8-in. quartz reactor. The *lower image* shows the stage in which the copper foil reacts with CH_4 and H_2 gases at high temperatures. **b** Roll-to-roll transfer of graphene films from a thermal release tape to a PET film at 120 °C. **c** A transparent ultra large-area graphene film transferred on a 35-in. PET sheet. **d** Screen printing process of silver paste electrodes on graphene/PET film. The *inset* shows 3.1 in. graphene/PET panels patterned with silver electrodes before assembly. **e** An assembled graphene/PET touch panel showing outstanding flexibility. **f** A graphene-based touch-screen panel connected to a computer with control software. Bae et al. [39] Copyright 2010. Reproduced with permission from Nature

of 8.8 Ω/□ at transmittance of 84 %, which outperformed the best values of ITO and other carbon-based materials, and a carrier density as high as 8.9×10^{14} cm^{-2} with a room temperature and carrier mean free path as large as ~0.6 μm. In addition, the $FeCl_3$-FLGs were stable in air up to 1 year, which was favorable in practical application.

Graphene-Conducting polymer hybrid transparent electrodes was reported by Lee's group [47]. They developed a new method for the transfer of CVD-synthesized graphene by using a popular conducting polymer, i.e. poly(3,4-ethylenedioxythiophene)/poly(styrenesulfonate) (PEDOT/PSS), which acted as a supporting layer. Since PEDOT/PSS is a well-known conducting polymer widely used as a hole-transfer layer or electrode in many organic electronic devices, unlike the conventional transfer (CT) method (Fig. 5a), the graphene/PEDOT/PSS bilayer electrodes could be used directly form any applications as a doping layer without

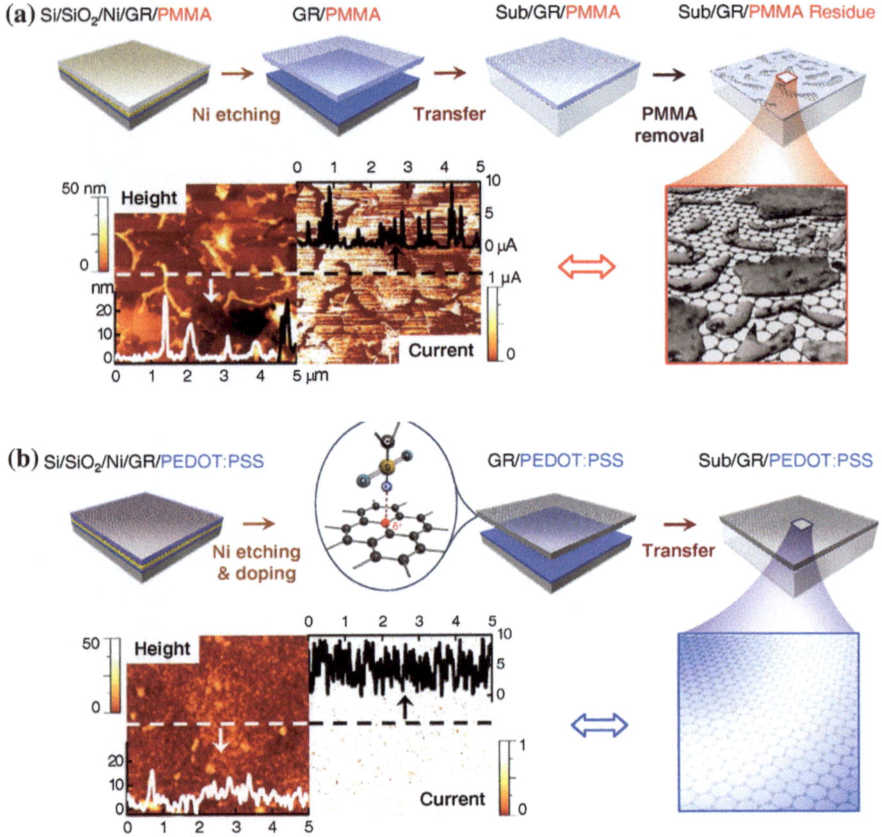

Fig. 5 Schematic illustrations and representative AFM image for the **a** conventional transfer (*CT*) and **b** doping transfer method (*DT*) using PEDOT/PSS. Lee et al. [47] Copyright 2014. Reproduced with permission from John Wiley & Sons

removing the supporting layer in the following processes. By using this doping transfer (DT) processes (Fig. 5b), the graphene/PEDOT:PSS hybrid electrodes showed less residue contamination compared with the CT method, and together with the achievement of stable doping, a better devices performance in bulk heterojunction (BHJ) polymer solar cells was achieved.

3 Photoluminescence of Graphene and Biosensing by Photoluminescence Quenching by Graphene

The luminescence of carbon materials have been studied by many researchers since the discovery of carbon dots in 2004 [48] due to the potential applications in bioimaging, photocatalyst, LEDs and other optoelectronic systems [49, 50]. Similarly, graphene has also been reported a promising material in these fields by sophisticated chemical tailoring. On account of the large conjugated system in the structure, perfect one-atom thick graphene is a zero-band gap semiconductor [3, 51] and is not of photoluminescence. However, photoluminescent emission was observed in the imperfect graphene when the electronic structure was changed. For example, photoluminescence was observed in GO, which contains a portion sp^3-hybridized carbon instead of the former sp^2-hybridized carbon [52–55]. Another strategy is to control the size of graphene smaller than 100 nm to obtain GQDs. GQDs possess quantum confinement and edge effect, as well as an energy band gap. These enable the observation of bright photoluminescence emissions in GQDs [56, 57].

3.1 *Synthesis of GQDs and Structure-Dependent Photoluminescent Properties*

Preparation of GQDs with expected photoluminescence in an efficient manner is significant for the applications in future devices. Great efforts have been devoted for developing strategies to prepare GQDs, and the reported methods can be divided into two categories: top-down and bottom-up methods. The main idea of top-down methods is cutting graphene into small pieces while the bottom-up is synthesizing GQDs from small-molecule precursors.

Among them, Pan et al. proposed that hydrothermal treatment was one of the easy-to-implement top-down methods for the synthesis of GQDs with blue photoluminescence. Graphene sheets that were prepared by thermal reduction of GO sheets were oxidized in concentrated H_2SO_4 and HNO_3 to introduce oxygen-containing functional groups at the edge and on the basal plane, including C=O/COOH, OH, and C–O–C. After that, the sheets were hydrothermal treated in a poly(tetrafluoroethylene) (PTFE)-lined autoclave at 200 °C for 10 h, during which, epoxy groups might break and form smaller graphene sheets (Fig. 6) [56].

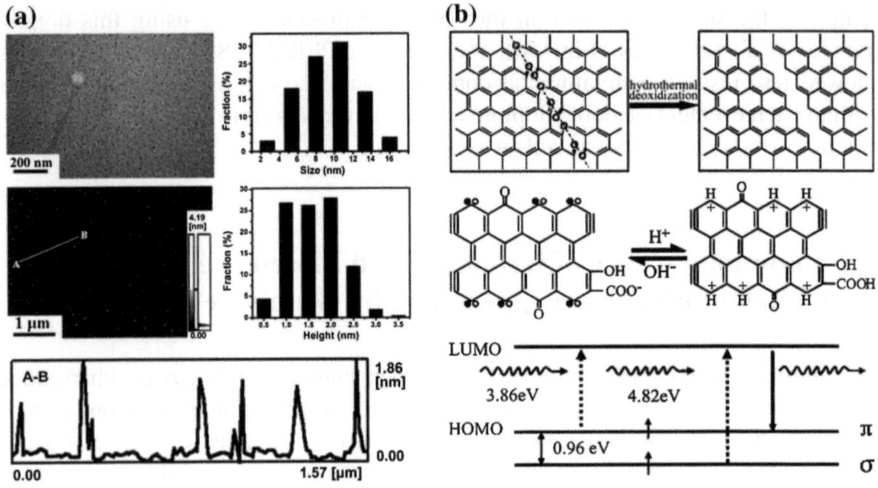

Fig. 6 **a** TEM/AFM images, size distribution and height profile of GQDs prepared by hydrothermal method. **b** Mechanism for the hydrothermal cutting of oxidized graphene sheets into GQDs, models of the GQDs in acidic (*right*) and alkali (*left*) media, and the typical electronic transitions of triple carbenes at zigzag sites observed in the optical spectra. The two models can be converted reversibly depending on pH. The pairing of σ (*filled circle*) and π (*open circle*) localized electrons at carbene-like zigzag sites and the presence of triple bonds at the carbyne-like armchair sites are represented. Pan et al. [56] Copyright 2010. Reproduced with permission from John Wiley & Sons

TEM and AFM images of the GQDs showed their diameters were narrowly distributed around 5–13 nm and the topographic heights of ca. 1–2 nm. The as-prepared GQDs illustrated a strong blue luminescence, which was ascribed to the free zigzag sites with a carbene-like triplet ground state described as $\sigma^1\pi^1$.

GQDs also can be synthesized from graphite using electrochemical methods. As a pioneer, a facile one-pot electrochemical exfoliation was applied to synthesize fluorescent carbon nanoribbons, nanoparticals, and graphene sheets from graphite electrode in ionic liquids. The fluorescence of the nanoparticles can be tuned from the ultraviolet to visible regions by changing the water/ionic liquid ratio in the electrolyte [58]. Li et al. reported that functional GQDs with a uniform size of 3–5 nm could be prepared by means of electrochemical approach in phosphate buffer solution. The obtained GQDs exhibited a green luminescence and were able to be well dispersed in water. Due to the high mobility of GQDs, it was used in the polymer photovoltaic devices in which act as an electronic-accepting material [59].

Apart from the popular approaches described above, other interesting strategies were also developed for top-down synthesis of GQDs. For example, photo-Fenton reaction was carried out to prepare GQDs in mass scale from GO. Fenton reagent could produce ·OH and reacted with GO under UV irradiation. The reaction involved the cleavage of C–C bonds and the subsequent formation of GQDs with plenty of carboxylic groups [60]. Ye et al. reported that coal was another abundant source for producing GQDs. Three different types of coal were selected and various GQDs with different morphologies were obtained by sonication of coals in concentrated sulfuric acid and nitric acid (Fig. 7a, b). Due to the unique structure of these coals over pure sp^2-carbon allotropes, the fabricated GQDs showed stable size-dependent (Fig. 7c) and pH-dependent photoluminescence in aqueous

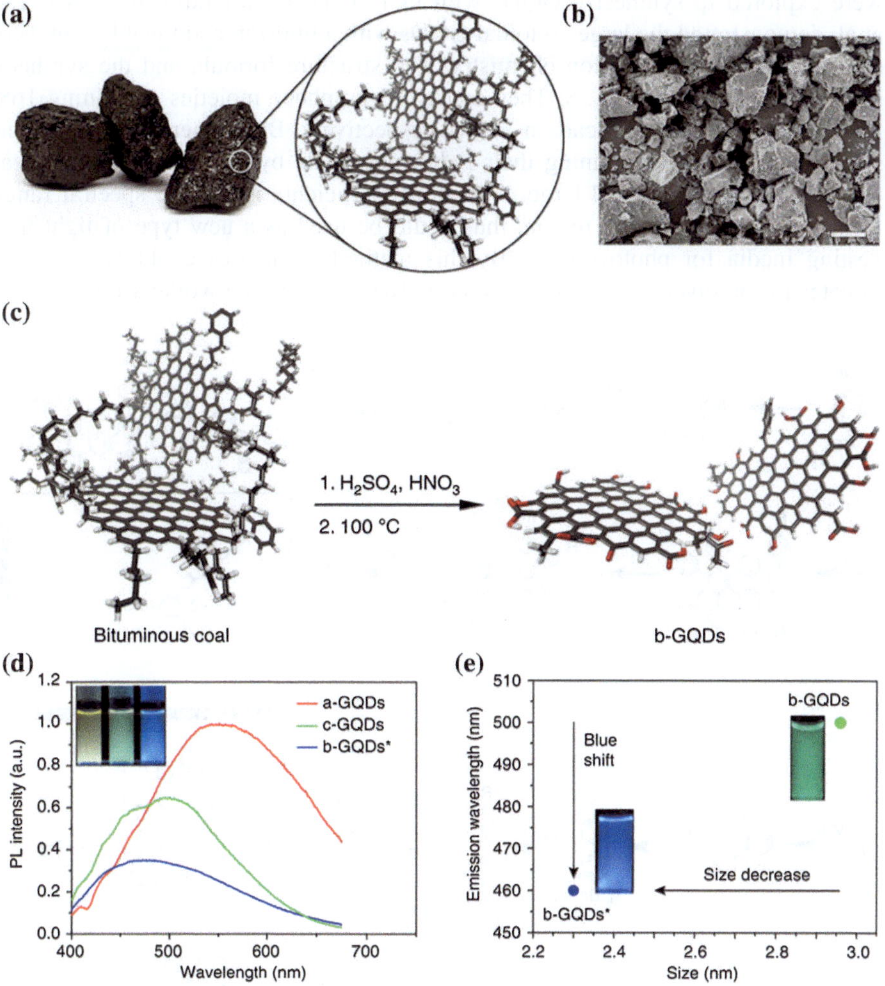

Fig. 7 **a** Macroscale image and simplified illustrative nanostructure of coal. **b** SEM image of ground bituminous coal with sizes ranging from one to hundreds of microns in diameter. *Scale bar*, 50 mm. **c** Schematic illustration of the synthesis of b-GQDs. Oxygenated sites are shown in *red*. **d** PL emission of GQDs excited at 345 nm. *Inset* is the photograph showing fluorescence of *yellow* (a-GQDs), *green* (c-GQDs) and *blue* (b-GQDs*). **e** PL emission wavelength versus the size of the GQDs, smaller GQDs lead to *blue* shift. Ye et al. [61] Copyright 2013. Reproduced with permission from Nature

solutions [61]. Besides, but not less important, the synthesized GQDs, produced in up to 20 % isolated yield from coal, were soluble and fluorescent in aqueous solution, providing promise for applications in areas such as bioimaging, biomedicine, photovoltaics and optoelectronics. Nevertheless, the reduction products of GQDs remain to be studied as well as their insertions into practical optical, electronic and structural composite materials.

In addition to the facile and efficient top-down methods, bottom-up methods were explored to synthesize GQDs with more concise structures and sizes. Yan et al. demonstrated the large colloidal GQDs with a uniform and tunable size were synthesized through solution chemistry. The structure formula and the synthetic steps were outlined in Fig. 8. They consist of graphene moieties containing 168, 132, and 170 conjugated carbon atoms, respectively. By further increasing their sizes of the GQDs and tuning their redox potentials by chemical functionalization, the GQDs illustrated large extinction coefficients in a wide spectral range from UV to near-infrared region, thus could be used as a new type of light harvesting media for photovoltaics. By this method, structures could be designed to cater to the diverse demands of devices [62]. Liu and co-workers reported the

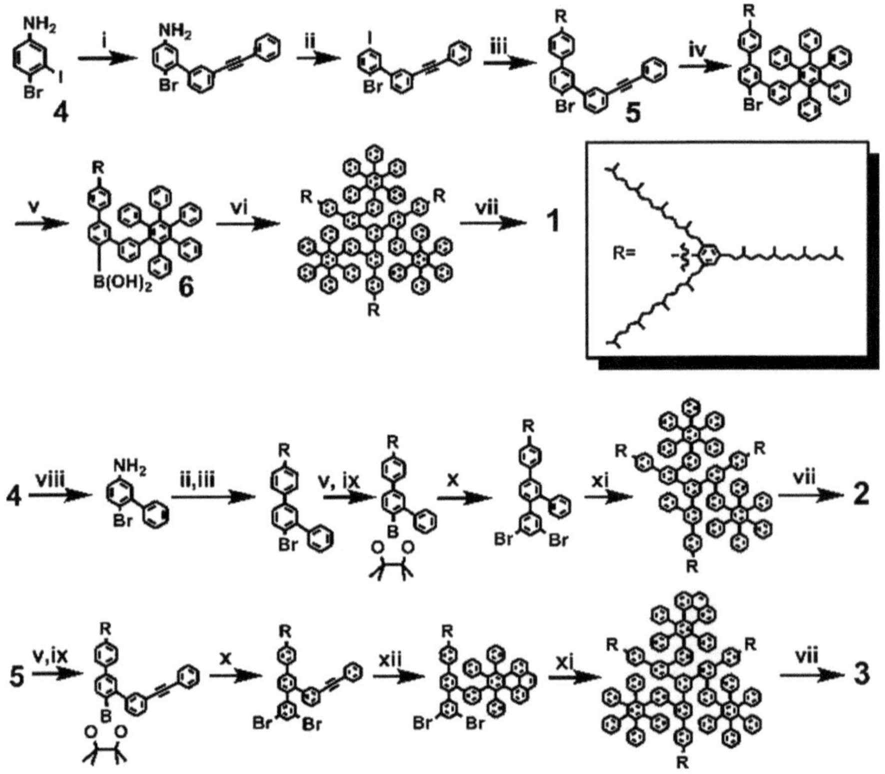

Fig. 8 Synthesis of GQDs in a bottom-up way. Yan et al. [62] Copyright 2010. Reproduced with permission from American Chemical Society

unsubstituted hexa-*peri*-hexabenzocoronene as the precursor of bottom-up fabricated GQDs. They obtained monodispersed GQDs with a size of ~60 nm diameter and 2–3 nm thickness, and multicolor photoluminescence was observed at different excitation wavelength [63]. GQDs prepared using glucose by microwave-assisted hydrothermal method was also reported. The GQDs possessed an average diameter of 1.65 nm (~5 layers) and exhibited ultraviolet emission of 4.1 eV. Unlike previously reported GQDs, the emission wavelength was independent of the size of GQDs from this method. These might be caused by the self-passivated layer on the surface of the GQDs. The as-prepared GQDs showed quantum yields of 7–11 % and were able to convert blue light to white light when coated onto a blue light emitting diode [64].

The tunable optical properties of GQDs are of significant importance for applications in optoelectronic devices. Some efforts were made to tailor the properties of GQDs. Zhu et al. [65] demonstrated that the functional groups on the surface of GQDs were modified or reduced, the original green luminescence would convert to blue luminescence. The mechanism was studied and attributed to the transformation between defect state emission and intrinsic state emission. Jin et al. further explored the possible mechanism by both experimental analyses and density functional theory (DFT). It was revealed that the charge transfer effect due to the chemical functionalization with amine groups (Fig. 9a) was responsible for the photoluminescent emission shift (Fig. 9b, c). Moreover, it was found that the photoluminescence emission of GQDs also changed upon modulate the pH of the

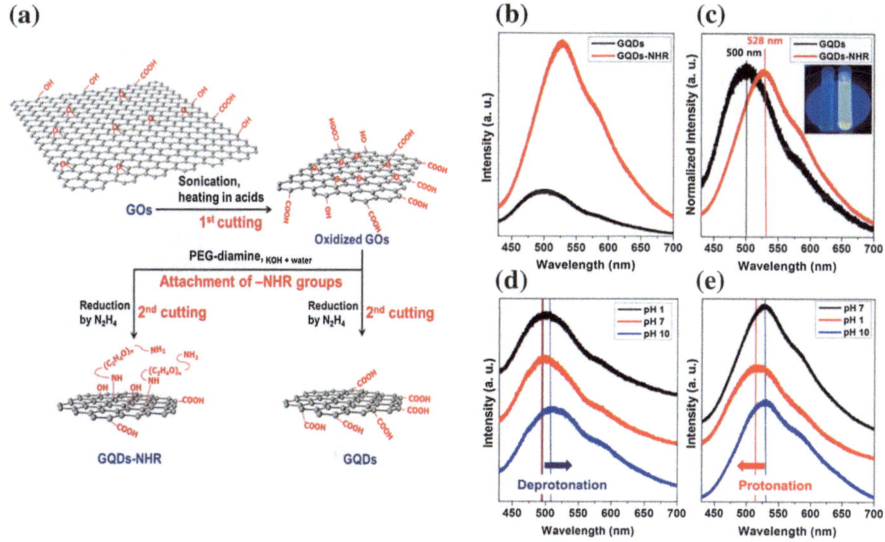

Fig. 9 **a** Schematics of GQDs and GQDs-NHR from GO, **b** PL, **c** normalized PL spectra of GQDs and GQDs-NHR in water (*inset* photograph of GQDs (*left*) and GQDs-NHR (*right*) taken under 355 nm laser excitation); pH-dependent PL spectra of **d** GQDs and **e** GQDs-NHR. Jin et al. [66] Copyright 2013. Reproduced with permission from American Chemical Society

solutions (Fig. 9d). These observations were also in line with the DFT calculations [66]. Therefore, it is feasible to prepare GQDs with task-specific photoluminescent properties by delicate chemical functionalization of the target GQDs and selecting proper solution conditions. It should be noted that the most GQDs that have been reported so far only emit blue light or the light with shorter wavelength. In this sense, it is highly anticipated to further tailor GQDs so as to broaden their photoluminescent emission in the visible/near-infrared light in the future studies.

Besides the surface chemistry, size and shape are also significant factors of the photoluminescent properties of GQDs. Kim et al. [67] reported the anomalous visible photoluminescence observed from GQDs in the size of 5–35 nm. The photoluminescence peak energies of these GQDs have a minimum at $d_a = \sim 17$ nm.

3.2 Photoluminescence of GO and RGO

Although GO possesses a larger planar size than GQDs, photoluminescence in ultraviolet to near-infrared region was observed from GO in many studies. Both liquid and solid GO emits fluorescence in visible with a long infrared emission tail under a proper excitation. Interestingly, Luo et al. reported that the photoluminescence of solid GO and the band gaps would be modulated to near-infrared region upon progressive chemical reduction using hydrazine vapor (Fig. 10) [54]. Mechanism investigations revealed that a Kekule pattern of bond distortions might account for the observed phenomenon. Near-UV-to-blue photoluminescence from chemically derived GO in a solution-processed way was reported by Eda et al. [55]. DFT calculation of the fused aromatic rings containing different atomic number showed smaller sp^2 clusters of few aromatic rings had a smaller energy gap (Fig. 11a). In this sense, the PL intensity could be manipulated by varying

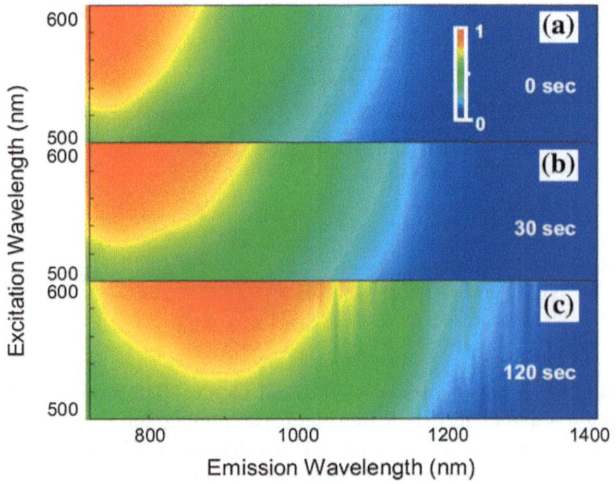

Fig. 10 Normalized PL excitation-emission maps for solid GO taken in transmission during hydrazine vapor exposure. Luo et al. [54] Copyright 2009. Reproduced with permission from American Institute of Physics

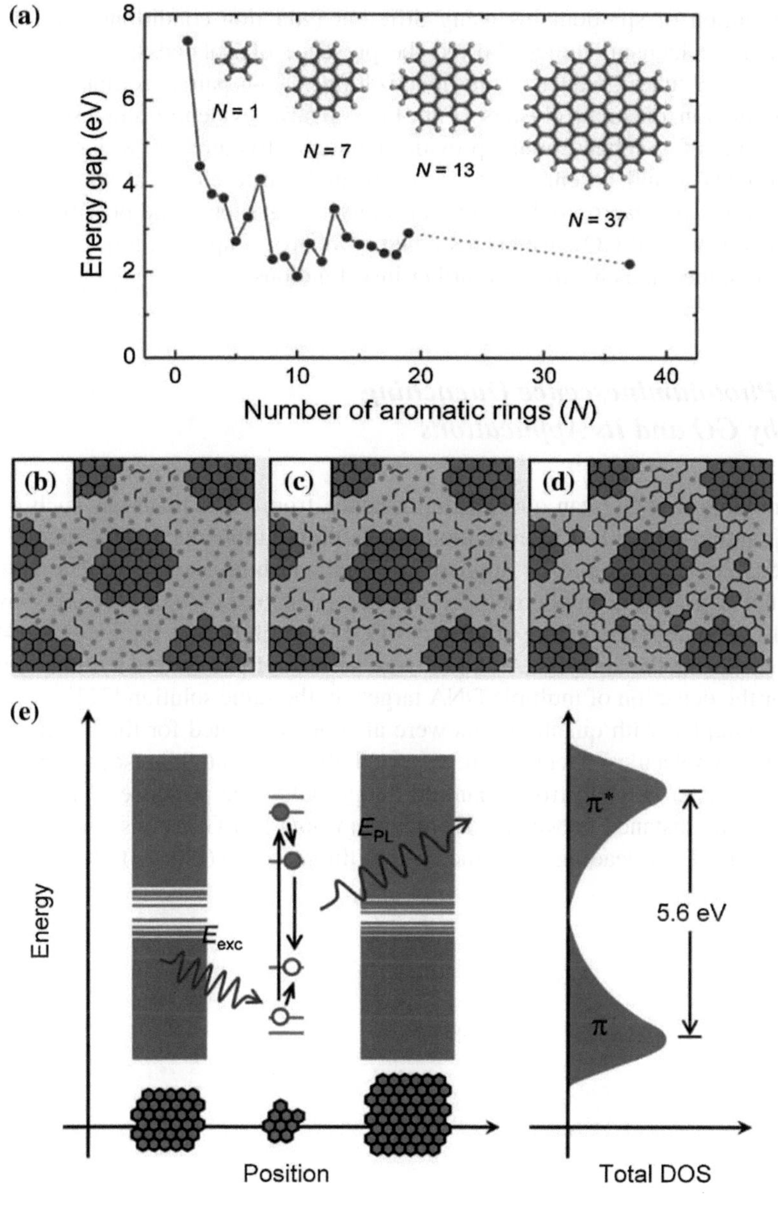

Fig. 11 **a** Energy gap of π–π* transitions calculated based on DFT as a function of the number of fused aromatic rings (N). *Inset* the structures of the graphene molecules used for calculation, **b–d** structural models of GO at different stages of reduction. **e** Representative band structure of GO. The energy levels are quantized with large energy gap for small fragments due to confinement. A photogenerated **e h** pair recombining radiatively is depicted. Eda et al. [55] Copyright 2010. Reproduced with permission from John Wiley & Sons

the evolution of sp^2 domains using different reduction conditions (Fig. 11b–d). Detailed mechanism studies showed the presence of isolated sp^2 clusters within the C–O sp^3 matrix leaded to the localization of e–h pairs, facilitating radiative recombination of small clusters (Fig. 11e). It also evidently suggested that the framework of graphene could provide a parent structure on which fluorescent components would be engineered sophisticatedly in a chemical way meanwhile maintaining the macroscopic structural integrity. As shown, the possible luminescent mechanism of GO have been studied by many groups [54, 55], but the incontrovertible mechanism still need further investigations.

3.3 Photoluminescence Quenching by GO and Its Applications

It is notable that GO can quench fluorescence from other materials such as dyes and quantum dots [68–70], even though it is photoluminescent. This unique property enables graphene to be applied in visualized biosensors [70, 71]. He et al. reported a graphene nanoprobe for the detection of DNA targets, as shown in Fig. 12. The high quenching efficiency of GO contribute to a high signal-to-background ratio in the analysis. And, the large surface of GO enable a multicolor sensor for the detection of multiple DNA targets in the same solution [72].

GO coupled with quantum dots were also demonstrated for the effective sensing of biomolecules. Dong et al. reported the efficient fluorescence resonance energy transfer (FRET) from quantum dots to GO. The presence of target would increase the distance between CdTe quantum dots and GO, consequently hindering the FRET and leading to the increase of fluorescence (Fig. 13) [70].

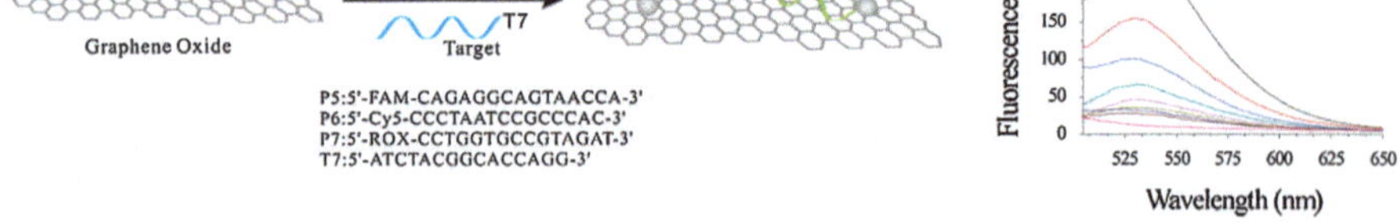

Fig. 12 **a** Scheme of the GO-based multicolor DNA analysis. Fluorescence spectra of mixture probes (*P5*, *P6*, *P7*) in the presence of different targets T5 (*blue*), T6 (*red*) and T7 (*orange*) with the excitation wavelengths of 494, 643, and 587 nm. **b** Fluorescence quenching of FAM of 50 nM in the absence (*black*) and presence of GO with a series of concentrations (*top* to *bottom* 5, 10, 15, 20, 25, 30, 35 mg/mL). FAM: carboxyfluorescein, Cy5: cyanine 5, ROX: 6-carboxy-X-rhodamine. He et al. [72] Copyright 2010. Reproduced with permission from John Wiley & Sons

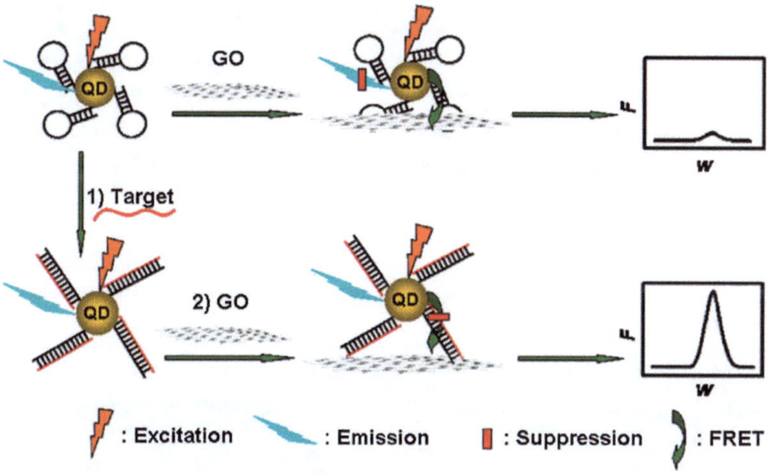

Fig. 13 Schematic representation of GO-induced fluorescence quenching of molecular beacon-QDs and biosensing mechanism. Dong et al. [70] Copyright 2010. Reproduced with permission from American Chemical Society

4 Graphene-Based Nanocomposites for Photocatalysis and Photoelectrochemical Solar Fuels

The utilization of solar energy is a research focus in recent years, for instance, solar cells, photocatalysis for hydrogen evolution and photocatalytic degradation of organic pollutions. However, the efficiency of solar energy conversion remains to be improved. As a 2D novel material, graphene was incorporated into semiconductors for the performance optimization due to the large surface area and high carrier mobility.

For instance, titanium dioxide (TiO_2) nanoparticles were loaded onto photocatalytic reduced GO to form rGO-TiO_2 composite, and the photocurrent generated from this composite was ~10 times larger than pure TiO_2 nanoparticles. This phenomenon was studied by both current transient analysis and electrochemical impedance spectroscopy (EIS). From current transient analysis, the electron lifetime of the composite was 4 times longer. And, EIS show that rGO enhances the conductivity within the composite as well as the conduction between the composite and FTO substrates [73]. Kim et al. fabricated the nanographene-coated TiO_2 composite which had a core/shell structure, and compared the photoelectrochemical properties between this composite and typical composite in which TiO_2 nanoparticles were loaded onto graphene sheets. These results of photocatalytic and photoelectrochemical measurements indicated that the nanographene-coated TiO_2 performed better than the latter one in hydrogen evolution due to the slower charge recombination and faster electron transfer [74]. Graphene was also applied to decorate TiO_2 nanotube array, and the fabricated photoelectrode showed higher

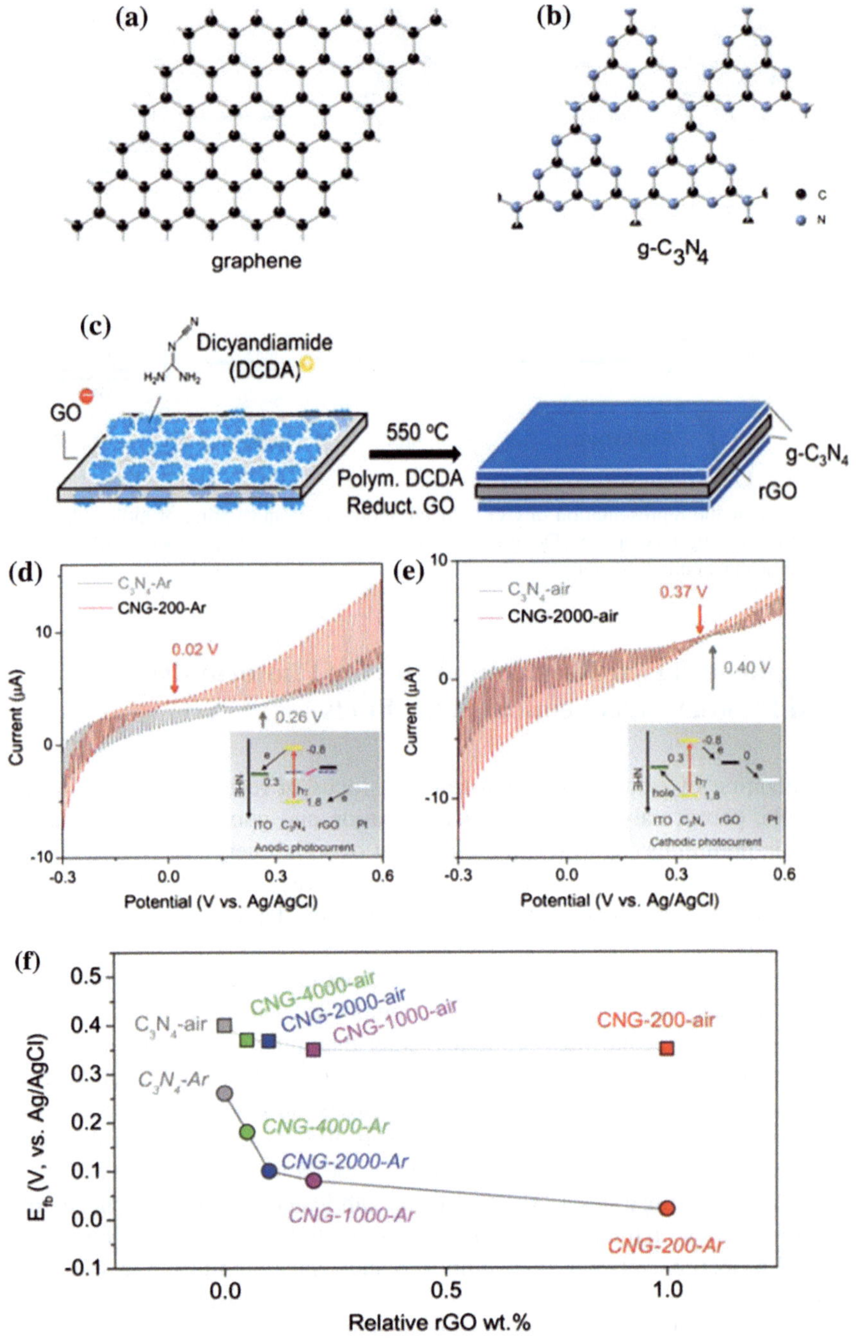
(a)
graphene
(b)
C
N
g-C3N4
(c)
Dicyandiamide
(DCDA)
GO
550 °C
Polym. DCDA
Reduct. GO
g-C3N4
rGO
(d)
C3N4-Ar
CNG-200-Ar
0.02 V
0.26 V
Current (μA)
Potential (V vs. Ag/AgCl)
ITO C3N4 rGO Pt
Anodic photocurrent
(e)
C3N4-air
CNG-2000-air
0.37 V
0.40 V
Current (μA)
Potential (V vs. Ag/AgCl)
ITO C3N4 rGO Pt
Cathodic photocurrent
(f)
C3N4-air
CNG-4000-air
CNG-2000-air
CNG-1000-air
CNG-200-air
C3N4-Ar
CNG-4000-Ar
CNG-2000-Ar
CNG-1000-Ar
CNG-200-Ar
Efb (V, vs. Ag/AgCl)
Relative rGO wt.%

◀ **Fig. 14** Idealized motif of graphene (**a**) and graphitic C_3N_4 (g-C_3N_4) sheet (**b**). Brief procedure of preparing rGO-intercalated g-C_3N_4 (**c**). Photo voltammograms of rGO-doped and pristine g-C_3N_4 synthesized in Ar (**d**) and in air (**e**) under chopped visible light (AM 1.5G) in 0.1 M KCl aqueous solution, scan rate: 10 mV s^{-1}, the *arrows* indicated the respective photocurrent onset potential, which was used to estimate E_{fb}, *inset* proposed mechanisms for anodic and cathodic photocurrent, respectively. The correlation of E_{fb} and relative rGO wt% for various rGO-doped g-C_3N_4 (**f**). Zhang et al. [81] Copyright 2011. Reproduced with permission from Royal Society of Chemistry

photocatalytic activity for the degradation of methyl blue compared with the pristine TiO_2 nanotube array [75].

Graphitic carbon nitride (g-C_3N_4) is an organic semiconductor which possesses 2D structure and tunable band gap [76, 77]. The low cost of mass-production and high thermal stability against oxidation make it a promising candidate in photovoltaic and photocatalytic fields [78–80]. Due to both 2D conjugated structure (Fig. 14a, b), through π–π stacking interaction, rGO was intercalated into g-C_3N_4 matrix to modulate the electronic structure of g-C_3N_4 (Fig. 14c). As a result, the flat band potential of the composites could be negatively or positively shifted depending on the π–π stacking interaction between rGO and g-C_3N_4, which could be facilely adjusted by their interlayer distance. As the thickness of GO would increase if it had more oxygen-containing groups, the interlayer distance was manipulated by doping g-C_3N_4 with rGO upon different reduction. In this case, both the cathodic and anodic photocurrent was enhanced in a controlled manner [81]. This work also suggested the utilization of graphene as the dopants for other semiconductors, which was very rarely reported previously. Hou et al. reported that nitrogen-doped graphene coupled with molybdenum disulfide (MoS_2) were introduced into g-C_3N_4 to form a ternary nanojunction. With the aid of the MoS_2, the composite can absorb more light and separate photogenerated charge more efficiently. Nitrogen-doped graphene act as an electron mediator for shuttling electrons-holes between g-C_3N_4 and MoS_2 sheets. The nanojunction show a higher photocurrent density and photocatalytic activity than pure g-C_3N_4 under the simulated sunlight irradiation [82].

5 Conclusion and Perspectives

In summary, graphene and graphene-based composites have shown potential applications in optoelectronics due to its unique optical and electric properties. The high transparency and conductivity of graphene make it promising for flexible transparent conductors, as well as the graphene-based nanocomposites for higher photoelectrochemical activity. Moreover, the photoluminescence and photoluminescence quenching ability of graphene also attracted tremendous attentions for the optoelectronics. Notably, the different size and structures of graphene are of significant to modulate their roles in different applications. For example, synthesis of graphene with expected scale and less defection efficiently, together with the transfer of graphene to a target substrate are still a challenge for the development of graphene as transparent conductors. Simultaneously, the luminescent properties can be tuned by the functionalization and size regulation of graphene, thus the structure-property relationship is remaining a research focus for the usage of graphene. Different structures of graphene also influence the performance of composites based on graphene. For this, the synthesis and fabrication of graphene and graphene-based nanocomposites with particular task-specific structure should be further studied. We anticipate optoelectronic devices based on graphene-based nanocomposites would change our life in the near future.

Acknowledgments This work was supported in part by the National Natural Science Foundation of China (21203023 and 91333110), NSF of Jiangsu province (BK2012317), and Fundamental Research Funds for the Central Universities, China for financial support.

References

1. Kroto H.W., Heath J.R., Obrien S.C., Curl R.F., Smalley R.E. (1985) C-60-buckminsterfullerene. Nature 318 (6042):162-163.
2. Iijima S. (1991) Helical Microtubules of Graphitic Carbon. Nature 354 (6348):56-58.
3. Novoselov K.S., Geim A.K., Morozov S.V., Jiang D., Zhang Y., Dubonos S.V., Grigorieva I.V., Firsov A.A. (2004) Electric Field Effect in Atomically Thin Carbon Films. Science 306 (5696):666-669.
4. Peierls R.E. (1935) Quelques Proprietes Typiques des Corpses Solides. Ann. Inst. H. Poincare 5:177-222.
5. Landau L.D. (1937) Zur Theorie der Phasenumwandlungen II. Phys. Z. Sowjetunion 11:26-35.
6. Venables J.A., Spiller G.D.T., Hanbucken M. (1984) Nucleation and Growth of Thin-films. Rep. Prog. Phys. 47 (4):399-459.
7. Bekyarova E., Itkis M.E., Ramesh P., Berger C., Sprinkle M., de Heer W.A., Haddon R.C. (2009) Chemical Modification of Epitaxial Graphene: Spontaneous Grafting of Aryl Groups. J. Am. Chem. Soc. 131 (4):1336-1337.
8. Balog R., Jorgensen B., Nilsson L., Andersen M., Rienks E., Bianchi M., Fanetti M., Laegsgaard E., Baraldi A., Lizzit S., Sljivancanin Z., Besenbacher F., Hammer B., Pedersen T.G., Hofmann P., Hornekaer L. (2010) Bandgap Opening in Graphene Induced by Patterned Hydrogen Adsorption. Nat. Mater. 9 (4):315-319.
9. Han M.Y., Oezyilmaz B., Zhang Y., Kim P. (2007) Energy Band-gap Engineering of Graphene Nanoribbons. Physical Review Letters 98 (20).
10. Rana K., Singh J., Ahn J.H. (2014) A graphene-based transparent electrode for use in flexible optoelectronic devices. J. Mater. Chem. C 2 (15):2646.
11. Bonaccorso F., Sun Z., Hasan T., Ferrari A.C. (2010) Graphene photonics and optoelectronics. Nat. Photonics 4 (9):611-622.
12. Cairns D.R., Witte R.P., Sparacin D.K., Sachsman S.M., Paine D.C., Crawford G.P., Newton R.R. (2000) Strain-dependent Electrical Resistance of Tin-doped Indium Oxide on Polymer Substrates. Appl. Phys. Lett. 76 (11):1425-1427.
13. Zhang M., Fang S.L., Zakhidov A.A., Lee S.B., Aliev A.E., Williams C.D., Atkinson K.R., Baughman R.H. (2005) Strong, Transparent, Multifunctional, Carbon Nanotube Sheets. Science 309 (5738):1215-1219.
14. Li J., Hu L., Wang L., Zhou Y., Gruner G., Marks T.J. (2006) Organic Light-emitting Diodes Having Carbon Nanotube Anodes. Nano Lett. 6 (11):2472-2477.

15. Tenent R.C., Barnes T.M., Bergeson J.D., Ferguson A.J., To B., Gedvilas L.M., Heben M.J., Blackburn J.L. (2009) Ultrasmooth, Large-Area, High-Uniformity, Conductive Transparent Single-Walled-Carbon-Nanotube Films for Photovoltaics Produced by Ultrasonic Spraying. Adv. Mater. 21 (31):3210-3216.
16. Kuila T., Bose S., Mishra A.K., Khanra P., Kim N.H., Lee J.H. (2012) Chemical Functionalization of Graphene and Its Applications. Prog. Mater. Sci. 57 (7):1061-1105.
17. Sun Z., Yan Z., Yao J., Beitler E., Zhu Y., Tour J.M. (2010) Growth of Graphene from Solid Carbon Sources. Nature 468 (7323):549-552.
18. Hernandez Y., Nicolosi V., Lotya M., Blighe F.M., Sun Z., De S., McGovern I.T., Holland B., Byrne M., Gun'Ko Y.K., Boland J.J., Niraj P., Duesberg G., Krishnamurthy S., Goodhue R., Hutchison J., Scardaci V., Ferrari A.C., Coleman J.N. (2008) High-yield Production of Graphene by Liquid-phase Exfoliation of Graphite. Nat. Nanotech. 3 (9):563-568.
19. Liu W.W., Wang J.N. (2011) Direct Exfoliation of Graphene in Organic Solvents with Addition of NaOH. Chem. Commun. 47 (24):6888-6890.
20. Paton K.R., Varrla E., Backes C., Smith R.J., Khan U., O'Neill A., Boland C., Lotya M., Istrate O.M., King P., Higgins T., Barwich S., May P., Puczkarski P., Ahmed I., Moebius M., Pettersson H., Long E., Coelho J., O'Brien S.E., McGuire E.K., Sanchez B.M., Duesberg G.S., McEvoy N., Pennycook T.J. et al. (2014) Scalable Production of Large Quantities of Defect-free Few-layer Graphene by Shear Exfoliation in Liquids. Nat. Mater. 13 (6):624-630.
21. Blake P., Brimicombe P.D., Nair R.R., Booth T.J., Jiang D., Schedin F., Ponomarenko L.A., Morozov S.V., Gleeson H.F., Hill E.W., Geim A.K., Novoselov K.S. (2008) Graphene-based Liquid Crystal Device. Nano Lett. 8 (6):1704-1708.
22. Lotya M., Hernandez Y., King P.J., Smith R.J., Nicolosi V., Karlsson L.S., Blighe F.M., De S., Wang Z., McGovern I.T., Duesberg G.S., Coleman J.N. (2009) Liquid Phase Production of Graphene by Exfoliation of Graphite in Surfactant/Water Solutions. J. Am. Chem. Soc. 131 (10):3611-3620.
23. Gao W., Alemany L.B., Ci L., Ajayan P.M. (2009) New Insights into the Structure and Reduction of Graphite Oxide. Nat. Chem. 1 (5):403-408.
24. Brodie B.C. (1860) Sur le Poids Atomique du Graphite. Ann. Chim. Phys. 59:466-472.
25. Staudenmaier L. (1898) Verfahren zur Darstellung der Graphitsäure. Ber. Dtsch. Chem. Ges. 31:1481-1487.
26. Hummers W.S., Offeman R.E. (1958) Preparation of Graphitic Oxide. J. Am. Chem. Soc. 80 (6):1339-1339.
27. Li D., Muller M.B., Gilje S., Kaner R.B., Wallace G.G. (2008) Processable Aqueous Dispersions of Graphene Nanosheets. Nat. Nanotech. 3 (2):101-105.
28. Wang X., Zhi L., Muellen K. (2008) Transparent, Conductive Graphene Electrodes for Dye-sensitized Solar Cells. Nano Lett. 8 (1):323-327.
29. Cote L.J., Kim F., Huang J. (2009) Langmuir-Blodgett Assembly of Graphite Oxide Single Layers. J. Am. Chem. Soc. 131 (3):1043-1049.
30. Wu J., Agrawal M., Becerril H.A., Bao Z., Liu Z., Chen Y., Peumans P. (2010) Organic Light-Emitting Diodes on Solution-Processed Graphene Transparent Electrodes. ACS Nano 4 (1):43-48.
31. Chang H., Wang G., Yang A., Tao X., Liu X., Shen Y., Zheng Z. (2010) A Transparent, Flexible, Low-Temperature, and Solution-Processible Graphene Composite Electrode. Adv. Funct. Mater. 20 (17):2893-2902.
32. Becerril H.A., Mao J., Liu Z., Stoltenberg R.M., Bao Z., Chen Y. (2008) Evaluation of Solution-processed Reduced Graphene Oxide Films as Transparent Conductors. ACS Nano 2 (3):463-470.
33. Xu Y., Bai H., Lu G., Li C., Shi G. (2008) Flexible Graphene Films via the Filtration of Water-soluble Noncovalent Functionalized Graphene Sheets. J. Am. Chem. Soc. 130 (18):5856-5857.
34. Reina A., Jia X., Ho J., Nezich D., Son H., Bulovic V., Dresselhaus M.S., Kong J. (2009) Large Area, Few-layer Graphene Films on Arbitrary Substrates by Chemical Vapor Deposition. Nano Lett. 9 (1):30-35.

35. Li X., Cai W., An J., Kim S., Nah J., Yang D., Piner R., Velamakanni A., Jung I., Tutuc E., Banerjee S.K., Colombo L., Ruoff R.S. (2009) Large-area Synthesis of High-quality and Uniform Graphene Films on Copper Foils. Science 324 (5932):1312-1314.
36. Zhang Y., Zhang L., Zhou C. (2013) Review of Chemical Vapor Deposition of Graphene and Related Applications. Accounts. Chem. Res. 46 (10):2329-2339.
37. Kim K.S., Zhao Y., Jang H., Lee S.Y., Kim J.M., Kim K.S., Ahn J.H., Kim P., Choi J.Y., Hong B.H. (2009) Large-scale Pattern Growth of Graphene Flms for Stretchable Transparent Electrodes. Nature 457 (7230):706-710.
38. Li X., Zhu Y., Cai W., Borysiak M., Han B., Chen D., Piner R.D., Colombo L., Ruoff R.S. (2009) Transfer of Large-area Graphene Films for High-performance Transparent Conductive Electrodes. Nano Lett. 9 (12):4359-4363.
39. Bae S., Kim H., Lee Y., Xu X., Park J.S., Zheng Y., Balakrishnan J., Lei T., Kim H.R., Song Y.I., Kim Y.J., Kim K.S., Ozyilmaz B., Ahn J.H., Hong B.H., Iijima S. (2010) Roll-to-roll Production of 30-inch Graphene Films for Transparent Electrodes. Nat. Nanotech. 5 (8):574-578.
40. Watcharotone S., Dikin D.A., Stankovich S., Piner R., Jung I., Dommett G.H.B., Evmenenko G., Wu S.E., Chen S.F., Liu C.P., Nguyen S.T., Ruoff R.S. (2007) Graphene-silica Composite Thin Films as Transparent Conductors. Nano Lett. 7 (7):1888-1892.
41. Tung V.C., Chen L.-M., Allen M.J., Wassei J.K., Nelson K., Kaner R.B., Yang Y. (2009) Low-Temperature Solution Processing of Graphene-Carbon Nanotube Hybrid Materials for High-Performance Transparent Conductors. Nano Lett. 9 (5):1949-1955.
42. Khrapach I., Withers F., Bointon T.H., Polyushkin D.K., Barnes W.L., Russo S., Craciun M.F. (2012) Novel Highly Conductive and Transparent Graphene-based Conductors. Adv. Mater. 24 (21):2844-2849.
43. Yin Z., Wu S., Zhou X., Huang X., Zhang Q., Boey F., Zhang H. (2010) Electrochemical Deposition of ZnO Nanorods on Transparent Reduced Graphene Oxide Electrodes for Hybrid Solar Cells. Small 6 (2):307-312.
44. Chen Z., Ren W., Gao L., Liu B., Pei S., Cheng H. (2011) Three-dimensional Flexible and Conductive Interconnected Graphene Networks Grown by Chemical Vapour Deposition. Nat. Mater. 10 (6):424-428.
45. Yan Z., Peng Z., Casillas G., Lin J., Xiang C., Zhou H., Yang Y., Ruan G., Raji A.R.O., Samuel E.L.G., Hauge R.H., Yacaman M.J., Tour J.M. (2014) Rebar Graphene. ACS Nano 8 (5):5061-5068.
46. Słoma M., Wróblewski G., Janczak D., Jakubowska M. (2014) Transparent Electrodes with Nanotubes and Graphene for Printed Optoelectronic Applications. J. Nanomater. 2014:1-7.
47. Lee B.H., Lee J.H., Kahng Y.H., Kim N., Kim Y.J., Lee J., Lee T., Lee K. (2014) Graphene-Conducting Polymer Hybrid Transparent Electrodes for Efficient Organic Optoelectronic Devices. Adv. Funct. Mater. 24 (13):1847-1856.
48. Xu X.Y., Ray R., Gu Y.L., Ploehn H.J., Gearheart L., Raker K., Scrivens W.A. (2004) Electrophoretic Analysis and Purification of Fluorescent Single-walled Carbon Nanotube Fragments. J. Am. Chem. Soc. 126 (40):12736-12737.
49. Li H., He X., Kang Z., Huang H., Liu Y., Liu J., Lian S., Tsang C.H., Yang X., Lee S.T. (2010) Water-soluble Fluorescent Carbon Quantum Dots and Photocatalyst Design. Angew. Chem. Int. Ed. 49 (26):4430-4434.
50. Sun Y.P., Zhou B., Lin Y., Wang W., Fernando K.A.S., Pathak P., Meziani M.J., Harruff B.A., Wang X., Wang H.F., Luo P.J.G., Yang H., Kose M.E., Chen B.L., Veca L.M., Xie S.Y. (2006) Quantum-sized Carbon Dots for Bright and Colorful Photoluminescence. J. Am. Chem. Soc. 128 (24):7756-7757.
51. Eda G., Chhowalla M. (2009) Graphene-based Composite Thin Films for Electronics. Nano Lett. 9 (2):814-818.
52. Sun X., Liu Z., Welsher K., Robinson J.T., Goodwin A., Zaric S., Dai H. (2008) Nano-Graphene Oxide for Cellular Imaging and Drug Delivery. Nano Res. 1 (3):203-212.

53. Chien C.T., Li S.S., Lai W.J., Yeh Y.C., Chen H.A., Chen I.S., Chen L.C., Chen K.H., Nemoto T., Isoda S., Chen M., Fujita T., Eda G., Yamaguchi H., Chhowalla M., Chen C.W. (2012) Tunable Photoluminescence from Graphene Oxide. Angew. Chem. Int. Ed. 51 (27):6662-6666.
54. Luo Z., Vora P.M., Mele E.J., Johnson A.T.C., Kikkawa J.M. (2009) Photoluminescence and Band Gap Modulation in Graphene Oxide. Appl. Phys. Lett. 94 (11):111909.
55. Eda G., Lin Y.Y., Mattevi C., Yamaguchi H., Chen H.A., Chen I.S., Chen C.W., Chhowalla M. (2010) Blue Photoluminescence from Chemically Derived Graphene Oxide. Adv. Mater. 22 (4):505-509.
56. Pan D., Zhang J., Li Z., Wu M. (2010) Hydrothermal Route for Cutting Graphene Sheets into Blue-luminescent Graphene Quantum Dots. Adv. Mater. 22 (6):734-738.
57. Ponomarenko L.A., Schedin F., Katsnelson M.I., Yang R., Hill E.W., Novoselov K.S., Geim A.K. (2008) Chaotic Dirac Billiard in Graphene Quantum Dots. Science 320 (5874):356-358.
58. Lu J., Yang J., Wang J., Lim A., Wang S., Loh K.P. (2009) One-Pot Synthesis of Fluorescent Carbon Nanoribbons, Nanoparticles, and Graphene by the Exfoliation of Graphite in Ionic Liquids. Acs Nano 3 (8):2367-2375.
59. Li Y., Hu Y., Zhao Y., Shi G., Deng L., Hou Y., Qu L. (2011) An Electrochemical Avenue to Green-luminescent Graphene Quantum Dots as Potential Electron-acceptors for Photovoltaics. Adv. Mater. 23 (6):776-780.
60. Zhou X., Zhang Y., Wang C., Wu X., Yang Y., Zheng B., Wu H., Guo S., Zhang J. (2012) Photo-Fenton Reaction of Graphene Oxide: A New Strategy to Prepare Graphene Quantum Dots for DNA Cleavage. ACS Nano 6 (8):6592-6599.
61. Ye R., Xiang C., Lin J., Peng Z., Huang K., Yan Z., Cook N.P., Samuel E.L., Hwang C.C., Ruan G., Ceriotti G., Raji A.R., Marti A.A., Tour J.M. (2013) Coal as an Abundant Source of Graphene Quantum Dots. Nat. Commun. 4:2943.
62. Yan X., Cui X., Li L.S. (2010) Synthesis of Large, Stable Colloidal Graphene Quantum Dots with Tunable Size. J. Am. Chem. Soc. 132 (17):5944-5945.
63. Liu R., Wu D., Feng X., Muellen K. (2011) Bottom-Up Fabrication of Photoluminescent Graphene Quantum Dots with Uniform Morphology. J. Am. Chem. Soc. 133 (39):15221-15223.
64. Tang L., Ji R., Cao X., Lin J., Jiang H., Li X., Teng K.S., Luk C.M., Zeng S., Hao J., Lau S.P. (2012) Deep Ultraviolet Photoluminescence of Water-Soluble Self-Passivated Graphene Quantum Dots. ACS Nano 6 (6):5102-5110.
65. Zhu S., Zhang J., Tang S., Qiao C., Wang L., Wang H., Liu X., Li B., Li Y., Yu W., Wang X., Sun H., Yang B. (2012) Surface Chemistry Routes to Modulate the Photoluminescence of Graphene Quantum Dots: From Fluorescence Mechanism to Up-Conversion Bioimaging Applications. Adv. Funct. Mater. 22 (22):4732-4740.
66. Jin S.H., Kim D.H., Jun G.H., Hong S.H., Jeon S. (2013) Tuning the Photoluminescence of Graphene Quantum Dots through the Charge Transfer Effect of Functional Groups. ACS Nano 7 (2):1239-1245.
67. Kim S., Hwang S.W., Kim M.K., Shin D.Y., Shin D.H., Kim C.O., Yang S.B., Park J.H., Hwang E., Choi S.H., Ko G., Sim S., Sone C., Choi H.J., Bae S., Hong B.H. (2012) Anomalous Behaviors of Visible Luminescence from Graphene Quantum Dots: Interplay between Size and Shape. ACS Nano 6 (9):8203-8208.
68. Treossi E., Melucci M., Liscio A., Gazzano M., Samori P., Palermo V. (2009) High-Contrast Visualization of Graphene Oxide on Dye-Sensitized Glass, Quartz, and Silicon by Fluorescence Quenching. J. Am. Chem. Soc. 131 (43):15576-15577.
69. Kim J., Cote L.J., Kim F., Huang J. (2010) Visualizing Graphene Based Sheets by Fluorescence Quenching Microscopy. J. Am. Chem. Soc. 132 (1):260-267.
70. Dong H., Gao W., Yan F., Ji H., Ju H. (2010) Fluorescence Resonance Energy Transfer between Quantum Dots and Graphene Oxide for Sensing Biomolecules. Anal. Chem. 82 (13):5511-5517.

71. Chang H., Tang L., Wang Y., Jiang J., Li J. (2010) Graphene Fluorescence Resonance Energy Transfer Aptasensor for the Thrombin Detection. Anal. Chem. 82 (6):2341-2346.
72. He S., Song B., Li D., Zhu C., Qi W., Wen Y., Wang L., Song S., Fang H., Fan C. (2010) A Graphene Nanoprobe for Rapid, Sensitive, and Multicolor Fluorescent DNA Analysis. Adv. Funct. Mater. 20 (3):453-459.
73. Bell N.J., Ng Y.H., Du A., Coster H., Smith S.C., Amal R. (2011) Understanding the Enhancement in Photoelectrochemical Properties of Photocatalytically Prepared TiO2-Reduced Graphene Oxide Composite. J. Phys. Chem. C 115 (13):6004-6009.
74. Kim H., Moon G., Monllor Satoca D., Park Y., Choi W. (2012) Solar Photoconversion Using Graphene/TiO2 Composites: Nanographene Shell on TiO2 Core versus TiO2 Nanoparticles on Graphene Sheet. J. Phys. Chem. C 116 (1):1535-1543.
75. Cheng X., Liu H., Chen Q., Li J., Wang P. (2014) Preparation of Graphene Film Decorated TiO2 Nano-tube Array Photoelectrode and Its Enhanced Visible Light Photocatalytic Mechanism. Carbon 66:450-458.
76. Zhang Y., Mori T., Ye J. (2012) Polymeric Carbon Nitrides: Semiconducting Properties and Emerging Applications in Photocatalysis and Photoelectrochemical Energy Conversion. Sci. Adv. Mater. 4 (2):282-291.
77. Deifallah M., McMillan P.F., Cora F. (2008) Electronic and Structural Properties of Two-dimensional Carbon Nitride Graphenes. J. Phys. Chem. C 112 (14):5447-5453.
78. Zhang Y., Antonietti M. (2010) Photocurrent Generation by Polymeric Carbon Nitride Solids: An Initial Step towards a Novel Photovoltaic System. Chem.-Asian J. 5 (6):1307-1311.
79. Wang X., Maeda K., Thomas A., Takanabe K., Xin G., Carlsson J.M., Domen K., Antonietti M. (2009) A Metal-free Polymeric Photocatalyst for Hydrogen Production from Water under Visible Light. Nat. Mater. 8 (1):76-80.
80. Yan S.C., Li Z.S., Zou Z.G. (2009) Photodegradation Performance of g-C3N4 Fabricated by Directly Heating Melamine. Langmuir 25 (17):10397-10401.
81. Zhang Y., Mori T., Niu L., Ye J. (2011) Non-covalent Doping of Graphitic Carbon Nitride Polymer with Graphene: Controlled Electronic Structure and Enhanced Optoelectronic Conversion. Energy Environ. Sci. 4 (11):4517-4521.
82. Hou Y., Wen Z., Cui S., Guo X., Chen J. (2013) Constructing 2D Porous Graphitic C3N4 Nanosheets/nitrogen-doped Graphene/layered MoS2 Ternary Nanojunction with Enhanced Photoelectrochemical Activity. Adv. Mater. 25 (43):6291-6297.

Graphene Filled Polymers in Photovoltaic

Dipankar Barpuzary and Mohammad Qureshi

Abstract Graphene—a two-dimensional lattice oriented monolayer of sp^2-hybridized carbon atoms—has taken up considerable attention leading to a growing scientific interest due to its exceptionally high electrical conductivity (orders of magnitude higher than copper), optical transparency (>90 %), chemical robustness (more than 500 °C) and mechanical stiffness (more than 1,000 GPa) as well as high specific surface area. Design and development of graphene incorporated polymer photovoltaics is one of the promising routes to harness the extraordinary properties of graphene for the generation of efficient solar-to-power conversion devices. Graphene as well as its chemically functionalized forms, graphene oxide (GO) and reduced-GO, are the smart materials for photovoltaic cells performing specific functions depending upon their intriguing properties. Herein we review the multifunctional and practical applicability of graphene and its composite materials as the electron acceptor, counter electrode and hole transport components of polymer solar cells. We conclude the chapter with the present scenario and challenges related to the stability and commercialization of graphene–polymer based photovoltaic devices.

Keywords Solar cell · Counter electrode · Photovoltaics · Electron-hole transport

1 Introduction

An emerging approach towards the high-technology photovoltaic applications employing graphene—a two-dimensional (2D) lattice oriented monolayer of sp^2-hybridized carbon atoms—as the smart material has led to a growing scientific interest due to its exceptionally high electrical conductivity, optical transparency,

D. Barpuzary · M. Qureshi (✉)
Materials Science Laboratory, Department of Chemistry, Indian Institute of Technology Guwahati, Guwahati 781039, Assam, India
e-mail: mq@iitg.ernet.in; mq@iitg.ac.in
URL: http://www.iitg.ernet.in/mq

K.K. Sadasivuni et al. (eds.), *Graphene-Based Polymer Nanocomposites in Electronics*, Springer Series on Polymer and Composite Materials,
DOI 10.1007/978-3-319-13875-6_7

chemical and mechanical robustness, high specific surface area as well as fracture strength [1–4]. Andre K. Geim and Konstantine S. Novoselov of the University of Manchester, UK, were awarded the Nobel Prize in Physics (2010) for their discovery to isolate "free" and "perfect" graphene as the single sheet of carbon atoms from Graphite demonstrating extraordinary electronic properties in 2004. As a result of their achievement, the current state-of-the art scenario of polymer nanocomposite based photovoltaics has been modified and improved dramatically. The unique property of graphene is the excellent electron transport with a minimum scattering or resistance as well as a high electron mobility value of ~20,000–25,000 cm^2/V s depending on the synthesis protocol and purity of the sample [5, 6]. One of the possible technical approaches to harness the outstanding properties of graphene for practical applications is to incorporate them into a suitable polymer or composite material thereby improving the physical properties of the host material even at a smaller loading. Particularly, the introduction of graphene has changed the landscape of graphene-based polymer nanocomposite materials and their recent advancements in photovoltaics. Moreover, the "wonder material" graphene and its derivatives, graphene oxide (GO) and reduced GO (RGO), have been extensively investigated as the promising candidates to enhance the performance parameters of energy-related devices [7]. In addition to the development of photovoltaics, graphene has significantly influenced the fabrication and implementation of a wide variety of potential device applications such as nanoelectronics, nanocomposites, sensors, batteries, supercapacitors etc [8–19]. In photovoltaic cells, graphene can be exploited in plentiful ways—electron/hole transport and electron acceptor materials, interfacial or buffer layers to retard the charge recombination within the components, transparent counter electrodes, etc. However considering the practical performances and low-cost of manufacturing, there needs a large-scale production of graphene in sufficiently pure quantity in its sheet form as well as their homogeneous distribution into the host/polymer materials. Even if the large-scale production of graphene nanosheets at low-cost is achieved, it is still a scientific challenge to determine how to apply these graphene nanosheets in the most effective way for efficient output. Herein the exfoliation of graphene or its chemical functionalization to produce GO and RGO to enhance the binding interactions and the intriguing properties of graphene–polymer composites comes into picture. Interestingly, compared to graphene, the use of GO is advantageous, wherein, it clutches the polymer composites owing to the presence of chemically reactive oxygen functionality (carboxylic, epoxy and hydroxyl groups) at the basal planes of GO [20]. The electrical conductivity and electron/hole mobilities of GO can be further enhanced by its reduction to produce RGO, but RGO offers a less interactive ability in contrast to GO [21]. It can be concluded that regardless the smaller or bigger size, the organic/polymer composite materials can be non-covalently attached to the graphitic surfaces of chemically modified graphene and covalently anchored to GO via the highly reactive oxygen functionalities thereby improving the customized properties for the optoelectronic applications. In this chapter, we focus the general methodologies to synthesize graphene and graphene–polymer composites by the exfoliation and chemical functionalization of graphene with an emphasis on the fabrication and production of graphene–polymer based photovoltaic devices in detail. We conclude with present scenario and challenges related to the stability and commercialization of graphene–polymer based devices.

2 Graphene Based Organic Polymer Solar Cells

Recent advancements in the polymer-based photovoltaic devices have changed the landscape of its current state-of-the art technological approach to challenge or overwhelm the ever-increasing global energy demand and sustainability issues. Since the discovery of dye-sensitized solar cells by Grätzel in 1991, tremendous efforts have been carried out by the researchers to develop and synthesize or modify the innovative materials to incorporate them in photovoltaic devices to achieve high efficiency employing a number of polymer materials, organic molecules or metal complexes [22]. If we look back at the practical acquirement of the natural solar energy in terms of electricity, silicon based solar cells have primarily gained the commercial enactment, however, the high cost of device fabrication and production limits its indoor–outdoor implementation at a larger scale [23]. As a replacement for the inorganic photovoltaic assemblies, the organic or polymer-based photovoltaic devices have been deemed to be one of the attractive and efficient alternative low-cost renewable energy resources concerning the scalability, compatibility and practical execution [24–29]. The organic or polymer solar cells (PSCs) are advantageous owing to the structural diversity and facile functionalization tunability of the organic or polymer components via chemical modifications to achieve the characteristic prerequisites for better performance of the devices, such as light absorption, charge separation and electron transport. In addition, the solution processed low-cost fabrication compatibility by screen printing, doctor blading, inkjet printing and spray deposition techniques in conjugation with the flexibility, transparency and light weight stimulate the potential role of PSCs to achieve high throughput over large-area production [30]. The PSCs incorporating the conjugated polymers as the electron donor materials blended with various electron acceptor organic molecules/polymers have demonstrated high PCEs up to ~8 % using single bulk-heterojunction (BHJ) device structures [31–34]. You et al. [35] have recently established an excellent power conversion efficiency (PCE) of ~10.6 % for a tandem PSC utilizing a bezothiadiazole based polymer having narrow band gap. The electron transport property of the polymers used in the fabricated device have been tuned via chemical modification of the molecular framework that altered the band gap alignments within the system leading to an enhanced light absorption as well electron transfer. The prime target to promote the development of PSCs is the facile molecular engineering of the conjugated polymers or their blends with the materials having highly efficient electron transport properties. In accordance to the importance of developing new strategies as the driving forces to improve the photovoltaic performance, the conjugated polymers should uphold three fundamental features: (i) a narrow band gap ensuring efficiently broad region solar light harvesting, (ii) suitable highest-occupied molecular orbital (HOMO) level and lowest-unoccupied molecular orbital (LUMO) level ascribing an effective charge separation with minimum energy loss and (iii) higher hole mobility improving the charge transport [36]. Even though, appropriate modification of the structural components and side groups of π conjugated polymers usually enhances their light absorptivity and band edge alignments, the combination of efficient electron acceptor/transport material to these polymers also needs sincere attention in order to enhance their charge carriers mobilities (~10^{-3}–10^{-4} cm^2/V s) and exciton diffusion lengths (~10 nm) to achieve high performance photovoltaic devices [37]. In this regard, an emerging tactic to accomplish the milestones of PCEs is to employ the ever-known thinnest smart material graphene having exceptional inherent physical properties. Theoretically, graphene has a zero-band gap with a small area overlap between its valence and conduction bands leading to the electrical conductivity anticipating the access to graphene-based organic photovoltaics [38]. Moreover by virtue of quantum confinement effect, the band gap of graphene can be fine-tuned via its size reduction up to the nano regime (less than 100 nm) forming graphene quantum dots (GQDs) and graphene nanoribbons, which carry unique semiconducting and photoluminescence properties to demonstrate a potential application in photovoltaics [39–42]. For the first time, Yong and Tour [43] have predicted an efficiency of ~12 % for the single-cell graphene-based organic photovoltaic cell. Their experimental evidences suggested that the bandgap and band position tunability as well as higher thermal, mechanical stability of graphene endowing it as a credible material for the fabrication of next-generation graphene-based flexible organic or PSCs.

Although appears to be very simple, working principle of PSCs needs to be understood and grasped to improve our ability to categorize the predictive materials based on their optimized chemical anonymity in the device architecture. In general, the primary design of organic solar cells comprises of an organic molecule or polymer material as the active layer sandwiched between the two electrodes as shown in Fig. 1. When the device is illuminated, photons are absorbed by the molecule in the active layer (donor) and electrons gets excited from its HOMO to the LUMO level generating electron–hole pairs, known as excitons,

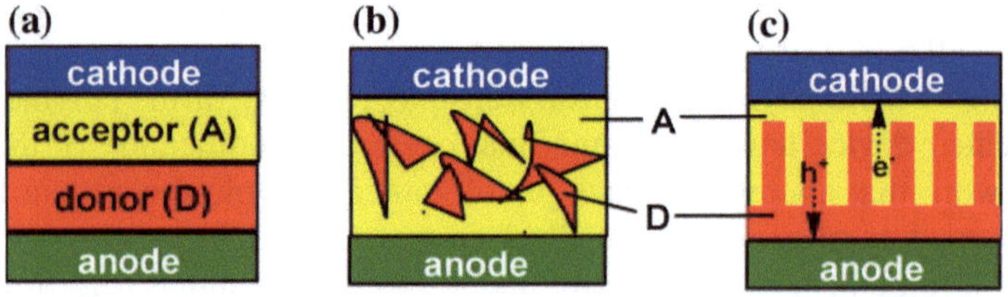

Fig. 1 Basic architectures for three donor–acceptor morphologies of organic solar cells: **a** double layer morphology, **b** blended D–A morphology and **c** ideal BHJ morphology [108]. Copyright 2012. Reproduced with permission from Elsevier Ltd.

which relaxes with a binding energy between ~1.0 and 0.14 eV [44–46]. The exciton binding energy of inorganic semiconductor materials remains within a range of few milli electron volts, while the same for the organic semiconductors are relatively larger due to their localized electron and hole wave functions and lower dielectric constants ($\varepsilon \approx 3$–4) thereby enhancing the Coulombic attraction between the electron and hole. In order to generate the photocurrent in the devices, these Coulombic bound electron–hole pair needs to get separated into its component electron and hole, if not then it eventually recombines and loses via the competitive processes like photon emission (fluorescence) or thermal decay (non-radiative recombination). In the case of PSCs, the exciton dissociation relies on gradients of the potential created across the interface of an electron donor (D) and electron acceptor (A). The photoinduced electron transfers from the donor material to the acceptor component, provided, the acceptor possesses a higher electron affinity (EA) and lower ionization potential (IP) than that of the donor material [47]. The donor–acceptor (D–A) heterojunction provides the driving force necessary to dissociate the electron–hole pair by means of the difference in EA and IP of the donor and acceptor materials. Additionally, an acceptor material should possess a HOMO level relatively lower than that of the donor material in order to facilitate the photo excited electron transfer from the donor to acceptor material leaving behind the hole in the valence band of the donor itself. A schematic of the working principle of a donor–acceptor type solar cell has been illustrated in Fig. 2.

$$\mathrm{D} + \mathrm{A} + hv \rightarrow \mathrm{D}^* + \mathrm{A}\left(\text{or } \mathrm{D} + \mathrm{A}^*\right) \rightarrow \mathrm{D}^{\cdot +} + \mathrm{A}^{\cdot -}$$

It can be envisioned that for the planar heterojunction type devices, where the active layer of the device is composed of separated donor and acceptor layers, the efficiency is governed by the exciton diffusion length of the organic counterparts. The average distance traveled by the excitons before undergoing recombination

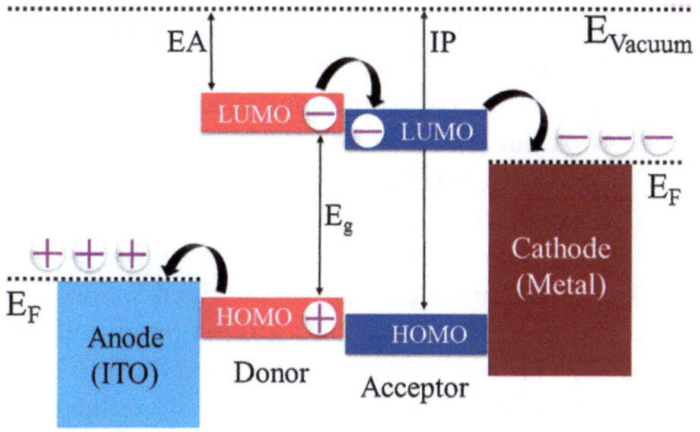

Fig. 2 Schematic drawing of the basic working principle of an organic or polymer solar cell. Processes involved in the conversion of light into electrical energy: (1) absorption of photons by the active material generating the exciton, (2) diffusion of excitons to an active donor–acceptor interface, (3) separation of the exciton into an electron and a hole at the donor–acceptor interface and (4) collection of the electrons and holes at the terminal cathode and anode, respectively

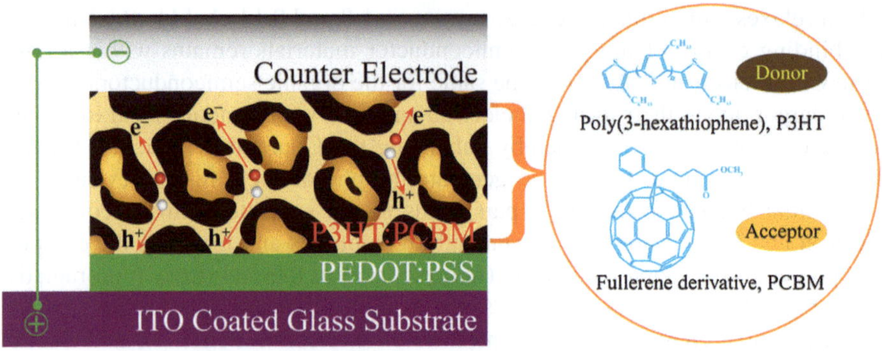

Fig. 3 Basic principle of the photon absorption to photocurrent generation in an organic bulk heterojunction solar cell, where ITO (indium tin oxide) is the transparent conductive electrode, PEDOT:PSS is the buffer/hole transport layer and P3HT:PCBM is the electron donor–acceptor layer

is ~3–10 nm for most of the organic semiconductors [48–50]. Once the photo generated excitons are dissociated into electrons and holes, the charge carriers are collected at the electrodes by means of a built-in-field drift and the diffusion process. This occurs within the lifetime of excitons to avoid the recombination process [51]. Notably, the exciton formation near the heterojunction is very crucial to have the maximum proficiency for light harvesting by the active material, which is practically problematic for the planar bilayer heterojunction cells due to the requirement of extremely thin photoactive layer. In contrast, the BJH structures are advantageous for the PSCs, where most of the excitons are separately generated at the heterojunction within the exciton diffusion length in an interpenetrating network of the donor and acceptor materials thereby creating an efficient solar harvesting via highly folded architectures. The simplest device showing the D–A interface is a BJH solar cell based on poly (3-hexylthiophene): [6,6]-phenyl C_{71} butyric acid methyl ester (P3HT:PCBM) blended films as shown in Fig. 3, where P3HT, PCBM and poly(3,4-ethylenedioxythiophene):poly(styrenesulfo nate) (PEDOT:PSS) acts as the electron donor, electron acceptor and hole transport layer, respectively. Although ITO/FTO coated rigid glass substrate based BHJ solar cells employing P3HT:PCBM blends have shown efficient power conversion efficiencies (η) up to ~6 %, the complete processing advantages of the organic materials can be fulfilled only by using the flexible substrates [52–56]. Potential use and roll-to-roll manufacturing process of the flexible substrates have stimulated rich interests in terms of manufacturing throughput and economy [57]. It should be noted that the observed performances of the flexible PSCs comprised of P3HT:PCBM are comparatively lower in contrast to those having rigid ITO/FTO electrodes depending on the sheet resistance [58–61].

The challenge to improve the ultimate efficiency of these BHJ solar cells basically includes the approaches to overcome the limiting factors in the devices such as—(i) charge transport explained by inefficient hopping (ii) structural traps hindering the electron transport, (iii) isolated domains and (iv) phase-segregation mediated incomplete pathways in the random percolation networks etc [62, 63]. In this regard, several attempts have been carried out to generate continuous pathways within the components of the BHJs in order to enhance the charge transport efficiently to the electrodes prior to the recombination process. One such approach to overcome the electron transport bottleneck is to substitute the zero-dimensional (0-D) fullerene based PCBM with other allotropes of carbon, such as one-dimensional (1D) carbon nanotube (CNT) or graphene, as the electron acceptors in the devices [64].

2.1 Graphene as the Electron Acceptor Material

As discussed earlier, the electron transport property of graphene can be utilized by applying graphene as an electron acceptor material in the polymer composites to enhance the charge transport in the photovoltaic devices. Liu et al. [65] have fabricated a PSC integrating P3HT as the electron donor and graphene-doped PCBM as the electron acceptor yielding a PCE of ~1.4 %, which was found to be 59 % enhanced PCE as compared to that of the identical graphene-free device. Enhanced performance of the graphene loaded device is accredited to the prolonged exciton diffusion area and faster electron transfer through graphene. However, Chen et al. [66] have unexpectedly observed the lower efficiency (<0.5 %) for the PSCs employing CNTs as the electron acceptor material owing to a few unfavorable factors of CNTs such as solubility, impurity and bundling structures eventually hindering the device performance. Indeed, the PSCs incorporating graphene in its solution-processable functionalized form as the electron acceptor material against poly(3-octylthiophene) (P3OT) or P3HT donor polymers have shown a maximum PCE of ~1.4 %, which is moderately comparable to the best performing PSCs employing non-fullerene based electron acceptors (Fig. 4) [16, 67].

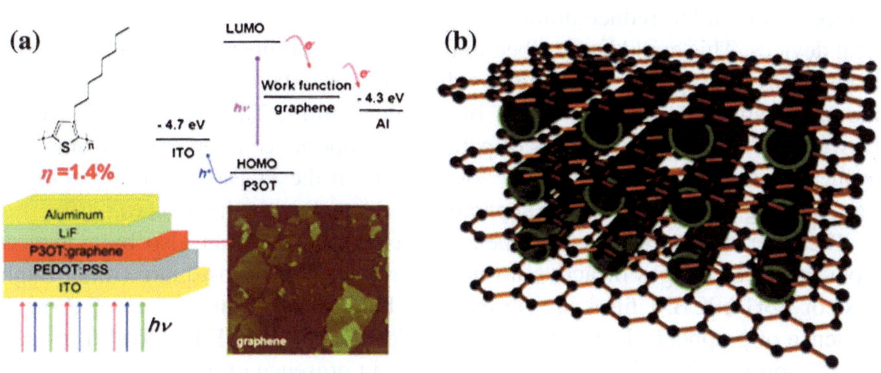

Fig. 4 **a** The scheme of an organic photovoltaic cell using graphene as the electron acceptor material. **b** The model of the 3-D hybrid graphene/CNT material [109]. Copyright 2012. Reproduced with permission from American Chemical Society

In their report, the electron transport from the electron donor materials to graphene, attributed to the higher electron mobility (~7×10^4 cm^2/V s) of graphene, was confirmed by significant quenching of fluorescence. The quenching in fluorescence emission of the donor-graphene composite is attributed to a faster decay of the excited state electron lifetime of the donor molecule in presence of graphene. Similarly, Liu et al. [68] have also reported an efficiency of ~1.4 % utilizing solution-processed functionalized graphene (SPF Graphene) in D–A type based PSC. Chen et al. [66] have revealed the combined performance of two carbon allotropes, namely, SPF Graphene and functionalized single walled CNTs (f-SWCNTs), in the active layer of the device ITO/PEDOT:PSS/P3HT-f-MWCNTs-SPF Graphene/LiF/Al. Even with a maximum efficiency of ~1.05 %, the device presented the mutual intriguing properties of SPF Graphene as the electron acceptor providing an electron percolation path as well as f-SWCNTs as the competent hole transporter. Another methodology to enhance of the device efficiency, as shown by Yu et al. [69], is to employ the PCBM–graphene hybrid as an electron acceptor, where in the chemical interaction of graphene with PCBM altering the HOMO–LUMO energy levels is expected. In the presented work, the fullerene-grafted graphene composite material, obtained using lithiation reaction, have been used in aP3HT based PSC to exhibit an efficiency of ~1.22 %. Interestingly in a very recent report, Chauhan et al. [70] have demonstrated a simple process of ultrasonic exfoliation of graphene flakes and the resultant thinner graphene sheets to be interacted with PCBM molecules via the π–π interactions to form their composite. These PCBM–graphene sheet composite was further used to fabricate a flexible BHJ solar cell exhibiting a promising efficiency of ~2.51 % using ITO-coated polyethylene naphthalate (PEN) substrates, P3HT electron donor and PEDOT as the hole transport material (Fig. 5).In the fabricated devices, the PCE of ~1.23 % for the graphene-free device (ITO/PEDOT/P3HT:PCBM/Al) has been observed to enhance to ~2.51 % for the graphene-based device (ITO/PEDOT/P3HT:PCBM–graphene sheet/Al). In the control device ITO/PEDOT/P3HT:PCBM–graphene flake/Al using simple blending of PCBM and graphene flake, the PCE performance was found to reduce dramatically even from that of the graphene-free identical devices. This could be ascribed to the absence of proper interfacial interaction in the PCBM–graphene flake blend, which impedes the charge migration due to the aggregation of graphene flakes in the blend films. The complete miscibility between PCBM and graphene is of paramount importance to ensure a strong electronic interaction between the two components in the PCBM–graphene composites resulting an efficient charge separation and transport in the devices.

The interactions between graphene sheets and PCBM can also be evidenced from the AFM images of active layers, where the root-mean-square (rms) roughness of P3HT:PCBM films are observed to be reduced from ~4.2 to ~3.2 nm in presence of graphene sheets. The typical AFM images of P3HT:PCBM films confirm the promoted packing of the active layer in presence of graphene sheets. The interaction between the P3HT:PCBM and graphene can also be established from the Raman spectroscopy by comparing the characteristic Raman peaks for P3HT (~1,380, 1,450 cm^{-1}), PCBM (~1,430, 1,465, 1,573 cm^{-1}) and graphene (D band

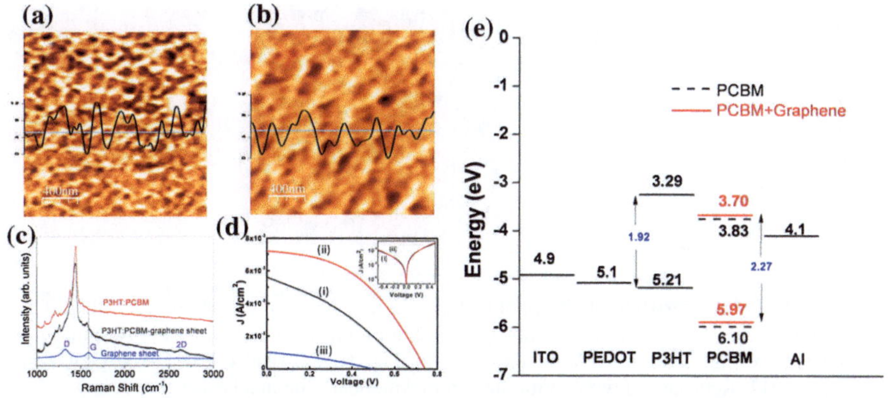

Fig. 5 AFM images of active layer blend films of **a** P3HT:PCBM and **b** P3HT:PCBM-graphene-sheets, **c** Raman spectra recorded for the films of pure graphene-sheets, P3HT:PCBM and P3HT:PCBM-graphene composite, **d** current–voltage characteristics for the bulk heterojunction solar cells fabricated using different active layers: (*i*) P3HT:PCBM, (*ii*) P3HT:PCBM-graphene-sheets and (*iii*) P3HT:PCBM-graphene-flakes. The inset shows the dark current for (*i*) P3HT:PCBM and (*ii*) P3HT:PCBM-graphene-sheets based devices, **e** energy level diagrams for the solar cell structures: ITO/PEDOT/P3HT:PCBM/Al and ITO/PEDOT/P3HT:PCBM-graphenesheets/Al [70]. Copyright 2014. Reproduced with permission from American Institute of Physics

for sp^3 carbon and G band for sp^2 carbon). The presence of strong electronic interaction between PCBM and graphene shows a downshift of the characteristic G band for graphene. This strong interaction between P3HT:PCBM and graphene sheets leads to an efficient dissociation of excitons at the D–A interface and collection of the separated charges at the electrodes. The energy difference between the excitonic state and charge separated excited state mediates the exciton dissociation and is associated with the difference in the LUMO levels of the donor and acceptor. Therefore it should be mentioned that in the presence of graphene sheets, LUMO level of PCBM–graphene composite is up fielded by ~0.13 eV towards the vacuum level, which should theoretically oppose the exciton dissociation. However, the lesser exciton dissociation triggered by graphene is compensated by the enhanced charge carrier mobility achieved by the improved morphology of the D–A interface due to the presence of graphene sheets. The closely packed structure of the P3HT:PCBM–graphene blend enriched with higher degree of ordering and phase separation offers the efficient charge transport within the device. Based on these effects, the current density–voltage (*J–V*) characteristics of the devices have shown higher short-circuit current (J_{sc}) and open-circuit voltage (V_{oc}), yielding higher PCE for the graphene sheet based device.

Interestingly, graphene in the quantum dot (QD) form also exhibits excellent optical and electrical properties in conjugation with high specific surface area and electron mobility as well as tunable band gap. The unique properties of GQDs can be explored as the electron acceptor material in solar cells. Gupta et al. [42]

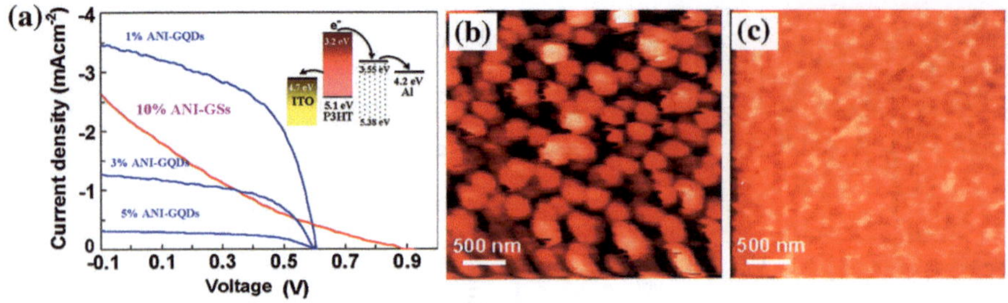

Fig. 6 **a** *J–V* characteristics of the photovoltaic devices based on ANI-GQDs with different GQDs content and ANI-GS (under optimized condition) annealed at 160 °C for 10 min, in AM 1.5 G 100 mW illumination. AFM images of **b** P3HT/ANI-GSs and **c** P3HT/ANI-GQDs [42]. Copyright 2011. Reproduced with permission from American Chemical Society

have demonstrated the application of aniline functionalized GQDs (ANI-GQDs) as the electron acceptor in the device ITO/PEDOT:PSS/P3HT:ANI-GQDs/LiF/Al and compared the performance with respect to the graphene sheet (ANI-GS) based control devices. A simple hydrothermal route was carried out to synthesize the ANI-GQDs and the effect of 0.5, 1.0, 3.0, 5.0 wt% GQD loading to P3HT was tested in the devices. A maximum PCE of ~1.14 % was accomplished with 1.0 wt% GQD loading to P3HT, which was higher than the identical device loaded with graphene sheets. The photovoltaic performance of the cells is governed by the transport of photo generated charge carriers through the P3HT/ANI-GQD blend. In contrast to the larger domains (~100–200 nm in diameter) representing a large scale phase separation in P3HT/ANI-GSs film, using AFM image, P3HT/ANI-GQDs offer a uniform film the phase separation at the nano regime enhancing the migration of excitons to the D–A interface with less resistance. However, the superior performance of GQDs was observed to decrease at the higher loading of GQD in the fabricated solar cells. A maximum J_{sc} of ~3.5 mA/cm^2 with a V_{oc} of ~0.61 V and a FF of ~53 %, affording the PCE of ~1.14 % were achieved by the GQD based cell as shown in Fig. 6. An alternate to the 2D graphene sheets, 0-D GQDs can also be electrochemically prepared in order to facilitate their application in the nanodevices. Based on this idea, Li et al. [40] have reported an electrochemical conversion technique of 2D to 0-D approach for direct formation of functional GQDs with a uniform size distribution of ~3–5 nm and stability of several months in water. A PCE of ~1.28 % was achieved using these green luminescent GQDs as the electron acceptors in the fabricated PSCs with structure ITO/PEDOT:PSS/P3HT:GQDs/Al.

The critical prerequisite to demonstrate the overwhelmed performance of graphene sheets or QDs as the additive/dopant to the acceptor material is the proper interaction of graphene to the donor material. In this context, a functionalized form of graphene can reinforce its binding interactions with the electron acceptor materials more efficiently. The chemical conversion of graphene to produce GO appears to be a convenient approach to introduce various functional groups (–COOH, –OH, epoxy etc.) to the basal planes of graphene, which enhances its

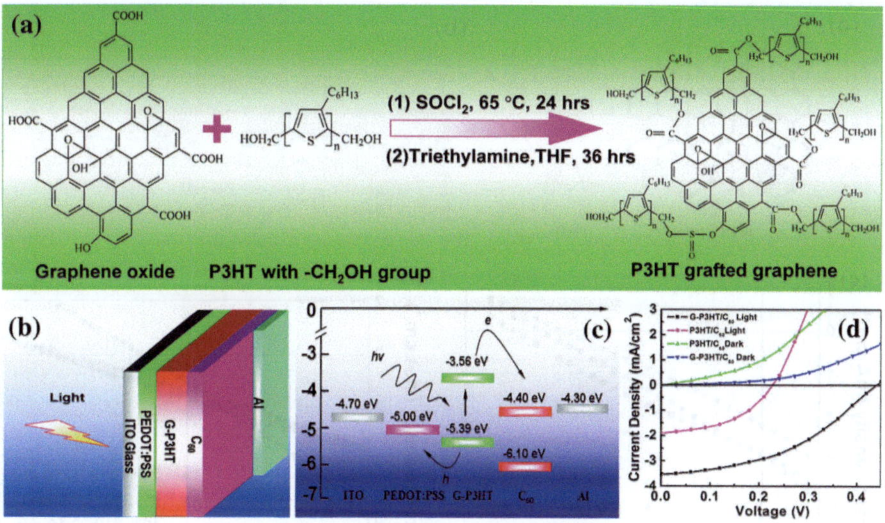

Fig. 7 Synthesis procedure for chemical grafting of CH_2OH-terminated P3HT chains onto graphene, which involves the $SOCl_2$ treatment of GO (step 1) and the esterification reaction between acyl-chloride functionalized GO and MeOH-terminated P3HT (step 2). **a** Schematic and **b** energy level diagram of an ITO/PEDOT:PSS/G-P3HT/C60/Al photovoltaic device, **c** current–voltage characteristics of the photovoltaic devices using P3HT/C60 or G-P3HT/C60 as the active layer [16]. Copyright 2010. Reproduced with permission from American Chemical Society

covalent anchoring ability with the acceptor molecules. As a result of this, GO can be incorporated to PCBM in the devices to achieve higher current and conductivity. Hill et al. (2011) have shown an enhancement in J_{sc} and conductivity upon GO binding to P3HT in the device ITO/PEDOT:PSS/P3HT:GO/Al, but the V_{oc} decreased [71]. In another report, Yu et al. (2010) have fabricated a bilayer photovoltaic device, as shown in Fig. 7, involving the P3HT covalently grafted GO sheets (GP3HT) heterostructures as the acceptor material and observed a PCE of ~0.61 % [16]. The solution processable GP3HT involves the chemical grafting of *regio*-regular P3HT onto GO sheets via the ester linkage between the –COOH groups of GO and $–CH_2OH$ terminal groups of P3HT. The ester linkage stimulates the blend to be highly soluble in organic solvents favoring the structure/property characterization and the solution processing. The bilayer device ITO/PEDOT:PSS/G-P3HT/C60/Al using solution-cast GP3HT and thermally evaporated C_{60} showed an increment of 200 % in contrast to the GO-free device. As illustrated in Fig. 7, the improved performance of the GP3HT based cell in presence of GO is attributed to the improved charge transport/collection concomitant with the GO sheets leading to an enhanced J_{sc} value from ~1.9 to ~3.5 mA/cm^2. Moreover, the reduction of the HOMO energy level of P3HT upon ester linked grafting to GO leads to an increase in the V_{oc} value from ~0.23 to ~0.43 V in the fabricated devices.

Similarly, Liu et al. [72] have also shown a PCE of ~1.1 % with a maximum J_{sc} of ~4.0 mA/cm^2 and V_{oc} of ~0.88 V using SPFGraphene–P3HT composite in the

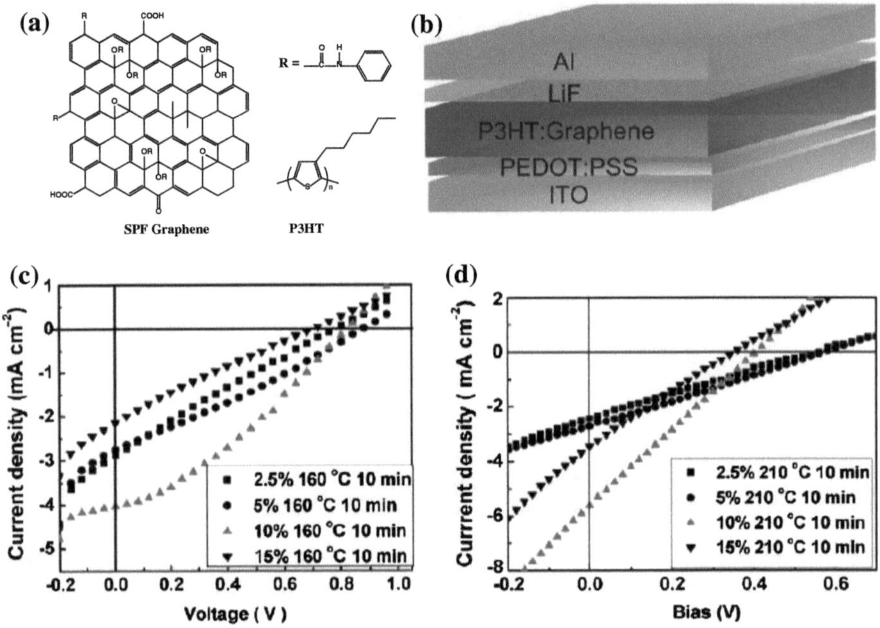

Fig. 8 **a** The schematic chemical structure of SPF Graphene and P3HT. **b** Schematic structure of the devices with the P3HT/SPF Graphene thin film as the active layer; ITO (~17 Ω/sq)/PEDOT:PSS (40 nm)/P3HT:SPF Graphene (100 nm)/LiF (1 nm)/Al (70 nm). **c**, **d** *J*–*V* characteristics of the P3HT/SPF Graphene-based photovoltaic devices with different graphene contents (2.5, 5, 10, and 15 wt%) under a simulated AM1.5G 100 mW illumination with annealing at 160 °C for 10 min and at 210 °C for 10 min [72]. Copyright 2009. Reproduced with permission from John Wiley & Sons

active layer of the device ITO/PEDOT:PSS/P3HT:graphene/LiF/Al as depicted in Fig. 8. In this study, the graphene based device performances have been tested by varying the amounts of graphene (in wt%) loaded to that of P3HT and observed a maximum PCE at 10 wt% graphene loading along with a post annealing treatment at 160 °C, while increasing the annealing temperature to 210 °C has shown a reduction in PCE up to ~0.57 %. The SPF Graphene–P3HT composite was synthesized by following the chemical functionalization route, wherein, the mixing of water-soluble graphene to the organic-soluble P3HT is promoted through a chemical reaction process, which is otherwise a challenging task due to the water–solvent immiscibility. Stankovich et al. (2006) have demonstrated an idea to achieve higher conductivity and low percolation threshold in a graphene–polystyrene composites via simple mechanical mixing [4]. Following the same, a chemical method was applied by Liu et al. (2009) to synthesize the P3HT–SPFGraphene composites in two steps—oxidation of graphene to produce GO, followed by its ester-linked organic functionalization. In the first step, graphene is oxidized using strong oxidizing agents (H_2SO_4, $KMnO_4$) following Hummers or modified-Hummers method, which cuts the graphene sheets into small pieces of several hundred nanometers along with the introduction of polar functional groups such as –COOH, –OH, epoxy, –C=O and –CH_2– into the basal planes of the graphene sheet [73, 74].

Certainly the presence of these highly reactive oxygen functionalities in GO is vital for further chemical functionalization with polymers; however, they introduce hydrophilicity in the GO sheets that hinders their direct application as the fillers in the hydrophobic polymer matrices which are soluble only in organic solvents. In order to anchor the GO sheets into water insoluble polymer matrices, the hydrophilic surface properties of GO can be altered into the hydrophobic nature using an organic isocyanate (phenyl isocyanate), which ultimately proclaims the solubility of GO in organic solvents such as 1,2-dichlorobenzene [75]. This technique offers a homogenous dispersion of graphene sheets in the organic matrices, particularly conjugated polymer molecules, to use the composites in the fabrication of solution-processable photovoltaic devices. It should be noted that a post annealing treatment at an optimum temperature (~160 °C for 10 or 20 min) is often performed after the deposition of the active layer in the typical solution-processable P3HT–graphene based PSCs. The post-annealing treatment usually removes the organic functional groups from the graphene sheet thereby enhancing the recovery of the π–π conjugation and charge carrier mobility of the graphene sheets within the composite of the device. However, prolonged annealing at an optimum temperature or annealing at higher temperature might alter the work function of graphene and raise its Fermi level, which eventually decreases the performance of the SPF Graphene–P3HT based devices because their V_{oc} is directly governed by the difference between the work function of graphene and the HOMO level of P3HT.

In the similar manner, another approach involving an organic synthetic route via the 1,3-dipolar cycloaddition reaction of azomethine ylide could also be utilized for the polymeric functionalization of RGO [76–78]. Herein, a highly soluble functional material incorporating the π conjugated polymer and RGO can be achieved from the reaction between the sarcosine, RGO and polymer having aldehyde groups. Li et al. (2012) have demonstrated the synthesis of a new solution-processable poly[(9-phenyl-9H-carbazole){4,7-di(thiophen-2-yl)benzo[c] [1,2,5] thiadiazole}(9,9-dihexyl-9H-fluorene)] (PCTF) polymer material covalently grafted with RGO to form the functional derivative PCTF-RGO (Fig. 9) [79]. The functional composite PCTF-RGO is advantageous due to its ability to offer the binary requisites for the device applications—(i) D–A character facilitated by the electron accepting benzothiadiazole unit and the electron donating carbazole, thiophene and fluorene moieties and (ii) RGO as the hole transporting material. The cell configuration ITO/PEDOT:PSS/PCTF-RGO/LiF/Al has shown a PCE of ~0.01 % probing the photovoltaic performance of PCTF-RGO. A blend of PCTF and [70]PCBM ([6]-phenyl C71 butyric acid methyl ester) in a 1:3 w/w ratio in the device has exhibited a PCE of ~1.14 %, whereas the identical device in presence of RGO shows a PCE up to 1.45 % especially showing a considerable enhancement in the photocurrent (J_{sc}) from ~3.94 to ~4.68 mA/cm^2. This elucidates the primitive role of RGO in the device performance directly through an enhancement in the number of photons absorbed by the active layer, favorable generation of mobile carriers due

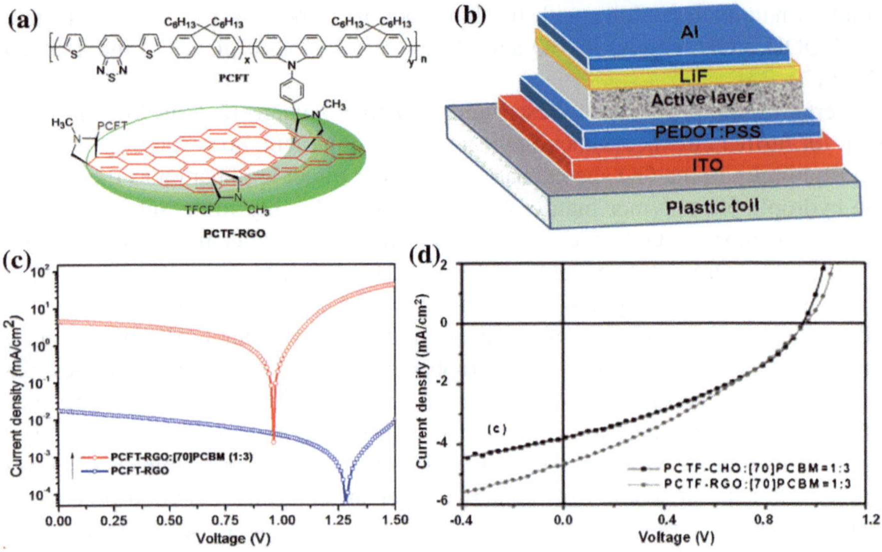

Fig. 9 **a** Molecular structure of PCFT-RGO; **b** the device structure; and **c** current density–voltage characteristics (*J–V*) of the PCTF-RGO-based photovoltaic device with and without [70]PCBM under a simulated AM1.5G 100 mW illumination, **d** *J–V* of the device using PCTF-CHO: [70]PCBM or PCTF-RGO: [70]PCBM as active layer [79]. Copyright 2012. Reproduced with permission from John Wiley & Sons

to D–A interfacial charge separation as well as the accelerated charge collection at the two electrodes. Thus an increment of ~259 % in the J_{sc} and ~144 % in the PCE has been accredited by the PCTF-RGO: [70]PCBM in contrast to the PCTF-RGO counterpart. The remarkable improvement in the J_{sc} value enhancing the PCE of the devices (with and without [70]PCBM) can be directly correlated to the higher number of photons absorbed by RGO and [70]PCBM.

In view of the solar energy conversion mechanism that has been understood so far, the most essential criteria to bring all the solution processable PSC technology one step further in terms of the efficiency improvement is the development of electron transport. It is thereby preferable to perform the possible modification to the electron transport layers mediated through a solution processable route for the development of printable PSCs. In this regard, a recent approach by Jayawardena et al. [80] have shown a solution processable metal oxides–RGO based electron transporting interfacial layer in the PSCs (Fig. 10). The performance of the ZnO–RGO and TiO_2–RGO interfacial layers has been explored in the PSC involving a low band gap [3,4-b] thiophene/benzodithiophene (PTB7) donor polymer and [70]PCBM acceptor. The metal oxide–RGO devices yielded a maximum PCE of ~7.5 %, whereas the similar RGO-free devices have shown a PCE of ~7.2 %. The enhancement in J_{sc} (exceeding ~15 mA/cm^2) observed in presence of RGO is attributed to the characteristic charge transport property provided by RGO as well as the reduced series resistance (R_s) of the metal oxide–RGO counterpart in

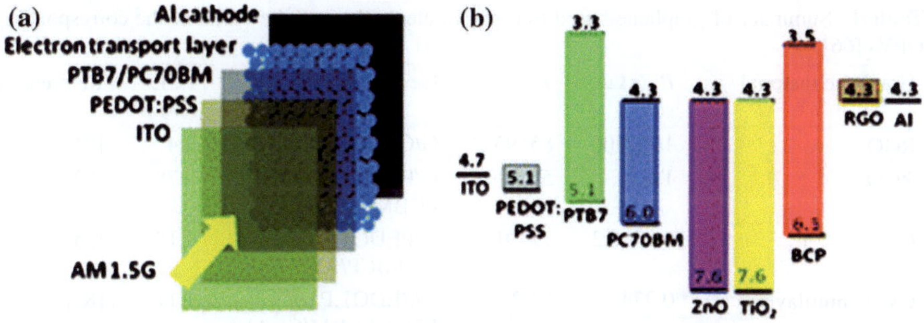

Fig. 10 **a** Schematic of the energy levels of ITO, PEDOT:PSS, PTB7, PC70B, ZnO, TiO_2, RGO, BCP and Al, **b** schematic of the device architecture employed. The *boxed areas* in (**b**) indicate the different solution processed layers used in the device [80]. Copyright 2013. Reproduced with permission from Royal Society of Chemistry

the device as compared to bare ZnO and TiO_2. Moreover, the metal oxide–RGO assisted improvement in the *FF* of the devices approaching ~68 % is ascribed to the balanced electron and hole mobilities. The core–shell type metal oxide–RGO composite also offers the Ohmic contact of RGO with the metal oxide, which improves the charge extraction properties between the metal oxide nanoparticles.

2.2 *Graphene as the Counter Electrode*

Ideal counter electrode for the photovoltaic device should possess high transparency (>80 %) over wider wavelength range, lower sheet resistance (R_s < 100 Ω/sq) and suitable work function of ~4.5–5.2 eV. Although at present ITO coated glass or flexible polymer substrates having transparency (>90 % at 550 nm) and R_s of ~10–30 Ω/sq are mostly being used as the counter electrodes for PSCs, efforts are being made to develop the new generation transparent electrodes based on graphene with low-cost, unique optical property, chemical robustness and high flexibility. The superior electron transportability of graphene can be exploited by using it as a transparent counter electrode material for the PSCs. A few of the recent advances in the organic solar cells using graphene as the counter electrode are shown in Table 1.

In principle, the transmittance and electrical conductivity of graphene sheets depend on the thickness of the film. In case of one-atom thick layer of graphene, the electrons in its individual sheet gets delocalized over the complete sheet providing a ballistic electron transport as well as a lower light absorption [81]. Chen et al. [66] have demonstrated a favorable transmittance of >80 % (400–1800 nm) for the single layer graphene sheets suitable for the device applications, but with a high R_s of ~10^2–10^3 Ω/sq as compared to that for the ITO electrodes. Unlikely for the practical applications, large area graphene film fabrication using solution processing route leads to an increase in the electrical resistance of the

Table 1 Summary of graphene-based transparent electrodes and the PCEs of the corresponding OPVs [66]

Graphene material	R_s (kΩ/sq)	T (%)	Device structure[a]	PCE (%)	References
RGO	100–500	85–95	G/CuPc/C60/BCP/Ag	0.4	[81]
RGO	1.6	55	G/PEDOT:PSS/P3HT:PCBM/TiO_2/Al	0.78	[83]
CVD	0.23–8.3	72–91	G/PEDOT:PSS/CuPc/C60/BCP/Al	1.18	[85]
CVD–multilayer	0.374	84.2	G/PEDOT:PSS/P3HT:PCBM/Ca:Al	1.17	[87]
CVD–$AuCl_3$ doped	0.5–0.3	97.1, 91.2	G/PEDOT:PSS/CuPc/C60/BCP/Ag	1.63	[88]
CVD–acid doped	0.08	90	G/MoO_3/PEDOT:PSS/P3HT:PCBM/LiF/Al	2.5	[89]
CVD–organic molecules doped	0.278	92.2	G/PEDOT:PSS/P3HT:PCBM/Ca:Al	2.58	[91]
RGO/CNTs	0.24	86	G/PEDOT:PSS/P3HT:PCBM/Ca:Al	0.85	[92]

[a] *G* represents graphene or its derivatives (including reduced graphene oxide (*RGO*), doped graphene, or multilayered graphene, etc.)-based transparent electrodes

material due to the generation of multiple grain boundaries, lattice defects and oxidative traps. The sheet resistance can be fine-tuned to minimize its value to a practical limit by making thicker layers or a multilayer stacking approach—while doing so, the transmittance of the graphene film is reduced. A photovoltaic cell based on graphene anode offered a PCE of only ~0.4 % (~0.84 % for ITO based identical device) owing to the high resistivity of graphene, provided the thicknesses of ~4 and 7 nm of graphene film showed a transmittance of ~95–85 % and a R_s of ~100–500 kΩ/sq, respectively. Even if the RGO sheets has the high resistivity as compared to the corresponding ITO substrates, it offers a unique advantage of excellent flexible nature for the practical applicability as the transparent flexible electrodes in PSCs. Yin et al. [82] have shown a transparent (~55 %) and conductive (R_s ~1.6 kΩ/sq) electrodes made of RGO on flexible polyethylene terephthalate (PET) substrate and achieved a PCE of ~0.78 % as shown in Fig. 11. Even though the observed PCE of the highly flexible device was inferior in contrast to the ITO counterpart, the device sustained a thousand cycles of bending at a tensile strain of ~2.9 %, whereas the flexible ITO-based devices cracked down upon bending.

In order to enhance the efficiency of the graphene electrode based cells, the resistivity of graphene could be further reduced by applying a chemical vapor deposition (CVD) method, which could yield high quality graphene in-build with high transmittance and electrical conductivity [83]. Typically the multi-layered smooth graphene films as the flexible transparent electrode can be fabricated by the CVD approach in large-area with respect to the device fabrication as reported by Arco et al. [84]. They have obtained continuous graphene films with a surface

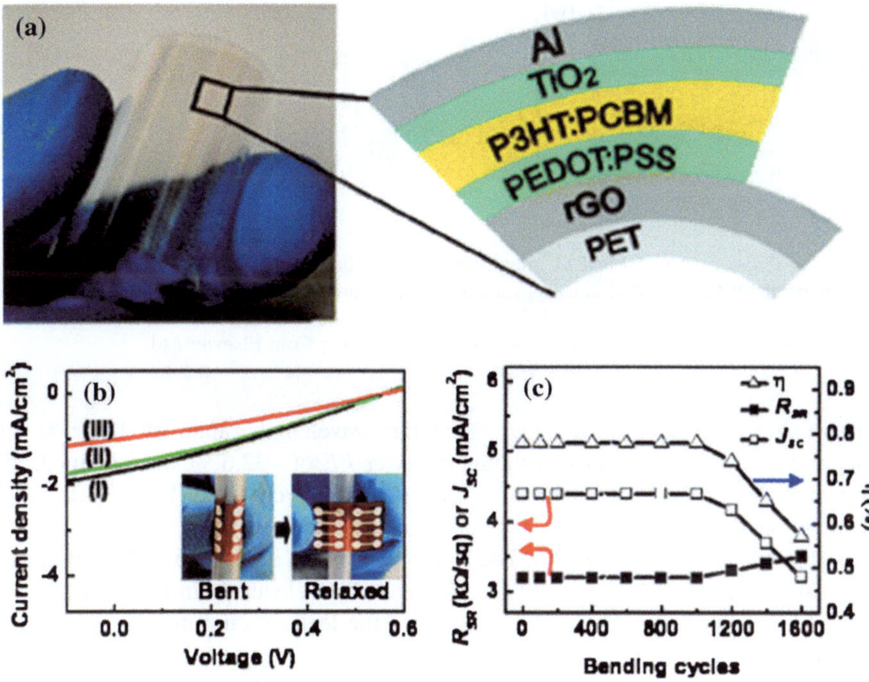

Fig. 11 **a** Photograph of RGO/PET and the schematic representation of the layer structure of the device, i.e., RGO/PEDOT:PSS/P3HT:PCBM/TiO_2/Al, with RGO as the transparent electrode; **b** current density–voltage (*J–V*) curves of the device after applying (*I*) 400, (*II*) 800, and (*III*) 1200 cycles of bending; **c** the short-circuit current density (J_{sc}), overall power conversion efficiency (η), and the sheet resistance (R_{SR}) of the device as a function of bending cycles [82]. Copyright 2010. Reproduced with permission from American Chemical Society

roughness of only ~0.9 nm. Notably, a comparable PCE of ~1.18 % in contrast to ITO based device (PCE ~1.27 %) has been demonstrated, where RGO shows a minimal resistivity (R_s) of ~230 Ω/sq at ~72 % transparency. In addition, the graphene-based device has shown a superior flexibility to operate even at a bending angle of 138°, while the ITO-based devices have cracked and irrevocably failed at a maximum bending tolerability of 60°. The sheet resistance of graphene is governed by the relation $R_s = (\sigma_{2D}N)^{-1}$, where σ_{2D} is the 2D sheet conductivity and N is the number of layers [85]. We have already discussed the possibility to increase the number of layers via stacking multilayers of graphene, keeping in mind the retention of an optimum transparency of the film. Alongside, the conductivity of graphene films can also be enhanced by chemical doping thereby increasing the carrier concentration. Following the same, Choi et al. [86] have reported a PCE of ~1.17 % for the device MLG/PEDOT:PSS/P3HT:PCBM/Ca/Al (Fig. 12) fabricated on the CVD-grown multilayer graphene (MLG) films having a relatively lower R_s of ~374 Ω/sq. The 4-layered MLG films, as confirmed by its characteristic G and 2D bands obtained from Raman spectra, applied in the devices have shown

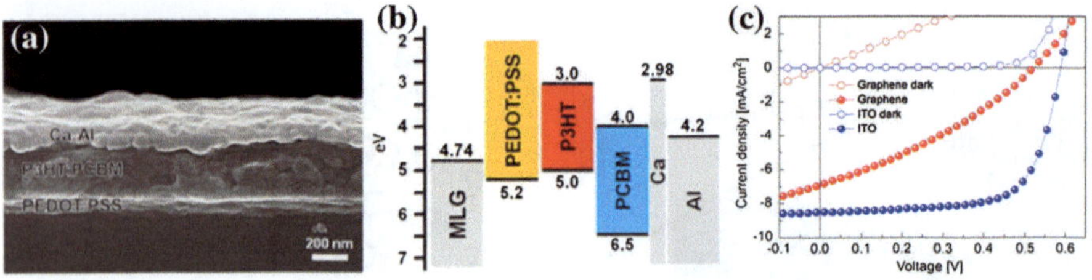

Fig. 12 **a** Cross-sectional FESEM image and **b** band diagram of a bulk-heterojunction OSC fabricated on MLG, **c** current density–voltage characteristics (AM 1.5 G with an incident light power intensity of 100 mW cm^{-2}) of OSCs fabricated on the MLG electrode and reference ITO electrodes [86]. Copyright 2012. Reproduced with permission from Elsevier Ltd.

a promising transparency of ~84.2 % at the wavelength range of 400–800 nm comparable to ITO substrates. Although a lower *FF* of ~32.6 % was obtained, the fabricated device has demonstrated a J_{sc} of ~6.9 mA/cm^2 and V_{oc} of ~0.521 V, whereas the similar device with ITO electrode has yielded a J_{sc} ~ 8.5 mA/cm^2, a V_{oc} ~ 0.589 V and a *FF* ~ 68.4 %. The lack of formation of smooth and wrinkle-free MLG in contrast to the cost-intensive sputtered ITO electrode in the devices generated lower shunt resistance, which produced the leakage current and lowered the overall PCE.

In an earlier report by Choe et al. [87] a maximum PCE of ~1.3 % has been achieved using CVD-grown (at 1,000 °C) MLG films as the transparent conductive electrodes in the device MLG/PEDOT:PSS/P3HT:PCBM/Al. In their work, the effect of CVD growth temperature for MLG films in the device performance has been explored. It has been observed that at lower temperatures of 800 and 900 °C, the MLG films show higher transparency, but at the same time lowering of the sheet resistance depending on the defect density of graphene is also been observed. This was confirmed from the Raman peak intensity ratio analysis for the MLG films grown at the varying temperatures of 800, 900 and 1,000 °C, which reflected the increased film thickness, reduced optical transparency, decreased defect density and lowered sheet resistance of graphene with an increase in its CVD growth temperature (Fig. 13). Indeed, the sheet resistances of ~1,730 (±600), ~990 (±400) and ~610 (±140) Ω/sq and mobility values of ~660 (±270), ~1,030 (±440) and 1,180 (±260) cm^2/V s were observed for the MLG films grown at 800, 900, and 1000 °C, respectively—however, the work function (~4.45 ± 0.04 eV) of the MLG films, measured using Kelvin probe measurement, did not show any significant variation with the change in the growth temperature. Typically, a J_{sc} of ~6.25 mA/cm^2, aV_{oc} of ~0.62 V and a *FF* of ~0.35 was obtained by the fabricated device, which could be further enhanced to achieve J_{sc} ~9.03 mA/cm^2, V_{oc} ~0.60 V and *FF* ~0.48, yielding a PCE of ~2.6 % by applying a hole-blocking TiO_x layer between the P3HT:PCBM active layer and Al cathode. The added TiO_x layer executes its dual role in enhancing the device performance by being—(i) an optical spacer facilitating the charge transport that improves the J_{sc} value and (ii) a hole-blocking layer reducing the contact

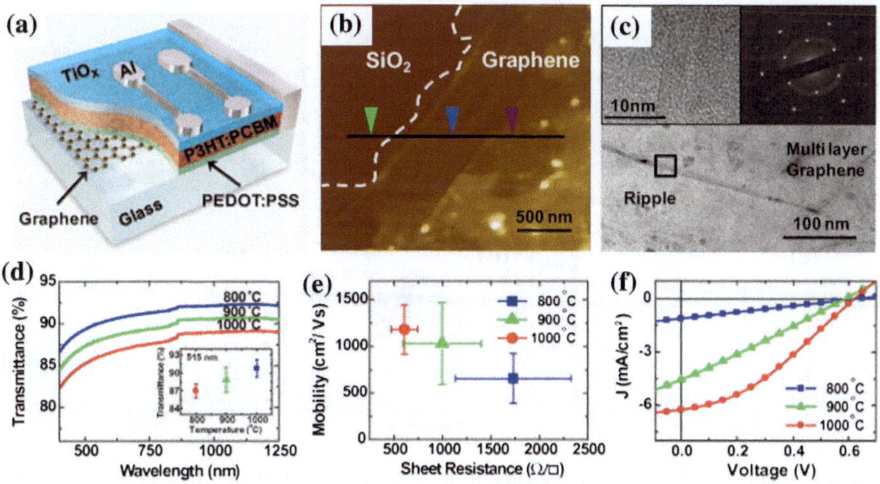

Fig. 13 **a** Schematic diagram of the photovoltaic device structure with MLG electrodes and a hole-blocking TiO_x layer. **b** AFM image of a MLG film showing the edge of the MLG layers (*dashed*) on a SiO_2 substrate. **c** TEM image of a region of MLG. The insets show an HRTEM image of the rippled regions **d** Transmission spectra of MLG films. The inset shows the transmittance of 90.7 ± 1.5, 88.8 ± 2.2, and 86.9 ± 1.2 % at 515 nm wavelength, 800, 900, and 1,000 °C, respectively. **e** Mobility versus sheet resistance of MLG films prepared at different growth temperatures (800, 900, and 1,000 °C). **f** *J–V* curves of photovoltaic devices with 1,000°C-grown MLG electrodes (*circles*) and with ITO electrodes (*diamonds*); the curves without the TiO_X layer (*filled symbols*) are compared to the ones with the TiO_X layer (*open symbols*) [87]. Copyright 2010. Reproduced with permission from Elsevier Ltd.

resistance between the active layer and Al counter electrode thereby increasing the *FF* of the devices.

As discussed earlier, a practical approach to improve the conductivity or reduce the resistivity is the stacking of multilayers of graphene or its chemical doping. An enhanced PCE of ~1.6–2.5 % has been recognized by Wang et al. [89] in the P3HT:PCBM based devices containing the acid-doped multilayer graphene as the anodes. Despite the use of HCl and HNO_3 doped 4-layer graphene multilayers having a R_s of ~80 Ω/sq at ~90 % transmittance (at 550 nm) has exhibited a PCE of ~2.5 %, which is ~83.3 % of the PCE of ITO-based cell, the devices suffered from the reliability issues. This is attributed to the fact that upon heating or electrical stressing, the acid dopants are released and ultimately diffuses through the P3HT:PCBM active layer thereby causing the stability problems in the device. In a similar device, reported by Lee et al. [90], top laminated 10-layer graphene electrode has been deposited using thermal releasing process simultaneously during the thermal annealing of the P3HT:PCBM layer and a PCE of ~2.5 % has been recorded. Apart from the acid dopants, organic molecule can also be chemically doped to graphene to enhance the conductivity of graphene electrode. For instance, Hsu et al. [91] have demonstrated a sandwiched type graphene/TCNQ/graphene (G/T/G) stacked films (Fig. 14), where

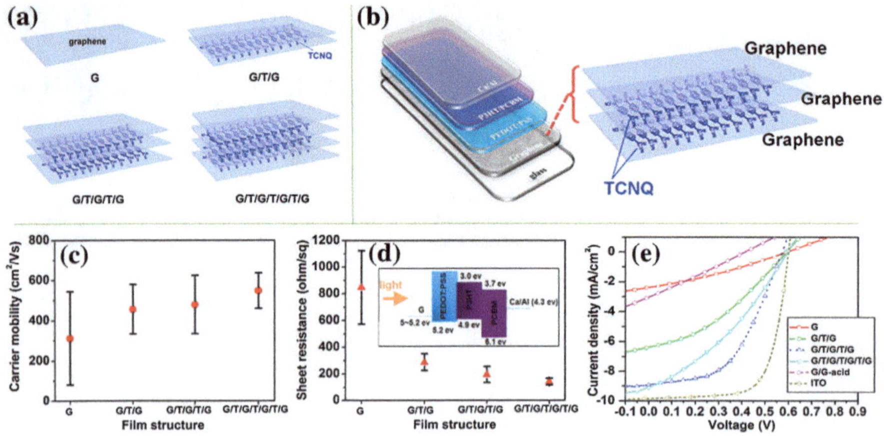

Fig. 14 **a** Four anode structures prepared using graphene and TNCQ, where G and T represent a graphene layer and a thermally evaporated TCNQ layer, respectively, **b** the schematic of the device architecture showing the G/T/G layers. Statistical Hall effect measurement results for these four anode films: **c** sheet resistance and **d** hole carrier mobility. Inset to **d** shows the scheme for the solar cell device with G as an anode, where the graphene anode is sitting on a glass substrate and PEDOT:PSS is spun on top of the graphene anode for enhancing hole transport. Ca/Al is used as the reflective cathode. A band diagram for the solar cell is also shown, **e** measured current density (*J*sc)–applied bias (*V*) curves of the fabricated devices [91]. Copyright 2012. Reproduced with permission from American Chemical Society

the tetracyanoquinodimethane (TCNQ) molecules were thermally evaporated onto the surfaces of graphene sheets in a controlled mode to obtain the layer-by-layer G/T/G films with tunable conductivities and investigated their role as the anode for the device performances. The intrinsic p-dopant character of TCNQ, due to the presence of highly electron deficient aromatic rings, induces an increased hole concentration in the multilayered graphene sheets and lowers the R_S value of graphene–TCNQ electrodes. As the number of layer-by-layer structure of graphene–TCNQ is increased, the R_S value of ~839 Ω/sq for monolayer graphene electrode is observed to decrease in the order ~629 Ω/sq (for G/T/G), ~278 Ω/sq (for G/T/G/T/G) and ~182 Ω/sq (for G/T/G/T/G/T/G). In view of the highest conductivity or the lowest R_S of the graphene–TCNQ films, the best performance is expected for the device involving G/T/G/T/G/T/G electrode, but the maximum PCE of ~2.58 % have shown by the device involving G/T/G/T/G—however, PCEs of ~1.37 and ~1.99 % has been achieved for G/T/G and G/T/G/T/G/T/G based devices. This could be attributed to the optimum optical transparency of ~92.2 % and an enhanced *FF* of ~0.48 with respect to the number of G/T layers.

In addition to the smaller molecules like TCNQ, poly(vinylidene fluoride-co-tri-fluoroethylene) (P(VDF-TrFE)) type ferroelectric polymers can also be used as the p-dopants of graphene [93]. The advantage of such polymers is the stability and durability under ambient conditions as well as the higher optical transparency of more than ~95 % from the visible to the near-infrared wavelength range due to the very low optical absorption of the ferroelectric polymer. Ni et al. [93] have reported a high loading capacity (~3×10^{13} cm^{-2}) of P(VDF-TrFE) as the dopant to the CVD grown monolayer graphene and achieved a low sheet resistance of ~120 Ω/sq.

In a different approach by Emmanuel et al. [94], laser-reduced GO (LRGO) films are deposited via spin coating on temperature sensitive substrates directly to fabricate the in situ flexible transparent conductive electrodes in one step. Conductive graphene films having a R_s value of ~700 Ω/sq at an optical transparency of ~44 % has been fabricated in a large area for the device applications. A PCE of ~1.1 % has been yielded by the flexible ITO-free device PET/LRGO/PEDOT:PSS/P3HT:PCBM/Al having LRGO film with R_s ~1.6 kΩ/sq at 70 % transparency.

2.3 Graphene as Hole Transporting Layer

The performance efficiency of PSCs is governed by their characteristic J_{sc}, V_{oc}, *FF* and cell life-time values, all the four are dependent on the interfacial properties between the active layer and electrodes as well as on the physical properties of the materials. The basic PSC device model encounters the direct electrical contact of the donor and acceptor layers with the respective electrodes, which leads to the possible carrier recombination and leakage current. In order to overwhelm such recombination and to suppress the current leakage, the electron blocking and hole transporting layers (HTLs), particularly the wide band gap p-type materials, are often deposited on the top of the transparent counter electrodes. The semiconducting polymer PEDOT:PSS is mostly employed as such anode interfacial layer that improves the anode contact as well as the hole collection. The solution processability of PEDOT:PSS is advantageous as it can reduce the surface roughness of ITO as well as the suitable alignment of the work functions between P3HT and ITO for smooth collection of holes. Notably, a few practical problems exists along with the use of PEDOT:PSS, namely, ITO corrosion at high temperatures facilitated by the highly acidic (pH ~ 1) aqueous suspensions of PEDOT:PSS, introduction of water into the active layer and the non-homogeneous electrical properties ensuing deprived long-term stability [95–97]. Even though several inorganic semiconductor metal oxides such as V_2O_5, MoO_3, NiO, etc. have also been tested as the HTL for the organic solar cells achieving PCEs of ~3–5 %, the requirement of cost-intensive vacuum deposition method has lead a way to search for the alternate solution processable materials for HTL. In this context, GO has received considerable attention as a robust material having mechanical flexibility and tunable electronic properties to use as a hole-transporting material [98–100]. Li et al. [101] have applied the chemically derived GO thin films (thickness of ~2 nm), deposited from the neutral aqueous suspensions, as the HTL in the devices wherein a maximum PCE of 3.5 % has been achieved with a J_{sc} ~11.4 mA/cm^2, a V_{oc} ~0.7 V and a *FF* ~0.54, while the identical device employing PEDOT:PSS showed a comparable PCE ~3.6 % and the GO-free device yielded a PCE of only ~1.8 %.

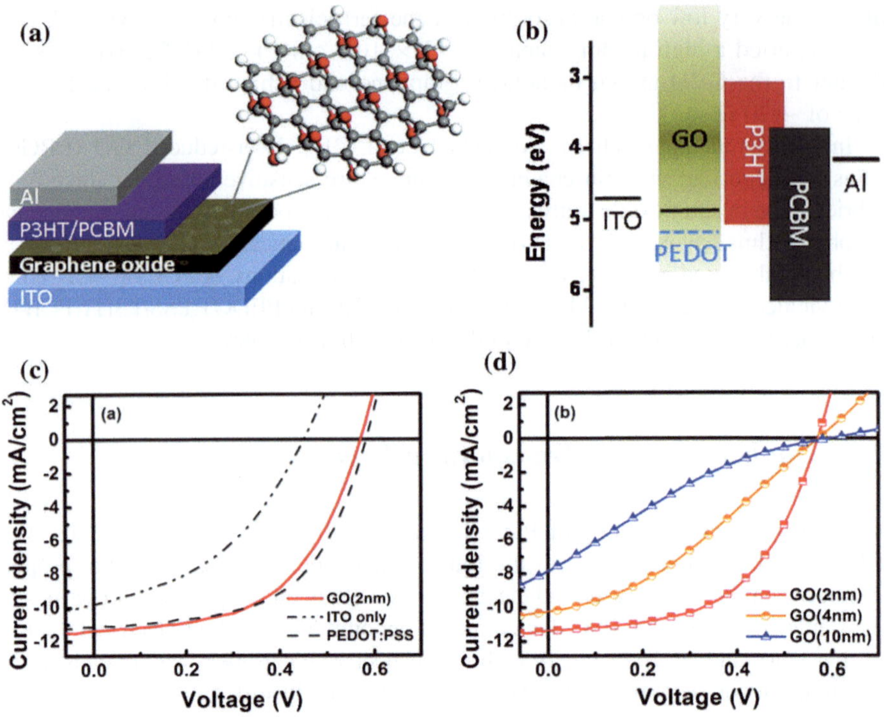

Fig. 15 **a** Schematic of the photovoltaic device structure consisting of the following: ITO/GO/P3HT:PCBM/Al, **b** energy level diagrams of the bottom electrode ITO, interlayer materials (PEDOT:PSS, GO), P3HT (donor), and PCBM (acceptor), and the top electrode Al, **c** current–voltage characteristics of photovoltaic devices with no hole transport layer (curve labeled as ITO), with 30 nm PEDOT:PSS layer, and 2 nm thick GO film, **d** current–voltage characteristics of ITO/GO/P3HT:PCBM/Al devices with different GO thickness. All of the measurements were under simulated A.M. 1.5 illumination at 100 mW/cm^2 [101]. Copyright 2010. Reproduced with permission from American Chemical Society

In the presented PSC with structure ITO/GO/P3HT:PCBM/Al as shown in Fig. 15, the work functions of GO and ITO matched well and the LUMO level of P3HT has also favored an effective hole transfer within the device. It should be mentioned that GO as the HTL has established longer stability in contrast to the PEDOT:PSS devices. The increase in the thickness of the GO layer from ~2 to ~10 nm has decreased the device performance, which could be ascribed to the increased sheet resistance as well as the optical transmittance along with the increased thickness. However, the spin coating deposition of smooth GO films with a thickness of ~2 nm resulting in single or bilayer GO film is a challenging task. On the other hand, the full coverage underlying ITO surface for a larger area device fabrication needs thicker GO films, which results in higher sheet resistance thereby lowering the J_{sc}, V_{oc} and FF of the devices. In this regard, Kim et al. [102] have introduced an idea to use the composites of GO and single-walled CNTs (SWCNTs) as the HTL for the PSC fabrication, where a PCE comparable to the PEDOT:PSS-based device has been recorded. The GO–SWCNT composite has significantly improved the vertical conductivity of GO without affecting its surface roughness even at the presence of a small amount of SWCNTs with a diameter of ~1 nm. Herein the present approach demonstrates the possibility to use slightly thicker and thus easier-to-make GO films as HTL via solution processed spin coating route. The substantial role of the SWCNTs in the device performance has been demonstrated by comparing the PCEs of the devices with the varying thickness of the GO films in presence and absence of SWCNTs. It was observed that the SWCNT-free devices have shown a decreased trend in the PCE from ~3.26 to ~2.36 % with the increase in the thickness of the GO layer from ~1 to ~3–4 nm, while the same involving the GO–SWCNT composite showed an efficiency decrease from ~3.66 to 3.13 %.

In another report, Yun et al. [103] have compared the performance of the photovoltaic devices ITO/AIL/P3HT:PCBM/Ca(Al) employing GO, a conventional RGO and a p-toluenesulfonyl hydrazide reduced GO (PRGO) films of different thicknesses as the anode interfacial layer (AILs) or HTLs as shown in Fig. 16. The maximum PCEs of ~3.63 and 3.53 % have been recorded for the PRGO based devices for ~2 and 5 nm thicker PRGO (HTL) film, respectively. In the fabricated devices, the PRGO films exhibited a higher conductivity values (R_s ~ 2.0–2.2 Ω cm^2) than the GO films (R_s ~ 16.5–483.4 Ω cm^2), while the conductivities of RGO films (R_s ~ 4.0–4.1 Ω cm^2) were found to be similar to PRGO. On the other hand, the film quality of PRGO films (rms roughness ~1.23–1.24 nm) was found to be better than that of RGO films (rms roughness ~2.65–3.39 nm) and similar to that of the GO films (rms roughness ~1.17–1.21 nm). Thus, depending on the combined effects of the sheet resistance as well as smoothness of the GO, RGO and PRGO films, a deprived PCE maxima of ~2.23 % for GO-based and ~2.83 % for RGO-based devices are observed in contrast to the PRGO-based devices (PCE ~ 3.53–3.63 %). It has been further investigated that apart from achieving a comparable performance efficiency, the PRGO-based PSCs have offered a superior long-term stability as compared to PEDOT:PSS counterpart.

Murray et al. [104] have investigated the role of an electronically tuned single layer GO credibly facilitating the device durability and replacing the traditional PEDOT:PSS as the HTL (Fig. 17). The fabricated GO based devices involving PTB7 as the donor and PC71BM as the acceptor materials have shown a promising PCE up to ~7.39 % with a J_{sc} of ~15.21 mA/cm^2, a V_{oc} of ~0.716 V and a *FF* of ~67.7 %, while the PEDOT:PSS device resulted in J_{sc} ~ 14.55 mA/cm^2, V_{oc} ~ 0.741 V, *FF* ~ 68.1 % yielding a PCE ~ 7.46 %, provided PTB7 is poly[[4,8-bis[(2-ethylhexyl)oxy]benzo[1,2-b:4,5-b′]dithiophene-2,6-diyl][3-fluoro-2-[(2-ethylhexyl)carbonyl]-thieno[3,4-b]thiophenediyl]] and PC71BM is [6]-phenyl-C71-butyric-acidmethyl-ester. Moreover, incorporating GO as the HTL in the devices have exhibited 5 times improvement in the thermal aging lifetime and 20 times enrichment in the humid ambient lifetime in contrast to the equivalent PEDOT:PSS-based devices. The observed rms roughness for PEDOT:PSS film was found to be ~3.5 nm, while the comparatively lower value of only ~0.7 nm for the Langmuir–Blodgett derived GO film reflects the lower

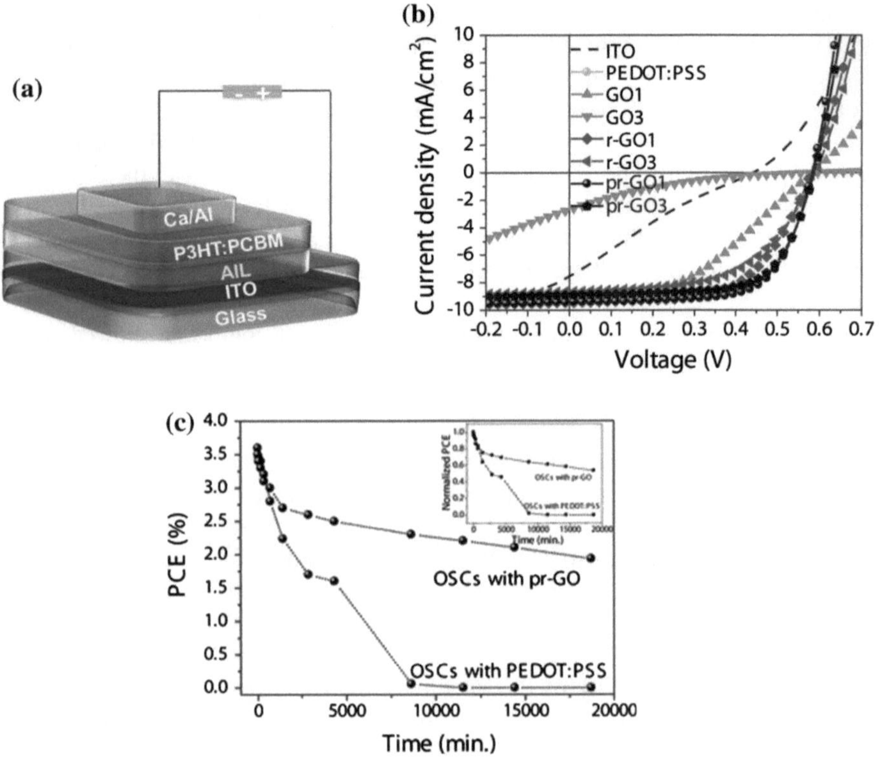

Fig. 16 **a** Schematic of the device structure ITO/AIL/P3HT:PCBM/Ca(Al), where AIL represents GO, RGO and PRGO. **b** The representative *J–V* curves for the devices with varying thicknesses of GO, RGO and PRGO films with the thicknesses measured as: GO1 (2 nm), GO3 (5 nm), RGO1 (4 nm), RGO3 (5 nm), PRGO1 (2 nm) and PRGO3 (5 nm). **c** Changes in PCE of a conventional PEDOT:PSS-based organic solar cell (OSC) and an OSC with the pr-GO anode interfacial layer during exposure to air [103]. Copyright 2011. Reproduced with permission from John Wiley & Sons

sheet resistance of the GO film. Also the GO films template PTB7 polymer face-on oriented π stacking more effectively as compared to PEDOT:PSS leading to an enhanced PCE for GO based cells.

2.4 Graphene Filled Hybrid Solar Cells

The growing interest of graphene based photovoltaic devices and cost-effective large-area synthesis of graphene composites upholds the current research attention on the photovoltaic technology one step forward by utilizing hybrid nanostructure based heterojunction assemblies. The unique physical properties of graphene in its pure, oxidized or reduced form can be incorporated in the inorganic–organic type

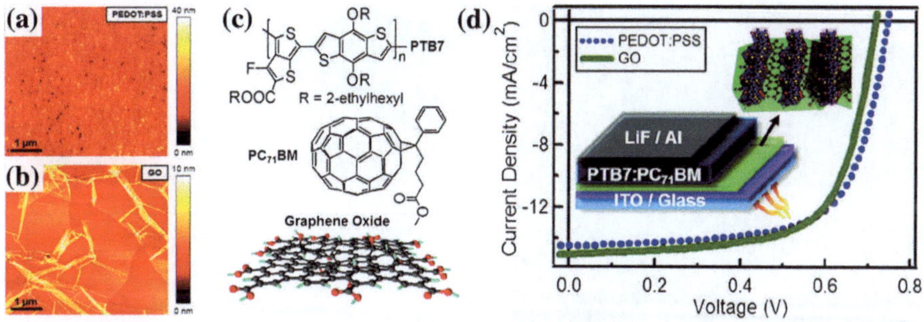

Fig. 17 AFM image of **a** a spin-cast and annealed PEDOT:PSS film with an rms roughness of 3.5 nm and **b** a GO film deposited by LB assembly, annealed, and treated with UV ozone (UVO) having an rms roughness of 0.7 nm, **c** chemical structures of the PTB7 donor polymer, PC71BM acceptor, and GO, **d** comparative photovoltaic performance of PTB7:PC71BM device with PEDOT:PSS or GO HTLs. Inset (*left*) shows the schematic of a standard OPV indicating the location of the IFLs. *Inset* (*right*) shows the structural model from GIXS data showing preferential face-on PTB7 π-stacking on ITO/GO substrates [104]. Copyright 2011. Reproduced with permission from American Chemical Society

hybrid solar cells. For instance, Park et al. [105] have shown a method to use the CVD grown graphene thin films (R_s ~ 300 ± 12 Ω/sq, transparency ~92.0 ± 0.4 % at 550 nm) on glass substrates as the conductive electrodes, onto which ZnO nanowires can be vertically grown using a low-temperature (90 °C) hydrothermal treatment (Fig. 18). Notably, the ZnO seed layer is needed to spin coat over the ITO or graphene sheet in order to provide a platform for the uniform growth of ZnO nanowires, but the lower surface free energy and hydrophobic nature of graphene causes poor wetting of the seed solution over the graphene surface. Therefore, the graphene surface is modified using two non-destructive conducting polymers (PEDOT:PEG and RG-1200), which eventually acted as the interfacial layers, over which ZnO seed layer is spin coated. Herein, the smoothness of the seed layer is the critical parameter for the uniform one dimensional growth of ZnO nanowires over the modified graphene surfaces. The graphene/PEDOT:PEG and graphene/PEDOT:PEG/ZnO films have shown the rms roughness of ~34 and ~24 nm, respectively. The graphene films were further smoothened when RG-1200 polymer was used resulting in the decreased rms roughness values of ~6 and ~2 nm for graphene/RG-1200 and graphene/RG-1200/ZnO films, respectively. Moreover, ZnO nanowires grown on the RG-1200 substrates have shown better nanowire alignment (order parameter, $S_{RG\text{-}1200}$ = 0.992) than PEDOT:PEG substrates ($S_{PEDOT:PEG}$ = 0.938) and a comparable alignment to the nanowire arrays grown on ITO (S_{ITO} = 0.997), which is in good agreement with the smoothness observed for the as-grown films. The P3HT based hybrid solar cells involving graphene/RG-1200/ZnO have shown a maximum PCE of ~0.5 % along with a J_{sc} of ~2.4 mA/cm^2, a V_{oc} of ~0.57 V and a FF of ~32.9 %, while analogous devices involving graphene/PEDOT:PEG/ZnO (J_{sc} ~ 1.9 mA/cm^2, V_{oc} ~ 0.49 V,

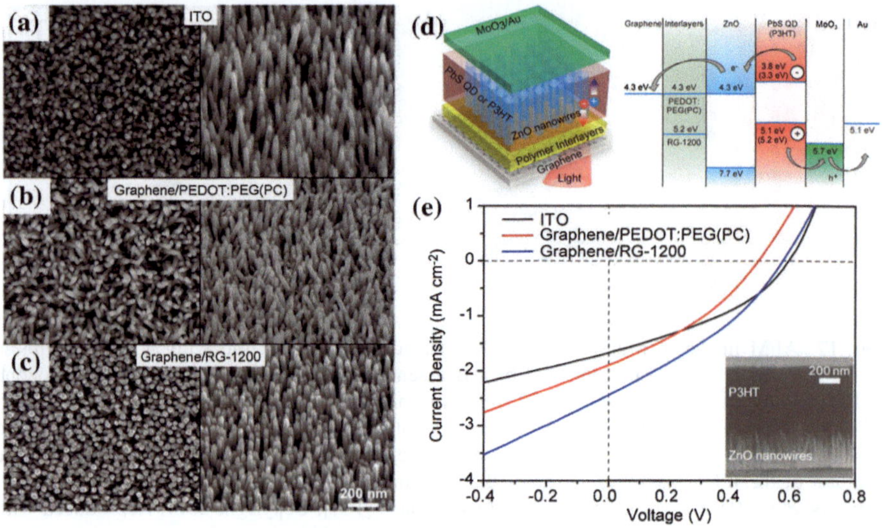

Fig. 18 SEM characterization of ZnO nanowire arrays grown on **a** ITO substrates, **b** graphene/PEDOT:PEG(PC), and **c** graphene/RG-1200, **d** *left* Schematic diagram of the graphene cathode hybrid solar cells: graphene deposited on quartz is covered by a polymer (PEDOT:PEG(PC) or RG-1200), followed by the ZnO seed layer and 400 nm long ZnO nanowires. The nanowires are then infiltrated and covered with P3HT (700 nm) and finally with MoO_3 (25 nm)/Au (100 nm) top electrodes. *Right* Flat-band energy level diagram of the device. **e** *J–V* characteristics of representative graphene-based P3HT devices with different polymer interlayers, compared with an ITO reference device. *Inset* to (**e**) shows SEM cross-section images of the complete device, showing substantial infiltration of P3HT into the nanometre-scale gaps between ZnO nanowires [105]. Copyright 2013. Reproduced with permission from American Chemical Society

FF ~ 33.8 %, PCE ~ 0.3 %) and ITO/ZnO (J_{sc} ~ 1.7 mA/cm^2, V_{oc} ~ 0.59 V, *FF* ~ 36.8 %, PCE ~ 0.4 %) have exhibited inferior photovoltaic performances.

Tong et al. [106] have recently shown another high performance inorganic–organic hybrid PSC demonstrating the integration of poly[[9-(1-octylnonyl)-9H-carbazole-2,7-diyl]-2,5-thiophenediyl-2,1,3-benzothiadiazole-4,7-diyl-2,5-thiophenediyl] (PCDTBT) polymer as the donor and a composite of core–shell CdSe/CdTe nanocrystals and RGO as the electron acceptor (Fig. 19). The homogeneous mixture of the components were achieved by introducing a chemical treatment with amines endorsing the compatibility of the interfaces. The treatment with oleylamine promotes the surface modification of the tetrapod shaped CdSe/CdTe nanocrystals by inducing hydrophobic properties. On the other hand, the acidic functional groups of RGO binds with the amine through acid–base complexation ensuring an enhanced solubility of RGO in organic solvents such as 1,2-dichlorobenzene. Thus, the amine treatment dramatically increases the homogeneous dispersion of the nanocrystal in RGO and the miscibility of the acceptor to the donor facilitating an excellent interfacial charge transfer. The significant role of amine treatment has been evidenced from more than 2-fold enhancement in the

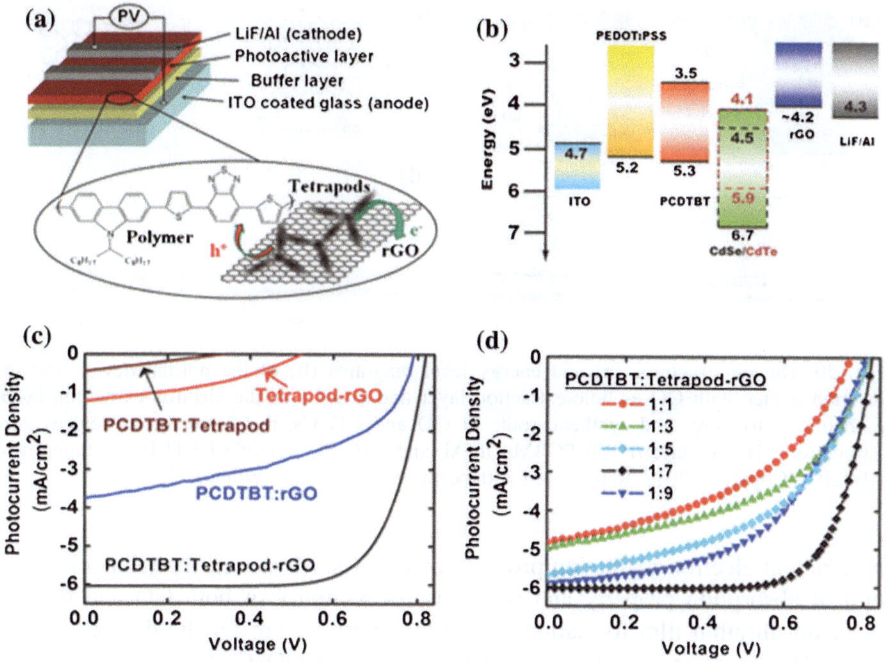

Fig. 19 **a** Photovoltaic device configuration used in the current study. **b** Schematic energy level diagram of the device. *J–V* characteristics of the devices with **c** different photoactive layer and **d** PCDTBT:Tetrapod-RGO in various blend ratio under light illumination intensity of 100 mW/cm^2 [106]. Copyright 2013. Reproduced with permission from John Wiley & Sons

PCEs of the devices with the hybrid active layer of PCDTBT:tetrapod–RGO in contrast to PCDTBT:tetrapod, tetrapod-RGO and PCDTBT:RGO. The ternary hybrid exhibited an outstanding device performance achieving a J_{sc} of ~6.0 mA/cm^2, a V_{oc} of ~0.82 V and a *FF* of ~59 %, affording a PCE of ~3.27 %, while the reference devices involving PCDTBT:tetrapod, tetrapod-RGO and PCDTBT:RGO have exhibited lower PCEs of ~0.04, 0.29, and 1.43 %, respectively.

In a different methodology, Liu et al. [107] have demonstrated a controlled functionalization GO involving a simple charge neutralization process using Cs_2CO_3, wherein the periphery –COOH groups present in GO are converted to –COOCs. As a result of this, the hole-extraction GO as well as its work function of ~4.7 eV are altered to produce electron-extraction Cesium-neutralized GO (GO–Cs) with an work function of ~4.0 eV as shown in the schematic of Fig. 20. The hole-extraction property of GO is administered by the presence of peripheral –COOH groups, which eventually dope the P3HT polymer to facilitate an Ohmic contact between GO and P3HT at the interface created by the similar/matched work function values of GO (~4.7 eV) and the HOMO level of P3HT. Once these peripheral –COOH groups are converted to –COOCs, the altered work function of GO–Cs (~4.0 eV) matches with the LUMO level of P3HT providing

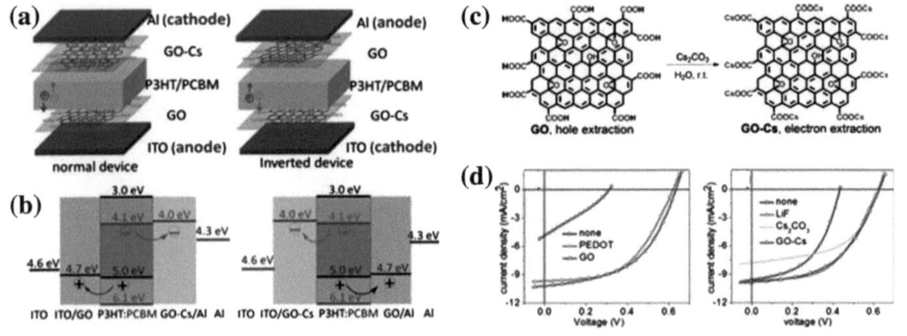

Fig. 20 Device structures (**a**) and energy level diagrams (**b**) of the normal device and the inverted device with GO as hole-extraction layer and GO–Cs as the electron-extraction layer, **c** chemical structures and synthetic route of GO and GO–Cs, **d** *J–V* curves of devices with -structures ITO/interlayer/P3HT:PCBM/LiF/Al and ITO/PEDOT:PSS/P3HT:PCBM/interlayer/Al) [107]. Copyright 2012. Reproduced with permission from John Wiley & Sons

an efficient electron-extraction property of GO–Cs. Moreover, the relatively weak optical absorption property and solution-processability of both GO and GO–Cs favor an ultrathin film formation with a thickness ~2 nm for the device applications. The measured conductivity of these GO and GO–Cs thin films are found to be ~3.8×10^{-3} and ~5.0×10^{-3} S m^{-1}, respectively. The normal configuration of the devices involving hole extraction GO layer has shown a maximum PCE of ~3.67 % with a J_{sc} of ~10.3 mA/cm^2, a V_{oc} of ~0.61 V and a *FF* of ~59 %, whereas the inverted structure of the devices using electron-extraction GO–Cs layer afforded a maximum PCE ~ 2.97 %, a J_{sc} ~10.6 mA/cm^2, a V_{oc} ~0.51 V and a *FF* ~54 %. In another hybrid solar cell approach, the promising role of GO as an electron acceptor to minimize the electron-hole recombination has also been explored in a ternary ZnO-CdS-GO composite based device affording a PCE of ~2.8%, while the GO-free device yielded a PCE of only ~1.9 % [110]

3 Conclusion and Outlook

Since the discovery of graphene in 2004, the substantial growth of graphene based materials has been continuously demonstrating its superior electronic properties in the BJH and hybrid photovoltaic assemblies. Recent advances in this area have witnessed the new pathways to develop high-performance photovoltaic assemblies. In the brief discussion on graphene based PSCs, in this article, we have focussed on the variety of interesting roles of graphene to enhance the device performance efficiencies. Excellent electronic as well as the surface properties of graphene could be employed in the devices via its single-layered, oxidized or reduced forms, which offer great opportunities to produce the functionalized and composite materials involving graphene. Coupling of graphene with polymer, organic or semiconductor materials presents a contemporary approach to design and develop wide variety of novel materials for the photovoltaic applications. Although pristine, oxidized or highly reduced forms of graphene have exhibited best prospects in the PSCs, experimental routes applied to incorporate graphene into the polymer hosts and the fabrication of highly conducting graphene based electrodes need sincere attention. Pristine graphene exhibits inferior device performance owing to its poor compatibility with the organic polymers. In this regard, graphene sheets seemed to be a promising choice to fulfil the conductivity requisites as well as the easy synthetic routes to accomplish well-dispersed blends with the polymer matrix, especially using the solution processed approaches. Chemical and surface modifications of graphene can also be carried out to improve the surface properties and electrical conductivity of graphene, thereby enriching the graphene–polymer electronic interactions for the best results. In addition to the advancement of the properties of active layer in PSCs, graphene based materials can likewise promote the role of counter electrodes to replace or modify the rigid electrodes due to their superior electrical conductivity, chemical robustness and high flexibility, consecutively succeeding the roll-to-role industrial processing applicability. Single- or multi-layered sheets of graphene have been explored as the counter electrodes that can outshine the performance of platinum based electrodes in terms of flexibility and comparable sheet resistance, optical transparency and mechanical strength. Nevertheless, it is still challenging to stabilize the large area production of high quality graphene films onto flexible substrates for the practical applications. One of the suitable applied routes is the solution processed fabrication of mono-layer graphene electrodes, but it induces a serious increase in the electrical resistance within the material. The resistivity of the electrodes could be further lowered to a minimum practical limit by introducing the thicker or multi-layer stacking approach—while doing so, optical transparency of the films is compromised. Indeed, a CVD technique can be employed as an effective alternate to overcome these issues producing transparent and high quality multi-layer graphene films with a very low resistivity or high electrical conductivity. The electrical conductivity of multi-layer graphene films could also be enhanced by introducing p-dopants, which eventually promotes the hole concentration thereby minimizing the sheet resistance in the system. Graphene in its oxidized or reduced forms have also been applied as the smart hole transport materials establishing a longer stability in contrast to the well-known material PEDOT:PSS. However, the fabrication or spin coating of smooth GO or RGO films with a thickness of ~1–2 nm ensuring the single or bilayer structure is a prime challenge. Similarly for the large scale applications, the widespread coverage of ITO surfaces with a GO/RGO HTL is anticipated—while doing so, a thicker GO/RGO layer is needed that encounters a higher sheet resistance, lowering the device performance. Rather exciting results are observed applying the monolayer GO as the HTL, electronically coupled through a π-stacked face-on microstructure of the active layer in the device, enriching the durability and lifetime comparable to PEDOT:PSS. Moreover, graphene and its chemically modified derivatives or their doped structures can be utilized to perform superior enhancement in the efficiencies of hybrid inorganic–organic solar cells. Thus far, using graphene based materials in PSCs have perceived a rapid upsurge in research and prolific results, keeping in mind that specific graphene material have specific properties– inherently dependent upon the synthesis route and fabrication process –and each specialized material performs specific functions in the solar cells. An advanced phase of innovative research, to involve these graphene materials in the PSC community progressively, is obligatory to study whether the applications briefed in this article can be carried over to produce modern graphene devices.

References

1. Zhu Y, Murali S, Cai W, Xuesong Li, Suk J W, Potts J R, Ruoff R S (2010) Graphene and graphene oxide: synthesis, properties, and applications. Adv Mater 22:3906–3924
2. Geim A K, Novoselov K S (2007) The rise of graphene. Nat Mater 6:183–191
3. Compton O C, Nguyen S B T (2010) Graphene oxide, highly reduced graphene oxide, and graphene: versatile building blocks for carbon-based materials. Small 6:711–723
4. Stankovich S, Dikin D A, Dommett G H B (2006) Graphene-based composite materials. Nat 442:282–286
5. Soldano C, Mahmood A, Dujardin E (2010) Production, properties and potential of graphene. Carbon 48:2127–2150
6. Katsnelson M I (2007) Graphene: carbon in two dimensions. Mater Today 10:20–27
7. Geim A K (2009) Graphene: status and prospects. Science 324:1530–1534
8. Novoselov K S, Jiang Z, Zhang Y, Morozov S V, Stormer H L, Zeitler U, Maan J C, Boebinger G S, Kim P, Geim A K (2007) Room-temperature quantum hall effect in graphene. Science 315:1379–1379
9. Stoller M D, Park S, Zhu Y, An J, Ruoff R S (2008) Graphene-based ultracapacitors. Nano Lett 8:3498–3502
10. Loh K P, Bao, Q L, Eda G, Chhowalla M (2010) Graphene oxide as a chemically tunable platform for optical applications. Nat Chem 2:1015–1024
11. Eda G, Fanchini G, Chhowalla M (2008) Large-area ultrathin films of reduced graphene oxide as a transparent and flexible electronic material. Nat Nanotechnol 3:270–274
12. Yu D S, Dai L (2010) Self-assembled graphene/carbon nanotube hybrid films for supercapacitors. J Phys Chem Lett 1:467–470
13. Wang X, Zhi L J, Mullen K (2008) Graphene electrodes for dye-sensitized solar cells. Nano Lett 8:323–327
14. Eda G, Chhowalla M (2009) Graphene-based composite thin films for electronics. Nano Lett 9:814–818
15. Yu D S, Dai L (2010) Voltage-induced incandescent light emission from large-area graphene films. Appl Phys Lett 96:143107(1–3)
16. Yu D S, Yang Y, Durstock M, Baek J-B, Dai L (2010) Soluble P3HT-grafted graphene for efficient bilayer-heterojunction photovoltaic devices. ACS Nano 4:5633–5640
17. Gilje S, Song H, Wang M, Wang K L, Kaner R B (2007) A chemical route to graphene for device applications. Nano Lett 7:3394–3398
18. Liu Y, Yu D, Zeng C, Miao Z, Dai L (2010) Biocompatible graphene oxide-based glucose biosensors. Langmuir 29:6158–6160
19. Qu L, Liu Y, Baek J, Dai L (2010) Nitrogen-doped graphene as efficient metal-free electrocatalyst for oxygen reduction in fuel cells. ACS Nano 4:1321–1326
20. Dreyer D R, Park S, Bielawski C W, Ruoff R S (2009) The chemistry of graphene oxide. Chem Soc Rev 39:228–240
21. Lightcap V, Kamat P V (2012) Fortification of CdSe quantum dots with graphene oxide. excited state interactions and light energy conversion. J Am Chem Soc 134:7109–7116
22. O'Regan B, Grätzel M (1991) A low-cost, high-efficiency solar cell based on dye-sensitized colloidal TiO_2 films. Nat 353:737–740

23. Imahori H, Umeyama T, Ito S (2009) Large π-aromatic molecules as potential sensitizers for highly efficient dye-sensitized solar cells. Acc Chem Res 42:1809–1818
24. Ye L, Zhang S, Huo L, Zhang M, HouJ (2014) Molecular design toward highly efficient photovoltaic polymers based on two-dimensional conjugated benzodithiophene. Acc Chem Res 47:1595–1603
25. Park S H, Roy A, Beaupre S, Cho S, Coates N, Moon J S, Moses D, Leclerc M, Lee K, Heeger A J (2009) Bulk heterojunction solar cells with internal quantum efficiency approaching 100%. Nat Photonics 3:297–302
26. Li G, Shrotriya V, Huang J, Yao Y, Moriarty T, Emery K, Yang Y (2005) High-efficiency solution processable polymer photovoltaic cells by self-organization of polymer blends. Nat Mater 4:864–868
27. Ma W, Yang C, Gong X, Lee K, Heeger A J (2005) Thermally stable, efficient polymer solar cells with nanoscale control of the interpenetrating network morphology. Adv Funct Mater 15:1617–1622
28. Peet J, Kim J Y, Coates N E, Ma W L, Moses D, Heeger A J, Bazan G C (2007) Efficiency enhancement in low-bandgap polymer solar cells by processing with alkane dithiols. Nat Mater 6:497–500
29. Chen H-Y, Hou J, Zhang S (2009) Polymer solar cells with enhanced open-circuit voltage and efficiency. Nat Photonics 3:649–653
30. Li G, Zhu R, Yang Y (2012) Polymer solar cells. Nat Photonics 6:153–161
31. He Z, Zhong C, Huang X, Wong W-Y, Wu H, Chen L, Su S, Cao Y (2011) Simultaneous enhancement of open-circuit voltage, short-circuit current density, and fill factor in polymer solar cells. Adv Mater 23:4636–4643
32. Small C E, Chen S, Subbiah J, Amb C M, Tsang S-W, Lai T-H, Reynolds J R, So F (2012) High-efficiency inverted dithienogermolethienopyrrolodione-based polymer solar cells. Nat Photonics 6:115–120
33. Chen H-Y, Hou J, Zhang S, Liang Y, Yang G, Yang Y, Yu L, Wu Y, Li G (2009) Polymer solar cells with enhanced open-circuit voltage and efficiency. Nat Photonics 3:649–653
34. Liang Y, Xu Z, Xia J, Tsai S-T, Wu Y, Li G, Ray C, Yu L (2010) For the bright future–bulk heterojunction polymer solar cells with power conversion efficiency of 7.4%. Adv Mater 22:E135–E138.
35. You J, Dou L, Yoshimura K, Kato T, Ohya K, Moriarty T, Emery K, Chen C-C, Gao J, Li G, Yang Y (2013) A polymer tandem solar cell with 10.6% power conversion efficiency. Nat Commun 4:1446(1–10)
36. Clarke T M, Durrant J R, (2010) Charge photogeneration in organic solar cells. Chem Rev 110:6736–6767
37. Pan Z, Gu H, Wu M-T, Li Y, Chen Y (2012) Graphene-based functional materials for organic solar cells. Opt Mater Express 2:814–824
38. Ren L, Qiu J, Wang S (2013) Photovoltaic properties of graphene nanodisk-integrated polymer composites. Compos: Part B 55:548–557
39. Acik M, Chabal Y J (2011) Nature of graphene edges: a review. Jpn J Appl Phys 50:070101(1–16)
40. Li Y, Hu Y, Zhao Y, Shi G, Deng L, Hou Y, Qu L (2011) An electrochemical avenue to green-luminescent graphene quantum dots as potential electron-acceptors for photovoltaics. Adv Mater 23:776–780
41. Yan X, Cui X, Li B, Li L S (2010) Large, solution-processible graphene quantum dots as light absorbers for photovoltaics. Nano Lett 10:1869–1873
42. Gupta V, Chaudhary N, Srivastava R, Sharma G D, Bhardwaj R, Chand S (2011) Luminscent graphene quantum dots for organic photovoltaic devices. J Am Chem Soc 133:9960–9963
43. Yong V, Tour J M (2010) Theoretical efficiency of nanostructured graphene-based photovol taics. Small 6:313–318
44. Hill I G, Kahn A, Soos Z G, Pascal R A J (2000) Charge-separation energy in films of π-conjugated organic molecules. Chem Phys Lett 327:181–188

45. Alvarado S F, Seidler P F, Lidzey D G, Bradley D D C (1998) Direct determination of the exciton binding energy of conjugated polymers using a scanning tunneling microscope. Phys Rev Lett 81:1082–1085
46. Kersting R, Lemmer U, Deussen M, Bakker H J, Mahrt R F, Kurz H, Arkhipov V I, Bässler H, Göbel E O (1994) Ultrafast field-induced dissociation of excitons in conjugated polymers. Phys Rev Lett 73:1440–1443
47. Xue J, Rand B P, Uchida S, Forrest S R (2005) A hybrid planar–mixed molecular heterojunction photovoltaic cell. Adv Mater 17:66–71
48. Scully S R, McGehee M D (2006) Effects of optical interference and energy transfer on exciton diffusion length measurements in organic semiconductors. J Appl Phys 100:034907(1–5)
49. Markov D E, Hummelen J C, Blom P W M, Sieval A B (2005) Dynamics of exciton diffusion in poly(p-phenylene vinylene)/fullerene heterostructures. Phys Rev B 72:045216(1–5)
50. Peumans P, Yakimov A, Forrestb S R (2003) Small molecular weight organic thin-film photodetectors and solar cells. J Appl Phys 93:3693–3723
51. Hoppe H, Sariciftci N S (2004) Organic solar cells: an overview. J Mater Res 19:1924–1945
52. Dennler G, Scharber M C, Brabec C J (2009) Polymer-fullerene bulk-heterojunction solar cells. Adv Mater 21:1323–1338
53. Kniepert J, Lange I, Kaap N J v d, Koster L J A, Dieter Neher (2014) A conclusive view on charge generation, recombination, and extraction in as-prepared and annealed P3HT:PCBM blends: combined experimental and simulation work. Adv Energy Mater 4:1301401(1–10)
54. Wang T, Pearson A J, Lidzey D G, Jones R A L (2013) Evolution of structure, optoelectronic properties, and device performance of polythiophene:fullerene solar cells during thermal annealing. Adv Funct Mater 21:1383–1390
55. Oklobia O, Shafai T S (2013) A study of donor/acceptor interfaces in a blend of P3HT/PCBM solar cell: effects of annealing and PCBM loading on optical and electrical properties. Solid-State Electron 87:64–68
56. Kasry A, Ashry M E, Nistor R A, Bola A A, Tulevskia G S, Martynaa G J, Newns D M (2012) High performance metal microstructure for carbon-based transparent conducting electrodes. Thin Solid Films 520:4827–4830
57. Søndergaard R, Hösel M, Angmo D, Larsen-Olsen T T, Krebs F C (2012) Roll-to-roll fabrication of polymer solar cells. Mater Today 15:36–49
58. Carle J E, Helgesen M, Madsen M V, Bundgaarda E, Krebs F C (2014) Upscaling from single cells to modules – fabrication of vacuum- and ITO-free polymer solar cells on flexible substrates with long lifetime. J Mater Chem C 2:1290–1297
59. Lee S, Yeo J-S, Ji Y, Cho C, Kim D-Y, Na S-I, Lee B H, LeeT (2012) Flexible organic solar cells composed of P3HT:PCBM using chemically doped graphene electrodes. Nanotechnol 23:344013
60. Na S-I, Kim S-S, Jo J, Kim D-Y (2008) Efficient and flexible ITO-free organic solar cells using highly conductive polymer anodes. Adv Mater 20:4061–4067
61. Ohzeki M, Fujii S, Arai Y, Yanagidate T, Yanagi Y, Okukawa T, Yoshida A, Kataura H, Nishioka Y (2014) Performance improvement of flexible bulk heterojunction solar cells using PTB7:$PC_{71}BM$ by optimizing spin coating and drying processes. J Appl Phys, Part 1 53:02BE04(1–5)
62. Bässler H (1994) Non-dispersive and dispersive transport in random organic photoconductors. Mol Cryst Liq Cryst Sci A 252:11–21;
63. Huynh W U, Dittmer J J, Paul A (2002) Hybrid nanorod-polymer solar cells. Science 295:2425–2427
64. Yin Z, Zhu J, He Q, Cao X, Tan C, Chen H, Yan Q, Zhang H (2014) Graphene-based materials for solar cell applications. Adv Energy Mater 4:1300574(1–19)
65. Liu Z, He D, Wang Y, Wu H, Wang J (2010) Graphene doping of P3HT:PCBM photovoltaic devices. Synth Met 160:1036–1039

66. Chen D, Zhang H, Liu Y, Li J (2013) Graphene and its derivatives for the development of solar cells, photoelectrochemical, and photocatalytic applications. Energy Environ Sci 6:1362–1387
67. Liu Q, Liu Z, Zhang X, Zhang N, Yang L, Yin S, Chen Y (2008) Organic photovoltaic cells based on an acceptor of soluble graphene. Appl Phys Lett 92:223303(1–3)
68. Liu Z, He D W, Wang Y, Wu H, Wang J (2010) Solution-processible functionalized graphene in donor/acceptor-type organic photovoltaic cells. Sol Energy Mater Sol Cells 94:1196–1200
69. Yu D, Park K, Durstock M, Dai L (2011) Fullerene-grafted graphene for efficient bulk heterojunction polymer photovoltaic devices. J Phys Chem Lett 2:1113–1118
70. Chauhan A K, Gusain A, Jha P, Koiry S P, Saxena V, Veerender P, Aswal D K, Gupta S K (2014) Graphene composite for improvement in the conversion efficiency of flexible poly 3-hexyl-thiophene:[6,6]-phenyl C_{71} butyric acid methyl ester polymer solar cells. Appl Phys Lett 104:133901(1–5)
71. Hill C M, Zhu Y, Pan S (2011) Fluorescence and electroluminescence quenching evidence of interfacial charge transfer in poly (3-hexylthiophene): graphene oxide bulk heterojunction photovoltaic devices. ACS Nano 5:942–951
72. Liu Q, Liu Z, Zhang X, Yang L, Zhang N, Pan G, Yin S, Chen Y, Wei J (2009) Polymer photovoltaic cells based on solution-processible graphene and P3HT. Adv Funct Mater 19:894–904
73. Jr W S H, Offeman R E (1958) Preparation of graphitic oxide. J Am Chem Soc 80:1339–1339
74. Becerril H A, Mao J, Liu Z, Stoltenberg R M, Bao Z, ChenY (2008) Evaluation of solution-processed reduced graphene oxide films as transparent conductors. ACS Nano 2:463–470
75. Stankovich S, Piner R D, Nguyen S B T, Ruoff R S (2006) Synthesis and exfoliation of isocyanate-treated graphene oxide nanoplatelets. Carbon 44:3342–3347
76. Zhang B, Liu G, Chen Y, Zeng L-J, Zhu C-X, Neoh K-G, Wang C, Kang E-T (2011) Conjugated polymer-grafted reduced graphene oxide for nonvolatile rewritable memory. Chem Eur J 17:13646–13652
77. Li P-P, Chen Y, Zhu J, Feng M, Zhuang X, Lin Y, Zhan H (2011) Charm-bracelet-type poly(N-vinylcarbazole) functionalized with reduced graphene oxide for broadband optical limiting. Chem Eur J 17:780–785
78. Zhang B, Chen Y, Liu G, Xu L-Q, Chen J, Zhu C-X, Neoh K-G, Kang E-T (2012) "Push-pull archetype of reduced graphene oxide functionalized with polyfluorene for nonvolatile rewritable memory. J Polym Sci, Part A: Polym Chem 2:378–387
79. Li Y, Pan Z, Fu Y, Chen Y, Xie Z, Zhang B (2012) Soluble reduced graphene oxide functionalized with conjugated polymer for heterojunction solar cells. J Polym Sci, Part A: Polym Chem 50(9):1663–1671
80. Jayawardena K D G I, Rhodes R, Gandhi K K, Prabhath M R R, Dabera G D M R, Beliatis m J, Rozanski L J, Henley S J, Silva S R P (2013) Solution processed reduced graphene oxide/metal oxide hybrid electron transport layers for highly efficient polymer solar cells. J Mater Chem A 1: 9922–9927
81. Wu J, Becerril H A, Bao Z, Liu Z, Chen Y, Peumans P (2008) Organic solar cells with solution-processed graphene transparent electrodes. Appl Phys Lett 92:263302(1–3)
82. Yin Z, Sun S, Salim T, Wu S, Huang X, He Q, Lam Y M, Zhang H (2010) Organic photovoltaic devices using highly flexible reduced graphene oxide films as transparent electrodes. ACS Nano 4:5263–5268
83. Bae S, Kim H, Lee Y, Xu X, Park J-S, Zheng Y, Balakrishnan J, Lei T, Kim H R, Song Y I, Kim Y-J, Kim K S, Özyilmaz B, Ahn J-H, Hong B H, Iijima S (2010) Roll-to-roll production of 30-inch graphene films for transparent electrodes. Nat Nanotechnol 5:574–578
84. Arco L G D, Zhang Y, Schlenker C W, Ryu K, Thompson M E, Zhou C (2010) Continuous, highly flexible, and transparent graphene films by chemical vapor deposition for organic photovoltaics. ACS Nano 4:2865–2873

85. Bonaccorso F, Sun Z, Hasan T, Ferrari A C (2010) Graphene photonics and optoelectronics. Nat Photonics 4: 611–622
86. Choi Y Y, Kang S J, Kim H- K, Choi W M, Na S-I (2012) Multilayer graphene films as transparent electrodes for organic photovoltaic devices. Sol Energy Mater Sol Cells 96:281–285
87. Choe M, Lee B H, Jo G, Park J, Park W, Lee S, Hong W-K, Seong M-J, Kahng Y H, Lee K, LeeT (2010) Efficient bulk-heterojunction photovoltaic cells with transparent multi-layer graphene electrodes. Org Electron 11:1864–1869
88. Park H S, Rowehl J A, Kim K K, Bulovic V, Kong J (2010) Doped graphene electrodes for organic solar cells. Nanotechnol 21:505204
89. Wang Y, Tong S W, Xu X F, Özyilmaz B, Loh K P (2011) Interface engineering of layer-by-layer stacked graphene anodes for high-performance organic solar cells. Adv Mater 23:1514–1518
90. Lee Y- Y, Tu K- H, Yu C- C, Li S-S, Hwang J-Y, Lin C-C, Chen K-H, Chen L-C, Chen H-L, Chen C-W (2011) Top laminated graphene electrode in a semitransparent polymer solar cell by simultaneous thermal annealing/releasing method. ACS Nano 5:6564–6570
91. Hsu C- L, Lin C- T, Huang J- H, Chu C-W, Wei K-H, Li L-J (2012) Layer-by-layer graphene/TCNQ stacked films as conducting anodes for organic solar cells. ACS Nano 6:5031–5039
92. Tung V C, Chen L M, Allen M J, Wassei J K, Nelson K, Kaner R B, Yang Y (2009) Low-temperature solution processing of graphene–carbon nanotube hybrid materials for high-performance transparent conductors. Nano Lett 9:1949–1955
93. Ni G- X, Zheng Y, Bae S, Tan C Y, Kahya O, Wu J, Hong B H, Yao K, Özyilmaz B (2012) Graphene–ferroelectric hybrid structure for flexible transparent electrodes. ACS Nano 6:3935–3942
94. Emmanuel K, Kyriaki S, Minas M S, Fotakis C, Stratakis E (2013) Flexible organic photovoltaic cells with in situ nonthermal photoreduction of spin-coated graphene oxide electrodes. Adv Funct Mater 23:2742–2749
95. M. Jørgensen, K. Norrman, F. C. Krebs, Stability/degradation of polymer solar cells. Sol Energy Mater Sol Cells 92:686–714
96. Kim Y- H, Lee S- H, Noh J, han S-H (2006) Performance and stability of electroluminescent device with self-assembled layers of poly(3,4-ethylenedioxythiophene)–poly(styrenesulfonate) and polyelectrolytes. Thin Solid Films 510:305–310
97. Lagemaat J, Barnes T M, Rumbles G, Shaheen S E, Coutts T J, Weeks C, Levitsky I, Peltola J, Glatkowski P (2006) Organic solar cells with carbon nanotubes replacing In_2O_3:Sn as the transparent electrode. Appl Phys Lett 8:233503 (1–3)
98. Matyba P, Yamaguchi H, Chhowalla M, Robinson N D, Edman L (2010) Flexible and metal-free light-emitting electrochemical cells based on graphene and PEDOT:PSS as the electrode materials. ACS Nano 5:574–580
99. Dreyer D R, Park S, Bielawski C W, Ruoff R S (2010) The chemistry of graphene oxide. Chem Soc Rev 39:228–240
100. Loh K P, Bao Q, Eda G, Chhowalla M (2010) Graphene oxide as a chemically tunable platform for optical applications. Nat Chem 2:1015–1024
101. Li S-S, Tu K-H, Lin C-C, Chen C-W, Chhowalla M (2010) Solution-processable graphene oxide as an efficient hole transport layer in polymer solar cells. ACS Nano 4:3169–3174
102. Kim J, Tung V C, Huang J X (2011) Water processable graphene oxide:single walled carbon nanotube composite as anode modifier for polymer solar cells. Adv Energy Mater 1:1052–1057
103. Yun J M, Yeo J S, Kim J, Jeong H-G, Kim D-Y, Noh Y-J, Kim S-S, Ku B-C, Na S-I (2011) Solution-processable reduced graphene oxide as a novel alternative to PEDOT:PSS hole transport layers for highly efficient and stable polymer solar cells. Adv Mater 23:4923–4928
104. Murray I P, Lou S J, Cote L J, Loser S, Kadleck C J, Xu T, Szarko J M, Rolczynski B S, Johns J E, Huang J, Yu L, Chen L X, Marks T J, HersamM C (2011) Graphene Oxide

Interlayers for Robust, High-Efficiency Organic Photovoltaics. J Phys Chem Lett 2:3006–3012
105. Park H, Chang S, Jean J, Cheng J J, Araujo P T, Wang M, Bawendi M G, Dresselhaus M S, Bulović V, Kong J, Gradečak S (2013) Graphene cathode-based ZnO nanowire hybrid solar cells. Nano Lett 13: 233–239
106. Tong S W, Mishra N, Su C L, Nalla V, Wu W, Ji W, Zhang J, Chan Y, Loh K P (2014) High-performance hybrid solar cell made from CdSe/CdTe nanocrystals supported on reduced graphene oxide and PCDTBT. Adv Funct Mater 24:1904–1910
107. Liu J, Xue Y, Gao Y, Yu D, Durstock M, Dai L (2012) Hole and electron extraction layers based on graphene oxide derivatives for high-performance bulk heterojunction solar cells. Adv Mater 24:2228–2233
108. Iwan A, Chuchmała A (2012) Perspectives of applied graphene: Polymer solar cells. Prog Polym Sci 37:1805–1828
109. Wan X, Huang Y, Chen Y (2012) Focusing on energy and optoelectronic applications: a journey for graphene and graphene oxide at large scale, Acc Chem Res 45(4):598-607
110. Barpuzary D, Qureshi M (2013) Enhanced photovoltaic performance of semiconductor-sensitized ZnO-CdS coupled with graphene oxide as a novel photoactive material. ACS Appl Mater Interfaces 5:11673–11682

Graphene Composites Based Photodetectors

Shichao Song, Long Wen and Qin Chen

Abstract Graphene is an amazing material with unique electrical and optical properties that have never been observed in conventional materials. Graphene can absorb light from ultraviolet to infrared and transit carriers at a speed of 1/300 of light, which make graphene an excellent candidate for optoelectronic applications. Graphene composites consisting of graphene and other materials combine the high carrier mobility property of graphene and the excellent light absorption properties of other semiconductors, which are ideal for development of next-generation optoelectronic devices, especially photodetectors. In this chapter, we review the recent progress of graphene composite photodetectors with significant performance improvement compared to the original graphene photodetectors and discuss its future developments. We consider that graphene composite photodetectors would play an important role in future optical interconnect and imaging systems.

Keywords Graphene · Photodetector · Composite

List of Abbreviations

CVD	Chemical vapor deposition
CNT	Carbon nanotube
OLED	Organic light emitting devices
NW	Nanowire
QDs	Quantum dots
FET	Field-effect transistor
MSM	Metal-semiconductor-metal

S. Song · L. Wen · Q. Chen (✉)
Key Lab of Nanodevices and Applications, Collaborative Innovation Center of Suzhou Nano Science and Technology, Suzhou Institute of Nano-Tech and Nano-Bionics, Chinese Academy of Sciences (CAS), Suzhou 215123, China
e-mail: qchen2012@sinano.ac.cn

K.K. Sadasivuni et al. (eds.), *Graphene-Based Polymer Nanocomposites in Electronics*, Springer Series on Polymer and Composite Materials,
DOI 10.1007/978-3-319-13875-6_8

IR	Infrared
UV	Ultraviolet
THz	Terahertz

1 Introduction

Graphene, the origin of all graphitic forms, can be described as a one-atom thick layer of graphite and is densely packed in a regular sp^2-bonded hexagonal pattern. Over the past decades, graphene has become one of the most exciting topics of research in nanoscience and nanotechnology [1]. It has a large specific surface area (2,630 m^2 g^{-1}), high intrinsic mobility (200,000 cm^2 v^{-1} s^{-1}) [2, 3], large Young's modulus (~1.0 TPa) [4], high thermal conductivity (~5,000 W m^{-1} K^{-1}) [5, 6], and broadband photoresponse (300 nm–6 μm) [7, 8], etc. The prominent properties that graphene possesses merit attention for applications in such as heat sinks [9–11], transistor [12, 13], photodetector [12, 14–17], modulator [18–21], polarization controller [22], transparent conductive electrode [23, 24], etc. To further exploit these properties in various kinds of applications, versatile and reliable synthetic routes have been developed to grow graphene, such as mechanical exfoliation [25], chemical vapor deposition (CVD) [26], liquid phase exfoliation [27], molecular assembly [28], synthesis on SiC [29], etc. Since Geim and Novoselov at University of Manchester first isolated single-layer graphene from bulk graphite [30–32], which were awarded the 2010 Nobel Prize in Physics, great progress in graphene fabrication has been achieved, for example, Sungkyunkwan University and Samsung have developed a novel fabrication technology for 30-in. graphene film massive production via CVD and roll-to-roll transferred methods in 2010 [24]. With all these advantages in preparation of graphene, it is desirable to harness the useful properties of graphene in composites, through the incorporation with various kinds of functional materials [33]. Till now, graphene-based composites have been successfully made with polymers [34–36], nanostructures [37, 38], CNTs [39, 40] and are intensively explored in applications such as supercapacitors [41, 42], transparent conductive electrodes [43], photovoltaic devices [44, 45], photodetectors [46, 47], sensors [48, 49], etc. There is no doubt that graphene transparent electrodes will be the first appearance outside laboratories for the applications in OLEDs, touch screen and so on. In a recent important review on graphene [50], graphene photodetector, one of the most actively studied devices, is considered to be the first graphene optoelectronic devices for optical interconnect applications before 2020. In this chapter, we will review recent development in graphene composites based photodetectors and discuss the possible urgent tasks to push this device to the market. The chapter is organized as follows: Sect. 2 gives a brief introduction of the graphene production method; Sect. 3 introduces the carrier transition and light absorption property in graphene; Sect. 4 gives a general introduction of graphene optoelectronic applications; Sect. 5 focuses on various graphene composite photodetectors; In Sect. 6, we will summarize the chapter.

2 Graphene Production Method

There are several production methods nowadays for graphene. High quality graphene was first produced by **mechanical exfoliation** of graphite [25]. It is a very simple way to get monolayer and few graphene layers by a scotch tape and provides the best quality graphene so far. However, it is not an efficient method to produce graphene on a large scale due to the less controllability and low repeatability. Therefore it is only used in academic research. **Liquid phase exfoliation** method is to exfoliate graphene in aqueous or non-aqueous solvents by ultrasonication [27]. Control of lateral size and number of layers can be achieved by separation in centrifugal fields. Therefore, it has advantages of low cost and scalability, and also it is a perfect method to fabricate graphene composite. However, these exfoliation methods are not ideal for the integration of graphene devices into the existing platforms. **CVD** is the most popular way to produce graphene as all the process steps occur in completely semiconductor compatible environment. Large-area uniform polycrystalline graphene films are usually grown by CVD on metal foils and then transferred onto other substrates [26], for example, 30-in. graphene films have been produced via CVD [24]. The issue of the CVD method is that most as-grown samples are multilayered [14].

Transferring graphene layer is a key step in all above methods. Extensive experiences of transfer process have been developed for CNTs, which are also applied to graphene. A single layer graphene and few layer graphene on SiO_2/Si were reported to be transferred to other substrates [51]. In this method, a layer of poly(methyl methacrylate) (PMMA) is spin coated on the graphene covered substrate. Then the substrate is subsequently detached by SiO_2 wet etching. After this, the PMMA/graphene membrane is placed on the target substrate and the PMMA is dissolved in acetone. It is worth to mention that a scaled transferring process of roll-to-roll method has been already reported to transfer graphene from copper foil onto plastic substrates [24].

3 Carrier Transition and Light Absorption in Graphene

Graphene, a zero-bandgap material or semimetal, has a linear dispersion relationship around the Dirac point as shown in Fig. 1, which causes an ultra-high intrinsic mobility and broadband absorption. The dielectric constant of graphene is usually taken as $1 + 4\pi i\sigma/\omega t$, where σ is the conductivity, ω is the angular frequency and t is the thickness of graphene [52]. The conductivity from both interband and intraband contributions for graphene can be calculated using the Kubo formula [53, 54],

$$\begin{aligned}\sigma_{inter}(\omega) = {} & \frac{\pi e^2}{4h}\left(\tanh\frac{\hbar\omega + 2\mu}{4k_B T} + \tanh\frac{\hbar\omega - 2\mu}{4k_B T}\right) + \\ & i\left(\frac{1}{\hbar\omega}\frac{2e^2}{h}\mu - \frac{e^2}{2h}\log\left|\frac{\hbar\omega + 2\mu}{\hbar\omega - 2\mu}\right|\right)\end{aligned} \tag{1}$$

$$\sigma_{\text{intra}}(\omega) = i\frac{2e^2}{h}\frac{2k_BT}{\hbar(\omega + i\tau^{-1})}\log\left(2\cosh\left(\frac{\mu}{2k_BT}\right)\right) \tag{2}$$

$$\sigma = \sigma_{\text{inter}} + \sigma_{\text{intra}} \tag{3}$$

where σ_{inter} and σ_{intra} are the interband and intraband contribution to conductivity, respectively. h and ħ are Planck and reduced Planck constant. μ is the chemical potential of graphene. k_B is Boltzmann constant, T is the thermodynamic temperature. e is the charge of the electron, τ is the average scattering time of free electrons.

As illustrated in the Fig. 1, in the visible and near infrared ranges, the light absorption of graphene is calculated to be independent of frequency and have a magnitude given by $\pi\alpha \approx 2.3\ \%$, where α is the fine constant [55]. And with a negligible reflectance (<0.1 %), this absorbance increases linearly with the layer numbers from 1 to 5 [55]. Such broadband absorbance, not affected by the details of the graphene band structure, is equivalent to a universal optical sheet conductivity $\sigma_0 = \pi G_0/4$, where $G_0 = e^2/(4\hbar) \approx 6.08 \times 10^{-5}\Omega^{-1}$ is the quantum conductance [56]. However, as shown in Fig. 1c, the deviation from this broadband absorption is found with the energy of incident light less than 0.5 eV (or wavelength larger than ~2,480 nm), which is attributed to the effects of finite-temperature, doping, and intraband transitions [54, 57].

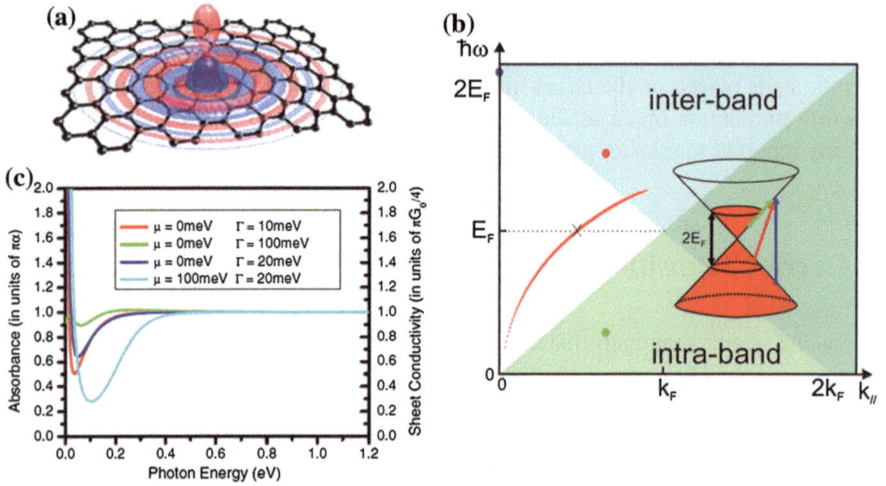

Fig. 1 **a** Schematic of a grapheme sheet and the near electric field produced by a perpendicular dipole situated 10 nm away from doped graphene. **b** The linear dispersion relation around the Dirac point and the interband/intraband transition. **c** Calculated graphene absorption spectra and sheet conductivity from 0 to 1.2 eV, including both the inter band and intra band contributions at a temperature of 300 K. **a**,**b** Reference [58], Copyright 2011. Reproduced with permission from American Chemical Society. **c** Reference [54], Copyright 2008. Reproduced with permission from American Physical Society

4 Graphene Optoelectronic Devices

Graphene can be seen as two dimensional electron gas, where electrons in graphene behave as massless two-dimensional particles, leading to a significant wavelength-independent absorption (~2.3 %) for normal incident light below about 3 eV [50]. Additionally, mono- and bi-layer graphene become nearly transparent when the optical energy is smaller than double the Fermi level due to Pauli blocking. These properties would suit many controllable photonic devices [50], such as transparent conductive electrodes, photodetectors, modulators, etc. In this section, we briefly introduce graphene optoelectronic devices.

4.1 *Transparent Conductive Electrodes*

Transparent conductive coatings, such as Indium Tin Oxides (ITO), Aluminum-doped Zinc Oxide (AZO), are widely used in optoelectronic devices such as touch screen displays, electronic paper and OLEDs, which require both low sheet resistance and high transmittance. Graphene monolayer provides a high transmittance of 97.7 % but with a large resistance of 2,000–5,000 Ω. Combined with metal meshes, a screen with ~90 % transparency and 20 Ω was achieved [59]. Graphene also has outstanding mechanical flexibility and chemical durability, which is very important for flexible devices. In contrast, traditionally used material, ITO, has a transmittance of 90 % and a resistance of about 50 Ω. But ITO is expensive due to indium scarcity, brittle and difficult to process [24, 50]. Therefore, graphene is very promising for the application as transparent conductive electrodes in OLEDs, organic photovoltaics, display, and so on.

4.2 *Optical Modulator*

Optical modulator is one of the key active components used to encode signals into light by altering phase, intensity and polarization [60]. There are various techniques to modulate light such as electro-optic effect [61], thermal-optic effect [62], acoustic-optic effect [63], electroabsorption [64], and so on. Graphene is a perfect material for optical modulation because its optical transmission largely depends on the Fermi level that could be dramatically modified by electrical gating [18, 19, 65] or chemical doping [66]. For photons with energy large than $2E_F$ (Fermi energy), there is no interband transition.

Graphene electroabsorption modulators have excellent performance in terms of large modulation depth and modulation speed, which are two important figure-of-merits of optical modulators. With some electromagnetic-confined structures, an operation bandwidth of more than 50 GHz has been suggested, using integrated dual graphene layers [19] to reduce the resistance in the RC delay time, offering a pathway to a regime of hundreds of gigahertz. The optical bandwidth of such modulators exceeds 12 THz with an operation speed ranging from 160 to 850 GHz and limited only by graphene resistance [67]. A record high electroabsorption modulation of 1 dB/μm was demonstrated by integrating a graphene monolayer onto a silicon waveguide [68]. What is more, this graphene waveguide modulator has a bandwidth of more than 500 nm due to the broadband photoresponse of graphene.

4.3 Photodetector

A photodetector absorbs light and converts it to electrical signal, which is usually manifested as a photocurrent. Generally, there should be at least two energy levels for electrons transition, where only the incident photons with energy larger than the energy difference between these two levels can be absorbed. Therefore, it has limited detecting wavelength range like semiconductor photodetectors, such as InGaAs-based photodetector for infrared and Si-based photodetector for visible. However, graphene can in principle be used for a wide spectral range from ultraviolet to infrared due to its linear dispersion around the Dirac point [55]. And the ultrahigh carrier mobility of graphene provides high speed of graphene-based photodetectors [12]. Photoresponse in graphene has been extensively investigated both theoretically [69] and experimentally [70]. Three main mechanisms including photovoltaic effect [71], thermoelectric effect [72] and bolometric effect [73] have been identified as the origins of photocurrent in graphene junctions. Furthermore, hot carrier transport and carrier multiplication are believed to play an important role [74]. Beside light absorption, carrier separation and collection are also very important in photodetector and it is usually realized by introducing an electric field by external bias or build-in field, where p–n junction and Schottky junction are usually applied. The recent develop progress of graphene composite photodetectors will be reviewed in Sect. 5 in details.

4.4 Graphene Plasmonics in Terahertz

The intraband transition in graphene is in the THz range resulting in various graphene THz devices. Furthermore, optical properties of graphene in the THz range are similar to the noble metals in the visible range [52]. Therefore, it is expected that graphene film or nanostructure graphene could provide plasmonic-like phenomena, for example, perfect absorption [75, 76], beam manipulating [77], spectral engineering [78, 79], and even THz generation [72, 80] and detection [70, 81].

5 Graphene Composite Photodetectors

In a recent review on graphene [50], a graphene photodetector is considered to be the first graphene optoelectronic devices for optical interconnect applications and it first prototype could be expected before 2020. Although both the transition-limited bandwidth up to 1.5 THz and the RC-limited bandwidth up to 640 GHz predict a high-speed performance, the photoresponsivity of current graphene photodetector is at a level of a few mA/W, which is quite low compared to the required 1 A/W for applications. There are three main reasons for the low responsivity of graphene photodetectors: (1) the weak light absorption of graphene monolayer; (2) the ps-timescale combination; (3) small junction region for charge carrier collection. Combining graphene with other nanomaterials or nanostructures is an effective way to improve light absorption and carrier collection. Conversely, graphene benefits the carrier transition and separation in other active nanomaterials. In this Section, we will review the recent progress of graphene composite photodetectors and the contents are arranged as follows: (Sect. 5.1) graphene-semiconductor film photodetectors; (Sect. 5.2) graphene-NW photodetectors; (Sect. 5.3) graphene-QDs photodetectors; (Sect. 5.4) graphene-CNT photodetectors; (Sect. 5.5) nanophotonic graphene photodetectors.

5.1 Graphene-Semiconductor Film Photodetectors

Since Xia demonstrated the first ultrafast (40 GHz) transistor-based graphene photodetector [12], lots of efforts have been put into the development of high-response, ultrafast, broadband graphene photodetectors [15, 70, 81–88]. The basic structure is as shown in the inset of Fig. 2a, which can be seen as a back gate FET. The photocurrent can be observed without a source-drain bias due to the build-in field near the Schottky junction. However, the small junction area and the fast carrier recombination limited the photoresponsivity of the first graphene photodetector to only ~0.5 mA/W [12]. To solve this issue, Xia proposed a MSM graphene photodetector with interdigitated electrodes as shown in Fig. 2b [15]. The photoresponsivity increased to 6.1 mA/W due to the better carrier collection and the larger junction area compared to the device in Fig. 2a. But the metal shielding effect and the low single path absorption still limited the photoresponsivity.

Alternatively, the low photoresponsivity can be simply increased by integration of multiple graphene layers [89–91]. As shown in Fig. 3a, p–i–n junctions are formed due to chemical doping or electrical gating. Each depleted intrinsic section (*i*) is formed at reverse bias, where the photocarriers induce the photocurrent in each graphene monolayer. Then the photogenerated electrons and holes propagate through the channel toward the p-section and n-section, respectively. In theory, the responsivity of the multiple-graphene-layer photodetectors with large number of graphene layers could be very high, for example 74 % for 20 layer graphene [89].

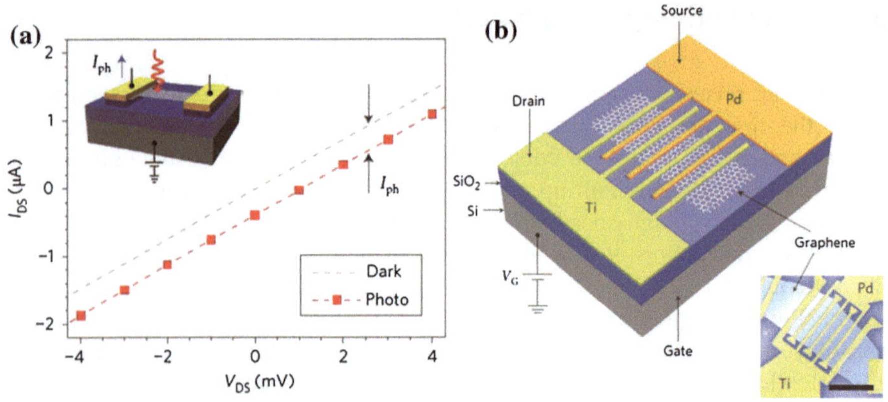

Fig. 2 **a** *I–V* curves of the graphene photodetector with and without light illumination. *Inset* is the schematic of the device. **b** Metal–graphene–metal photodetectors with asymmetric metal contacts. **a** Reference [12], Copyright 2009. Reproduced with permission from Nature Publishing Group. **b** Reference [15], Copyright 2010. Reproduced with permission from Nature Publishing Group

The optimized detectivity with different graphene layers, shown in Fig. 3b, gets its limit order of 10^{13} cm $Hz^{1/2}$ W^{-1} (Jones) at 77 K and 10^9 Jones at 300 K due to the dark current limit. However, the photoresponsivity increases with the increasing number of the graphene layer, and gets its maximum 227 A/W when the number is 100. Although the photoresponsivity is improved, the fabrication process becomes very complex, resulting in low production.

Different from the pure graphene photodetector above where light is absorbed in graphene with generated electron-hole pairs, using graphene as a junction material has been considered. Graphene has near-zero band gap and high conductivity, and thus it can be regarded as metal. Semiconductor/graphene interface therefore forms a semiconductor/metal-like Schottky contact. An et al. [92] demonstrated a graphene-silicon Schottky detector at visible range, where silicon forms an Ohmic contact with the bottom Gold/Titanium electrode and a Schottky contact with the top graphene layer as shown in Fig. 4. Compared to graphene FET [70, 82] and MSM [15] photodetectors, there is no metal shadow effect due to the high transparency of graphene. Furthermore, the junction area only depends on the size of graphene but not limited by the metal/graphene interface. Therefore, charge carriers are generated at the whole graphene/silicon interface and separated by the build-in field of the Schottky barrier. Because of the vertical transport of charge carriers and the thin thickness of graphene, the recombination of carrier is suppressed. The photocurrent responsivity reaches 225 mA/W at 2 V bias, which is 1–2 orders of magnitude higher than those of previously graphene-based photodetectors, such as the reported ~6.1 mA/W by Mueller et al. [15]. Similar device structure was applied to develop a graphene-Si detector in the near infrared range [93]. An internal quantum efficiency as high as 30 % was obtained, which is much higher than 6–16 % in Ref. [12]. Compared to an all-Si-based photodetector [94],

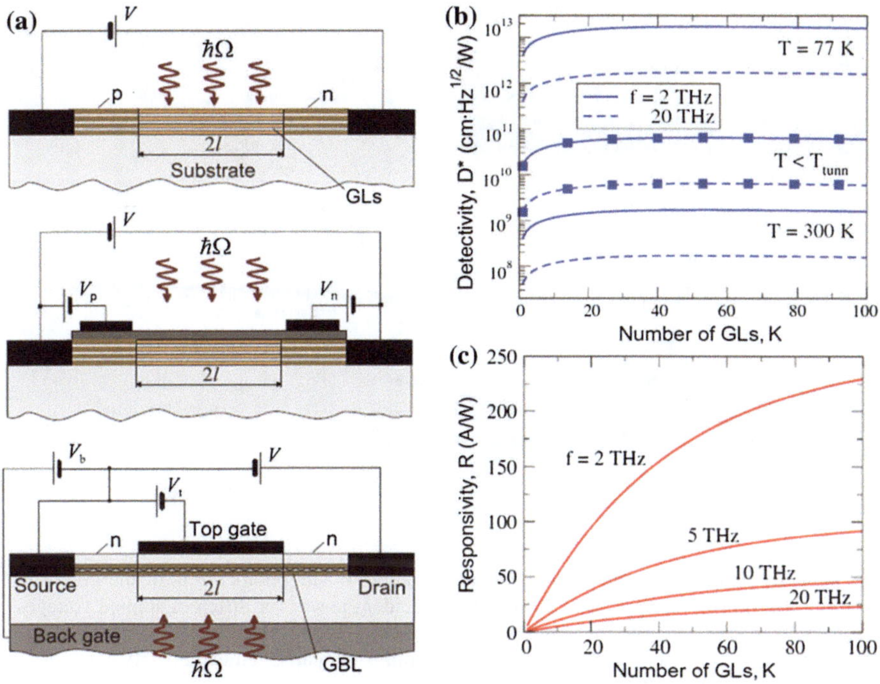

Fig. 3 **a** A gradual optimized multiple-graphene-layer p–i–n photodetector. **b** Responsivity of multiple-graphene-layers photodetector versus numbers of graphene layers with the radiation frequencies ranging from 2 to 20 THz. **c** The detectivity of the photodetector versus number of graphene layers with different temperatures. Reference [90], Copyright 2011. Reprinted with permission from Elsevier

the photoresponse of 9.9 mA/W of the graphene-Si Schottky detector is 123 times higher. As seen, the graphene-Si Schottky detectors show broadband characteristics as graphene photodetectors.

Encouraging by the simple device fabrication process of graphene-silicon detectors, other semiconductor materials were also investigated to construct a Schottky detector with graphene. A graphene-Ge Schottky photodetector showed a responsivity of 51.8 mA/W and a detectivity of 1.38×10^{10} Jones under 1.55 μm light illumination with power of 18 W/cm^2 [95]. A Schottky junction can be formed between graphene and CdSe, owing to the work function difference between graphene (~4.5 eV) and CdSe (4.2 eV), which is a direct bandgap material. By constructing a Schottky photodetector with graphene and CdSe, high response photodetection was achieved in the visible range [96, 97]. As shown in Fig. 5, CdSe nanobelts were fabricated on graphene sheet coated SiO_2/Si substrate (Fig. 5a) and polyethylene terephthalate (PET) substrate (Fig. 5d). Under 633 nm light illumination, the photoresponsivity of such graphene-CdSe photodetectors are about 10.2 and 8.7 A/W, and the on/off ratios are 3.5×10^5 and 1.2×10^5, respectively, which are much higher than the graphene-silicon one [93]. Although

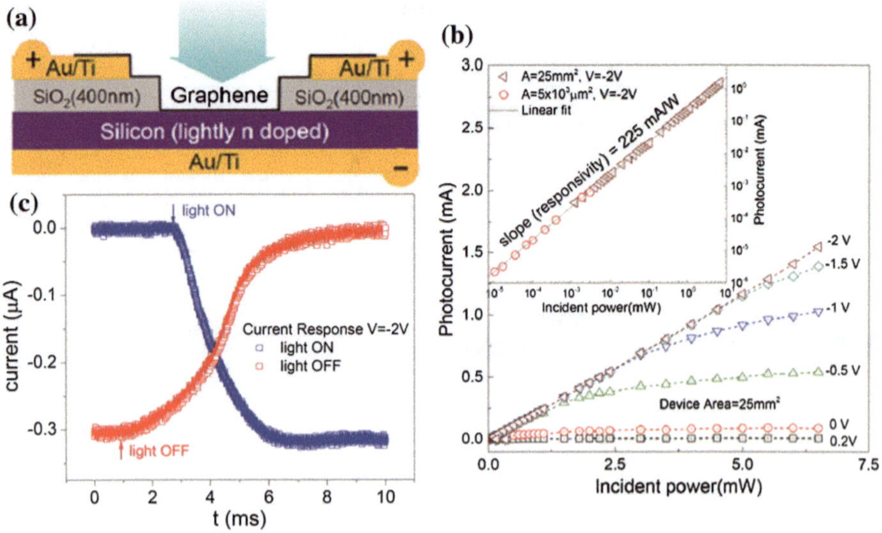

Fig. 4 **a** Schematic of the graphene-Si Schottky diode. **b** The photocurrent of the graphene-Si Schottky junction photodetector as a function of incident power for different applied voltages. **d** Transient photocurrent response of the photodetector at −2 V reverse applied voltage. Reference [92], Copyright 2013. Reprinted with permission from American Chemical Society

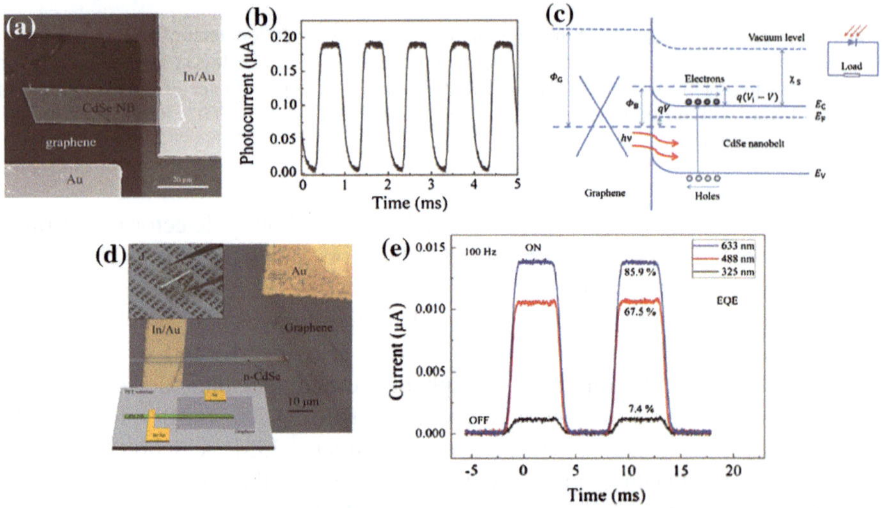

Fig. 5 Schematics and microscope photos of the graphene/CdSe nanobelt photodetector on SiO_2/Si substrate (**a**) and PET (**d**, **e**), respectively. **b** Photocurrent response of the photodetector (**a**) under a 1,000 Hz light switching frequency with 633 nm laser illumination. **c** The energy band diagrams of the photodetector under light illumination. **e** The photocurrent responses of the device (**d**) with the calculated external quantum efficiencies under different photon wavelengths with fixed photon density (~1.3×10^{17} s^{-1} cm^{-2}). **a**, **b** Reference [96], Copyright 2012. Reprinted with permission from Royal Society of Chemistry. **c**–**e** Reference [97], Copyright 2013. Reprinted with permission from Royal Society of Chemistry

the success in improving the photoresponsivity, these graphene-semiconductor junction photodetectors may lose the high-speed property because of the limited carrier mobility of semiconductors.

5.2 Graphene-Nanowire Photodetectors

Semiconductor nanowires with excellent photoelectric properties are important building blocks for ultracompact photodetectors [98]. For example, ZnO UV photodetectors show great potential in high sensitivity and fast response [99]. Generally, Schottky nanowire photodetectors have much higher sensitivity than photoconductive detectors. However, the metal electrodes shield the light and dramatically reduce the absorption efficiency. The good transparency and high carrier mobility make graphene a promising candidate as transparent electrodes. What's more, graphene-nanowire is expected to form a Schottky barrier for affiliating the charge carrier separation and collection for fast response. It is particularly important for UV photodetectors as the traditional transparent conductive electrode, for example ITO [100], has significant absorption in UV range.

Several graphene-nanowire photodetectors have been recently reported based on CdS [101], ZnO [102, 103], GaN [104] nanowires and TiO_2 nanotubes [105]. In these graphene-nanowire configurations, graphene can be used as Schottky contacts [85], transparent electrodes [102] and even just a growth template [106]. In Ref. [102], ZnO nanowires were grown by electrodeposition technique with high residual n-type doping and the graphene sheet consists of four graphene monolayer as shown in Fig. 6. As a result, the work function of ZnO is very close to that of graphene electrode. Due to the low resistance contact formed between graphene and nanowires, Ohmic-like behavior was observed, indicating that graphene acts as transparent electrode. The excellent light-trapping effect of nanowire array and the high transparency of graphene electrode provide responsivity as high as 104 A/W in the near UV range with a long rise/fall time above 100 s. Similar work was demonstrated by a graphene-GaN nanowire composite photodetector [104], where a room-temperature responsivity of 25 A/W at 357 nm was demonstrated.

In the above examples [102, 104], the Ohmic contact was formed at the graphene-ZnO nanowire interface due to the low resistance contact for heavily doped nanowires. If a high resistance contact, i.e., a Schottky contact, could be formed, it provides better carrier separation and collection than the Ohmic contact. As shown in Fig. 7, ZnO nanowire forms Schottky contacts at both top and bottom interfaces with graphene layers [85], where ZnO nanowires were synthesized via a chemical vapor deposition technique and the graphene sheet is monolayer resulting in high resistance contacts. The photoconductive gain of the UV photodetector was calculated to be ~1.6 at the bias of 3 V and under a UV laser power of 50 $\mu W/\mu m^2$. The recovery time is approximately 0.5 s, much faster than the previously reported ZnO nanowire photodetector with Ohmic contact (several seconds to minutes). The graphene-ZnO nanowire Schottky junctions provide fast electron-hole pair

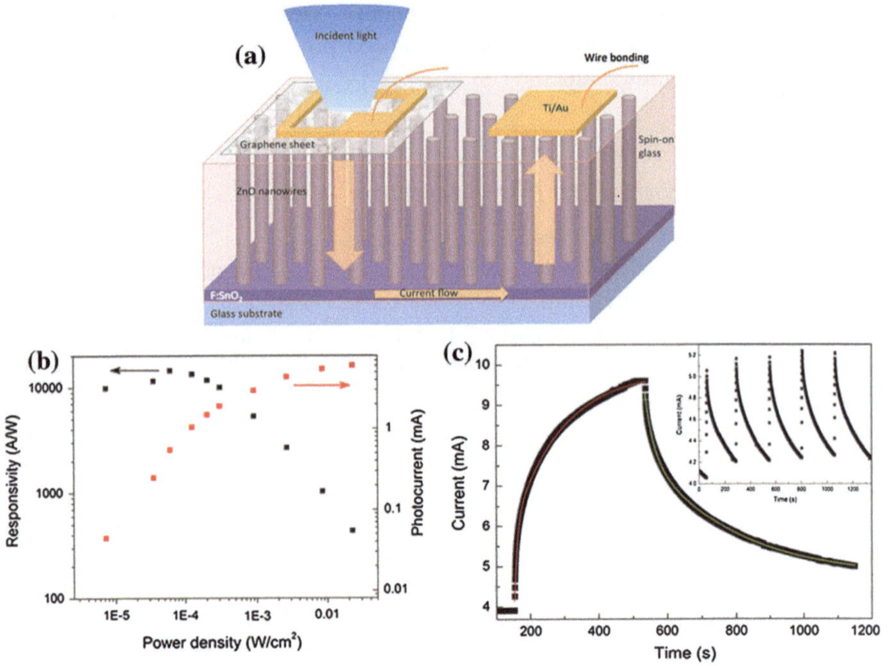

Fig. 6 **a** Schematic of the UV photodetector. **b** Dependence of the photoresponsivity and photocurrent on the incident power density. **c** Current transient under UV illumination during 450 s and relaxation during 600 s. *Inset* shows the response to multiple illumination cycles with 5 s on and 240 s off time. Reference [102], Copyright 2013. Reprinted with permission from AIP Publishing LLC

separation, resulting in a decrease of the recombination rate and an increase of the carrier lifetime. It is believed that the increased carrier density in ZnO raises the Fermi level and reduces the work function of ZnO, resulting in lower and thinner Schottky barriers, i.e., fast recovery time. In addition, the vertical sandwiched dual Schottky barrier structure has larger light absorption than the conventional one with metal electrodes due to the high transparency of graphene. Besides the single nanowire-graphene composite photodetector, nanorod array combined with graphene were also developed for Schottky junction photodetectors as shown in Fig. 8 [103]. Due to the formation of Schottky junction, the strong light trapping effect of the ZnO nanorod arrays and the ultra-high mobility of graphene, a high photoresponsivity of 113 A/W was obtained and the response speed was extremely high (rise/fall time: 0.7/3.6 ms), much quicker than that of similar UV photodetectors (order of 1–10 s) [107, 108].

Most recently, a ZnO MSM photodetector as shown in Fig. 9 was fabricated on graphene coated oxidation silicon substrate [106]. The crystalline quality of ZnO was greatly improved due to the smaller difference of the lattice constant between

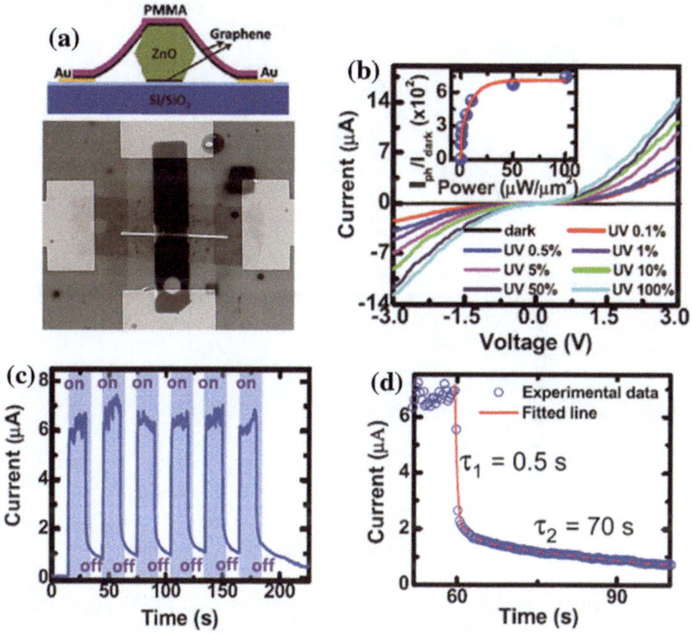

Fig. 7 **a** Schematic and SEM image of the graphene/ZnO NW/graphene Schottky junction UV photodetector. **b** *I–V* characteristic of the graphene/ZnO NW/graphene device in dark and under illumination of a 325 nm. **c** Time dependence of the photocurrent with switching on/off the 325 nm UV laser illumination with the 2 V bias voltage. **d** Magnification of the detailed behavior of the photocurrent decay. Reference [85], Copyright 2012. Reproduced with permission from AIP Publishing LLC

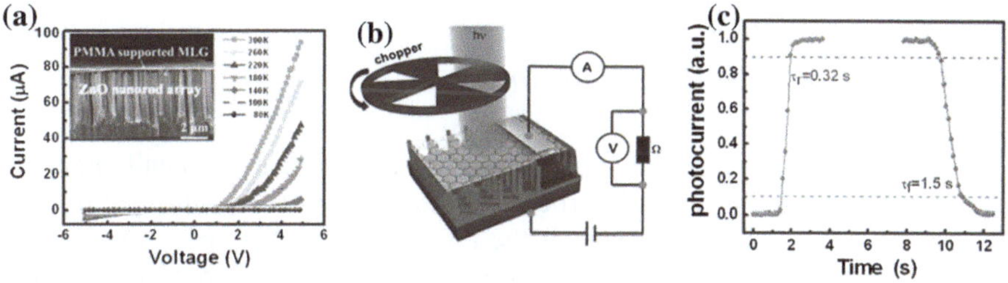

Fig. 8 **a** *I–V* characteristics of multilayer graphene film/ZnO nanorod array Schottky diode at temperatures ranging from 80 to 300 K, *inset* shows the cross-sectional SEM image of the UV photodetector. **b** Schematic illustration of the experimental setup for studying the time response of the UV photodetector. **c** The rise and decay edges of the device. Reference [103], Copyright 2013. Reprinted with permission from John Wiley & Sons, Inc.

graphene and ZnO. Both the photoluminescence spectra and the photoresponsivity show a performance improvement by the introduction of graphene. With 1 V bias voltage, the highest responsivity is 188 A/W for the 370 nm-incident light, and average responsivity is 0.02 A/W for an incident visible light wavelength from 410

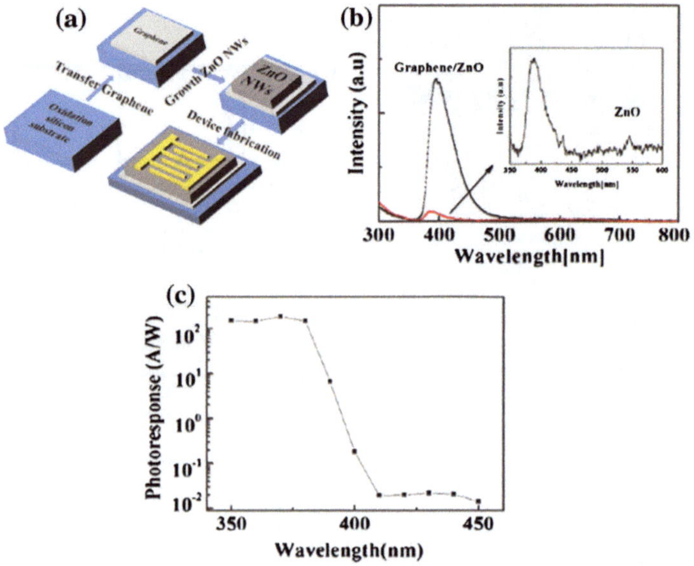

Fig. 9 **a** The fabrication steps and a schematic diagram of ZnO MSM photodetector. **b** Room temperature photoluminescence spectra of ZnO NWs grown on a graphene layer and on a silicon dioxide substrate. **c** Spectral photoresponse of the ZnO NW MSM photodetector at 1 V bias voltage. Reference [106], Copyright 2014. Reproduced with permission from IOP Publishing

to 450 nm. The surface plasmon at the interface of ZnO nanowires and graphene was believed to enhance the response by the strong near field confinement.

5.3 Graphene-QDs Photodetectors

QDs are nanocrystals made of semiconductor materials that are small enough to exhibit quantum mechanical properties. QD photodetectors attract broad research interests due to its large tunability of the response band, high photoresponsivity, high operation temperature and low noise. But it also suffers from the poor carrier transition property due to the non-continuous distribution and surface defects, which degenerates both photoresponsivity and transient response. Therefore, it is highly desired to improve the carrier transition in QD photodetectors.

Recently, a graphene-ZnO QD UV photodetector was proposed and demonstrated a significant enhancement of the photocurrent [83], where the photocurrent increases from 3.36×10^{-5} to 5.15×10^{-3} A for the devices without and with graphene, respectively (Fig. 10). As shown in Fig. 10a, graphene is embedded between aluminum electrode and the hybrid layer of ZnO QD and poly-N-vinylcarbazole (PVK). The graphene layer with a high electrical conductivity attached to the hybrid layer facilitates the electron transportation to aluminum electrode as shown in Fig. 10b. This high carrier transition also improves the response speed of the

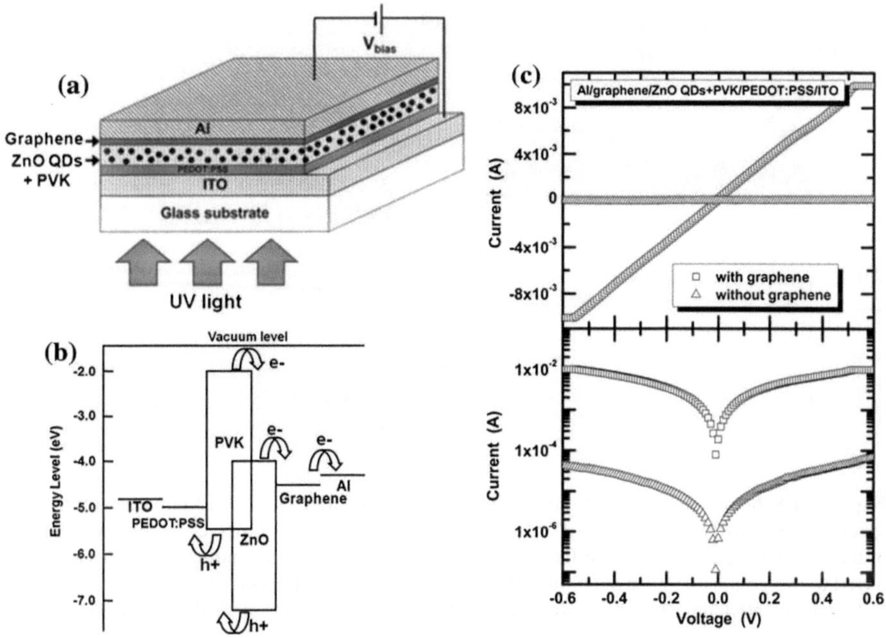

Fig. 10 **a** Schematic of a graphene-ZnO QDs UV photodetector. **b** Schematic diagram of the energy bands. **c** Currents as functions of the applied voltage for the UV photodetector with (*rectangles*) and without (*triangles*) graphene under illumination by a 365 nm UV light at a power of 6 W. Reference [83], Copyright 2010. Reprinted with permission from Elsevier

photodetectors. In Ref. [49], a fast transient response on the order of 10 ms together with a large photoresponsivity of 247 A/W was achieved in a graphene-ZnO QDs composite photodetector by using atomic layer deposition method.

A recorded high photoresponsivity of 10^7 A/W was obtained in graphene-PbS QDs composite photodetectors [109]. As shown in Fig. 11, the graphene flake is deposited onto a Si/SiO_2 structure and then coated with PbS QDs. Light absorption by a sensitizing layer of QDs results in exciton generation. Holes are transferred to graphene and then transit to electrode with a timescale of $\tau_{transit}$, but electrons are trapped in QDs ($\tau_{lifetime}$). The ultrahigh gain is caused by the high carrier mobility of graphene and the recirculation of charge carriers during the long trapped-charge lifetime in QDs, which is given by $G = \tau_{lifetime}/\tau_{transit}$. It is the reason that we see a significant enhancement of photoresponsivity in such a graphene-PbS QDs composite photodetector.

Differ from the simple graphene-QD composite structure, coupling quantum dots (QDs) on graphene in a PbSe–TiO_2-Graphene hybrid photodetector [110], shown in Fig. 12a, potentially allow the excited state of the QDs to be transformed into a charge-transfer state, and the charge injection to graphene is expected to occur on a timescale faster than exciton recombination [111]. Meanwhile, the synergistic effects of these QDs on graphene also cause a broadband photoelectric

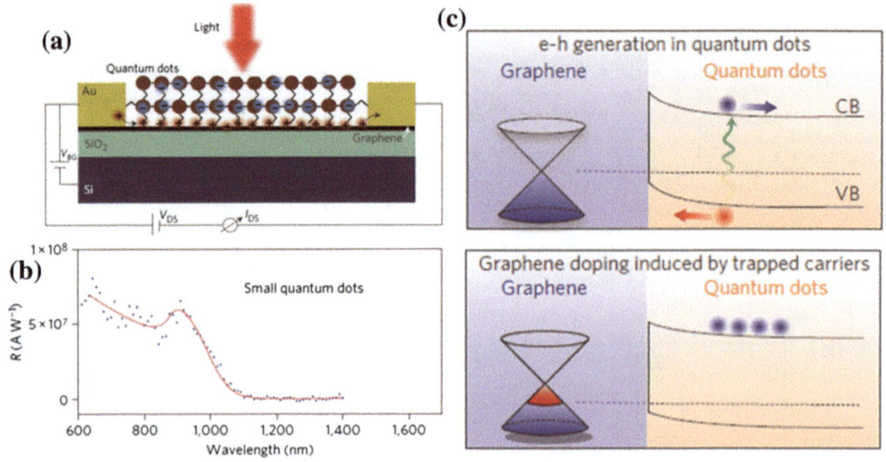

Fig. 11 **a** Schematic of the cross section of graphene-quantum dot hybrid photodetector. **b** The spectral responsivity of graphene of PbS QDs composited photodetectors. **c** Energy level diagram of the graphene/quantum dot interface. On photo-excitation of PbS quantum dots, holes are transferred to the graphene under the built-in field, leaving electrons trapped in the PbS quantum dots. Reference [109], Copyright 2012. Reprinted with permission from Nature Publishing Group

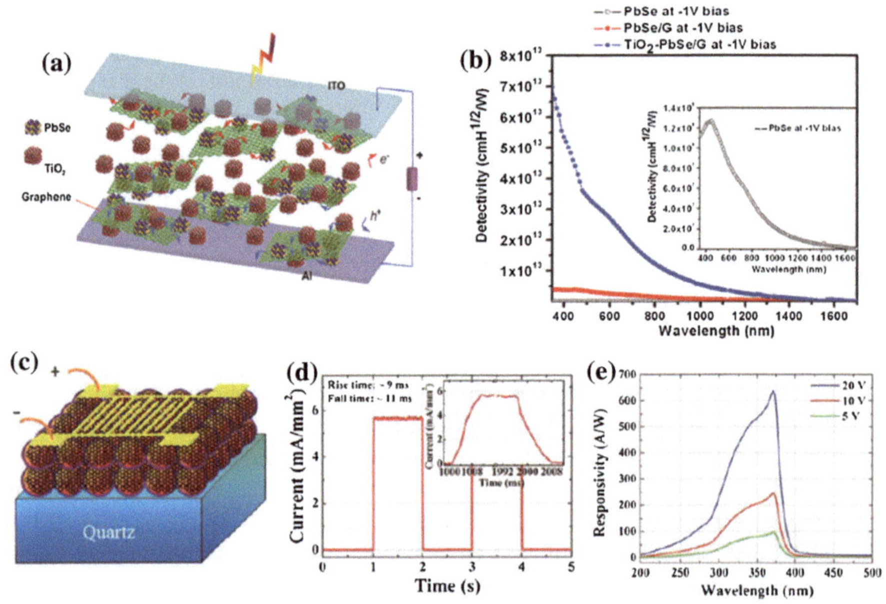

Fig. 12 **a** Schematic of a PbSe–TiO$_2$–Graphene hybrid photodetector. **b** The Detectivities of different components as a function of wavelength measured at −1 V applied bias. **c** Schematic of a ZnO nanoparticles-Graphene core-shell photodetector. **d** Transient response of the UV photodetector. **e** Responsivity of the photodetector. **a**, **b** Reference [110], Copyright 2012. Reprinted with permission from John Wiley & Sons, Inc.; **c**–**e** Reference [113], Copyright 2013. Reprinted with permission from Royal Society of Chemistry

conversion [112], thus, the detectivities in the visible and IR range can reach ~3×10^{13} Jones and ~5.7×10^{12} Jones, respectively. Alternatively, a ZnO nanoparticle-graphene core-shell structure [113] may be another method to improve the photoresponsivity of the photodetectors, as shown in Fig. 12c. Warping graphene onto the ZnO nanoparticles makes graphene to better contact with each nanoparticle, which contributes to a high responsivity and fast transient response. A large photoresponsivity of 640 A/W with an order of 10 ms transient response was achieved in this core-shell photodetector.

It is worth to note that light absorption and charge-transport processes are decoupled in the graphene-QDs composite photodetectors. Different from graphene field-effect phototransistors [12], light absorbers are semiconductor QDs rather than graphene. Therefore, we might not expect the broadband and high speed photoresponse in such devices as in a graphene photodetector. For example, only 10 ms transient response was achieved in [113]. It is also expected in the graphene-semiconductor film Schottky junction detectors in Sect. 5.1 and the graphene-nanowires detectors in Sect. 5.2.

5.4 Graphene-CNT Photodetectors

Graphene can also form a junction photodetector with CNT [40]. The graphene-CNT hybrid suspension as fabricated by mixing graphene flakes and CNTs. After vacuum filtration, the films were transferred onto an oxide coated silicon substrate as shown in Fig. 13a. Graphene may form heterojunctions on CNTs with band edge offset. As shown in the inset of Fig. 13b, the work function of multiwall CNT is larger than 4.5 eV with a diameter of several tens of nm. Significantly enhanced photoresponsivity in near-IR region was achieved in the graphene/multi-walled nanotubes (MWCNT) nanohybrid IR photodetectors as compared to the best achieved so far on MWCNT IR photodetectors. In addition, the photodetectivity increases by a factor of 500 % to 1.5×10^7 Jones. The improvement attributes to a novel exciton dissociation mechanism through heterojunctions self-assembled at the graphene-multiwall CNT interfaces.

5.5 Nanophotonic Graphene Photodetectors

All the above sections focus on the photodetectors consisting of graphene and various semiconductor materials, in which the performance improvement relies on the decrease of charge carrier recombination, the increase of junction area, and the enhanced carrier separation and collection in the graphene composite. There is another important issue in all kinds of graphene photodetectors, i.e., the weak light absorption. In this section, we review the techniques and results of nanophotonic graphene photodetectors, in which nanostructures rather than nanomaterials are applied to improve the photoresponsivity.

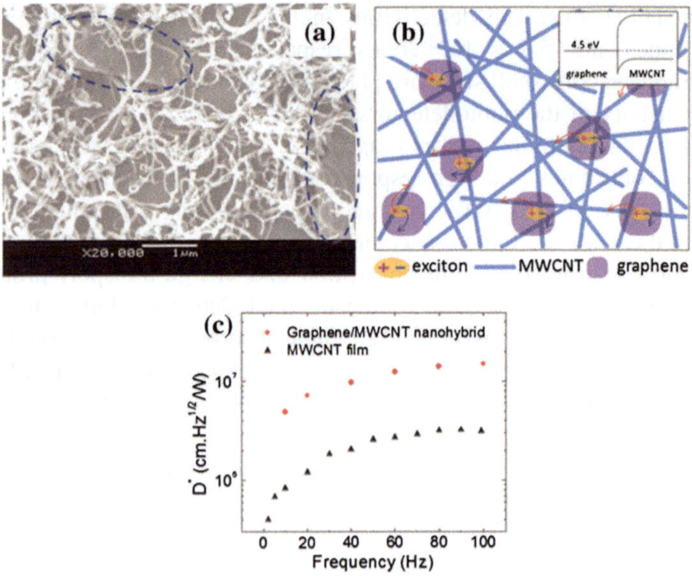

Fig. 13 **a** SEM image of graphene/MWCNT nanohybrid structure. **b** Schematic of exciton dissociation in graphene/MWCNT nanohybrid. The *inset* shows the band diagram of graphene/MWCNT nanohybrid structure. **c** Comparisons of photodetectivity as a function of frequency for graphene/MWCNT nanohybrid and reference MWCNT films with $\sim$0.3 mW/mm^2 NIR intensity and 2 mA bias current. Reference [40], Copyright 2013. Reprinted with permission from American Chemical Society

5.5.1 Microcavity-Based Graphene Photodetectors

Pristine graphene monolayer has light absorption of 2.3 % from UV to IR for a single pass. A direct way to increase the photocurrent is to use multilayer graphene as shown in Fig. 3 [89–91]. Alternatively, we can modify the light pathway to allow multiple pass through graphene monolayer. It is actually typical photon recycling in nanophotonics. A planar cavity with active material sandwiched between two reflectors is the easiest way to realize the photon recycling, for example resonant cavity enhanced photodetectors [114–118]. Besides the recycling effect, a microcavity supports highly confined resonant modes, which significantly increase the overlap between active materials and light leading to larger absorption and photocurrent. Furthermore, the compact size of the cavity-based photodetector enables high-speed operation due to the reduced RC parameters. In addition, the resonant mechanism provides wavelength selectivity, which is very useful in spectroscopy and spectral imaging [117, 118].

Figure 14a shows such a dielectric microcavity with two distributed Bragg reflector (DBR) mirrors [119]. Optimization of the cavity structures ensures a maximum of field distribution at the position of graphene and the high reflection of the DBRs minimizes the transmission and reflection loss at the resonant

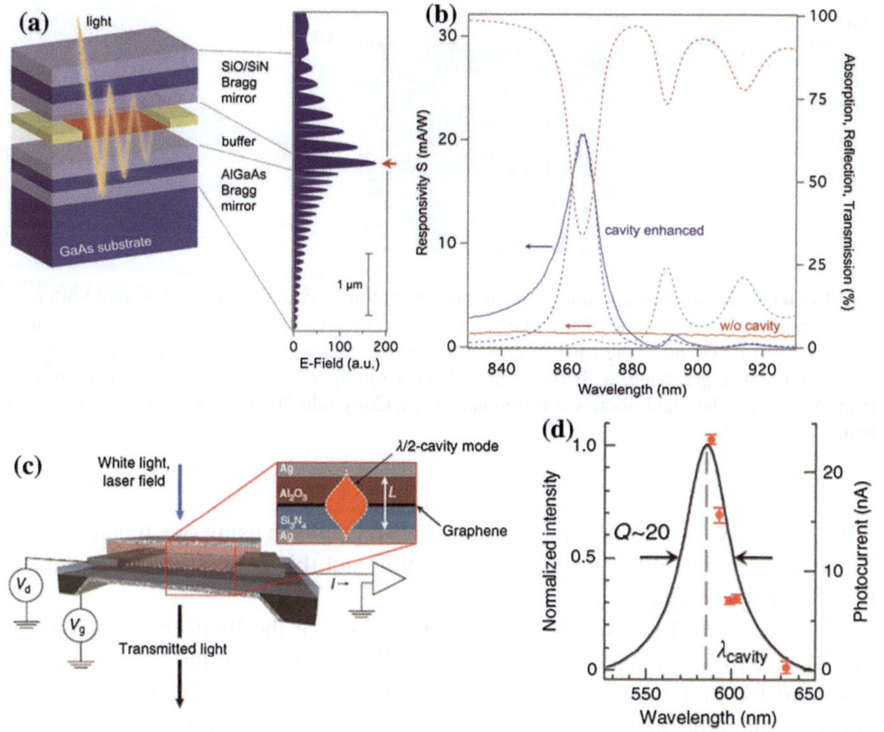

Fig. 14 **a** Schematic of dielectric microcavity graphene photodetector. **b** Performance comparison for devices in (**a**) with and without microcavity. The *dashed lines* are calculation results for the microcavity device: reflection (*red*), transmission (*green*), and absorption (*blue*). The *solid lines* show the photoresponse of the cavity-enhanced device (*solid blue*) and conventional (*without cavity*) photodetector (*solid red*). **c** Schematic of metallic microcavity graphene photodetector. **d** Optical transmission spectrum (*black line*) of the device in (**c**) measured with white light illumination. And the measured laser-induced photocurrent amplitude (*red dots*) samples the spectral profile of the optical cavity resonance. **a**, **b** Reference [119], Copyright 2012. Reprinted with permission from American Chemical Society; **c**, **d** Reference [120], Copyright 2012. Reprinted with permission from Nature Publishing Group

wavelength, which is tunable and determined by the cavity structures. In theory the graphene absorption could be up to 100 % and the measured result researched 60 %, 26 times higher than that of pristine graphene monolayer. The photoresponsivity of 21 mA/W was achieved as shown in Fig. 14b, indicating a positive effect of the cavity. In contrast to the dielectric DBR cavity, metallic Fabry-Pérot cavity as shown in Fig. 14c was proposed [120], where metal reflectors rather than dielectric ones are placed at both sides of graphene. An increase of photocurrent by twice was also observed in experiment. Comparing these two cavities, both have pros and cons. Dielectric one has a higher improvement of photocurrent due to the lossless cavity structure but suffers from complex fabrication process and poor electrical properties, for example gating graphene is difficult through dielectric

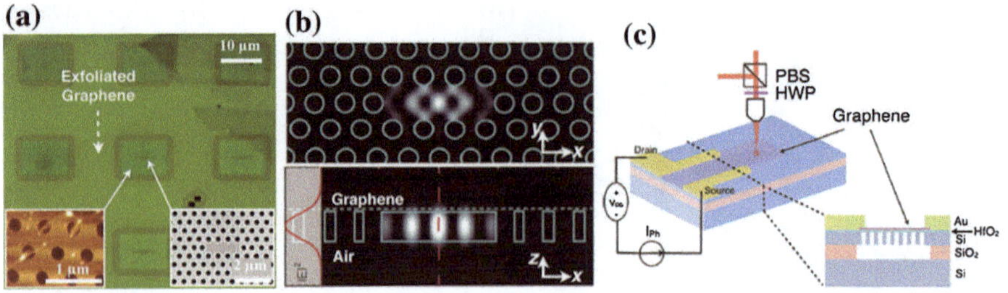

Fig. 15 **a** Optical microscope image of the planar photonic crystal cavities integrated with exfoliated graphene. **b** Schematic of the planar photonic crystal cavity integrated graphene photodetector. **c** SEM image and schematic of cross section of a planar photonic crystal cavity integrated with single-layer graphene. **a**, **b** Reference [121], Copyright 2012. Reprinted with permission from American Chemical Society **c** Reference [122], Copyright 2013. Reprinted with permission from AIP Publishing LLC

DBRs. Metallic one allows an easy way to gate graphene using one metal reflector as the gate electrode but the surface reflection and the metal attenuation reduced the photoresponsivity

Photonic crystal cavities are also used to enhance the light-graphene interaction [121–123]. A linear defect cavity in a 2D photonic crystal concentrates the light into the high-quality slow light mode as shown in Fig. 15. The interaction between the graphene and the evanescent field dominates the cavity loss and gives estimated 45 % light absorption. The measured reflection spectra show a 20 dB more attenuation in a cavity with graphene compared to the same cavity without graphene [121]. By fabricating source and drain electrodes on the graphene covered photonic crystal, the measured photocurrent shows an eight-fold enhancement [122]. Most recently, Piper and Fan [124] proposed another technique based on photonic crystal to improve the light absorption in graphene. By placing graphene onto a high-index photonic crystal slab, the incident light is expected to be coupled into the guided resonance mode in the slab and be absorbed in graphene during the lateral propagation.

5.5.2 Surface Plasmon Enhanced Graphene Photodetectors

Surface plasmon has been widely used in optoelectronic devices to improve light radiation [125], detection [126], modulation [127], and so on. Especially, its strong near field and super-scattering effects show significant improvement of photocurrent in plasmonic solar cells. Similarly, plasmonic or metamaterial nanostructures can enhance the photoresponsivity of graphene photodetectors [17, 69, 74].

As shown in Fig. 16a, the gold nanostructures were fabricated on graphene film [16]. By excitation of surface plasmon resonance assisted by the metallic nanostructures, the near field at the nanostructure significantly enhanced the light-graphene interaction, resulting in a photo responsivity of 10 mA/W. The strong confined electromagnetic field in the nanogap between gold particles was also used to improve the photoresponsivity as shown in Fig. 16b [128]. However, the responsivity is as small as 3 μA/W due to the very limited effective area in the nanogap. Large scale surface plasmon resonance enhanced graphene photodetector was obtained by annealing thin film gold layer on top of graphene as shown in Fig. 16c [17]. The fabricated nanoparticles increase the forward scattering and enhance the local electric fields. As a result, a responsivity of 2.2 mA/W is achieved, *i.e.*, an average more than 400 % enhancement over the one of 0.42 mA/W for the device without gold nanoparticles.

To further increase the photoresponsivity by surface plasmon, optimization of nanostructures with a strong Fano resonance was investigated and show a maximum responsivity of ~8 mA/W in a gold heptamer array sandwiched between two layers of graphene monolayer as shown in Fig. 16d [74]. The significant enhancement of photoresponsivity is due to the strong near field of Fano resonance and the

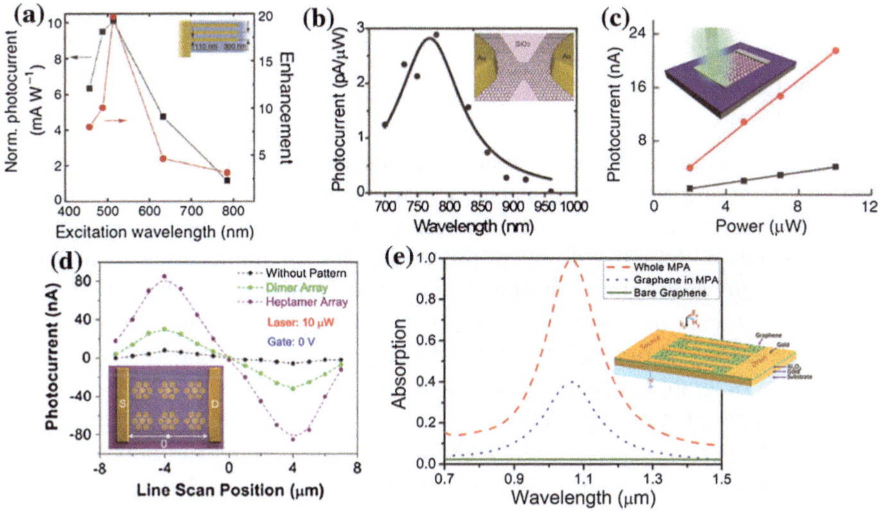

Fig. 16 **a** Normalized photocurrent and maximum enhancement coefficient for a graphene photodetector with metallic finger structures as shown in the *inset*. **b** Wavelength dependence of the photocurrent of a gold nanogap graphene detector as shown in the *inset*. **c** Photocurrent generated as a function of laser power under 514 nm light illumination in a nanoparticle covered graphene photodetector as shown in the *inset*. The *red* and *black lines* indicate the response of the device with and without Au nanoparticles, respectively. **d** Photocurrent measurements with the line scan position in the graphene photodetector, in which the two layers of graphene monolayer sandwich the gold heptamer array. **e** Absorption spectrum of a graphene-metamaterial photodetector as shown in the *inset*. The absorption spectra of the whole MPA and a bare graphene monolayer are shown for comparison. **a** Reference [16], Copyright 2011. Reprinted with permission from Nature Publishing Group; **b** Reference [128], Copyright 2011. Reprinted with permission from American Chemical Society; **c** Reference [17], Copyright 2011. Reprinted with permission from Nature Publishing Group; **d** Reference [74], Copyright 2012. Reprinted with permission from American Chemical Society; **e** Reference [69], acknowledgements to be used by RSC authors

conformal contact between the gold nanostructures and graphene. The hot-carrier effect was also claimed to contribute to the photocurrent.

We also proposed a novel graphene-metamaterial photodetector [69], where the graphene monolayer is embedded in a metamaterial perfect absorber as shown in Fig. 16e. The metamaterial perfect absorber consists of a periodically nanopatterned top metal layer and a continuous bottom metal film separated by a dielectric layer. The graphene is placed between the top metal nanostructure layer and the dielectric layer. By optimization this sandwich structure, we can obtain a perfect impedance matching between the structure and the air and therefore suppress the reflection loss at a resonance. At the same time, the transmission is zero for an optically thick bottom metal film. As a result, a perfect absorption occurs at a resonance of 1.07 μm and 40 % comes from graphene as shown in Fig. 16e. Besides of the 17 times enhancement of the light absorption, the device structure we proposed natively matches with the transistor, where the bottom metal layer can be used as the back gate and the top metal nanostructures can be used source and drain electrodes. Due to the impedance match, no metal shadow effect has to be worried.

6 Conclusion

Graphene is a unique two dimensional nanomaterial in the sense that it combines many superior properties, from mechanical to electronic. Its high carrier mobility, broadband optical absorption and ultra-thin properties have potential applications in high-speed and broadband optical processing and communication such as photodetectors. However, graphene monolayer has very weak light absorption due to the sub-1 nm-scale thickness, resulting in very low photoresponsivity, especially in photodetectors. Various methods have been reported to improve the photoresponsivities of graphene-based photodetectors. In this chapter, we reviewed the recent progress of graphene photodetectors and presented the promising results on graphene-nanomaterial and graphene-nanostructure composite photodetectors. Although different mechanisms for improving the photoresponsivity, all devices focus on increasing the light/graphene interaction, enlarging the junction area and improving the carrier separation and collection. It is found that both nanomaterials (QDs and nanowires) and nanostructures (microcavity and surface plasmon) can significantly increase the light absorption. One or two orders of magnitude high photocurrents have been achieved for devices from UV to IR range. Although it is still lower than the traditional semiconductor photodetectors, it shows a possibility to further improve the performance by optimization of the composite material and structures. Graphene composites consisting of graphene and other materials combine the high carrier mobility property of graphene and the excellent light absorption properties of other semiconductors. Therefore graphene composite photodetectors are one promising candidate for development of next-generation photodetectors. It is worth to note that we may lose the broadband and high-speed properties of graphene if all the focus is on improving the photoresponsivity, for example in graphene-QDs detectors [109] or graphene nanostructure detectors [129]. Of course, another important step is developing fabrication techniques for high-quality graphene devices at low cost and a large scale. Operation speed, power consumption and chip integration should be also carefully considered. With the fast continuous developments of material preparation, device design and fabrication, we believe that graphene composite photodetectors would play an important role of the future optoelectronic chips and systems.

Acknowledgments This work is supported by the grants from the National Natural Science Foundation of China (No. 11274344), the Hundred Talents Program of Chinese Academy of Sciences, the Scientific Research Foundation for the Returned Overseas Chinese Scholars and Suzhou Science and Technology Development Program Foundation (No. ZXG201425).

Reference

1. Geim, A.K., Novoselov, K.S.: The rise of graphene. *Nat. Mater.* **6**(3), 183-191 (2007). doi:10.1038/Nmat1849
2. Morozov, S.V., Novoselov, K.S., Katsnelson, M.I., Schedin, F., Elias, D.C., Jaszczak, J.A., Geim, A.K.: Giant intrinsic carrier mobilities in graphene and its bilayer. *Phys. Rev. Lett.* **100**(1), 016602 (2008). doi:10.1103/Physrevlett.100.016602
3. Mayorov, A.S., Gorbachev, R.V., Morozov, S.V., Britnell, L., Jalil, R., Ponomarenko, L.A., Blake, P., Novoselov, K.S., Watanabe, K., Taniguchi, T., Geim, A.K.: Micrometer-Scale Ballistic Transport in Encapsulated Graphene at Room Temperature. *Nano Lett.* **11**(6), 2396-2399 (2011). doi:10.1021/Nl200758b
4. Lee, C., Wei, X.D., Kysar, J.W., Hone, J.: Measurement of the elastic properties and intrinsic strength of monolayer graphene. *Science* **321**(5887), 385-388 (2008). doi::10.1126/science.1157996
5. Balandin, A.A., Ghosh, S., Bao, W.Z., Calizo, I., Teweldebrhan, D., Miao, F., Lau, C.N.: Superior thermal conductivity of single-layer graphene. *Nano Lett.* **8**(3), 902-907 (2008). doi:10.1021/Nl0731872
6. Balandin, A.A., Ghosh, S., Nika, D.L., Pokatilov, E.P.: Thermal Conduction in Suspended Graphene Layers. *Fuller. Nanotub. Car. N.* **18**(4-6), 474-486 (2010). doi:10.1080/1536383x.2010.487785
7. Choi, H., Borondics, F., Siegel, D.A., Zhou, S.Y., Martin, M.C., Lanzara, A., Kaindl, R.A.: Broadband electromagnetic response and ultrafast dynamics of few-layer epitaxial graphene. *Appl. Phys. Lett.* **94**, 172102 (2009). doi:10.1063/1.3122348
8. Zhao, W.S., Shi, K.F., Lu, Z.L.: Greatly enhanced ultrabroadband light absorption by monolayer graphene. *Opt. Lett.* **38**(21), 4342-4345 (2013). doi:10.1364/Ol.38.004342
9. Liao, A.D., Wu, J.Z., Wang, X.R., Tahy, K., Jena, D., Dai, H.J., Pop, E.: Thermally Limited Current Carrying Ability of Graphene Nanoribbons. *Phys. Rev. Lett.* **106**(25), 256801 (2011). doi:10.1103/Physrevlett.106.256801
10. Bae, M.H., Islam, S., Dorgan, V.E., Pop, E.: Scaling of High-Field Transport and Localized Heating in Graphene Transistors. *ACS Nano* **5**(10), 7936-7944 (2011). doi:10.1021/nn202239y
11. Pop, E., Varshney, V., Roy, A.K.: Thermal properties of graphene: Fundamentals and applications. *Mrs. Bull.* **37**(12), 1273-1281 (2012). doi:10.1557/mrs.2012.203
12. Xia, F.N., Mueller, T., Lin, Y.M., Valdes-Garcia, A., Avouris, P.: Ultrafast graphene photodetector. *Nat. Nanotechnol.* **4**(12), 839-843 (2009). doi:10.1038/Nnano.2009.292
13. Koybasi, O., Childres, I., Jovanovic, I., Chen, Y.P.: Graphene field effect transistor as a radiation and photo detector. *Proc. SPIE* **8373**, 83730H (2012). doi:10.1117/12.919628

14. Xia, F.N., Mueller, T., Golizadeh-Mojarad, R., Freitag, M., Lin, Y.M., Tsang, J., Perebeinos, V., Avouris, P.: Photocurrent Imaging and Efficient Photon Detection in a Graphene Transistor. *Nano Lett.* **9**(3), 1039-1044 (2009). doi:10.1021/Nl8033812
15. Mueller, T., Xia, F.N.A., Avouris, P.: Graphene photodetectors for high-speed optical communications. *Nat. Photonics* **4**(5), 297-301 (2010). doi:10.1038/Nphoton.2010.40
16. Echtermeyer, T.J., Britnell, L., Jasnos, P.K., Lombardo, A., Gorbachev, R.V., Grigorenko, A.N., Geim, A.K., Ferrari, A.C., Novoselov, K.S.: Strong plasmonic enhancement of photovoltage in graphene. *Nat. Commun.* **2**, 458 (2011). doi:10.1038/ncomms1464
17. Liu, Y., Cheng, R., Liao, L., Zhou, H.L., Bai, J.W., Liu, G., Liu, L.X., Huang, Y., Duan, X.F.: Plasmon resonance enhanced multicolour photodetection by graphene. *Nat. Commun.* **2**, 579 (2011). doi:10.1038/Ncomms1589
18. Liu, M., Yin, X.B., Ulin-Avila, E., Geng, B.S., Zentgraf, T., Ju, L., Wang, F., Zhang, X.: A graphene-based broadband optical modulator. *Nature* **474**(7349), 64-67 (2011). doi:10.1038/Nature10067
19. Kim, K., Choi, J.Y., Kim, T., Cho, S.H., Chung, H.J.: A role for graphene in silicon-based semiconductor devices. *Nature* **479**(7373), 338-344 (2011). doi:10.1038/Nature10680
20. Sensale-Rodriguez, B., Fang, T., Yan, R.S., Kelly, M.M., Jena, D., Liu, L., Xing, H.L.: Unique prospects for graphene-based terahertz modulators. *Appl. Phys. Lett.* **99**(11), 113104 (2011). doi:10.1063/1.3636435
21. Lu, Z.L., Zhao, W.S.: Nanoscale electro-optic modulators based on graphene-slot waveguides. *J. Opt. Soc. Am. B* **29**(6), 1490-1496 (2012). doi:10.1364/JOSAB.29.001490
22. Bao, Q.L., Zhang, H., Wang, B., Ni, Z.H., Lim, C.H.Y.X., Wang, Y., Tang, D.Y., Loh, K.P.: Broadband graphene polarizer. *Nat. Photonics* **5**(7), 411-415 (2011). doi:10.1038/Nphoton.2011.102.
23. Wang, X., Zhi, L.J., Mullen, K.: Transparent, conductive graphene electrodes for dye-sensitized solar cells. *Nano Lett.* **8**(1), 323-327 (2008). doi:10.1021/nl072838r
24. Bae, S., Kim, H., Lee, Y., Xu, X.F., Park, J.S., Zheng, Y., Balakrishnan, J., Lei, T., Kim, H.R., Song, Y.I., Kim, Y.J., Kim, K.S., Ozyilmaz, B., Ahn, J.H., Hong, B.H., Iijima, S.: Roll-to-roll production of 30-inch graphene films for transparent electrodes. *Nat. Nanotechnol.* **5**(8), 574-578 (2010). doi:10.1038/Nnano.2010.132.
25. Novoselov, K.S., Jiang, D., Schedin, F., Booth, T.J., Khotkevich, V.V., Morozov, S.V., Geim, A.K.: Two-dimensional atomic crystals. *P. Natl. Acad. Sci. USA* **102**(30), 10451-10453 (2005). doi:10.1073/pnas.0502848102
26. Li, X.S., Cai, W.W., An, J.H., Kim, S., Nah, J., Yang, D.X., Piner, R., Velamakanni, A., Jung, I., Tutuc, E., Banerjee, S.K., Colombo, L., Ruoff, R.S.: Large-Area Synthesis of High-Quality and Uniform Graphene Films on Copper Foils. *Science* **324**(5932), 1312-1314 (2009). doi:10.1126/science.1171245
27. Hernandez, Y., Nicolosi, V., Lotya, M., Blighe, F.M., Sun, Z.Y., De, S., McGovern, I.T., Holland, B., Byrne, M., Gun'ko, Y.K., Boland, J.J., Niraj, P., Duesberg, G., Krishnamurthy, S., Goodhue, R., Hutchison, J., Scardaci, V., Ferrari, A.C., Coleman, J.N.: High-yield production of graphene by liquid-phase exfoliation of graphite. *Nat. Nanotechnol.* **3**(9), 563-568 (2008). doi:10.1038/nnano.2008.215
28. Cai, J.M., Ruffieux, P., Jaafar, R., Bieri, M., Braun, T., Blankenburg, S., Muoth, M., Seitsonen, A.P., Saleh, M., Feng, X.L., Mullen, K., Fasel, R.: Atomically precise bottom-up fabrication of graphene nanoribbons. *Nature* **466**(7305), 470-473 (2010). doi:10.1038/Nature09211
29. Lin, Y.M., Dimitrakopoulos, C., Jenkins, K.A., Farmer, D.B., Chiu, H.Y., Grill, A., Avouris, P.: 100-GHz Transistors from Wafer-Scale Epitaxial Graphene. *Science* **327**(5966), 662-662 (2010). doi:10.1126/science.1184289
30. Novoselov, K.S., Geim, A.K., Morozov, S.V., Jiang, D., Zhang, Y., Dubonos, S.V., Grigorieva, I.V., Firsov, A.A.: Electric field effect in atomically thin carbon films. *Science* **306**(5696), 666-669 (2004). doi:10.1126/science.1102896
31. Novoselov, K.S., Geim, A.K., Morozov, S.V., Jiang, D., Katsnelson, M.I., Grigorieva, I.V., Dubonos, S.V., Firsov, A.A.: Two-dimensional gas of massless Dirac fermions in graphene. *Nature* **438**(7065), 197-200 (2005). doi:10.1038/Nature04233

32. Geim, A.K.: Nobel Lecture: Random walk to graphene. *Rev. Mod. Phys.* **83**(3), 851-862 (2011). doi:10.1103/RevModPhys.83.851.
33. Huang, X., Qi, X.Y., Boey, F., Zhang, H.: Graphene-based composites. *Chem. Soc. Rev.* **41**(2), 666-686 (2012). doi:10.1039/C1cs15078b
34. Qi, X.Y., Pu, K.Y., Li, H., Zhou, X.Z., Wu, S.X., Fan, Q.L., Liu, B., Boey, F., Huang, W., Zhang, H.: Amphiphilic Graphene Composites. *Angew. Chem. Int. Edit.* **49**(49), 9426-9429 (2010). doi:10.1002/anie.201004497
35. Qi, X.Y., Pu, K.Y., Zhou, X.Z., Li, H., Liu, B., Boey, F., Huang, W., Zhang, H.: Conjugated-Polyelectrolyte-Functionalized Reduced Graphene Oxide with Excellent Solubility and Stability in Polar Solvents. *Small* **6**(5), 663-669 (2010). doi:10.1002/smll.200902221
36. Yang, H.F., Zhang, Q.X., Shan, C.S., Li, F.H., Han, D.X., Niu, L.: Stable, Conductive Supramolecular Composite of Graphene Sheets with Conjugated Polyelectrolyte. *Langmuir* **26**(9), 6708-6712 (2010). doi:10.1021/La100365z
37. Wu, S.X., Yin, Z.Y., He, Q.Y., Huang, X.A., Zhou, X.Z., Zhang, H.: Electrochemical Deposition of Semiconductor Oxides on Reduced Graphene Oxide-Based Flexible, Transparent, and Conductive Electrodes. *J. Phys. Chem. C* **114**(27), 11816-11821 (2010). doi:10.1021/Jp103696u
38. Yin, Z.Y., Wu, S.X., Zhou, X.Z., Huang, X., Zhang, Q.C., Boey, F., Zhang, H.: Electrochemical Deposition of ZnO Nanorods on Transparent Reduced Graphene Oxide Electrodes for Hybrid Solar Cells. *Small* **6**(2), 307-312 (2010). doi:10.1002/smll.200901968
39. Tung, V.C., Chen, L.M., Allen, M.J., Wassei, J.K., Nelson, K., Kaner, R.B., Yang, Y.: Low-Temperature Solution Processing of Graphene-Carbon Nanotube Hybrid Materials for High-Performance Transparent Conductors. *Nano Lett.* **9**(5), 1949-1955 (2009). doi:10.1021/Nl9001525
40. Lu, R.T., Christianson, C., Weintrub, B., Wu, J.Z.: High Photoresponse in Hybrid Graphene-Carbon Nanotube Infrared Detectors. *ACS Appl. Mater. Inter.* **5**(22), 11703-11707 (2013). doi:10.1021/Am4033313
41. Shi, W.H., Zhu, J.X., Sim, D.H., Tay, Y.Y., Lu, Z.Y., Zhang, X.J., Sharma, Y., Srinivasan, M., Zhang, H., Hng, H.H., Yan, Q.Y.: Achieving high specific charge capacitances in Fe3O4/reduced graphene oxide nanocomposites. *J. Mater. Chem.* **21**(10), 3422-3427 (2011). doi:10.1039/C0jm03175e
42. Li, F.H., Song, J.F., Yang, H.F., Gan, S.Y., Zhang, Q.X., Han, D.X., Ivaska, A., Niu, L.: One-step synthesis of graphene/SnO2 nanocomposites and its application in electrochemical supercapacitors. *Nanotechnology* **20**(45) (2009). doi:10.1088/0957-4484/20/45/455602
43. Cai, W.W., Zhu, Y.W., Li, X.S., Piner, R.D., Ruoff, R.S.: Large area few-layer graphene/graphite films as transparent thin conducting electrodes. *Appl. Phys. Lett.* **95**(12) (2009). doi:10.1063/1.3220807
44. Yin, Z.Y., Sun, S.Y., Salim, T., Wu, S.X., Huang, X.A., He, Q.Y., Lam, Y.M., Zhang, H.: Organic Photovoltaic Devices Using Highly Flexible Reduced Graphene Oxide Films as Transparent Electrodes. *ACS Nano* **4**(9), 5263-5268 (2010). doi:10.1021/Nn1015874
45. Li, X.M., Zhu, H.W., Wang, K.L., Cao, A.Y., Wei, J.Q., Li, C.Y., Jia, Y., Li, Z., Li, X., Wu, D.H.: Graphene-On-Silicon Schottky Junction Solar Cells. *Adv. Mater.* **22**(25), 2743 (2010). doi:10.1002/adma.200904383
46. Shao, D.L., Sawyer, S., Hu, T., Yu, M.P., Lian, J.: Photoconductive Enhancement Effects of Graphene Quantum Dots on ZnO Nanoparticle Photodetectors. *P. IEEE Les. Eastm.* (2012)
47. Radoi, A., Dragoman, M., Cismaru, A., Konstantinidis, G., Dragoman, D.: Light-Harvesting Using Metallic Interdigitated Structures Modified with Au Sputtered Graphene. *International Semiconductor Conference* **2**, 117-120 (2012)
48. Chang, H.X., Sun, Z.H., Ho, K.Y.F., Tao, X.M., Yan, F., Kwok, W.M., Zheng, Z.J.: A highly sensitive ultraviolet sensor based on a facile in situ solution-grown ZnO nanorod/graphene heterostructure. *Nanoscale* **3**(1), 258-264 (2011). doi:10.1039/C0nr00588f
49. Shao, D.L., Sun, X., Xie, M., Sun, H.T., Lu, F.Y., George, S.M., Lian, J., Sawyer, S.: ZnO quantum dots-graphene composite for efficient ultraviolet sensing. *Mater. Lett.* **112**, 165-168 (2013). doi:10.1016/j.matlet.2013.09.031.

50. Novoselov, K.S., Fal'ko, V.I., Colombo, L., Gellert, P.R., Schwab, M.G., Kim, K.: A roadmap for graphene. *Nature* **490**(7419), 192-200 (2012). doi:10.1038/Nature11458
51. Reina, A., Son, H.B., Jiao, L.Y., Fan, B., Dresselhaus, M.S., Liu, Z.F., Kong, J.: Transferring and Identification of Single- and Few-Layer Graphene on Arbitrary Substrates. *J. Phys. Chem. C* **112**(46), 17741-17744 (2008). doi:10.1021/Jp807380s
52. Koppens, F.H.L., Chang, D.E., de Abajo, F.J.G.: Graphene Plasmonics: A Platform for Strong Light-Matter Interactions. *Nano Lett.* **11**(8), 3370-3377 (2011). doi:10.1021/Nl201771h
53. Gusynin, V.P., Sharapov, S.G., Carbotte, J.P.: Sum rules for the optical and Hall conductivity in graphene. *Phys. Rev. B* **75**(165407), 165407 (2007). doi:10.1103/Physrevb.75.165407
54. Mak, K.F., Sfeir, M.Y., Wu, Y., Lui, C.H., Misewich, J.A., Heinz, T.F.: Measurement of the Optical Conductivity of Graphene. *Phys. Rev. Lett.* **101**(19), 196405 (2008). doi:10.1103/Physrevlett.101.196405
55. Nair, R.R., Blake, P., Grigorenko, A.N., Novoselov, K.S., Booth, T.J., Stauber, T., Peres, N.M.R., Geim, A.K.: Fine structure constant defines visual transparency of graphene. *Science* **320**(5881), 1308 (2008). doi: 10.1126/science.1156965
56. Kuzmenko, A.B., van Heumen, E., Carbone, F., van der Marel, D.: Universal optical conductance of graphite. *Phys. Rev. Lett.* **100**(11), 117401 (2008). doi:10.1103/Physrevlett.100.117401
57. Li, Z.Q., Henriksen, E.A., Jiang, Z., Hao, Z., Martin, M.C., Kim, P., Stormer, H.L., Basov, D.N.: Dirac charge dynamics in graphene by infrared spectroscopy. *Nat. Phys.* **4**(7), 532-535 (2008). doi:10.1038/Nphys989
58. Weiss, N.O., Zhou, H.L., Liao, L., Liu, Y., Jiang, S., Huang, Y., Duan, X.F.: Graphene: An Emerging Electronic Material. *Adv. Mater.* **24**(43), 5782-5825 (2012). doi:10.1002/adma.201201482
59. Brody, H.: Graphene. *Nature* **483**(7389), S29-S29 (2012). doi:10.1038/483S29a.
60. Reed, G.T., Mashanovich, G., Gardes, F.Y., Thomson, D.J.: Silicon optical modulators. *Nat. Photonics* **4**(8), 518-526 (2010). doi:10.1038/nphoton.2010.179
61. Datta, S., Das, B.: Electronic Analog of the Electrooptic Modulator. *Appl. Phys. Lett.* **56**(7), 665-667 (1990). doi:10.1063/1.102730
62. Gosciniak, J., Bozhevolnyi, S.I., Andersen, T.B., Volkov, V.S., Kjelstrup-Hansen, J., Markey, L., Dereux, A.: Thermo-optic control of dielectric-loaded plasmonic waveguide components. *Opt. Express* **18**(2), 1207-1216 (2010). doi:10.1364/OE.18.001207
63. Gordon, E.I.: A Review of Acoustooptical Deflection and Modulation Devices. *Appl. Optics* **5**(10), 1629 (1966). doi:10.1364/Ao.5.001629
64. Kuo, Y.H., Chen, H.W., Bowers, J.E.: High speed hybrid silicon evanescent electroabsorption modulator. *Opt. Express* **16**(13), 9936-9941 (2008). doi:10.1364/Oe.16.009936.
65. Liu, M., Yin, X.B., Zhang, X.: Double-Layer Graphene Optical Modulator. *Nano Lett.* **12**(3), 1482-1485 (2012). doi:10.1021/Nl204202k
66. Yan, K., Wu, D., Peng, H.L., Jin, L., Fu, Q., Bao, X.H., Liu, Z.F.: Modulation-doped growth of mosaic graphene with single-crystalline p-n junctions for efficient photocurrent generation. *Nat. Commun.* **3**, 1280 (2012). doi:10.1038/Ncomms2286
67. Gosciniak, J., Tan, D.T.H.: Theoretical investigation of graphene-based photonic modulators. *Sci. Rep.* **3**, 1897 (2013). doi:10.1038/Srep01897
68. Chenran, Y., Khan, S., Zhuo Ran, L., Simsek, E., Sorger, V.J.: λ-Size ITO and Graphene-Based Electro-Optic Modulators on SOI. *Selected Topics in Quantum Electronics, IEEE Journal of* **20**(4) (2014). doi:10.1109/JSTQE.2014.2298451
69. Song, S.C., Chen, Q., Jin, L., Sun, F.H.: Great light absorption enhancement in a graphene photodetector integrated with a metamaterial perfect absorber. *Nanoscale* **5**(20), 9615-9619 (2013). doi:10.1039/C3nr03505k
70. Vicarelli, L., Vitiello, M.S., Coquillat, D., Lombardo, A., Ferrari, A.C., Knap, W., Polini, M., Pellegrini, V., Tredicucci, A.: Graphene field-effect transistors as room-temperature terahertz detectors. *Nat. Mater.* **11**(10), 865-871 (2012). doi:10.1038/Nmat3417
71. Chang, C.W., Wang, D.Y., Tan, W.C., Huang, I.S., Wang, I.S., Chen, C.C., Yang, Y.J., Chen, Y.F.: Enhanced performance of photodetector and photovoltaic based on carrier reflector and

back surface field generated by doped graphene. *Appl. Phys. Lett.* **101**(7), 073906 (2012). doi:10.1063/1.4746763

72. Prechtel, L., Song, L., Schuh, D., Ajayan, P., Wegscheider, W., Holleitner, A.W.: Time-resolved ultrafast photocurrents and terahertz generation in freely suspended graphene. *Nat. Commun.* **3**, 646 (2012). doi:10.1038/ncomms1656
73. Vora, H., Kumaravadivel, P., Nielsen, B., Du, X.: Bolometric response in graphene based superconducting tunnel junctions. *Appl. Phys. Lett.* **100**(15) (2012). doi:10.1063/1.3703117
74. Fang, Z.Y., Liu, Z., Wang, Y.M., Ajayan, P.M., Nordlander, P., Halas, N.J.: Graphene-Antenna Sandwich Photodetector. *Nano Lett.* **12**(7), 3808-3813 (2012). doi:10.1021/Nl301774e
75. Thongrattanasiri, S., Koppens, F.H.L., de Abajo, F.J.G.: Complete Optical Absorption in Periodically Patterned Graphene. *Phys. Rev. Lett.* **108**(4), 047401 (2012). doi:10.1103/Physrevlett.108.047401
76. Xu, J.L., Li, X.L., He, J.L., Hao, X.P., Wu, Y.Z., Yang, Y., Yang, K.J.: Performance of large-area few-layer graphene saturable absorber in femtosecond bulk laser. *Appl. Phys. Lett.* **99**(26), 261107 (2011). doi:10.1063/1.3672213
77. Wang, Z.F., Liu, F.: Manipulation of Electron Beam Propagation by Hetero-Dimensional Graphene Junctions. *ACS Nano* **4**(4), 2459-2465 (2010). doi:10.1021/Nn1001722
78. Ju, L., Geng, B.S., Horng, J., Girit, C., Martin, M., Hao, Z., Bechtel, H.A., Liang, X.G., Zettl, A., Shen, Y.R., Wang, F.: Graphene plasmonics for tunable terahertz metamaterials. *Nat. Nanotechnol.* **6**(10), 630-634 (2011). doi:10.1038/Nnano.2011.146
79. Lee, S.H., Choi, M., Kim, T.T., Lee, S., Liu, M., Yin, X., Choi, H.K., Lee, S.S., Choi, C.G., Choi, S.Y., Zhang, X., Min, B.: Switching terahertz waves with gate-controlled active graphene metamaterials. *Nat. Mater.* **11**(11), 936-941 (2012). doi:10.1038/Nmat3433
80. Ramakrishnan, G., Chakkittakandy, R., Planken, P.C.M.: Terahertz generation from graphite. *Opt. Express* **17**(18), 16092-16099 (2009). doi:10.1364/OE.17.016092
81. Mittendorff, M., Winnerl, S., Kamann, J., Eroms, J., Weiss, D., Schneider, H., Helm, M.: Ultrafast graphene-based broadband THz detector. *Appl. Phys. Lett.* **103**(2), 021113 (2013). doi:10.1063/1.4813621
82. Ryzhii, V., Ryzhii, M.: Graphene bilayer field-effect phototransistor for terahertz and infrared detection. *Phys. Rev. B* **79**(24) (2009). doi:10.1103/Physrevb.79.245311
83. Yang, H.Y., Son, D.I., Kim, T.W., Lee, J.M., Park, W.I.: Enhancement of the photocurrent in ultraviolet photodetectors fabricated utilizing hybrid polymer-ZnO quantum dot nanocomposites due to an embedded graphene layer. *Org. Electron.* **11**(7), 1313-1317 (2010). doi:10.1016/j.orgel.2010.04.009
84. Singh, R.S., Nalla, V., Chen, W., Wee, A.T.S., Ji, W.: Laser Patterning of Epitaxial Graphene for Schottky Junction Photodetectors. *ACS Nano* **5**(7), 5969-5975 (2011). doi:10.1021/Nn201757j
85. Fu, X.W., Liao, Z.M., Zhou, Y.B., Wu, H.C., Bie, Y.Q., Xu, J., Yu, D.P.: Graphene/ZnO nanowire/graphene vertical structure based fast-response ultraviolet photodetector. *Appl. Phys. Lett.* **100**(22), 223114 (2012). doi:10.1063/1.4724208
86. Zhang, W.J., Chuu, C.P., Huang, J.K., Chen, C.H., Tsai, M.L., Chang, Y.H., Liang, C.T., Chen, Y.Z., Chueh, Y.L., He, J.H., Chou, M.Y., Li, L.J.: Ultrahigh-Gain Photodetectors Based on Atomically Thin Graphene-MoS2 Heterostructures. *Sci. Rep.* **4**, 3826 (2014). doi:10.1038/Srep03826
87. Liu, C.H., Chang, Y.C., Norris, T.B., Zhong, Z.: Graphene photodetectors with ultra-broadband and high responsivity at room temperature. *Nat. Nanotechnol.* **9**(4), 273-278 (2014). doi:10.1038/nnano.2014.31
88. Li, J., Niu, L., Zheng, Z., Yan, F.: Photosensitive Graphene Transistors. *Adv. Mater.* **26**(31), 5239-5273 (2014). doi:10.1002/adma.201400349
89. Ryzhii, V., Ryzhii, M., Mitin, V., Otsuji, T.: Terahertz and infrared photodetection using p-i-n multiple-graphene-layer structures. *J. Appl. Phys.* **107**(5), 054512 (2010). doi:10.1063/1.3327441
90. Ryzhii, V., Ryzhii, M., Ryabova, N., Mitin, V., Otsuji, T.: Terahertz and infrared detectors based on graphene structures. *Infrared Phys. Techn.* **54**(3), 302-305 (2011). doi:10.1016/j.infrared.2010.12.034

91. Ryzhii, M., Otsuji, T., Mitin, V., Ryzhii, V.: Characteristics of p-i-n Terahertz and Infrared Photodiodes Based on Multiple Graphene Layer Structures. *Jpn. J. Appl. Phys.* **50**(7), 070117 (2011). doi:10.1143/Jjap.50.070117
92. An, X.H., Liu, F.Z., Jung, Y.J., Kar, S.: Tunable Graphene-Silicon Heterojunctions for Ultrasensitive Photodetection. *Nano Lett.* **13**(3), 909-916 (2013). doi:10.1021/Nl303682j
93. Amirmazlaghani, M., Raissi, F., Habibpour, O., Vukusic, J., Stake, J.: Graphene-Si Schottky IR Detector. *IEEE J. Quantum Electron.* **49**(7), 589-594 (2013). doi:10.1109/Jqe.2013.2261472
94. Casalino, M., Coppola, G., Iodice, M., Rendina, I., Sirleto, L.: Near-Infrared All-Silicon Photodetectors. *International J. Photoenergy* **2012**, 139278 (2012). doi:10.1155/2012/139278
95. Zeng, L.H., Wang, M.Z., Hu, H., Nie, B., Yu, Y.Q., Wu, C.Y., Wang, L., Hu, J.G., Xie, C., Liang, F.X., Luo, L.B.: Monolayer Graphene/Germanium Schottky Junction As High-Performance Self-Driven Infrared Light Photodetector. *ACS Appl. Mater. Inter.* **5**(19), 9362-9366 (2013). doi:10.1021/Am4026505
96. Jin, W.F., Ye, Y., Gan, L., Yu, B., Wu, P.C., Dai, Y., Meng, H., Guo, X.F., Dai, L.: Self-powered high performance photodetectors based on CdSe nanobelt/graphene Schottky junctions. *J. Mater. Chem.* **22**(7), 2863-2867 (2012). doi:10.1039/C2jm15913a
97. Gao, Z.W., Jin, W.F., Zhou, Y., Dai, Y., Yu, B., Liu, C., Xu, W.J., Li, Y.P., Peng, H.L., Liu, Z.F., Dai, L.: Self-powered flexible and transparent photovoltaic detectors based on CdSe nanobelt/graphene Schottky junctions. *Nanoscale* **5**(12), 5576-5581 (2013). doi:10.1039/C3nr34335a
98. Soci, C., Zhang, A., Bao, X.Y., Kim, H., Lo, Y., Wang, D.L.: Nanowire Photodetectors. *J. Nanosci. Nanotechnol.* **10**(3), 1430-1449 (2010). doi:10.1166/jnn.2010.2157
99. Soci, C., Zhang, A., Xiang, B., Dayeh, S.A., Aplin, D.P.R., Park, J., Bao, X.Y., Lo, Y.H., Wang, D.: ZnO nanowire UV photodetectors with high internal gain. *Nano Lett.* **7**(4), 1003-1009 (2007). doi:10.1021/Nl070111x
100. Bugallo, A.D., Tchernycheva, M., Jacopin, G., Rigutti, L., Julien, F.H., Chou, S.T., Lin, Y.T., Tseng, P.H., Tu, L.W.: Visible-blind photodetector based on p-i-n junction GaN nanowire ensembles. *Nanotechnology* **21**(31), 315201 (2010). doi:10.1088/0957-4484/21/31/315201
101. Lee, H., Heo, K., Park, J., Park, Y., Noh, S., Kim, K.S., Lee, C., Hong, B.H., Jian, J., Hong, S.: Graphene-nanowire hybrid structures for high-performance photoconductive devices. *J. Mater. Chem.* **22**(17), 8372-8376 (2012). doi:10.1039/C2jm16565a
102. Zhang, H., Babichev, A.V., Jacopin, G., Lavenus, P., Julien, F.H., Egorov, A.Y., Zhang, J., Pauporte, T., Tchernycheva, M.: Characterization and modeling of a ZnO nanowire ultraviolet photodetector with graphene transparent contact. *J. Appl. Phys.* **114**(23), 234505 (2013). doi:10.1063/1.4854455
103. Nie, B.A., Hu, J.G., Luo, L.B., Xie, C., Zeng, L.H., Lv, P., Li, F.Z., Jie, J.S., Feng, M., Wu, C.Y., Yu, Y.Q., Yu, S.H.: Monolayer Graphene Film on ZnO Nanorod Array for High-Performance Schottky Junction Ultraviolet Photodetectors. *Small* **9**(17), 2872-2879 (2013). doi:10.1002/smll.201203188
104. Babichev, A.V., Zhang, H., Lavenus, P., Julien, F.H., Egorov, A.Y., Lin, Y.T., Tu, L.W., Tchernycheva, M.: GaN nanowire ultraviolet photodetector with a graphene transparent contact. *Appl. Phys. Lett.* **103**(20), 201103 (2013). doi:10.1063/1.4829756
105. Wang, M.Z., Liang, F.X., Nie, B., Zeng, L.H., Zheng, L.X., Lv, P., Yu, Y.Q., Xie, C., Li, Y.Y., Luo, L.B.: TiO2 Nanotube Array/Monolayer Graphene Film Schottky Junction Ultraviolet Light Photodetectors. *Partical & Particle Systems Characterization* **30**(7), 630-636 (2013). doi:10.1002/ppsc.201300040
106. Xu, Q., Cheng, Q.J., Zhong, J.X., Cai, W.W., Zhang, Z.F., Wu, Z.Y., Zhang, F.Y.: A metal-semiconductor-metal detector based on ZnO nanowires grown on a graphene layer. *Nanotechnology* **25**(5), 5 (2014). doi:10.1088/0957-4484/25/5/055501
107. Jin, Y., Wang, J., Sun, B., Blakesley, J.C., Greenham, N.C.: Solution-Processed Ultraviolet Photodetectors Based on Colloidal ZnO Nanoparticles. *Nano Lett.* **8**(6), 1649-1653 (2008). doi:10.1021/nl0803702

108. Tsai, D.S., Lin, C.A., Lien, W.C., Chang, H.C., Wang, Y.L., He, J.H.: Ultra-High-Responsivity Broadband Detection of Si Metal-Semiconductor-Metal Schottky Photodetectors Improved by ZnO Nanorod Arrays. *ACS Nano* **5**(10), 7748-7753 (2011). doi:10.1021/Nn203357e
109. Konstantatos, G., Badioli, M., Gaudreau, L., Osmond, J., Bernechea, M., de Arquer, F.P.G., Gatti, F., Koppens, F.H.L.: Hybrid graphene-quantum dot phototransistors with ultrahigh gain. *Nat. Nanotechnol.* **7**(6), 363-368 (2012). doi:10.1038/Nnano.2012.60
110. Manga, K.K., Wang, J.Z., Lin, M., Zhang, J., Nesladek, M., Nalla, V., Ji, W., Loh, K.P.: High-Performance Broadband Photodetector Using Solution-Processible PbSe-TiO2-Graphene Hybrids. *Adv. Mater.* **24**(13), 1697-1702 (2012). doi:10.1002/adma.201104399
111. Manga, K.K., Wang, S., Jaiswal, M., Bao, Q.L., Loh, K.P.: High-Gain Graphene-Titanium Oxide Photoconductor Made from Inkjet Printable Ionic Solution. *Adv. Mater.* **22**(46), 5265-5270 (2010). doi:10.1002/adma.201002939
112. Gur, I., Fromer, N.A., Geier, M.L., Alivisatos, A.P.: Air-stable all-inorganic nanocrystal solar cells processed from solution. *Science* **310**(5747), 462-465 (2005). doi:10.1126/science.1117908
113. Shao, D.L., Yu, M.P., Sun, H.T., Hu, T., Lian, J., Sawyer, S.: High responsivity, fast ultraviolet photodetector fabricated from ZnO nanoparticle-graphene core-shell structures. *Nanoscale* **5**(9), 3664-3667 (2013). doi:10.1039/C3nr00369h
114. Kishino, K., Unlu, M.S., Chyi, J.I., Reed, J., Arsenault, L., Morkoc, H.: Resonant Cavity-Enhanced (Rce) Photodetectors. *IEEE J. Quantum Electron.* **27**(8), 2025-2034 (1991). doi:10.1109/3.83412
115. Unlu, M.S., Strite, S.: Resonant-Cavity Enhanced Photonic Devices. *J. Appl. Phys.* **78**(2), 607-639 (1995)
116. Jervase, J.A., Zebda, Y.: Characteristic analysis of resonant-cavity-enhanced (RCE) photodetectors. *IEEE J. Quantum Electron.* **34**(7), 1129-1134 (1998). doi:10.1109/3.687854
117. Bugajski, M., Muszalski, J., Mroziewicz, B., Reginski, K., Ochalski, T.J.: Resonant cavity enhanced photonic devices. *Opt. Appl.* **31**(2), 273-288 (2001).
118. Bugajski, M., Muszalski, J., Ochalski, T., Katcki, J., Mroziewicz, B.: Resonant cavity enhanced photonic devices. *Acta. Phys. Pol. A* **101**(1), 105-118 (2002).
119. Furchi, M., Urich, A., Pospischil, A., Lilley, G., Unterrainer, K., Detz, H., Klang, P., Andrews, A.M., Schrenk, W., Strasser, G., Mueller, T.: Microcavity-Integrated Graphene Photodetector. *Nano Lett.* **12**(6), 2773-2777 (2012). doi:10.1021/Nl204512x
120. Engel, M., Steiner, M., Lombardo, A., Ferrari, A.C., Löhneysen, H.v., Avouris, P., Krupke, R.: Light–matter interaction in a microcavity-controlled graphene transistor. *Nat. Commun.* **3**, 906 (2012). doi:10.1038/ncomms1911
121. Gan, X.T., Mak, K.F., Gao, Y.D., You, Y.M., Hatami, F., Hone, J., Heinz, T.F., Englund, D.: Strong Enhancement of Light-Matter Interaction in Graphene Coupled to a Photonic Crystal Nanocavity. *Nano Lett.* **12**(11), 5626-5631 (2012). doi:10.1021/Nl302746n
122. Shiue, R.J., Gan, X.T., Gao, Y.D., Li, L.Z., Yao, X.W., Szep, A., Walker, D., Hone, J., Englund, D.: Enhanced photodetection in graphene-integrated photonic crystal cavity. *Appl. Phys. Lett.* **103**(24) (2013). doi:10.1063/1.4839235
123. Gan, X.T., Shiue, R.J., Gao, Y.D., Assefa, S., Hone, J., Englund, D.: Controlled Light-Matter Interaction in Graphene Electrooptic Devices Using Nanophotonic Cavities and Waveguides. *IEEE J. Sel. Top. Quant.* **20**(1), 600311 (2014). doi:10.1109/Jstqe.2013.2273412
124. Piper, J.R., Fan, S.: Total Absorption in a Graphene Monolayer in the Optical Regime by Critical Coupling with a Photonic Crystal Guided Resonance. *ACS Photonics* **1**(4), 347-353 (2014). doi:10.1021/ph400090p
125. Barnes, W.L., Dereux, A., Ebbesen, T.W.: Surface plasmon subwavelength optics. *Nature* **424**(6950), 824-830 (2003). doi:10.1038/Nature01937
126. Liedberg, B., Nylander, C., Lundstrom, I.: Surface-Plasmon Resonance for Gas-Detection and Biosensing. *Sensor Actuator* **4**(2), 299-304 (1983). doi:10.1016/0250-6874(83)85036-7

127. Nikolajsen, T., Leosson, K., Bozhevolnyi, S.I.: Surface plasmon polariton based modulators and switches operating at telecom wavelengths. *Appl. Phys. Lett.* **85**(24), 5833-5835 (2004). doi:10.1063/1.1835997
128. Shi, S.F., Xu, X.D., Ralph, D.C., McEuen, P.L.: Plasmon Resonance in Individual Nanogap Electrodes Studied Using Graphene Nanoconstrictions as Photodetectors. *Nano Lett.* **11**(4), 1814-1818 (2011). doi:10.1021/Nl200522t
129. Zhang, Y.Z., Liu, T., Meng, B., Li, X.H., Liang, G.Z., Hu, X.N., Wang, Q.J.: Broadband high photoresponse from pure monolayer graphene photodetector. *Nat. Commun.* **4**, 1811 (2013). doi:10.1038/Ncomms2830

Polymer/Nanographite Composites for Mechanical Impact Sensing

Maris Knite and Artis Linarts

Abstract The purpose of this chapter is to give a review of the polymer/nanographite composite (PNGC) materials specially developed for applications in mechanical strain and pressure sensors that can be used for design of flexible sensing systems. Our recent achievements in design, processing, and investigation of physical properties of elastomer and nanostructured carbon composites as prospective materials for mentioned sensors are also presented. In the beginning, theoretical principles of tunneling percolation theory and piezoresistivity have been described. We discuss the most suitable polymer matrices and electrically conductive nanographite fillers for sensitive PNGC. Preparation methods of mechanically sensitive PNGC have been considered. Different particularly produced and tested polymer/nanographite composites are overhauled and possible advantages and disadvantages of PNGC in different possible applications are analyzed.

Keywords Strain · Pressure · Sensors · Electronics

1 Introduction

The purpose of this chapter is to give a review of the polymer composite materials developed directly for application in strain and pressure sensors by using different nanographite structures as filler. We understand with the term "nanographite" the following fillers: extra-conductive highly structured carbon black (HSCB), carbon nanotubes (CNT), thermally exfoliated graphite (TEG) as well as the recently developed graphene. All of them have sp^2-hybridized crystal structure like graphite, and at least one dimension is smaller than 100 nm. Extra-conductive carbon black can be attributed to nanographite because its primary nanoparticle faces

M. Knite (✉) · A. Linarts
Institute of Technical Physics, Riga Technical University, 14/24 Azenes Street, Riga LV-1007, Latvia
e-mail: Maris.Knite@rtu.lv

K.K. Sadasivuni et al. (eds.), *Graphene-Based Polymer Nanocomposites in Electronics*, Springer Series on Polymer and Composite Materials,
DOI 10.1007/978-3-319-13875-6_9

consist of graphene platelets [1] or have "graphitic like organization" [2]. Our recent achievements in design, processing, and investigation of physical properties of elastomer and nanographite composites as prospective materials for mechanical (pressure, strain) indicators are also presented in this chapter.

Piezoelectric ceramics or constantan-chromium-alloy-based strain gauges are widely used as mechanical impact (MI) sensors. Such sensors are separate units from the monitored material or structure. There is a demand for new flexible large area sensors that could be easy embedded in different integrities and used for sensing multiple locations. High elasticity (hyperelasticity) polymer matrix based materials are still of interest. All of flexible polymer composite materials developed for MI sensing can be generally classified as electrically active (electronic response) or optically active (photonic response) materials. An example of a photonic response polymeric composite transducer for tactile sensing is based on optical fiber with Bragg grating embedded in polydimethilsiloxane [3]. The basic principle of this transducer lies in the monitoring of the wavelength shift of the returned Bragg-signal as a function of the strain or force. Comprehensive picture of current status of micro- and nanostructured flexible optical fiber sensors with particular reference to surface plasma resonance fiber sensors and photonic crystal fiber sensors is given by Fortes et al. [4], Yan et al. [5] and Lee et al. [6].

In the present chapter, we concentrate on smart polymer/nanographite composites (PNGC) that can give considerable electrical response to mechanical impact.

In the first section, the theoretical aspects of tunneling percolation as well as mechanical impact sensing mechanisms for PNGC based on quantum tunneling effect are going to be discussed. Thermodynamic force–response aspects of piezoresistivity are also presented.

In the second section, the most suitable nanographite filler materials, polymer matrix materials, and methods of preparation as well as principle of design of most sensitive PNGC are analyzed.

The third section informs about the most interesting PNGCs developed and investigated somewhere for mechanical impact sensing.

Our recent achievements in development of polymer/nanographite composite sensor-element systems for mechanical impact sensing as well as survey of particular technical solutions for sensing systems are given in the fourth section.

2 Theoretical Aspects of Mechanical Impact Sensing by Polymer/Nanographite Composites

2.1 Principles of Tunneling PercolationTheory

Conductive polymer composite for strain sensing can be obtained when particles of good conductors (carbon black, graphite powder, particles of metals, carbon nanotubes, graphene i.e.) are randomly inserted into an insulating polymer matrix. A continuous insulator–conductor transition is observed in two-component systems at gradual increase of the number of randomly dispersed conductor particles in an insulator matrix. Most often such transitions called percolation transitions are described by the model of statistical percolation [7, 8]. The volume concentration of conductor particles ϕ_C at which the transition proceeds is called percolation threshold or critical point. According to the statistical model, conductor particles, in the vicinity of ϕ_C, assemble in clusters and the correlation radius ζ (average distance between two opposite particles of a cluster) diverges as

$$\xi \sim |\phi - \phi_c|^{-\nu} \quad \text{if } \phi > \phi_c \tag{1}$$

upon approaching ϕ_C (ν—critical indices) [7].

In the vicinity of percolation threshold, electric conductivity σ of the composite changes as:

$$\sigma \sim |\phi - \phi_c|^{t} \quad \text{if } \phi > \phi_c \tag{2}$$

here t—critical index [8]. Balberg et al. [9] developed a theory for percolation-like behavior for polymer and conductive nanoparticle composites where interparticle charge tunneling considerably prevailed against direct (geometrical) interparticle contacts. It was found that Eq. (2) can still be used for such tunneling supported percolation system only the nonuniversal behavior of critical index t should be taken into account [9]. It means that the experimentally obtained values of t remarkably differ from the "universal" value $t = 2$. When such conductive composites are mechanically stressed, then the both ξ and σ change correspondingly. This is the reason of piezoresistance (the resistance changes vs. mechanical strain) effects. Changes of electric resistance with strain and pressure can be simply explained on microscopic level as a result of the percolation structure change of conductive particles network (destruction or formation of conductive micro- and nanochannels).

New interesting properties are expected in case when the composite contains dispersed nanosize extra-conducting particles [10]. Polymer/electroconductive nanostructure composites offer attractive alternatives for developing new generation of flexible large-size sensors because of their superior mechanical and electrical properties.

2.2 Charge Tunneling Models of Piezoresistivity

The phenomenon where the electrical resistivity of a material changes due to applied mechanical stress is called piezoresistivity. Correct theoretical description of piezoresistivity on macroscopic or phenomenological level has been done by Rocha et al. [11]. They developed piezoresistive coefficient tensor that describes all possible mechanical force effects on the electric resistivity of a composite with only one term "piezoresistivity". In the scientific literature term, "tensoresitivity" has also been used by some authors [10, 12]. If the resistivity changes under tensile force such material exhibits the tensoresistivity effect. For better understanding the particular manifestation of piezoresistivity in PNGCs, in this chapter, we propose to use terms "pressure coefficient of resistance" or "tension coefficient of resistance" like "temperature coefficient of resistance" used for characterization of temperature dependence of resistance on materials. So, it is possible to distinguish four different experimentally observed cases of resistivity change under MI forces:

1. negative pressure coefficient of resistance (NPR) when the resistivity decreases with compressive force [13, 14];
2. positive pressure coefficient of resistance (PPR) when the resistivity increases with compressive force [10, 14, 15];
3. positive tension coefficient of resistance (PTR)—resistivity rises by tensile strain [10]
4. negative tension coefficient of resistance (NTR) if the resistivity under stretching force decreases [16].

Generally speaking all of the above-mentioned effects of mechanical impact on mechanoelectrically active polymer composites can be explained by changes of nanodimensional structure of electroconductive nanoparticle grid inside the polymer matrix. If the inter-particle distance rises or the number of conductive channels diminishes, the resistivity of the sample rises (PPR and PTR). If the interparticle distance decreases or the number of conductive channels increases, the resistivity of sample decreases (NPR and NTR). Further, in this section, we report the more exhaustive theoretical explanations of all four effects on microscopic level based on quantum charge tunneling.

Theoretical description of NPR effect experimentally observed in polymer/metallic microparticle composites has been done by Zhang et al. [13] based on early developed theory for tunneling conductivity between dissimilar electrodes separated by a thin insulating film [17].

Knite et al. [10] modified the previously mentioned charge tunneling theory for experimentally observed reversible PTR effect in polyisoprene/HSCB composites. To explain the large PTR and PPR effects Knite et al. [10] assumed that bonds between the nanostructured agglomerates of carbon black and the hyperelastic polymer chains are stronger compared to the bonds between carbon nanoparticles themselves. Possibly, they are bonded by the free radicals of the chains thermally activated at vulcanization. At stretching of the composite, the carbon agglomerates remain covalently bonded to the polymer chains, as they are dragged along with the polymer chains and pulled apart. In case of poorly structured or microsize carbon filler, the bonding between carbon particles and polymer chains is weak and the filler makes irreversible (rigid and rather immobile) clusters in the insulating matrix. The effects are irreversible. In case of (HSCB) like PRINTEX-XE2, after the stretch is released, the contacts between carbon agglomerates are restored as the polymer chains return practically to its original positions [10].

The total electrical resistance of conductor-filled polymer composite is a function of both the resistance of each conducting particle and of the polymer matrix. As the conductivity of the conducting particles is very large in comparison to that of the polymer matrix, the resistance across the particles may be neglected. When particles are separated far enough from each other, no current flows through the composite. If the distances separating particles are small, tunneling currents may arise.

According to the model derived by Zhang et al.[13], the total electrical resistance R of the composite is calculated as:

$$R = \left(\frac{n}{N}\right)\left(\frac{8\pi hs}{3a^2\gamma e^2}\right)\exp(\gamma s), \tag{3}$$

where n is the number of particles forming a single conducting path, N—the number of conducting paths, h—Plank's constant, s—the least distance between conductive particles, a^2—the effective cross-section, where tunneling occurs, e—the electron charge, and γ is calculated as:

$$\gamma = \frac{4\pi(2m\phi)^{0.5}}{h}, \tag{4}$$

where m is the electron mass and ϕ—the height of potential barrier between adjacent particles.

If stress is applied to a composite sample, the resistance will be altered due to the change of particle separation. Assuming that under applied stress, the particle separation changes from s_0 to s, the relative resistance (R/R_0) is given by

$$\frac{R}{R_0} = \left(\frac{s}{s_0}\right)\exp\left[\gamma(s - s_0)\right] \tag{5}$$

where R_0 is the initial resistance, and s_0—the initial particle separation [13]. In case of elastomer composite, the separations under tensile strain is calculated as

$$s = s_0(1+\varepsilon) = s_0\left[1 + \left(\frac{\Delta l}{l_0}\right)\right] \tag{6}$$

where ε is the tensile strain of the elastomer matrix, Δl—deformation of the composite sample, and l_0—initial length of the sample. Substitution of Eq. (6) into Eq. (5) yields

$$\ln R = \ln R_0 + \ln\left[1 + \left(\frac{\Delta l}{l_0}\right)\right] + A_0\left(\frac{\Delta l}{l_0}\right) \tag{7}$$

where $A_0 = \gamma s_0$.

It was shown [10] that the model of tunneling currents quite well describes the experimental data at small deformations $\Delta l/l_0 < 0.1$ with $A_0 = 6.491$ and $R_0 = 3.770 \times 10^5\ \Omega$. Knite et al. [10] proposed that the high rate of the increase of R/R_0 at larger deformations $\Delta l/l_0$ is related to destruction of the conducting network, i.e., with decrease of the number of conducting paths N:

$$N = \frac{N_0}{\exp\left[A_1\left(\frac{\Delta l}{l_0}\right) + B\left(\frac{\Delta l}{l_0}\right)^2 + C\left(\frac{\Delta l}{l_0}\right)^3 + D\left(\frac{\Delta l}{l_0}\right)^4\right]} \tag{8}$$

where N_0 is the initial number of conducting path, A_1, B, C, and D all are constants [10].

The substitution of Eq. (8) into Eqs. (3) and (7) yields

$$\ln\left(\frac{R}{R_0}\right) = \ln\left[1+\left(\frac{\Delta l}{l_0}\right)\right] + A\left(\frac{\Delta l}{l_0}\right) + B\left(\frac{\Delta l}{l_0}\right)^2 + C\left(\frac{\Delta l}{l_0}\right)^3 + D\left(\frac{\Delta l}{l_0}\right)^4 \tag{9}$$

Equation (9) comprises both mechanisms—change of tunneling currents and disruption of tunneling currents in the conductive nanochannels. A good agreement between the theoretical and experimental curves have been achieved at $A = 8.206$; $B = -90.979$; $C = 873.911$; $D = -1{,}333.339$, and $R_0 = 3.835 \times 105\ \Omega$ for polyisoprene/HSCB composite [10].

At the first sight surprising surprising PPR effect in PNGC observed first by Knite et al. [10] has been later quantitatively explained by Zavickis et al. [18] by taking into account the transversal slippage that causes strain in the direction perpendicular to the applied axial pressure, which in turn provokes the tunneling barrier thickness increase at small deformations and subsequential rupture of the conductive channels in the direction perpendicular to the applied pressure at relatively larger strain. The exceptionally high structure of the conductive carbon black filler provides extremely entangled conductive grid structure and prevents new conductive pathways to develop in the direction parallel to the strain applied. Based on this, Zavickis et al. [18] adopted a more advanced theoretical model that describes the dependence of electrical resistivity directly on operational pressure for piezoresistive PNGC. This was done to simplify future calculations, since to characterize piezoresistive properties of PNGC the pressure is the primary argument instead of deformation. In case of compression of the elastomer matrix, authors considered the transversal elongation of the sample, thus the separations between adjacent particles can be calculated as

$$s = s_0\left(1+\varepsilon^{\perp}\right) = s_0\left[1+\left(\frac{\Delta L'}{L'_0}\right)\right], \tag{10}$$

where $\varepsilon^{\perp}$ is the transversal strain of the matrix in the perpendicular direction to applied axial pressure force, $\Delta L'$—transversal deformation of the sample, and L_0'—transversal length of the sample. Zavickis et al. [18] proposed that the relative elongation of the sample in the direction perpendicular to applied axial pressure $\Delta L'/L'_0$ can be expressed as follows:

$$\frac{\Delta L'}{L'_0} = -\left(-\nu\frac{\Delta L}{L_0}\right) = \nu\frac{P}{\lambda}, \tag{11}$$

where ν—Poisson ratio, λ—elastic modulus of composite, P—compressive pressure on the sample, and $-\Delta L/L_0$ is the negative elongation (reduction) in the direction parallel to applied compressive force. After taking into account Eqs. (10) and (11), Zavickis et al. [18] got equations similar to Eqs. (7) and (9), respectively:

$$\ln R = \ln R_0 + A_0' P \tag{12}$$

for small values of pressure (small deformations) and

$$\ln R = \ln R_0 + A'P + B'P^2 + C'P^3 + D'P^4 \tag{13}$$

for large values of pressure (large deformations), where

$$A_0' = A_0 \frac{\nu}{\lambda}, \quad A' = A \frac{\nu}{\lambda}, \quad B' = B\left(\frac{\nu}{\lambda}\right)^2, \quad C' = C\left(\frac{\nu}{\lambda}\right)^3 \quad \text{and} \quad D' = D\left(\frac{\nu}{\lambda}\right)^4.$$

To verify the assumptions in the theoretical description of the PPR effect, Zavickis et al. [18] fitted the experimental data points of the first upgoing cycle using Eqs. (12) and (13) and near perfect agreement was found between the two theoretical curves and the corresponding experimental values (Fig. 1). This was done for the operational pressure range of up to 4.1 bar where, according to approximate calculations, relative unidimensional axial compressive deformation of the sample is considered to be less than 10 %, which should correspond to the elastic deformation region.

Developed by Knite et al. [10], the combined piezoresistance model based on both the tunneling effect and the destruction of conducting paths have been modified and used also by other authors for theoretical description of different piezoresistance effects in some PNGCs [15, 19–21]. In two cases, our combined piezoresistance model was successfully applied for elastomer/graphite nanosheet composites [15, 20].

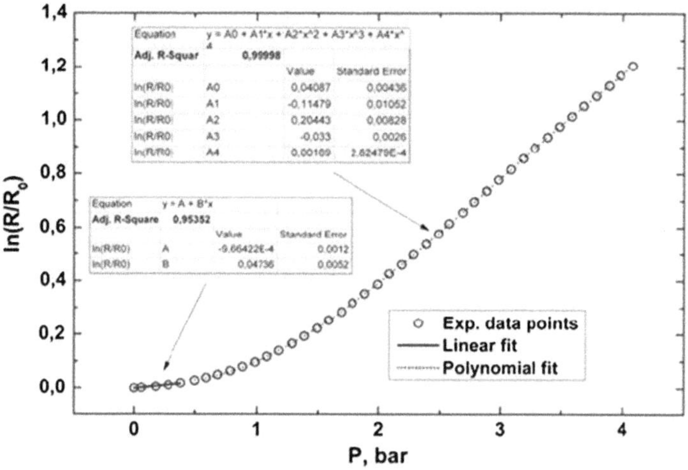

Fig. 1 A linear (12) and a fourth-order polynomial (13) fit to the experimentally observed positive piezoresistivity of a completely flexible sensor prototype [18]. Copyright 2011. Reproduced with permission from Elsevier Ltd

Wang Luheng et al. [19] generalized Eq. (9) for both PPR and NPR effects observed experimentally in the same kind of silicone rubber/HSCB composites but with different concentration of HSCB. The PPR effect has been observed in PNGCs with low mass ratio of HSCB to silicone rubber F ($0.08 \leq F \leq 0.09$) but NPR—in PNGCs with ($0.14 \leq F \leq 0.24$). These experimental results were explained by analyzing the changes due to the applied pressure in the gap sizes in the existing effective conductive paths and in the number of effective conductive HSCB paths in the composite samples [19].

Surprising at first, a NTR effect has been observed and qualitatively (non-numerically) explained by Flandin et al. [16] in case of ethylene-octene elastomer (EO)/HSCB composite. Authors (Flandin 2001 Polym) reported that the EO filled with HSCB exhibits a reversible decrease in resistivity under stretching with up to 30 % strain. In addition, these characteristics have been observed only in high-structure carbon black, as composites of the EO elastomer with low-structure carbon black or carbon fibers exhibited conventional PTR behavior [16]. An explanation was given by taking into account the unique features of both filler and matrix as follows. In contrast to covalently cured (chemically cross-linked) elastomers, the structural basis for EO elastomers is provided by a network of flexible chains with fringed micellar crystals (physically cross-linked). Authors believe that upon stretching, the fringed micellar junctions of EO elastomers do not remain fixed, but slide by a process of detachment–attachment, which can also be considered as partial melting. Due to local stress concentrations, the actual strain in the vicinity of the carbon particle can be noticeably higher than the average macroscopic strain. Therefore, the crystalline junctions can "melt" and flow out from the space between nearby particles. Either an existing pathway is improved by a rise in the charge tunneling conductivity or a new electrical pathway is thus created. Particularly interesting for strain gauge applications was a reversible and strain-rate independent decrease in resistivity with up to 30 % strain [16].

2.3 *The Shift of Percolation Threshold Under External Thermodynamic Forces*

From the macroscopic point of view, the principle of strain sensing is based on the shift of the percolation threshold due to some thermodynamic forces, for example, under tensile strain as shown in Fig. 2.

One can see that due to the shift of percolation threshold under 40 % strain the electric resistance changes more than 10^4 times for composite with 10 phr (parts per hundred rubber) of nanostructured carbon black. The maximum strain sensitivity is identified in the percolation region (9–11 mass parts of filler) for relaxed PNGC (Fig. 2). Thus, one may expect the maximum sensitivity of PNGC materials to the external thermodynamic forces in filler concentration region slightly above the percolation threshold of electric conductivity as demonstrated by Knite et al. [10].

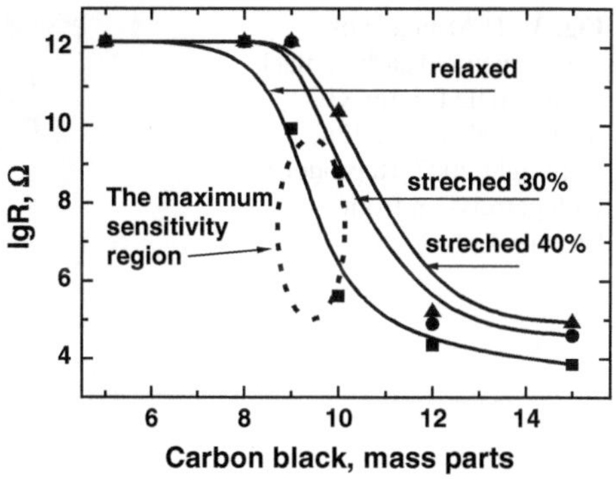

Fig. 2 Shift of the percolation threshold under tensile strain in the polyisoprene/high-structure carbon nanoparticles composite [22]. Copyright 2007. Reproduced with permission from Elsevier Ltd

3 Principles of Design and Characterization of Polymer/Nanographite Composites for Mechanical Impact Sensing

3.1 Most Suitable Filler Materials

In this chapter, we describe only such sp^2 hybrid carbon nanostructures (nanographites) that can be attributed to 2D nanostructures and we are going to pass the 1D ones like carbon nanotubes (CNT). One can find useful information about polymer/CNT composites developed for mechanical impact sensing in review papers [23, 24].

3.1.1 Extra-Conductive Carbon Black as Nanographite Filler

HSCB PRINTEX XE2 (ECB) is a commercially provided nanographite filler that is used in most cases for elaboration the PNGC [10, 25]. Several authors [1, 2] have reported a graphite-like organization of carbon atoms in HSCB nanoparticles near surface. In Fig. 3 one can see the polyhedron-like shape of primary particles of HSCB that is an indirect verification of the nanographite structure of HSCB. We believe that the faces of primary carbon nanoparticle consist of graphene nanoplatelets. Another indirect evidence for this is the high value of electric conductivity in comparison with other technical carbon blacks.

3.1.2 Thermochemically Exfoliated Graphite

Graphene is a single layer of sp^2-bonded carbon atoms that can be thought of as an individual atomic plane extracted from graphite [27]. Recent studies [28–30] have demonstrated that several stacked graphene layers, which essentially represent partially exfoliated graphite, can be applied successfully as fillers for polymeric

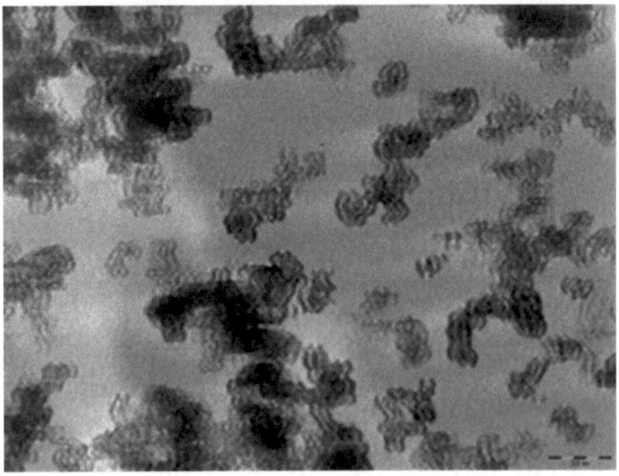

Fig. 3 TEM image of high-structured carbon black nanoparticle PrintexXE2 . Scale mark 200 nm [26]. Copyright 2007. Reproduced with permission from Elsevier Ltd

matrices. Thermochemically exfoliated graphite (TEG) produced from graphite subjected to thermal shock is shown lower on Fig. 10.

3.2 Most Suitable Matrix Materials

The polymer elastomers are the most useful matrix materials for the development of sensing elements intended for recurrent detection of large-scale mechanical impact there for they require reversible piezoresistive effect. For this purpose, the natural polyisoprene rubber has been found as the most promising matrix [10]. Silicone rubber also has been used as hyperelastic matrix for PNGC development [15, 19].

General principles of designing the structure of materials with the highest sensitivity that we adapted in this chapter for obtaining most sensitive multifunctional elastomer/nanographite composites are as follows:

1. Polyisoprene (natural rubber) with the best hyperelastic properties has to be chosen as the matrix material;
2. High-structured graphitized carbon nanoparticles (HSNP) (for example Printex XE2) providing a fine branching structure and a large surface area (better adhesion to polymer chains compared to low-structure graphitized nanoparticles (LSNP); short CNTs or small-size graphene platelets should be taken as filler. Because of a higher mobility of HSNP compared with LSNP, the electroconductive network in the elastomer matrix in this case is easily destroyed by very small tensile or compressive strain. We suppose this feature makes the elastomer–HSNP composite an option for more sensitive tactile elements.
3. The highest sensitivity is expected in the filler concentration region slightly above percolation threshold of a relaxed polyisoprene composite. The smallest mechanical strain or swelling of the composite matrix remarkably and

reversibly increases the resistance of such a composite. The sharper is the percolation transition of insulator/conductive particle composite, the higher should be the compressive stress sensitivity of sensing element [10].

3.3 Preparation Methods of Composites

Piezoresistive composites composed by an insulating matrix and conductive filler have been widely studied in the past years. Various fillers, polymers, and production methods have been used to produce continuous electrically conductive networks throughout the insulating polymer matrix. Electrical conductivity parameters of these heterogeneous composites have been found to be strongly dependant on filler concentration, filler geometry, and even composite preparation methods.

Depending on whether or not specific solvents have been used to lower the viscosity of the polymer matrix, all composite production methods can be divided into two groups:

1. Use of conventional polymer/rubber processing equipment that does not uses solvents—the filler is mechanically mixed into polymer, includes usage of various mixing mills, kneaders, banbury mixers, and even extruders. The polymer viscosity is generally high (Fig. 4);
2. Usage of certain solvents to dissolve the matrix and the filler is added to the mixture afterward. Polymer viscosity is much lower (Fig. 5).

In general, conventional polymer/filler mixing equipment offers high processing amounts and rates however the homogeneity of the composite structure is far worse compared to the polymer dissolution methods. Figure 5 shows typical scheme of piezoresistive composite production. At first, polymer is dissolved in adequate solvent then filler is added to the mixture and stirred for a certain period, afterward the solvent has to be evaporated. Long durations and usage of sometimes harmful solvents are the only drawbacks for this method. Ultrasound homogenizer can be used to produce suspensions/colloids of fillers and solvents before mixing with dissolved polymer. However even after the ultrasound treatment particles tend to aggregate due to the high specific surface of the nanosize fillers intermolecular Van Der Waals forces. Functionalization of particle surface might overcome this problem; however, the surface modification degrades the electrical properties of fillers.

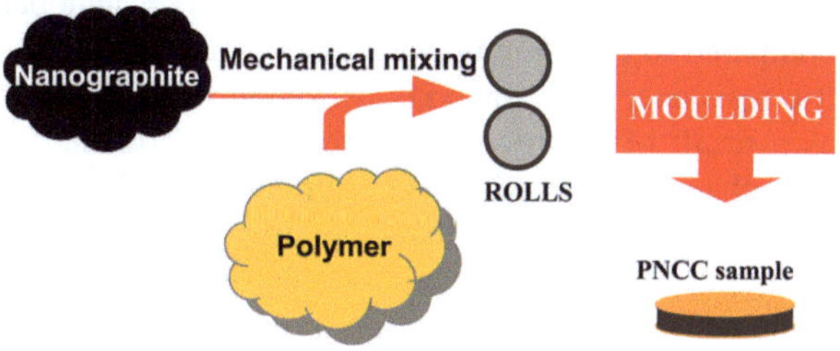

Fig. 4 Composite production scheme using conventional rubber processing equipment [25]. Copyright 2011. Reproduced with permission from lietuvos mokslų akademija

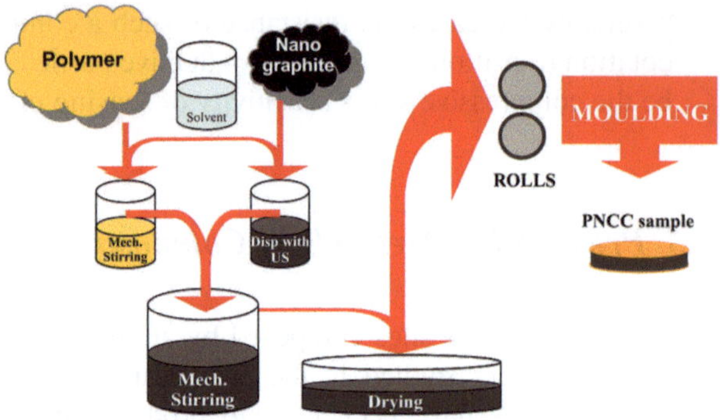

Fig. 5 Composite production scheme when the matrix is dissolved and then mixed with conductive filler [25]. Copyright 2011. Reproduced with permission from lietuvos mokslų akademija

4 Produced and Tested Polymer/Nanographite Composites for Mechanical Impact Sensing

4.1 Polymer/Extra Conductive (Graphitized) Carbon Nanoparticle Composites

Significant and reversible both PTR and PPR effects first were observed in polyisoprene/HSNP composites have been reported by Knite et al. [10]. Plates of 12 mm diameter were cut for piezoresistivty effect measurements from 1 mm thick 20 cm × 20 cm sheets vulcanized at high pressure. For studying the tensoresistivity effect 15 cm × 1.5 cm samples were cut. Copper foil electrodes were glued on both sides at sample ends and each pair of electrodes was shortcut by copper wiring. On a relaxed sample the distance between electrodes did not exceed 50 mm. Sandpaper was glued on the electrodes to fasten the samples in the stretching machine. Authors reported that of all the composites examined, the best results were obtained on samples with 10 phr of carbon nano-particles, which apparently belonged to the region of percolation phase transition. Electrical resistance of the samples increases by more than 4 orders upon a 40 % stretch (PTR effect) and more than 3 orders upon a 0.30 Mpa pressure (PPR effect) as seen in Fig. 6. Resistance practically returns to its previous value after the samples are relaxed (reversibility). The reversibility and the significant changes of electric resistance under both tensile and compressive strain were explained due to comparatively higher mobility of high structure nanoparticles compared to low structure particles as well as stronger adhesion of carbon nanoparticles to the polymer matrix compared to cohesion between nanoparticles themselves [10]. The growth of electric resistance with uniaxial stretching as well as with pressure can be explained as a result of destruction of the structure of the carbon electro-conductive nano size channel network. At low stretching deformation the experimental data has good coincidence with model of tunneling conductance [10], see also Eq. 7 in Sect. 2.2. The AC conductivity measurements also verify the tunneling model of conductance [10]. For description of the experimental results at high deformation

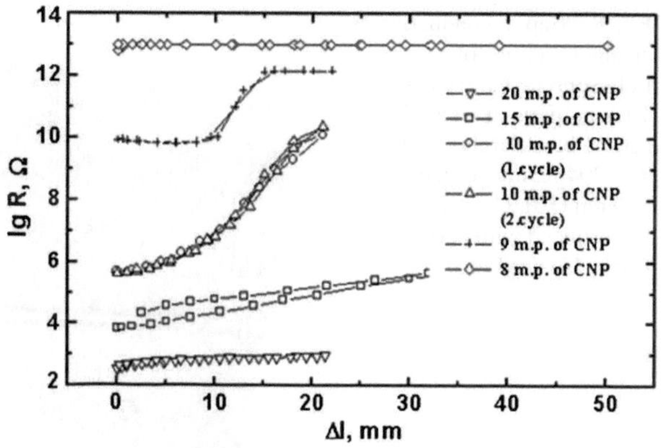

Fig. 6 Electrical resistance *R* of the nanocomposite as function of stretching deformation Δl at different phr (m.p.) of HSCB. On a relaxed sample the distance between electrodes was $l0 = 50$ mm, $T = 293$ K [10]. Copyright 2004. Reproduced with permission from Elsevier Ltd

the destruction of conductive network and decrease of conducting path have to be taken into account [10], see also Eq. 9 in Sect. 2.2. In case of uniaxial pressure (Fig. 7) destruction of electrically conductive channels is caused by deformation of the polyisoprene nanocomposite perpendicular to the direction of pressure (Eqs. 12 and 13, Sect. 2.2).

A tapping mode AFM Nanoscope III (Dimensions 3,000, Digital Instruments) was used to investigate the local nano-size properties of the polymers. The local conductivity patterns on the surface were studied on a modified conductive AFM. The AFM had standard silicon nitride cantilever tips. The AFM tip and cantilever were coated with a 5 nm thick Cr adhesive layer and a 15 nm thick Au layer. With a contact mode conductive atomic force microscope authors succeeded in obtaining a topographic picture of the sample surface and a nanoscale map of cross-sections of the electro-conductive channels and the insulating matrix of the same local spot. For example, it is seen in Fig. 8 that the extra-conductive carbon black nano-particles agglomerate during vulcanization process in clusters of size around 100 nm forming conductive channels throughout the whole sample. The blue regions in Fig. 8 represents insulating polyisoprene matrix and the red

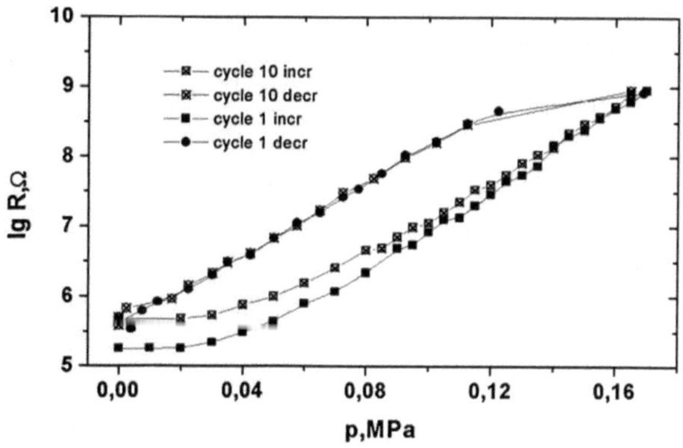

Fig. 7 Electrical resistance *R* of the 5-sheet nanocomposite block as function of uniaxial pressure *p*. The deformation of nanocomposite is constrained in the direction perpendicular to the acting force. $T = 291$ *K*. 293 K [10]. Copyright 2004. Reproduced with permission from Elsevier Ltd

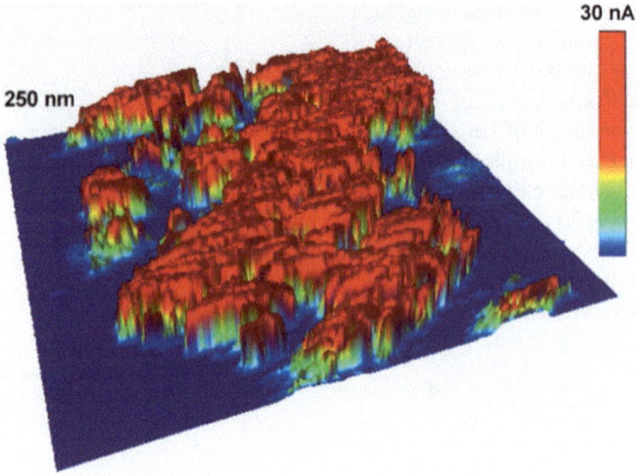

Fig. 8 Map of conductive channel sections of Polyisoprene matrix with 10 mass parts of nano-size carbon black. Relaxed state, $T = 294$ K, image size 250 nm [10]. Copyright 2004. Reproduced with permission from Elsevier Ltd

regions the electrically conductive HSNP channel sections on the surface of the PNGC sample [10].

The effect of a plasticizer on the change of the electrical resistance under deformation at strain was studied in polyisoprene composites containing dispersed HSCB at concentrations slightly above the percolation threshold [31, 32]. The addition of plasticizer to the initial materials increases the strain sensitivity of the composite and broadens the HSCB concentration interval of the percolation threshold. The observed improvements of the tensoresistive response are explained by decrease of cohesion forces between carbon nanoparticles and a higher mobility of the carbon nano-particles in the elastomer matrix in the presence of the plasticizer [31, 32].

Based on the PPR effect data Zavickis et al. [18] developed a completely soft matter hyper-elastic pressure (SHP) sensor prototype without any solid state (metallic) details. Sensor element is made using functional gradient multilayer approach, when elementary layers of PNGC with different conductive filler concentration are vulcanized together and forms uniform sensor body with integrated soft (hyper-elastic) electrodes as seen in Fig. 9.

Fig. 9 The SEM picture of the interface between completely vulcanized sensor (*S*) and electrode (*E*) PNGC layers. Picture made from surface of the sample broken in liquid nitrogen [18]. Copyright 2011. Reproduced with permission from Elsevier Ltd

The SHP sensor element was made from 3 functional layers of PNGC, each with different concentration of HSCB: The pressure sensitive middle layer was chosen with 10 phr of HSCB that is slightly above previously obtained percolation threshold [10], to achieve maximal piezoresistivity of the structure. The electrode layers on both sides of the sensitive layer were chosen with comparably larger HSCB concentration, to promote good electrical conductivity. All the conductive parts were incorporated into dielectric natural rubber shell without any HSCB filler. Thin brass wires were used only for easy connection of the measuring equipment, but according to SHP concept, the electric wiring although can be made from conductive rubber strips, incorporated into dielectric shell. Corresponding components were pre-shaped using incomplete vulcanization in reduced vulcanization temperature and shorter curing times in hot steel mould, using high temperature resistive polymer film inserts to prevent sticking to mould during sample removal. Afterwards the pre-vulcanized components were assembled according to sequence mentioned above and cured until complete vulcanization [18]. The measured PPR effect in this prototype is shown on Fig. 1 and is theoretically explained by quantum tunneling effect (Eqs. 12 and 13, Sect. 2.2).

Zheng et al. [21] recently investigated the piezoresistive behavior of PNGC cast films and pressed films made from high-density polyethylene and PRINTEX XE-B. The mechanical and piezoresistive response were measured under axial tensile test and the PTR effect was observed for all samples tested. Authors stated that the normalized electrical resistance of the cast film was almost constant with strain in the elastic region of stress-strain curve and starts to increase with strain in the plastic region. Authors described these experimental results by successfully modifying the Eq. (9) (combined piezoresistance model based of both tunneling effect and destruction of conducting paths). It was also concluded that cast PNGC films could be potentially used as a strain sensor to identify the elastic and plastic deformation regions of the films [21].

4.2 Polymer/Graphite Nanosheet Composites and Polymer/Graphene Composites

In the latest review papers particulary devoted to tactile sensors [14, 33] as well as to development and application of graphene based polymer composites [34–36] only few research papers were analyzed regarding to use polymer/graphene composites in field of mechanical impact sensing. In this section we are going to attempt to fulfill this space. The use of graphene for the development of a strain and damage sensor to be utilized in structural health monitoring of fiber reinforced polymers was evaluated and modeled by Chiacchiarelli et al. [29]. An Epikote 862 diglycidyl ether of bisphenolF(DGEBF)epoxy resin, kindly supplied by Hexion, was used as a matrix. Diethyltol-uenediamin e (DETDA), supplied by Lonza, was used as curing agent (26.4 phr). Graphenenanoplatelets (GNPs) were supplied by Cheap Tube Inc. (Grade 2). According to the manufacturer, the

GNPs had a surface area of 100 m^2/g, an average thickness of around 10 nm and an average diameter of 25 μm. They were used as supplied by the manufacturer. Autors obtained colloidal suspension of GNPs in chloroform after sonication with a Vibracell® VC750 tip sonicator for 1 h at an amplitude of 30 % (225 W). Then, the epoxy monomer was added and the mixture was further sonicated for 1 h at identical conditions. Afterwards, in order to fully remove the solvent, the solution was heated on a hot plate (Infrared spectroscopy was used to corroborate solvent evaporation). Finally, the hardener was added and the system was magnetically stirred for 5 min. This reactive mixture was then used to create a coating (50 × 6 mm 2 area) onto the carbon fiber/epoxy composite (CFRC) specimens with previously prepared Al electrodes. Finally, the coated specimens were cured at130 °C for 5 h. The PTR effect of sensitive composite layer was measured CFRC bending test cycles. Authors observed reversible and irreversible behavior of sensor tensoresistivity. Authors explained both these behaviors by modifying Eqs. (7) and (8) respectively (Sect. 2.2). It was experimentally shown that both the stress and damage of the composite can be detected by a simple measurement of the sample electrical resistance change.

Lu et al. found that in high-density polyethylene graphite nanosheet (5–20 μm; 30–80 nm) composites there is a critical pressure at which the piezoresistive behavior changes from PPR effect to NPR effect. The HDPE and GNs were first wet-mixed to achieve a uniform dispersion and afterward mixed on a two-roll mill. The critical pressure value increases with increasing filler concentration and generally up to 10 Mpa pressure, the effect is negative due to a decrease of interparticle separation but at even higher pressures (up to 40 Mpa) the PPR effect has been stated reason for this is the large deformation of polymer matrix that leads to destruction of conductive channels [37]. However, the NPR effect at low pressures might be attributed to increased/better electrical contact area when pressure is applied since sensors' resistivity is measured between steel electrodes.

Chen et al. produced finger pressure-sensitive composite based on silicone rubber graphite nanosheets (5–20 μm; 30–80 nm) [15]. Percolation threshold for such composition was determined to be 0,9 vol%, a composite containing 1,36 vol% conductive filler was found to be the most sensitive to external pressure reaching R/R_0 values around 800,000 under 0.7 MPa pressure. Authors described the piezoresistive behavior by the tunneling models as well as used the models for change of the number conducting paths that are similar to model developed for polyisoprene high-structure carbon black composites [10]. In the same way as Knite et al. [10] stated for nanostructured carbon black, the authors [15] made conclusions that stronger adhesion of the GNs to the polymer matrix compared to the cohesion between the nanosheets themselves lead to excellent piezoresistive properties of composite.

Alkyl-functionalized graphite oxide nanosheet polydimethylsilicone composites with low percolation threshold (0.63 vol%) have been prepared by Hou et al. [38] through wet mixing method. Low percolation threshold is attributed to high aspect ratio and homogeneous nanosheet dispersion in the polydimethylsilicone. The thickness of functionalized graphene oxide nanosheets was 2.7 nm.

Composites containing 1.19 vol% conductive filler showed the highest piezoresistive sensitivity ($R/R_0 > 400$).

Pressure-sensitive nanocomposites have been made by Soltani et al. [39] from room temperature vulcanizing silicone rubber, graphite nanosheets, and interfacial compatibilizer alkyl ammonium salt. Use of compatibilizer decreased the percolation threshold from 3 to 1 wt% as well as piezoresistive sensitivity from more than 100,000 to around 100 R/R_0 . The decrease of piezoresistive sensitivity is attributed to the high dispersion state of filler throughout the composite leading to insufficient distance between particles to make conductive paths under external pressure.

Sampo Tuukkanen et al. [40] report about the fabrication and characterization stretchable CNT and graphene nanocomposite electrodes on different rubber substrates. Authors used commercially available graphene ink (P3014 Graphene Screen Printing Ink from Innophene Co., Thailand) as well as custom-made CNT ink wich was prepared by mixing CNTs and cellulose derivatives with ultrasonication (obtained from Morphona Ltd., Finland). Both inks were deposited as stretchable electrodes using blade-coating method on four different types of rubber substrates: (1) blend of natural rubber (NR) and butadiene rubber (BR)—NR/BR with 15 phr clay, (2) NR/BR with 25 phr carbon black, (3) nitrile-butadiene rubber NBR unfilled, and (4) was chlorosulfonated polyethylene CSM. These substrates were chosen due to their different chemical structure and mechanical properties. Before the electrode deposition, the substrate rubber samples were stretched reversibly up to 20 % elongation to remove tensile stress softening. After deposition, the graphene film-like electrodes were dried for 7 min and the CNT electrodes for 5 min at 403 K. To obtain the sheet resistances approximately the same for both materials and in such way to make the analysis of the results more consistent, the final thicknesses of the graphene films were approximately 3–5 μm and 1 μm of the CNT films. The electrical resistance of the fabricated sheet electrodes was measured using four-point probe method. Accordingly to classification given in Sect. 2.1., the PTR effect has been found for both types of sheet electrodes. The CNT ink sheets showed good properties for all four rubber substrates. Their electrical resistance before and after the stretching was not changed. In contrary, the graphene ink was highly dependent on the substrate material. High-quality electrodes with low resistances were obtained in the cases of high surface energy rubbers 3 and 4. The graphene ink samples on substrates 1 have some cracks already before stretching but in case of substrate 2, the graphene electrodes contained some bubbles. Authors [40] believe that these faults caused higher resistances both before and after stretching. Authors also conclude that the relative resistance change during the stretching is larger and rubber relaxation after the stretching affects less to the electrical resistance in the case of CNT electrodes. This makes the CNT films more suitable for sensor applications where varying stress is applied to the element. On the other hand, the smaller relative resistance change during the stretching of the graphene electrodes would make them applicable to be used as stretchable electrodes.

Kumar et al. [41] have stated NPR effect for poly(isobutylene-co-isoprene) (IIR)/reduced graphene oxide (RGO) composites with 5 wt% RGO. The RGO was synthesized from graphite flakes using Improved Graphene Synthesis [42] with subsequent thermal reducing at 473 K temperature for 30 min. (IIR/RGO) composite samples with 5 wt% RGO were prepared by the solution mixing process in tetrahydrofuran (THF) using subsequent ultrasonic treatment for RGO suspension and mechanical mixing of RGO suspension with IIR in THF for 3 h. Thereafter, the mixture was dried at 333 K in vacuum oven until it achieves a constant weight and samples were molded at 433 K for optimum curing time determined by rheometer. SEM, AFM, and XRD investigations showed homogeneous dispersion of RGO in IIR whereas IIR/expanded graphite (EG) composites tested for comparison indicated poor dispersion of EG. The thickness of the sensor sample was 0.1 mm and relative resistance change of the sensor was measured while applying force in the range of 0.05–0.1 kN. Unfortunately, the comprehensive theoretical explanation of the piezoresistivity effect in this material is not given. Authors [41] mentioned only that the resistivity of the composite can change with external pressure due to the construction and destruction of conductive networks as explained in [39].

Very promising results for future development of highly sensitive tactile sensors have got Hodlur and Rabinal [43] by preparing polyurethane foam (PM) with self-assembled graphene layers inside of pores on the surface of polymer. They used the simple method as follows: First, the flexible PF was impregnated with hydrazine hydrate for a minute and washed with distilled water. Home-synthesized graphite oxide (GO) sample was dispersed in distilled water (0.5 mg in 1 ml water) to get GO colloidal solution. Second, the PF was soaked in GO colloidal solution and left to dry for an hour. Third, the PF was saturated with hydrazine hydrate to reduce GO to graphene. At the end it was washed as well as sonicated for few times in distilled and dried in room conditions. Authors [43] chosen for testing PF with approximately 3 wt% graphene. The conductivity of this sample changes by more than 5 orders of magnitude at applied voltage of 1 V if pressure rises from 1.01 to 1.52 atmospheres. So the negative pressure coefficient was stated for this material. Authors believe that such high value of pressure sensitivity can be explained by vertically oriented stacked graphene layers on polyurethane pores inside surface. FTIR analysis and SEM measurements indicates strong chemical bonding between PF surface and graphene flakes in vertically tilt configuration. In our opinion, the evidence for this are a little bit sophisticated because the high-resolution SEM recordings should be limited by the fact that high-energy electron beam is going to burn the PF. In our opinion, more convincing evidences for existence of vertically oriented stacked graphene layers on the inside surface of polyurethane pores should be found in future. We also hope that authors [43] will elaborate more comprehensive numerical model for mechanism of pressure sensing for this very promising material.

Vera Goncalves et al. [44] studied piezoresistive properties of electrically conductive polymer composites made from porous polyether block amide (PEBA 4,033 from "Atofina Chemicals") and various grades of graphene platelets obtained from "XG Sciences." Altogether five graphene platelet grades where used: C-750 (length 1–2 μm; thickness 2 nm; surface area 750 m^2/g), M5 (length

5 μm; thickness 6–8 nm; surface area 120–150 m^2/g), M15 (length 15 μm; thickness 6–8 nm; surface area 120–150 m^2/g), M25 (length 25 μm; thickness 6–8 nm; surface area 120–150 m^2/g), and H5 (length 5 μm; thickness 15 nm; surface area 50 m^2/g). The composites were prepared similarly to polymer dissolution method as described previously in Sect. 3. Porous morphology of electrically conductive polymer composite structures was achieved using all graphene platelets except C-750. The reason for this result is not explained in the following paper as well as the authors studied the influence of the remaining various graphene platelet grades on the piezoresisitive properties of their produced porous composites only at one specific concentration of conductive fillers—15 vol%. Therefore it is our responsibility to notify that the further discussed results might be inaccurate for determination of the most suitable graphene platelet grade for production of piezoresistive polymer/nanographite composites since piezoresistivity and conductivity highly depends on the conductive filler geometry and concentration. Composites containing H5 grade graphene exhibited the lowest electrical resistivity; however it did not exhibit any pressure sensitivity. The most promising piezoresistive response was achieved from composites containing M5 grade filler, which as authors highlight exhibited almost linear negative piezoresistive response to pressure up to 0.254 MPa on a log(R)–log(P) plot, but in cyclic loading/unloading conditions sensor exhibited significant hysteresis and resistivity drift due to poor mechanical properties of porous morphology. In Table 1, piezoresistive behavior and sensitivity of the previously mentioned polymer/graphite nanosheet composites are compared.

4.3 Hybrid Polymer/Nanographite Composites

Kim et al. [45] report about the preparation and investigation of nano smart hybrid material based on graphene. However, there is not explanation given why authors named the epoxide/graphene composite (EGC) a hybrid material. The appropriate amounts of graphene (KITECH) were incorporated into the epoxy (KUKDO CHEMICAL Co., YD-128) by direct dispersion into aqueous epoxy using ultrasonic homogenizer. A curing agent, Jeffamine (KUKDO CHEMICAL Co., A-230) was added to the dispersed graphene/epoxy suspension which was then located in a vacuum oven to remove the air. Afterwards the mixed suspension was poured into a silicon mold and cured at room temperature for 24 h and at 80 °C and 120 °C for 2 and 3 h respectively. The sensing element electrodes were made from conductive silver epoxy. The prepared sensor was tightly bonded on a steel beam with superglue to transfer the tensile strain from steel beam to a prepared epoxy/graphene sensor element [45]. Percolation threshold was found to be 3 wt% graphene. It is not shown by authors how percolation threshold was determined and the investigated piezoresistive sample also contains 3 wt% graphene. It is odd since, according to the definition, at percolation thresholds the composite merely starts to become conductive and its resistivity is too large for real piezoresistivity sensor application. The electrical resistance of EGC layer has been measured

Table 1 Comparison of piezoresistive sensitivity of polymer composites based on the graphite nanosheets

Filler	Matrix	Percolation threshold	Piezoresistive behavior	Piezoresistive sensitivity
Alkyl-functionalized graphite oxide (thickness 2.7 nm)	Silicone rubber	0.63 vol%	Positive	$R/R_0 > 400$ under 1.2 MPa [38]
Graphite nanosheets (5–20 μm; 30–80 nm)	Silicone rubber	0.9 vol%	Positive	$R/R_0 = 800{,}000$under 0.7 MPa [15]
Graphite nanoplatelets (thickness 10 nm)	Acrylnitrile butadiene rubber	0.5 phr	Positive	$R/R_0 > 100$ under 6 MPa [20]
Graphite nanosheets (thickness 30–80 nm)	Silicone rubber	3 wt%	Negative	$R/R_0 = 0.00001$ under 1 MPa [39]
Graphite nanosheets (5–20 μm; 30–80 nm)	High-density polyethylene		Negative till 12 MPa	$R/R_0 = 0.15$ under 12 MPa;
			Positive after 12 MPa	$R/R_0 = 800$ under 40 Mpa [37]
Reduced graphite oxide (nanoscale)	poly(isobutylene-co-isoprene)		Negative	$\Delta R/R_0 = 0.5$ under 0.05 kN [41]
Graphite oxide	Polyurethane foam		Negative	R/R_0 ~0.000001 under 52.7 kPa [43]
Graphene platelets (5 μm; 6–8 nm)	Porous polyether block amide		Negative	$R/R_0 = 0.42$ under 245 kPa [44]

versus steel beam deflection. The beam deflection was converted to strain by means of beam theory. In our classification the PTR effect was investigated. Except for some inaccuracies, authors [45] obtained promising results—exactly linear piezoresistivity in the range of 1,000 micro-strain as well as comparatively large gauge factor of 11.4 in comparison with epoxy/MWCNT composite—2.9. Authors also presented interesting explanation of such large gauge factor. If the strain sensor is subjected to tension, the contact resistance between adjacent graphene in conductive channel is increased mainly due to the reduction of the contact area of these overlapping adjacent graphene nanosheets. Based on this hypothesis, authors supposed that higher strain sensitivity of graphene composites can be explained as follows. The larger intercontact area among the graphene nanofillers due to their 2D structure may induce larger contact resistance change than other 1D structure carbon nanotubes. Charge tunneling between adjacent conductive nanoparticles was not discussed in this case.

Very interesting graphene-functionalized carbon nanotube and PVDF matrix composite strain-sensing materials were developed by Varrla Eswaraiah et al. [46]. These materials can be related to the class of hybrid composites. Authors prepared graphene-wrapped CNTs (GWCNTs) from fine ground powders of graphene oxide and alloy hydride (MmNi3) in a quartz tube in a furnace in subsequent argon, hydrogen, argon, acetylene, and argon atmospheres. The prepared graphene-wrapped CNTs were purified by refluxing in concentrated HNO3. Using the ultrasonic-assisted solvent casting method, GWCNTs were dispersed in PVDF matrix as follows: Definite quantity of GWCNTs as well as the corresponding amount of PVDF was dispersed separately in DMF with the help of ultrasonicator for 1 h at room temperature. These two solutions were mixed together by ultrasonicating for 1 h and the mixed solution was transferred into shear mixer and stirred at room temperature at 4,000 rpm for 2 h and at 80 °C for 30 min. The composite solution was taken out from the mixer and transferred into a Petri dish and kept in an oven at 80 °C in vacuum for 6 h to remove the solvent. Automatically peeled off polymer composite films were collected and cut into the required dimensions for further measurements. GWCNTs-based PVDF composite film was attached to one side of the aluminum (Al) specimen using high-strength epoxy to make a perfect bonding between the Al and the nanocomposite film; on the other side, a conventional metallic strain gauge was attached using glue. The specimens were investigated under tensile strain and the PTR linear effect was experimentally determined. Authors [46] compared results with insulating hydroxyl-group-functionalized CNT-based PVDF nanocomposites to show the superior performance of the graphene wrapped over CNT-based PVDF composites. A strain gauge factor of ≈ 20 has been obtained with 3 wt% conducting graphene wrapped over CNT-based PVDF composites whereas the gauge factor is ≈ 2 with insulating-molecule-functionalized CNT-based PVDF composites. Authors mentioned three main possible reasons for the piezoresistivity effect: (1) loss of conductive interconnections; (2) a tunneling effect between neighboring fillers, and (3) conductivity change from the deformed graphene-wrapped CNT hybrid. Authors concluded that since the conductance of the composite mainly comes from the nanofiller used and its distribution in the polymer matrix, a change in the number of conductive nanofillers or the loss of contacts in the polymer matrix is responsible in large for the change in resistance of the composite.

5 Development of Polymer/Nanographite Composite Sensor Element Systems for Mechanical Impact Sensing

Most of the pressure sensing systems are created as stiff structures made from brittle materials and therefore their usage is usually limited by ability to endure impact, vibrations, or large deformations. On the other hand, industries like civil and medical engineering, automotive as well as robotics are interested in cheap, reliable sensors without these limitations. In the above-described research, PNGC was proposed as a potential material for pressure and strain sensor elaboration. In this section, the influence of various nanographite fillers on the piezoresistive properties of PNGC is investigated and based on these results we chose the most suitable compositions for the development of relatively large-scale hyperelastic pressure sensor system (HPSS). Afterward the piezoresistive response of the developed HPSS was measured under external 0.1 and 1 MPa pressure in cyclic loading-unloading conditions; in addition, we determined the processing pressure influence on the piezoresistive sensitivity and behavior of HPSS under 0.1 MPa of pressure.

Two different types of nanographite fillers have been used:

1. 0-dimensional nanostructures—highly structured carbon black Printex XE2 obtained from Degussa®—specific surface area 950 m^2/g, average primary particle diameter 30 nm, DBP absorption 380 ml/100 g;
2. 2-dimensional nanostructures—thermally exfoliated graphite obtained from Kyiv National Taras Shevchenko University. Authors from this university showed that plastic deformation of graphite-epoxy/TEG composites under mechanical loading leads to irreversible changes in the value of normalized resistance R/R_0. The rate of these changes is determined by graphite filler content in epoxy matrix, method of preparation, and composite matrix porosity [47].

Depending on the filler type, two different raw PNGC production methods were used. In the case of HSCB filler, the polyisoprene (PI) and the necessary vulcanization ingredients (sulfur, stearic acid, zinc oxide, and N-Cyclohexyl-2-BenzothiazoleSulfenamide) were mixed with various high-structure carbon black concentrations using roll mixing. The obtained composition further in text is refered to as PICB. In contrast, for TEG filler, a multistep solution mixing method was used:

1. PI with curing ingredients was stirred and dissolved at room temperature for 24 h;
2. dispersion of TEG filler in chloroform (dispersed using ultrasonication with specific power 1 W $\times$ 5 min/1 ml to reduce the particle size Fig. 10) was added to the PI solution and stirred for 24 h;
3. obtained mixture was poured into Petri dishes and left for 24 h in drying chamber for chloroform to evaporate;
4. films were homogenized using cold rolling. Obtained compositions are abbreviated as PITEG.

To determine electrical as well as piezoresistive properties for each PNGC composition flat, round-shaped samples (diameter of 18 mm and average thickness of 1 mm) with brass foil electrodes were made by vulcanizing the raw rubber composites in hot stainless steel mold using Rondol thermostated press for 15 min under 3 MPa of pressure at 150 °C. The optimal curing conditions were determined using Monsanto 100 dynamic rheometer. After curing, the samples were shelf aged at room temperature for at least 24 h before any measurements were made.

The electrical conductivity of each PNGC sample was measured using Keithley 6,487 Picoammeter/Voltage source. The piezoresistive effect was determined

Fig. 10 Scanning electron images of TEG (*left*) and TEG after ultrasound treatment (*right*) particles

using Zwick/Roell Z2.5 universal material-testing machine coupled with Agilent 34970A data acquisition/switch unit. Due to the technical limitations of this measuring equipment, PNGC samples with conductivity lower than 10–8 S/m were not tested for piezoresistivity.

First the electrical conductivity of PNGC samples was determined. Figure 11 shows nanographite filler influence on the percolation transitions of PNGC compositions. The concentration of the conductive fillers is expressed in parts per hundred rubber (phr). PICB composition shows the steepest percolation curve; therefore, theoretically it could be expected for this composition to have the best piezoresistive behavior. Also it should be kept in mind that PICB composites were prepared with a different production method. TEG, on the other hand, is more or less a 2D nanostructure therefore the highest percolation transition in this case could be explained with difficulty to form a continuous conductive network of TEG particles throughout the matrix.

Fig. 11 The electrical percolation transition of PNCC

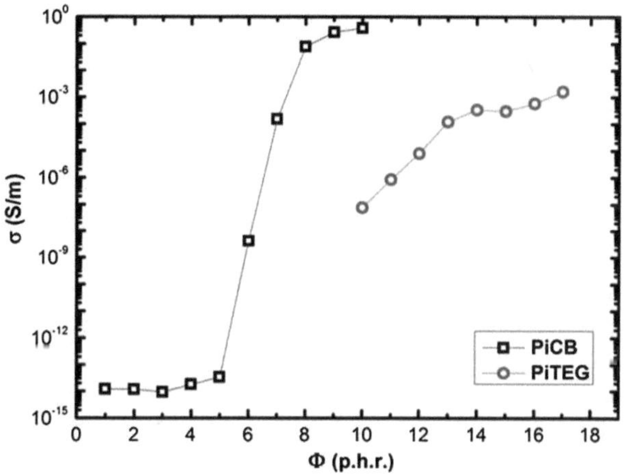

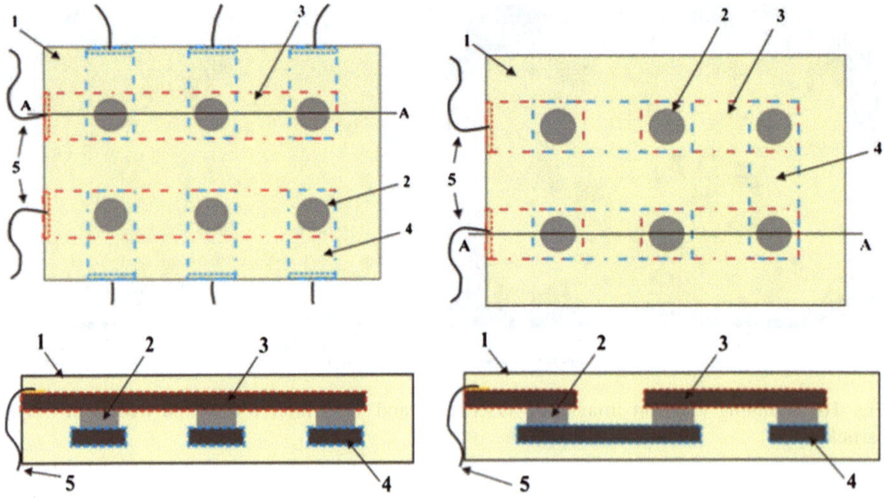

Fig. 12 **a** (left) schematic view of AHPS and schematic AA cross section of AHPS; **b** (right) schematic view of EHPS and schematic AA cross section of EHPS; consisting of : *1* non-conductive outer shell, *2* piezoresistivePiCB, *3* upper layer of conductive PICB, *4* lower layer of conductive PICB, *5* wires with soldered small brass foil plates. The geometrical dimensions of both designed systems were 100 × 70 × 5 mm

The hyperelastic pressure sensor system was made using layered composite design where pressure-sensitive elements were jointed with electrode elements and incorporated into protective nonconductive natural rubber shell (Fig. 12). As the pressure-sensitive elements, PICB with 8 phr CB was used since the piezoresistive sensitivity of this composition under 0.1 MPa was found to be the highest. Since the specific electrical conductivity for the composition PICB with 10 phr CB was high enough, it was used as hyperelastic electrode elements on both sides of sensitive elements. The placement of electrode layers in HPSS was designed to insure that:

1. each of the sensitive elements could be monitored separately (AHPS—addressed hyperelastic pressure sensor system);
2. the sensitive elements were connected in series to provide better piezoresistive sensitivity under external pressure (EHPS—enhanced hyperelastic pressure sensor system);

All HPSS elements were first separately partially vulcanized for 11 min under 3 MPa of pressure at 140 °C to ensure that they could maintain their shape during the final vulcanization when all elements where assembled in designated positions and cured together under 3 MPa of pressure at 150 °C for 20 min. To determine the influence of the vulcanization pressure on the piezoresistive behavior of the sensor, the EHPS elements were semi-vulcanized and the final product was vulcanized into one solid block under 0.5, 1, and 2 MPa of pressure (EHPS 0.5, EHPS 1, and EHPS 2). To connect HPSS to measuring equipment, small wires with soldered brass foil extensions were added to the side electrode layers. Cross-section of the EHPS system is shown in Fig. 13.

The piezoresistive behavior of EHPS under 0.1 MPa and 1 MPa operational pressures was determined (Fig. 14). As can be seen in Fig. 14, the piezoresistive behavior is very similar under 0.1 and 1 MPa of cyclic operational pressure; however, the piezoresistive sensitivity under 0.1 MPa of pressure is comparatively low (less than 2 %) as well as the piezoresistive behavior under repeated cyclic loading tends to decrease gradually. This can be explained with different speed of electrical relaxation for separate structural PICB elements of EHPS and therefore leads to the decrease of total piezoresistive effect in both ranges of operational pressure.

Figure 15 shows the piezoresistive behavior of EHPS made using different processing pressures under operational 0.1 MPa pressure. One can see that EHPS, which was made using highest processing pressure—3 MPa, appears to be the most piezosensitive. As one can see for small operational pressures (0.1 MPa), the sensitivity of EHPS drops considerably when processing pressure is reduced—this can be explained with improved mobility of electrically conductive particles in vulcanization process, leading to more electrically conductive channels in the composite structure.

Likewise, the piezoresistivity was determined for each AHPS each sensitive element up to 0.1 MPa pressure (Fig. 16). Variance of each element sensitivity at 0.1 MPa pressure is observable.

Fig. 13 Cross-section image of EHPS

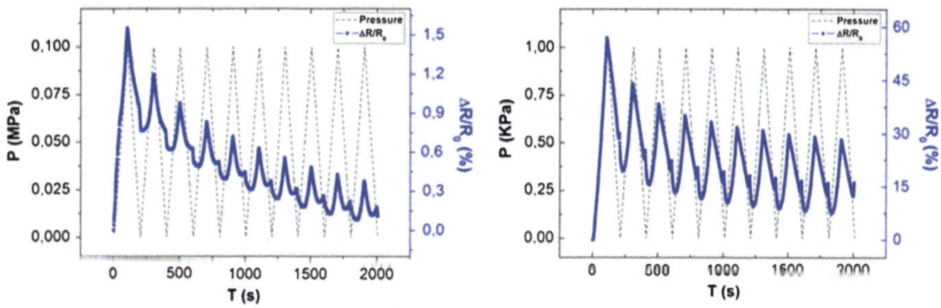

Fig. 14 Piezoresistive behavior of EHPS under cyclic operational pressure up to 0.1 MPa (*left*) and 1 MPa (*right*)

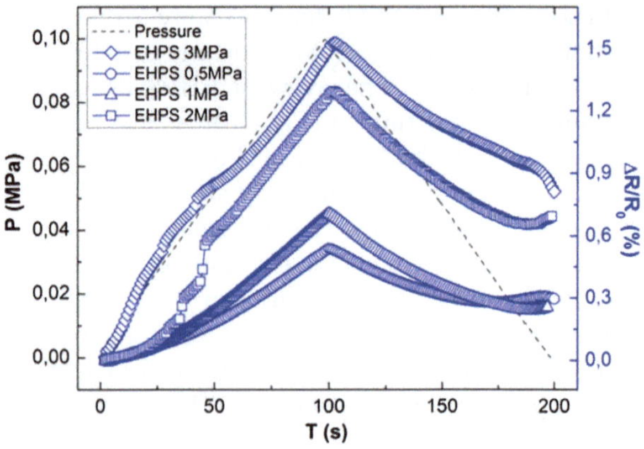

Fig. 15 Piezoresistive behavior of EHPS with different processing pressures under operational pressure of up to 0.1 MPa

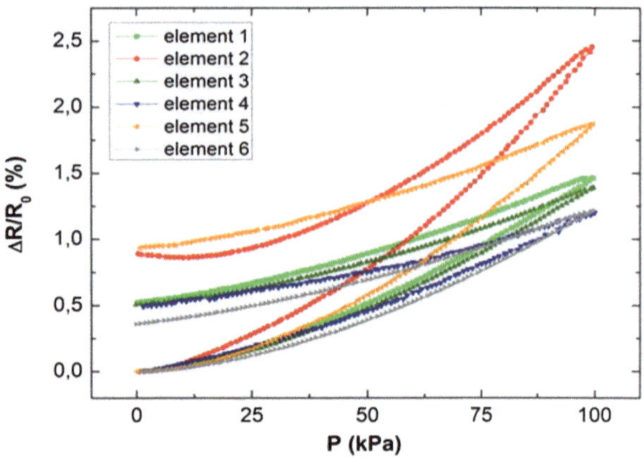

Fig. 16 Piezoresistive behavior of AHPS each sensitive element under operational pressure of up to 0.1 MPa

Similarly to HPSS manufacture, hyperelastic pressure sensor (HPS) with only one sensitive element (HPS) was developed. In Fig. 17 one can see a schematic image of HPS structure. As the pressure-sensitive elements in HPS, the PICB with 8 phr CB or the PITEG with 15 phr TEG were used since the piezoresistive sensitivities of these compositions were found to be the highest. As hyperelastic electrode elements on both sides of sensitive elements PICB with 10 phr filler was used. Small wires with soldered brass foil were added to the hyperelastic electrodes for resistivity monitoring. Finally, all of this was incorporated into protective nonconductive natural rubber shell.

The piezoresistive behavior of HPS with both sensitive elements was determined up to 0.1 MPa pressure (Fig. 18). As seen from Fig. 18, hyperelastic pressure sensor with PITEG sensitive layer shows noticeably higher pressure sensitivity. HPS with PICB-sensitive layer exhibits better effect reversibility. This could be explained as follows—in PITEG and PICB composites conductivity is possible due to tunneling currents between nanographite particles. Since intensity of tunneling currents is highly dependent on the average distance between conductive particles, piezoresistive behavior can be observed under external influences

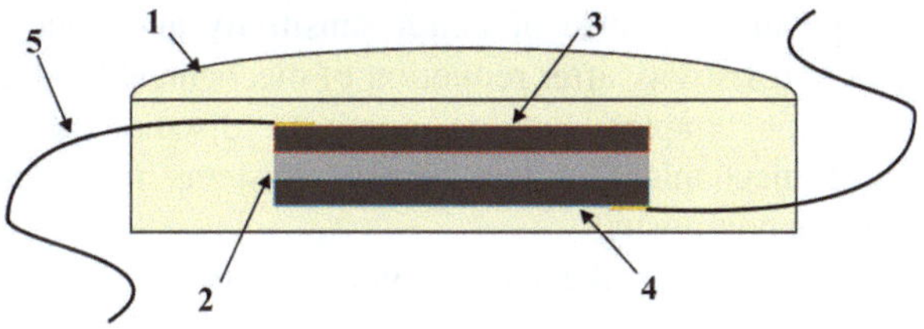

Fig. 17 Schematic view of HPS consisting of: *1* nonconductive outer shell, *2* piezoresistive PiCB or PITEG, *3* upper layer of conductive PICB, *4* lower layer of conductive PICB, *5* wires with soldered small brass foil plates. The geometrical dimensions of designed system was 18 mm in diameter and 5 mm thickness

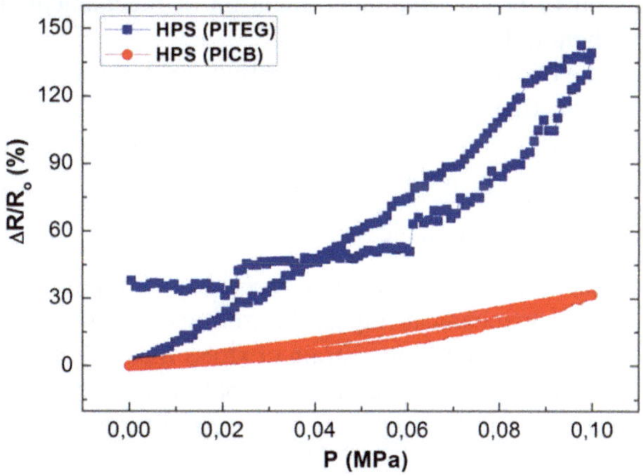

Fig. 18 Piezoresistive behavior of HPS with PITEG- and PICB-sensitive elements under operational pressure of up to 0.1 MPa

(pressure, strain) of PNGC. HSCB particles are believed to have a very complex surface characteristics possibly leading to more than one electrical contact point between two particles when forming conductive channels in PICB composites; therefore, electrical percolation is observed at lower filler concentrations compared to PITEG composites (Fig. 11). However, PITEG composites exhibits higher piezoresistive sensitivity since two TEG particles can form only one electrical contact point between them (due to geometrical nature of two planes) in PITEG-conductive channels leading to reduced electrical shunting of conductive channels when the whole structure is deformed.

6 Summary and Conclusions

Comparatively with other nanographite fillers, both graphene and graphene platelets are rather little used as fillers to elaborate the nanocomposite for mechanical impact sensing; however, in the last 2 years, the number of original papers in this field is rising very sharply [34–36, 40, 41, 43, 44]. One can see from the literature analysis that elastomers are mostly used as matrix material in such kind of sensitive nanocomposites, because of better sensitivity and comparatively faster recovery of electrical resistivity after reduction of mechanical loading.

At the end of this chapter, we are going to discuss the main strengths and weaknesses of mechanical impact sensing systems made on the basis of polymer/nanographite composites.

As main strengths of rubber-like sensors, we consider:

1. the durability against direct and strong mechanical impact in comparison with brittle piezoceramics or plastic metallic materials;
2. the durability against influence of definite environment factors like humidity, oxidation, etc.;
3. the possibility to adjust the chemical composition of composite material to cover broad dynamical diapason of mechanical stress sensing from few Pa to hundreds of MPa;
4. the cost of materials, relatively simple preparation of sensor material as well as detection of sensors output
5. the possibility to fabricate large and continuous sensing elements integrated in different provable systems.

As significant weaknesses one should mention:

1. the long-drawn recovery processes of small reorientations of polymer chains after mechanical impact that seriously limits the use of such materials for very accurate mechanical strain or stress measurement;
2. as follows from the previous, polymer nanographite composites at the state of the art can be used only for registration or counting of mechanical impacts on sensing element.

The future perspective as well as the challenges in this field we see as follows:

1. development of polymer nanographite composites with permanently aligned graphene 2D particles to improve the sensitivity;
2. elaboration of porous elastomer nanographite composites to improve the sensitivity; and
3. creation of hybrid polymer composites with mixed different allotropes of nanographite, where the sensor material properties will be improved due to the synergy between these fillers.

Acknowledgments This work was partly supported by ESF Grants Nr. 1DP/1.1.1.2.0/13/APIA /VIAA/030 and Nr. 1DP/1.1.1.2.0/13/APIA/VIAA/021

References

1. Hoffman W, Gupta H (2001) Kautchuk und Gummi Kunststoff. Doktor Gupta Verlag, Ratingen.
2. Donnet JB (2003) Nano and microcomposites of polymers elastomers and their reinforcement. Compos Sci Technol; 63:1085-1088.

3. Heo JS, Chung JH, Lee JJ (2006) Tactile sensor arrays using fiber Bragg grating sensors. Sensor Actuat A-Phys 126:312–327.
4. Fortes LM, Gonçalves MC, Almeida RM (2011) Flexible photonic crystals for strain sensing. Opt Mater 33:408-412.
5. Yan C, Ferraris E, Geernaert T, Berghmans F, Reynaerts D (2012) Characterisation of tactile sensors based on fibre Bragg gratings towards temperature independent pressure sensing. Procedia Engineering 47:1402-1405.
6. Lee B, Roh S, Park J (2009) Current status of micro- and nanostructured fiber sensors. Opt Fiber Technol 15:209-221.
7. Staufer D, Aharony A (1992) Introduction into percolation theory. Taylor & Francis, Washington.
8. Roldughin VI, Vysotskii VV (2000) Percolation properties of metal-filled polymer films, structure and mechanisms of conductivity. Prog Org Coat 39:81-100.
9. Balberg I, Azulay D, Toker D, Millo O (2004) Percolation and tunneling in composite materials. Int J Mod Phys B 18(15):2091-2121.
10. Knite M, Teteris V, Kiploka A, Kaupuzs J (2004) Polyisoprenecarbon black nanocomposites as strain and pressure sensor materials. Sensor Actuat A-Phys 110:142–149.
11. Rocha JG, Paleo AJ, Ferrie WJ, van Hattum WJ, Lanceros-Mendez S (2013) Polypropylene-carbon nanofiber composites as strain-gauge sensor. IEEE Sens J 13(7):2603-2609.
12. Laukhin V, Laukhina E, Lebedev V, Pfattner R, Rovira C, Veciana J (2011) Flexible all-organic highly tenzo-resistive bi layer films as weightless strain and pressure sensors for medical devices. In: Proceedings of the Second International Conference on Sensor device Technologies and Applications, SENSORDEVICES 2011, p 151.
13. Zhang XW, Pan Y, Zheng Q, Yi XS (2000) Time dependence of piezoresistance for the conductor filled polymer composites. J Polym Sci Pol Phys 38:2739-2749.
14. Stassi S, Cauda V, Canavese G, Pirri CF (2014) Flexible Tactile Sensing Based on Piezoresistive Composites: A Review. Sensors 14:5296-5332.
15. Chen L, Chen G, Lu L (2007) Piezoresistive Behavior Study on Finger Sensing Silicone Rubber/Graphite Nanosheet Nanocomposites. Adv Funct Mater 17:898-904.
16. Flandin L, Hiltner A, Baer E, (2001) Interrelationships between electrical and mechanical properties of a carbon black filled ethylene-octene elastomer. Polymer 42:827-8238.
17. Simmons JG (1963) Electric tunnel effect between dissimilar electrodes separated by thin insulating film. J Appl Phys 34(9):2581-2590.
18. Zavickis J, Knite M, Podins G, Linarts A, Orlovs R (2011) Polyisoprene – nanostructured carbon composite – a soft alternative for pressure sensor application. Sensor Actuat A-Phys 71:38-42.
19. Luheng W, Tianhuai D, Peng W (2009) Influence of carbon black concentration on piezoresistivity for carbon black filled silicone rubber composite. Carbon 47:3151-3157.
20. Al-solamya FR, Al-Ghamdib AA, Mahmoud WE (2012) Piezoresistive behavior of graphite nanoplatelets based rubber nanocomposites. Polym Advan Technol 23:478-482.
21. Zheng S, Deng J, Yang L, Ren D, Huang S, Yang W, Liu Z, Yang M (2014) Investigation on the piezoresistive behavior of high density polyethylene/carbon black films in the elastic and plastic regimes. Compos Sci Technol 97:34-40.
22. Knite M, Klemenok I, Shakale G, Teteris V, Zicans J (2007) Polyisoprene-carbon nano-composites for application in multifunctional sensors. J Alloy Compd 434–435:850-853.
23. Kanoun O, Müller C, Benchirouf A, Sanli A, Dinh TN, Al-Hamry A, Bu L, Gerlach C, Bouhamed A (2014) Flexible Carbon Nanotube Films for High Performance Strain Sensors. Sensors 14:10042-10071.
24. Bauhofer W, Kovacs JZ (2009) A review and analysis of electrical percolation in carbon nanotube polymer composites. Compos Sci Technol 69:1486 1498.
25. Zavickis J, Linarts A, Knite M (2011) The downshift of the electrical percolation threshold in polyisoprene nanostructured carbon composites. Energetika 8:44-49.
26. Knite M, Tupureina V, Fuith A, Zavickis J, Teteris V (2007) Polyisoprene – multi-wall carbon nanotube composites for sensing strain. Mat Sci Eng C-Biomim 27:1125-1128.

27. Park S, Ruoff RS (2009) Chemical methods for the production of graphenes. Nat Nano 4(4):217-224.
28. Grehov V, Kalnacs J, Matzui L, Knite M, Murashov A, Vilken A (2013) Nitrogen Ad-sorption by Thermoexfoliated Graphite. Latvian Journal of Physics and Technical Sciences 1:58-65.
29. Chiacchiarelli LM, Rallini M, Monti M, Puglia D, Kenny JM, Torre L (2013) The role of irreversible and reversible phenomena in the piezore-sistive behavior of graphene epoxy nanocomposites applied to structural health monitoring. Compos Sci Technol 80:73-79.
30. Knite M, Zavickis J, Sakale G, Ozols K, Linarts A (2013) Advanced smart polymer/nanographite composites for environmental pollution control. Green Design, Materials and Manufacturing Processes - Proceedings of the 2nd International Conference on Sustainable Intelligent Manufacturing, SIM 2013 p 587-592.
31. Knite M, Teteris V, Kiploka A (2003) The effect of plasticizing agent on strain induced change of electric resistivity of carbon polyisoprene nanocomposites. Mater Sci Eng C 23(6–8):787-790.
32. Knite M, Hill AJ, Pas SJ, Teteris V, Zavickis J (2006) Effects of plasticizer and strain on the percolation threshold in polyisoprene carbon nanocomposites: positron an-nihilation lifetime spectroscopy and electric resistance measurements. Mater Sci Eng C 26:771-775.
33. Knite M, Zavickis J (2009) Prospective polymer composite materials for applications in flexible tactile sensors. In: Rodic AD (ed) Contemporary robotics – challenges and solutions, In-The, India, p 99-128.
34. Kumar SK, Ponnamma D, Thomas S, Grohens Y (2014) Evolution from graphite to graphene elastomer composites. Prog Polym Sci 39:749-780.
35. Garima M, Vivek D, Kyong YR, Soo-Jin P, Wi RL (2014) A review on carbon nanotubes and graphene as fillers in reinforced polymer nanocomposites. J Ind Eng Chem in press: http://dx.doi.org/10.1016/j.jiec.2014.03.022.
36. Das TK, Prusty S (2013) Graphene-based polymer composites and their applications. Polym-Plast Technol 52:319-331.
37. Lu J, Chen X, Lu W, Chen G (2006) The piezoresistive behaviors of polyeth-ylene/foliated graphite nanocomposites. Eur Polym J 42:1015-1021.
38. Hou Y, Wang D, Zhang XM, Zhao H, Zha JW, Dang ZM (2013) Positive piezoresistive behavior of electrically conductive alkyl-functionalized graphene/ polydime-thylsiliconenanocomposites. J Mater Chem C 1:515-521.
39. Soltani R, Katbab AA (2010) The role of interfacial compatibilizer in controlling the electrical conductivity and piezoresistive behavior of the nanocomposites based on RTV silicone rubber/graphite nanosheets. Sensor Actuat A-Phys 163:213-219.
40. Tuukkanen S, Hoikkanen M, Poikelispaa M, Honkanen M, Vuorinen T, Kakkonen M, Vuorinen J, Lopo D (2014) Streching of solution processed carbon nanotube and graphene nanocomposite films on rubber substrates. Synt Met 191:28-35.
41. Kumar SK, Castro M, Saiter A, Delbreilh L, Feller JF, Thomas S, Grohens Y (2013) Development of poly(isobutylene-co-isoprene)/reduced graphene oxide nanocomposites for barrier, dielectric and sensing applications. Mater lett 96:109-112.
42. Marcano DC, Kosynkin DV, Berlin JM, Sinitskii A, Sun Z, Slesarev A, Alemany LB, Lu W, Tour JM (2010) Improved Synthesis of Graphene Oxide. ACS Nano 4(8):4806-4814.
43. Hodlur RM, Rabinal MK (2014) Self assembled graphene layers on polyurethane foam as a highly pressure sensitive conducting composite. Compos Sci Technol 90:160-165.
44. Goncalves V, Brandao L, Mendes A (2014) Development of porous polymer pressure sensors incorporating graphene platelets. Polym Test 37:129-137.
45. Kim YJ, Cha JY, Ham H, Huh H, So DS, Kang I (2011) Prepartion of piezoresistive nano smart hybrid material based on graphene. Curr Appl Phys 11:350-352.
46. Eswaraiah V, Aravind SSJ, Balasubramaniam K, Ramaprabhu S (2013) Graphene functionalized carbon nanotubes for conducting polymer nanocomposites and their improved strain sensing properties. Macromol Chem Physic 214:2439-2444.
47. Vovchenko L, Lazarenko A, Matzui L, Zhuravkov A (2012) The effect of mechanical stress on electric resistance of nanographite-epoxy composites. Physica E 44:940-943.

Graphene Filled Polymers for Vapor/Gas Sensor Applications

Tran Thanh Tung, Mickael Castro, Jean Francois Feller and Tae Young Kim

Abstract With their unique and excellent properties such as high carrier mobility and high surface area, graphene-base materials have shown great promise as efficient sensing materials for highly sensitive and low noise sensors. Graphene offers some important advantages over other carbon-based materials such as carbon nanotubes (CNTs), which includes enhanced sensitivity and low inherent electrical noise. These merits mainly comes from their structural features, as it is composed of all surface carbon atoms with large and flat geometry enabling high sensitivity and low contact resistance. Moreover, their surface can be functionalized with organic molecules (e.g., polymers, nanocrystalline, bio-molecular), and surface molecules on graphene surface can also be used as gas/vapor sensing materials that promote the sensing capability of overall composites. This has sparked interests in the development of highly sensitive and selective gas/vapor sensors based on graphene-based materials and their polymer composites. In this review, recent progress on graphene and its composites will be discussed in the context of their use in sensors. It mainly focuses on how engineering graphene with other functional molecules can affect their ability to detect a number of different gas/vapor. It also emphasizes achievements made with graphene-filled polymer composites for gas/vapor sensor applications.

Keywords Volatile organic compound · Gas · Sensing · Hybrid composites

T. ThanhTung (✉)
Carbon Nanostructure and Catalyst Group, Institute of Chemistry and Processes for Energy, Environment and Health (ICPEES), CNRS-Strasbourg University, Strasbourg, France
e-mail: thanhtungtran77@gmail.com

M. Castro · J.F. Feller
Smart Plastic Groups, LIMATB, University of South Brittany, Lorient, Bretagne, France

T.Y. Kim
Department of Bionanotechnology, Gachon University, 1342 Seongnamdaero, Sujeong-Gu, Seongnam-Si, Gyeonggi-Do, 461-701 Seongnam, South Korea

K.K. Sadasivuni et al. (eds.), *Graphene-Based Polymer Nanocomposites in Electronics*, Springer Series on Polymer and Composite Materials,
DOI 10.1007/978-3-319-13875-6_10

1 Introduction

Vapor/gas sensor is a device that transforms vapor/gas information, (e.g., composition, concentration, activity, pressure, etc.), into analytically useful signal which can be easily monitored. In recent years, vapor/gas sensors have attracted scientific interests in a variety of applications including human health, public safety, environmental monitoring, petroleum industry, and so on [1–6]. Nanostructured materials with novel physicochemical properties are considered as promising candidates for high performance sensing applications, and are expected to potentially replace the conventional gas analysis instruments based on semiconductors and gas chromatography mass spectrometry (GC-MS) in the future. One dimensional (1D) carbon nanostructured materials such as carbon nanotubes (CNTs) and nanofibers (CNFs) are well-known chemical sensing materials as their electrical conductance is sensitive to gases. These materials have shown to adapt conditions: easy-to-use processing, low cost and room temperature operation [7, 8]. The functionalization of CNTs for their application in gas/vapor sensors have been widely studied during the last decades [9–13]. However, their safety, rigid cylindrical structure, coupled with massive aspect ratios and relative inertness is still being an open question [14–17]. Recently, graphene, one-atom thick two-dimensional (2D) carbon nano-sheet, has also been the focus of recent research due to its remarkable properties such as high value of mechanical strength (fracture strength, 125 GPa), thermal conductivity (5,000 W $m^{-1}K^{-1}$), mobility charge carriers at room temperature (200,000 cm^2 $V^{-1}s^{-1}$) with a carriers density of 10^{12} cm^{-2}, and specific surface area (theoretical value, 2,630 m^2 g^{-1}) [18–20]. These excellent properties make graphene attractive as sensing materials for vapor/gas sensors. Previous reports described that graphene, with its unique 2D crystal lattice and high electron mobility, outperforms CNTs for vapor/gas sensing applications in term of low inherent electrical noise at room temperature. The large and planar geometry of graphene make them easier for making electrical contact as well [21–24]. Still a number of drawbacks, however, such as the dependence on specialized equipment for the fabrication of devices, the low solubility of graphene in most solvents, and the limited stability of graphene suspension, in particular specificity and reversibility monitoring, hinder the applications of these materials in the field of sensors. Nevertheless, these challenges possibly overcome by functionalizing graphene platelets with organic molecules (e.g., polymers, conducting polymers, bio-molecular) [25, 26]. The organic molecules introduced on the surface of the graphene platelets can function as effective stabilizer for graphene, preventing their agglomeration in liquid phase and thereby facilitating their processing for device fabrication [27, 28]. Moreover, those molecules can also be used as gas/vapor sensing materials that promote the sensing capability of overall composites. Hence, the combination of graphene and organic molecules has becomes an important strategy to achieve improved sensing performance, where the graphene provide electrically conductive path and organic molecules promotes adsorption of targeted analytes. This has attracted interests in the development of

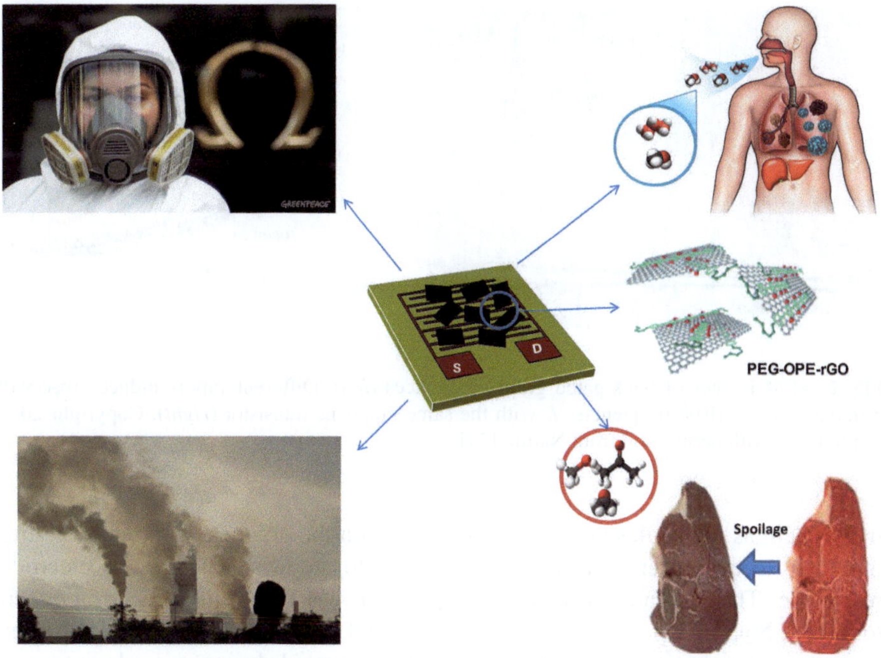

Fig. 1 Schematic illustrating the use of chemical vapor sensors based on graphene composite for detection of gas and vapor and their possible applications in public safety, human health, environmental monitoring, and food safety. Copyright 2010 and 2012. Reproduced with permission from John Wiley & Sons and American Chemical Society, respectively

highly sensitive and selective gas/vapor sensors based on graphene composites, and it holds a great potential for practical applications where sensors are widely used to monitor early indications. Figure 1 illustrates gas/vapor sensor devices based on graphene composite, which extends their potential applications in the area of human health, public safety, environmental monitoring, food packaging, petroleum industry and so on.

2 Preparation of Graphene and Its Derivatives for Using in Vapor/Gas Sensors

Single- to few-layered graphene can be prepared by a variety of methods such as micromechanical exfoliation [29], epitaxial growth [30], chemical vapor deposition [31], and chemical exfoliation route (i.e., sequential oxidation-exfoliation-reduction route) [32–34]. Pioneering studies was reported in 2007 by Geim and co-workers, where mechanically cleaved single-layer-graphene was used for gas sensors [35]. In this work, graphene-based gas sensor has shown to detect

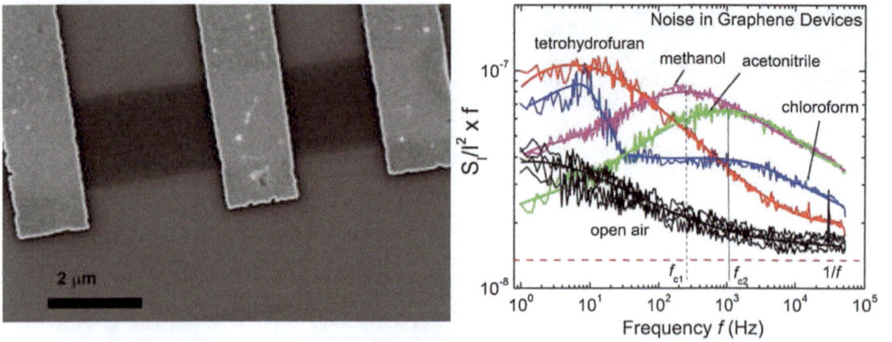

Fig. 2 SEM images of back-gated graphene devices (*left*). Different vapors induce noise with different characteristic frequencies f_C with the same graphene transistor (*right*). Copyright 2009. Reproduced with permission from Nature [34]

individual gas molecules adsorbed or desorbed on the graphene surface by changing their local carrier concentration, leading to stepwise changes in electrical resistance. The gas-induced changes depend on gas type, either electron acceptors (e.g., N_2O, H_2O) or electron donors (e.g., NH_3, CO, ethanol). The Hall-effect measurements showed that extra carriers were created during gas adsorption on their sensor device, and gas adsorption can increase the number of holes if the gas in an acceptor or increase the number of electrons if the gas is a donor. Modulation of carrier concentration by gas adsorption/desorption is basic mechanism that governs the operation of graphene-based gas sensors. They also claimed that this graphene-based gas sensor had a limit of detection (LOD) as low as parts per billion (ppb), and this value is comparable to those of the most sensitivity gas sensors ever reported. Balandin and co-wokers reported that a single pristine layer graphene transistor can possibly detect different chemical vapors by producing distinguishably different effects on the low-frequency Lorientzian noise spectra [36]. Upon gas exposure, the gas molecules can create specific traps and scattering centers in graphene, leading to the carrier fluctuation (due to fluctuation of traps or the scattering) and/or the kinetics of the molecules adsorption and desorption contributing to noise. Figure 2 shows that a different vapor induces noise with different characteristic frequencies.

The limitation of pristine graphene for gas sensor application is its incontrollable size and shape which hampers the fabrication of devices, and therefore it has been used for theoretical study and proof-of-concept demonstration. Recent studies have focused on the use of graphene grown by epitaxial growth and chemical vapor deposition (CVD) methods for the fabrication of sensors [37]. Such methods are most promising techniques for bottom-up synthesis of high-quality graphene sheets. In the epitaxial growth method, SiC substrate is heated under ultrahigh vacuum, so that silicon atoms sublimated from the substrate and carbon atoms rearranged into graphene layers. The thickness of graphene layers depend on annealing temperature and time [38, 39]. Recently, Giesbers et al. reported a high

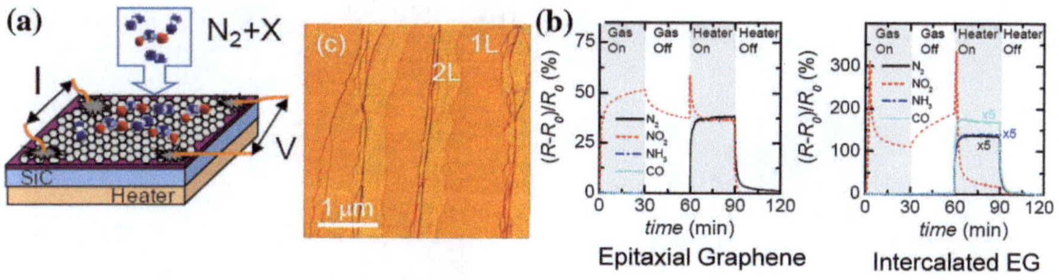

Fig. 3 **a** Schematic illustration of the sensing experiment where the sensing gas X diluted in N_2 carrier gas flows over a graphene-on-SiC sample while the resistance change of the graphene is measured, AFM phase image showing the single layer graphene areas (*dark*) and double layer graphene areas (*bright*) on top of the terraces. Chemiresistive gas sensors of epitaxial graphene (EG) and intercalated epitaxial graphene exposed to pure N_2, 100 ppb NO_2 (for EG), 40 ppb NO_2 (for QFEG), 300 ppm NH_3 and 3,000 ppm CO during annealed at 150 °C [37]. Copyright 2013. Reproduced with permission from American Institute of Physics

sensitive quasi freestanding epitaxial graphene gas sensor on 6H-SiC [40]. They have shown that the electrical response to NO_2, N_2, NH_3, and CO for both epitaxial graphene and quasi freestanding epitaxial graphene (hydrogen intercalated graphenes) on 6H-SiC substrate at 1 parts per million (ppm) and it can be possibly down to sub-ppm level. Specifically, the quasi freestanding epitaxial graphene shows a 6-fold increase in sensitivity towards NO_2 compared to epitaxial graphene (Fig. 3). The drastic increase in sensitivity of quasi freestanding epitaxial graphene can be explained by a Fermi-energy close to the Dirac point, leading to a strongly surface doping dependent sample resistance.

The CVD technique has led to the preparation of graphene film on metal substrates such as Ni, Cu, Co under decomposition of hydrocarbon gas source at moderate temperature (<1,000 °C) [41]. The resulting graphene is then transferred to other substrates for fabrication of gas sensor devices. The high conductivity, transparency and high-yield not only make the CVD-graphene desirable for the high performance electronic sensors, but also enable the mass production of devices with high flexibility.

It has been reported that nano-sized detector based on graphene could detect atomic oxygen with a LOD of 1.25 % [42]. Peter Boggild and co-workers showed that nanopatterned CVD-graphene using a spherical block copolymer etch mask presents a high defect density between the mesh and holes, achieving sensitivities for NO_2 concentrations as low as 300 ppt with an ultimate LOD of tens of ppt [43]. The dramatic improvement in the gas sensitivity is likely due to the high adsorption site density, from the combination of edge sites and points defect sites. Koratkar et al. [44] have developed another novel approach, which exploited a macroscopic three-dimensional (3D) network of graphene sheets, referred to as graphene foam (GF) for gas sensors. In this work, a scaffold of porous nickel foam was used as a template for the deposition of graphene. Then, nickel skeleton was etched by chemical etchants. The walls of the resulting foam are comprised of few-layer-graphene (FLG) sheets. The FLG grow uniformly on the entire surface

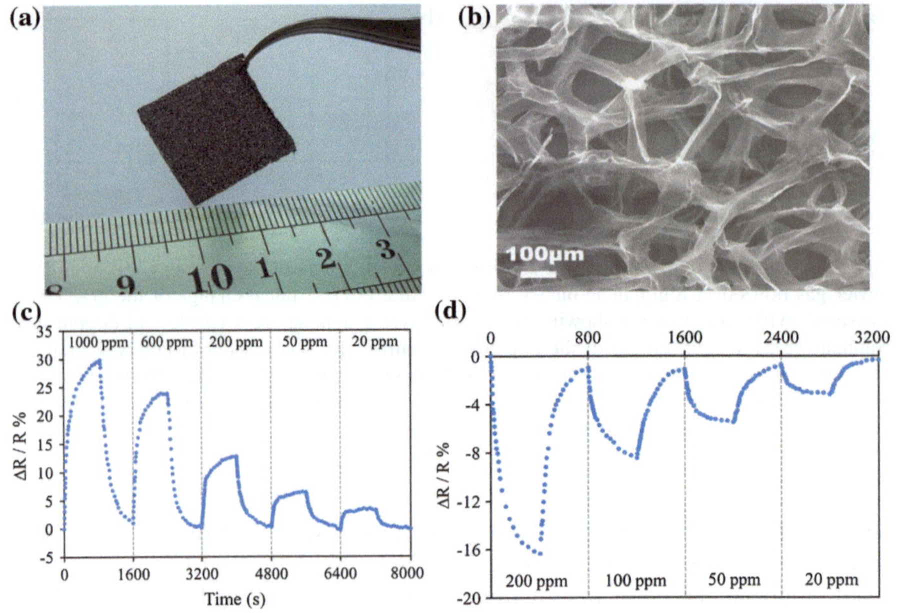

Fig. 4 **a** Photograph. **b** SEM image of the microporous 3D interconnected graphene grown on porous nickel foam by CVD method. Normalized change in electrical resistance of the GF as a function of time indicating the real-time detection of **c** NH_3 and **d** NO_2 at different concentrations [41]. Copyright 2009. Reproduced with permission from Elsevier Ltd

of the nickel foam and make them interconnected into each other, such that there is no interface or physical breaks in the network. This facilitates efficient electron transport though the FG and thus high sensing performance (Fig. 4). They demonstrated that the GF-based sensor can detect NH_3 and NO_2 at ~20 ppm level in air at room temperature.

However, both of epitaxial graphene and CVD-graphene entail additional transfer process for devices fabrication, which typically results in the surface contaminations and destruction of graphene intact. In addition, pristine graphene described above have limitative reversibility. Alternatively, chemical exfoliation route (i.e., sequential oxidation-exfoliation-reduction route) shows distinct advantages in terms of yield and cost [45–47]. This chemical method typically produces bulk quantities of reduced graphene oxide (rGO), which resembles graphene but with some oxygen groups and structural defects. Therefore graphene-based sensors can be mass-producible at cost [48, 49]. The certain amount of oxygen functional groups exist in the corresponding rGO due to incomplete reduction and structural defects and these can act as adsorption sites for interaction of gas molecules [50, 51]. Fowler et al. [52] reported a chemoresistance sensor made with a single-layer rGO coated on interdigitated electrode. This rGO-based sensors function as p-type semiconductor, in which the electron-withdrawing NO_2 cause a decrease in resistance, while the electron-donating NH_3 leads to an increase

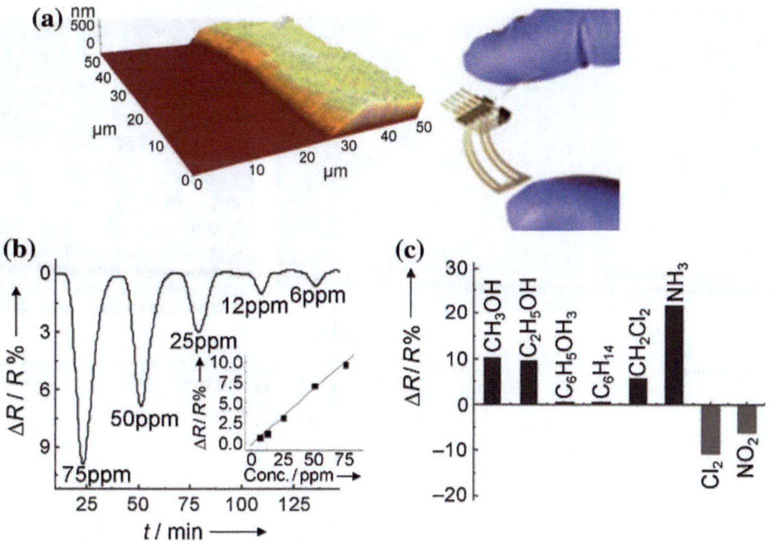

Fig. 5 **a** Digital image of inkjet-printed rGO/PET four-probe sensor. **b** Vapor sensing by inkjet-printed rGO/PET obtained by ascorbic acid reduction of dispersed GO when the film was exposed to a Cl_2 vapor at different concentrations ranging from 6 to 75 ppm, and **c** selectivity of sensor towards saturated organic vapors, NH_3 (100 ppm), NO_2 (100 ppm), and Cl_2 (100 ppm) [48]. Copyright 2008. Reproduced with permission from American Chemical Society

in resistance. This chemoresistor enable detection of 2, 4-dinitrotoluene (DNT), a relatively volatile component found in the trinitroluene (TNT) explosive at the ppb level. Dua et al. [53] developed a flexible chemiresistor composed of overlapped rGO film printed onto poly(ethyleneterephthalate) (PET) using inkjet printing technique. The rGO has both high electrical conductivity and chemically defect sites, making it a promising candidate for gas sensing. This sensor allows for reversible operation and shows selectivity towards chemical vapors such as NO_2, Cl_2 at concentration ranging from 100 ppm to 500 ppb. Electron-withdrawing vapors such as NO_2 sharply increase the conductivity of the inkjet-printed rGO/PET films with an increase in the charge carrier density (Fig. 5).

In general, graphene has high sensitivity toward diverse gases even at low gas concentration. However, there are still some challenges facing graphene-based sensors, for example particular specificity, reversibility monitoring, and reliability, which limits its use in practical application. In order to overcome these challenges many efforts have been made including functionalization of graphene with nanocrystalines or organic macromolecules. These hybrid materials display not only the individual properties of each component, but also additional novel properties due to the synergistic effect between them [54–57]. For example, Star et al [58] developed field-effect transistor (FET) devices using metal nanoparticle-decorated rGO as active materials for H_2 gas detection. In this work, enzymatic oxidation generated holes within the basal plane of honey graphene oxide (hGO),

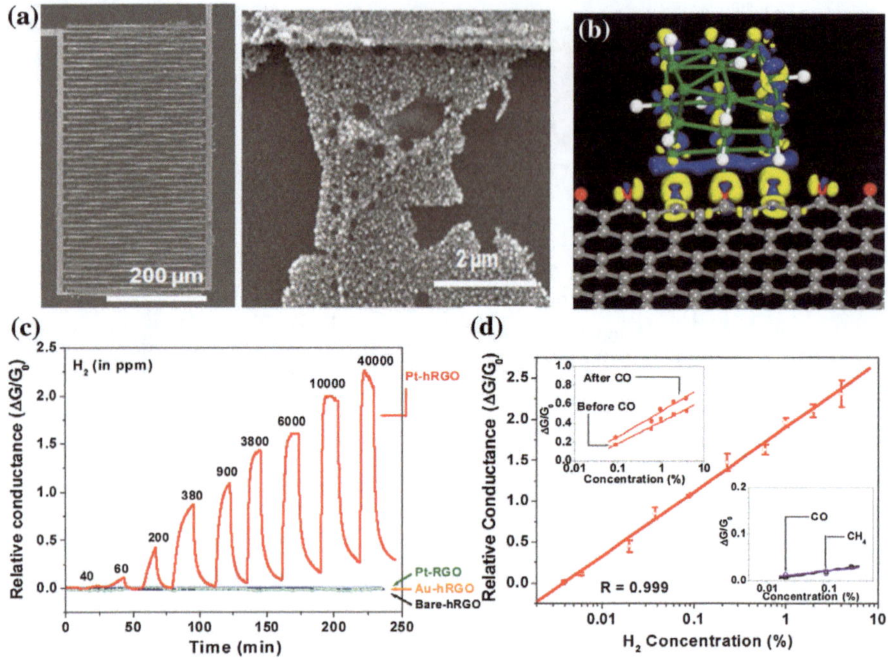

Fig. 6 **a** SEM images of Pt-decorated hrGO deposited on pre-patterned Au interdigitated microelectrodes. **b** Theoretical modeling on hydrogen adsorption on Pt-hrGO surface. Relative conductance ($\Delta G/G_0$) versus time curves for H_2 concentrations 40–40,000 ppm (in N_2) for bare-hrGO, Au-hrGO, Pt-hrGO, and Pt-rGO. Calibration curve of Pt-hrGO for response to H_2 gas, left inset shows the device response to H_2 before and after exposure to CO (0.25 % in N_2) and right inset shows the response to CO (0.05–0.25 % in N_2) and CH_4 (0.4–4 % in N_2) [58]. Copyright 2011. Reproduced with permission from American Chemical Society

and after reduction with hydrazine, reduced graphene oxide (hrGO) was formed. When Pt nanoparticles was decorated on the surface of hrGO by electrochemical method, the resulting hybrid materials exhibited a large and selective electronic response toward hydrogen gas. Based on the experimental results and theoretical modeling, they proposed that the increased edge-to-plane ratio, oxygen moieties, and Pt nanoparticle decoration were responsible for the improved gas sensing performance of hrGO nanostructures (Fig. 6).

In other work, Pd-functionalized multilayer graphene nanoribbons were fabricated and used for H_2 sensing [59]. Intrinsic graphene is a semimetal with zero band gap. However, graphene could exhibit band gaps useful for transistor operations, by engineering the graphene structure into a nanoribbon. This Pd-functionalized graphene nanoribbons with high specific surface area facilitates efficient functionalization and gas adsorption. These networks show high sensitivity to hydrogen at ppm concentration levels at room temperature with a fast response and recovery time.

Recently, we have demonstrated high performance chemical sensor that can reversibly detect volatile organic compounds (VOCs), based on the graphene decorated with silver nanoparticles [60]. The one-pot synthesis of the Ag-rGO hybrid materials is carried out by the in situ thermal reduction of Ag precursor and GO in the presence of poly(ionic liquids) (PIL). The PIL of poly(1-vinyl-3-ethylimidazolium) salt not only serves as a dispersant or capping agent for GO and Ag microspheres, but also acts as a linker for coupling rGO with Ag nanoparticles. In addition, PIL within the Ag-rGO hybrids can be used to transfer Ag-rGO hybrids between water phase and organic solvent phase by simply exchanging anions of PIL. Based on these Ag-rGO/PIL hybrids, we constructed the chemical vapor sensors to detect various VOCs and evaluate their performances in terms of response/recovery time, selectivity, and reliability. The improved sensing performance of Ag-rGO/PIL sensors was attributed to the synergistic effect of hybrid sensing materials. The Ag-rGO/PIL hybrid materials showed 3D porous structures enabled by Ag nanoparticles acting as spacers for graphene sheets, thus allowing for large surface area and a range of surface active sites. This translated into a large amount of organic vapour molecules adsorbed onto the sensing material. Therefore, a rapid and sensitive current response was achieved with Ag-rGO/PIL-based sensor by effective electron transfer between sensors and detection molecules.

Besides, there are studies on the effect of noble metal and metal oxide-based catalysts, e.g., Pt, Au, SnO_2, and Cu_2O, on gas sensing performance of graphene-based sensors [61–64]. They have demonstrated that the synergistic effect between them will play the role for improved gas sensing performance. At the same time, ternary systems with graphene and two distinct nanocrystals (NCs) have also been proposed and showed improved sensing performance as compared to pure graphene and graphene-NC binary systems [65].

Functionalization of graphene with organic molecules (capture agents) also enables the specific binding of target gases to the surface of graphene. Shi et al. [66] reported chemiresistive sensor based on sulfonate-modified rGO (S-G) and ethylenediamine-modified rGO (EDA-G) for selective detection of NO_2 gas. They have shown that the presence of functional groups played an important role in the sensing process. The electron-withdrawing NO_2 molecules tend to adsorb on electron-rich sites such as the lone-pair electrons of S or O atoms in those functional groups. The sensors were tested upon exposure to 50 ppm different vapor of NO_2, NH_3, H_2O and toluene, in which sensor has relatively good selectivity to NO_2 by high response (conductance change) whereas less reponse to NH_3 or no detectable can be observed for toluene or water. In other work, Jonhson et al. [67] showed one route towards improving the ability of graphene to work as a chemical sensor by using single stranded DNA (ssDNA) as a sensitizing agent. Graphene transistors were constructed using exfoliated graphite onto silicon substrate, then functionalized with a self-assembled layer of ssDNA. The current-gate voltage (I-Vg) characteristic of an individual device shows how the hole and electron mobilities change by carriers scattering on the surface of sensors. The sensor devices showed fast response times, complete and rapid recovery to initial state at room temperature, and discrimination between several similar vapor analytes (Fig. 7).

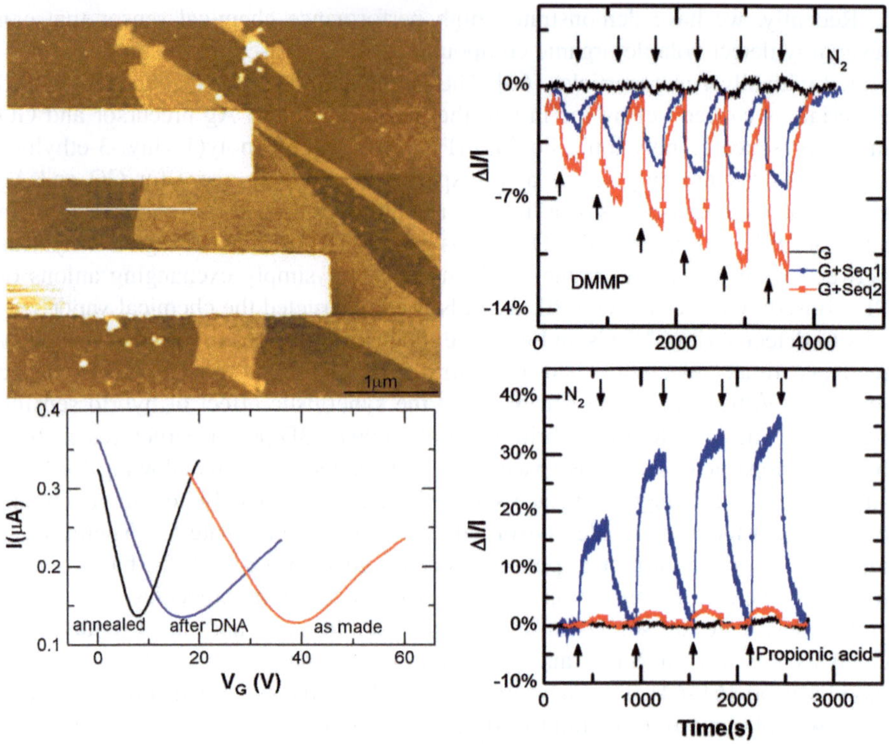

Fig. 7 AFM line scans of ssDNA on graphene, I-V_G characteristics for a graphene device through the steps of functionalization showing the expected doping shifts due to ssDNA application. Normalized changes in current versus time for comparison of clean graphene device (*black curve*) and ssDNA-functionalized graphene device (*red* and *blue curve*) when exposed to DMMP at concentrations of 20, 40, 60, 80, 100, 120 ppm (*top*) and propionic acid at concentrations of 90, 220, 435 ppm (*bottom*) [57]. Copyright 2012. Reproduced with permission from Elsevier Ltd

3 Hybrid Materials of Graphene and Conducting Polymers for Gas/Vapor Sensors

Conducting polymers (CP) can be used an active sensing material in chemical sensors since they exhibit physical changes occurring upon the exposure to various chemicals [68, 69]. In addition, they offer great design flexibility due to light weight, mechanical flexibility, and tunable opticaltransmittance [70–72]. Intrinsically conducting polymers such as poly(3,4-ethylenedioxythiophene) (PEDOT), polyalinine (PANi) and polypyrrole (PPy) are the most widely used due to their excellent properties such as gas sensing ability, response to a wide range of VOCs and good environment stability [73–77]. Therefore, the sensitivity and selectivity of graphene-based chemical sensors can be improved through their hybridization with conducting polymers. Various attempts have been made

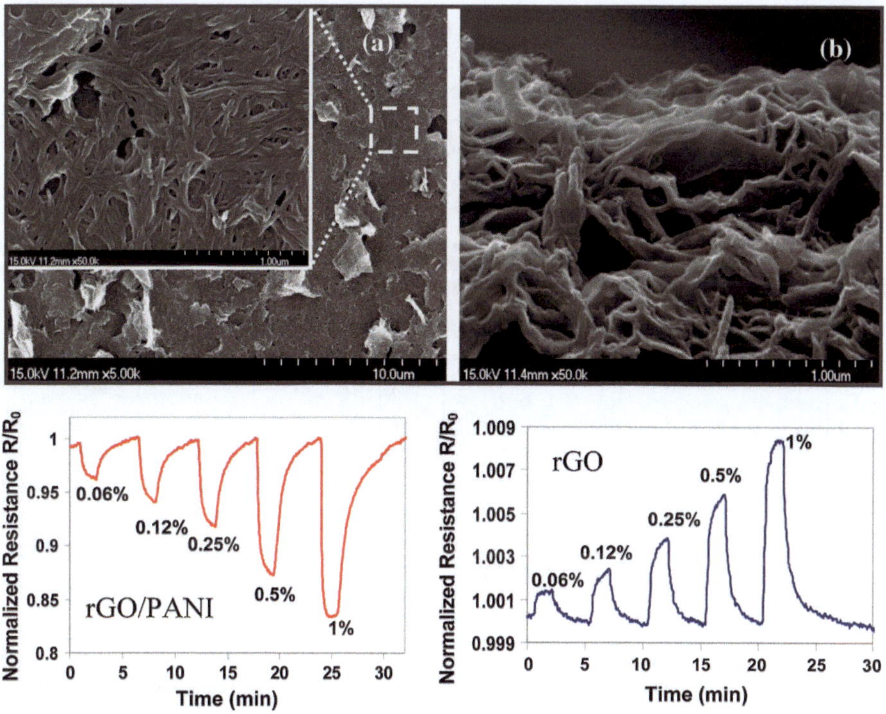

Fig. 8 SEM images of graphene/PANi nanocomposite **a** top view, **b** tilted view showing that PANi nanofibers have been grown on the surface of the graphene nanosheets. Dynamic responses of the H_2 gas sensors with sensitive layers of graphene/PANi and graphene only for low concentrations of less than 1 % H_2 gas [67]. Copyright 2010. Reproduced with permission from American Institute of Physics

to combine graphene or its derivatives with conducting polymers. For example, Al-Mashat et al. [78] developed a graphene/PANi nanocomposite for hydrogen gas sensor. Graphene was prepared using a chemical synthetic route, and it was then mixed with aniline monomer and ammonium persulfate and polymerized subsequently to form a graphene/PANi composite. A rapid reaction promotes homogeneous nucleation growth of PANi into uniform nanofibers which is then deposited on the surface of graphene. Gas sensors based on graphene/PANi were tested toward 1 % concentration of H_2 gas balanced in synthetic air and their sensing performance was compared with PANi nanofiber and graphene only devices. The results showed that the normalized resistance (R/R_0) of graphene/PANi composite was up to 16.7 %, which is higher than the sensitivities of PANi-based sensor (9.83 %) and only graphene (0.83 %) due to the higher porosity of the graphene/PANi composite (Fig. 8).

In other work, Wu et al. [79] reported a HCl-doped graphene/PANi composites synthesized by an in situ chemical oxidation polymerization. When graphene was hybridized with PANi, the π electrons cloud of graphene and PANi may overlap

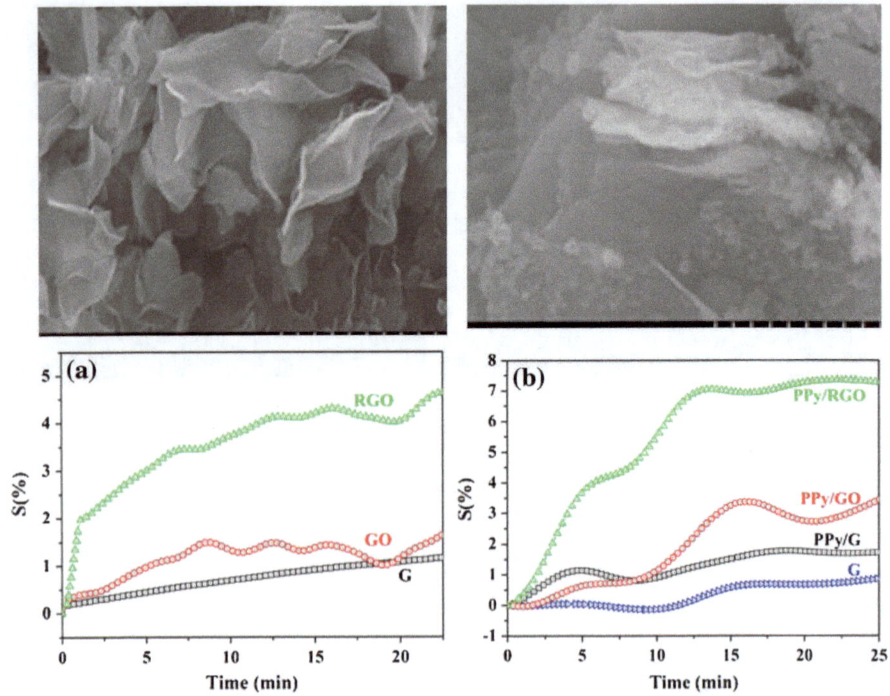

Fig. 9 SEM micrographs and ammonia gas sensitivity of graphitic materials (*left*, **a**) and nanocomposites of PPy and graphitic materials (*right*, **b**). PPy coating on the surface of graphitic materials plays an important role in the NH_3 sensing [69]. Copyright 2007. Reproduced with permission from Multidisciplinary Digital Publishing Institute

and form an extended π–π conjugation system. The extended π electron clouds of graphene/PANi composites promotes charge interaction with NH_3 gas molecules, achieving improved sensing performance as compared to PANi at the same NH_3 gas concentration. In addition, as graphene is homogeneously dispersed as a nanofiller in PANi, the surface area of graphene/PANi composites increases, enhancing the sensitivity of graphene/PANi sensor for NH_3 gas.

Jang et al. [80] prepared a nanocomposite of rGO and PPy using in situ polymerization method and used this composites to fabricate a NH_3 gas sensor with high sensitivity and rapid response time (Fig. 9). The improved response for NH_3 gas was mainly due to the effective electron change transfer between PPy and NH_3 as described below

$$PPy^{+} + NH_3 \rightarrow PPy^{0} + NH3^{+} \text{ adsorption}$$
$$PPy^{0} + NH_3^{+} \rightarrow PPy^{+} + NH3 \text{ desorption}$$

The PPy/rGO composite-based gas sensor also showed the excellent reproducibility during the recovery process at lower temperature of 373 K.

Hu et al. [81] has also reported a NH_3 gas sensors based on rGO and pyrrole, showing excellent responsive sensitivity and selectivity to ammonia (NH_3) gas. These rGO sensors showed resistance change as high as 2.4 % and response time as fast as 1.4 s when the concentration of NH_3 gas was as low as 1 ppb. Furthermore, the rGO sensors could rapidly recover to their initial states with IR illumination. The devices also showed excellent repeatability and selectivity to NH_3. These rGO sensors have advantages such as low cost, low power, easy fabrication, and scalable properties, showing great potential for ultrasensitive detection of NH_3 gas in a wide variety of fields. Shi et al. [82] prepared graphene oxide/conducting polymer (GO/CP) composite hydrogels including GO/PPy, GO/ PEDOT and GO/PANi by in situ chemical polymerization of corresponding aromatic monomers in aqueous dispersions of GO sheets. Among them, GO/PPy composite hydrogels showed high sensitivity towards ammonia gas by increasing 40 % resistance within 600 s upon exposure to 800 ppm NH_3 (about 7 % higher than pure PPy). The resistance increase is likely associated with dedoping of PPy by NH_3. The high performance of this sensor is partially attributed to the ultrathin PPy layers on the graphene sheets.

Our works have focused on graphene/PEDOT composite-based chemical sensors. We have developed a method to yield an interconnected 3D graphene network in conducting polymer matrix [83]. In this work, we used poly(ionic liquid)s (PIL) of poly(1-vinyl-3-ethylimidazolium) or poly($ViEtIm^+$) salts, as they can function to stabilize rGO platelets in the suspension and provide an effective linkage for the polymerization of PEDOT molecules on the rGO surface. Interestingly, these rGO-PIL complex and rGO-PIL/PEDOT composites were easily transferred from aqueous into organic solvent through an anion exchange of PILs adsorbed on the rGO surfaces. In comparison with pristine rGO sensor, the rGO-PIL and rGO-PIL/PEDOT-based sensors can detect VOCs such as methanol, ethanol, acetone, methyl acetate, dimethylsulfide and toluene with enhanced sensitivity and selectivity, especially at low VOC concentration in the range of several ppm. They also showed fast response characteristics as a steady signal was typically observed within 3 s (Fig. 10).

The electro-sorption behavior can be studied using the Langmuir-Henry-Clustering (LHC) model [Eq. (1)]. This model can be used not only to evaluate the amount of analyte in the environment of the chemical sensors from the amplitude of response, but also to identify the diffusion regimes taking place in the transducer.

$$A = \frac{b_L(f'' - f')f}{1 + b_L f} + k_H f + B(f - f')f^{n'} \qquad (1)$$

where b_L the Langmuir affinity constant, k_H the Henry diffusion parameter, n' the average number of solvent molecules per cluster, B the extra clustering coefficient, f' the solvent fraction over which clustering takes place and f'' the vapour fraction over which Langmuir's diffusion is replaced by Henry's diffusion. Based on the shape of the electro-sorption curves, it can be deduced that for graphene based sensors plotted for methanol and methyl acetate, the diffusion follows a Henry

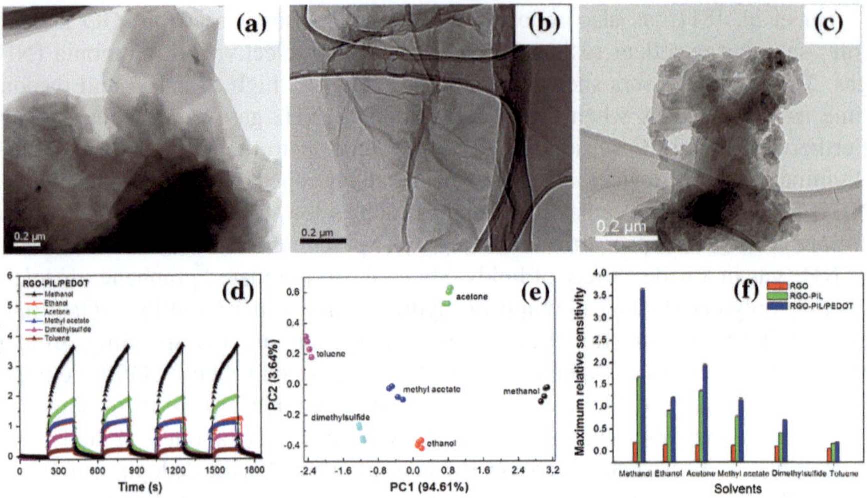

Fig. 10 TEM images of **a** rGO, **b** rGO-PIL, **c** rGO-PIL/PEDOT nanocomposite; **d** cyclic response of the nanocomposites with different VOC vapors, **e** a principal component analysis (PCA) indicated that sensors were capable of identifying types of VOCs, **f** sensitivities of graphene-only and functionalized graphene sensors for different VOCs [72]. Copyright 2005. Reproduced with permission from IEEE Xplore

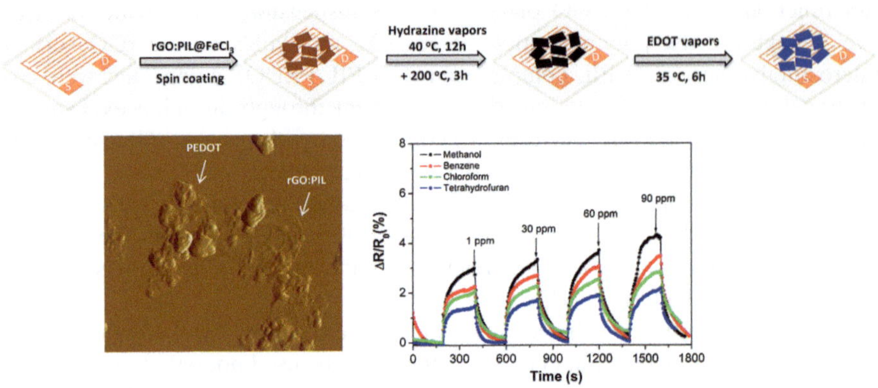

Fig. 11 Schematic of vapor-phase polymerization of PEDOT on PIL-modified rGO surface. The topographic AFM of the rGO:PIL/PEDOT composite on IDA substrate, in which PEDOT is seen as agglomerated granule on rGO sheets with an average thickness of 30 nm. Normalized electrical resistance change to different VOC vapor at 1 ppm concentration and their response and recover times [73]. Copyright 2005. Reproduced with permission from John Wiley & Sons

regime corresponding to a linear evolution, whereas the evolution is exponential for the rGO-PIL/PEDOT sensor exposed to methanol, suggesting a clustering regime of diffusion.

In other work, we prepared hybrid thin films of rGO and PEDOT using a vapor phase polymerization of PEDOT onto rGO platelets, and used them as active materials for chemiresistor to detect and monitor trace levels of different VOCs [84]. For hybridization of rGO with PEDOT, rGO platelets were decorated with ferric ions (Fe^{3+}) and modified by PILs. In this case, Fe^{3+} ions on the rGO platelets function to initiate the polymerization of EDOT, while PILs act as a bridging molecule for rGO and PEDOT to faciliated hybridization between rGO and PEDOT. For the fabrication of sensing devices, GO suspension containing oxidative Fe^{3+} ions and PIL in acetonitrile was spin-coated onto an inter-digitated electrode, followed by reduction of GO through both chemical and thermal methods. Finally, the as-prepared rGO:PIL was exposed to EDOT vapors to form hybrid film of rGO:PIL/PEDOT composites. The chemical sensors based on this hybrid film showed capability of detecting different VOC vapors at a level of ppm concentration with high sensitivity, selectivity and fast response (Fig. 11).

The high performance of the sensor arrays based on rGO:PIL/PEDOT is attributed to the synergistic effect of hybrid sensing materials—rGO and PEDOT.

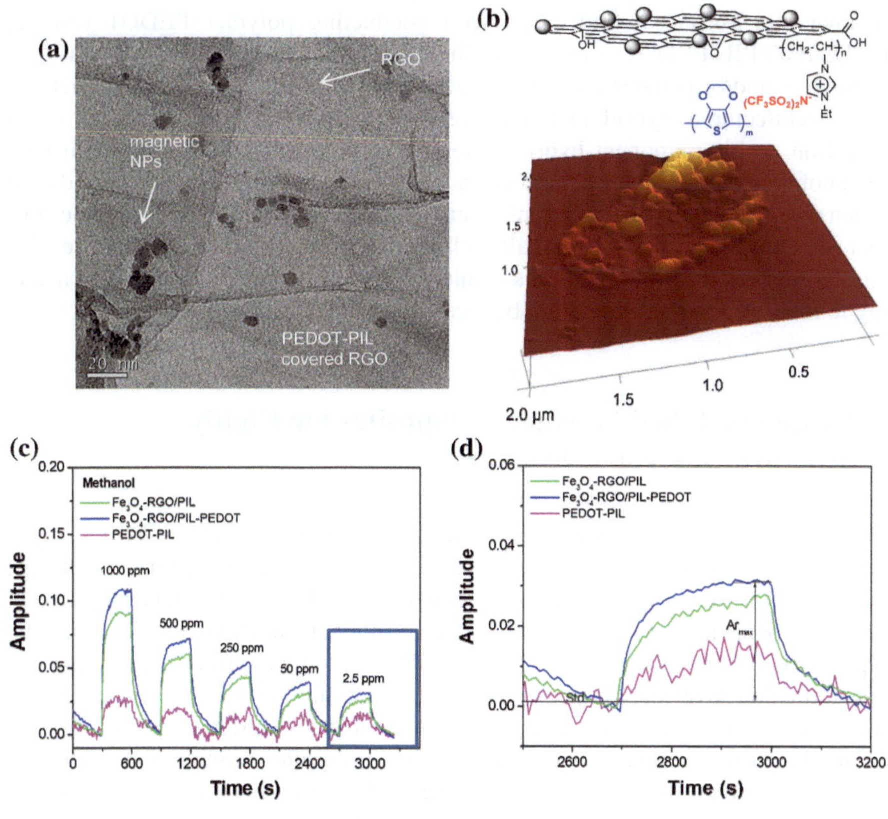

Fig. 12 **a** Representative TEM micrograph of the rGO Fe_3O_4/PIL PEDOT composites showing that magnetite NPs were decorated on rGO surface and then covered by layer of PEDOT. **b** AFM image shows a map of topography. **c** Representative sensor responses as a step of function of concentration for methanol. **d** Noise signals from sensors response at 2.5 ppm [74]. Copyright 2007. Reproduced with permission from John Wiley & Sons

Graphene-based material is well known to have a greatest possible surface per unit volume and extremely high electron mobility enabling high sensitivity and highly low noise sensing at room temperature. On the other hand, PEDOT functions as an absorbent layer that concentrates analyte molecules at the surface of graphene composites. The VOC analyte vapors can be dopants for rGO, which controls its hole/electron transport and carrier mobility, and yields a significant change in resistance. In addition, VOC vapors can also induce conformational change or doping of PEDOT chains, both of which are responsible for a change in electrical resistance. Therefore, hybrid film of rGO and PEDOT exhibits the combinatorial sensing characteristics from each active component with a synergistic contribution to the high sensitivity and fast response.

In order to explore full potential of graphene, we have also developed a hybrid materials consisting of graphene and nanocrystals of magnetic nanoparticles (e.g., Fe_3O_4 NPs) [85]. Once again, PIL was used as a versatile coupling agent to combine all the components together. Then the Fe_3O_4-rGO/PIL was used as a template for further hybridization with conducting polymer PEDOT resulting Fe_3O_4-rGO/PIL-PEDOT composites. Throughout the reaction, PILs serve as ionic stabilizers and/or counter ions of conducting polymers to create a multifunctional nano-architectured hybrid material. We demonstrated that the chemical sensor based on multi-component hybrid materials was capable of detecting both polar (ethanol, methanol, acetone, water) and non-polar (chloroform, styrene, dichlorobenzene, toluene) volatile organic compounds (VOC) (Fig. 12). It is underlined that the sensor's responses are still well defined at the ppm or sub-ppm level as attested by a SNR around 10. This result suggests its application in smart packaging to monitor food's degradation by the analysis of released VOC.

4 Graphene Filled Polymer Composites for Highly Sensitive Gas/Vapor Sensors

The 'perfect' graphene without any dangling bonds on its surface is unlikely to be very sensitive as vapor sensors, because gaseous molecules cannot readily get absorbed onto its surface. During the course of studying the graphene-based gas sensing, Johnson et al. [86] found that conventional nanolithographic processing typically leaves a polymer photoresist (e.g., PMMA) on the surface of graphene. The contamination layer degrades the electronic properties of the graphene. However, this layer can chemically dopes the graphene, enhances carrier scattering, and acts as an absorbent layer that concentrates analyte molecules at the graphene surface, thereby enhancing the sensor response. Figure 13 shows the graphene sensor with PMMA layer residue exhibits a strong electrical response towards vapor in ppm level, whereas a cleaning process that verifiably removes the contamination on the device structure made the intrinsic chemical responses of graphene to be insensitive.

Shi et al. [87] reported a GO/polypyrene (GO/PPr) composite films prepared by electrochemical co-deposition of GO and PPr in organic electrolyte. Blending of

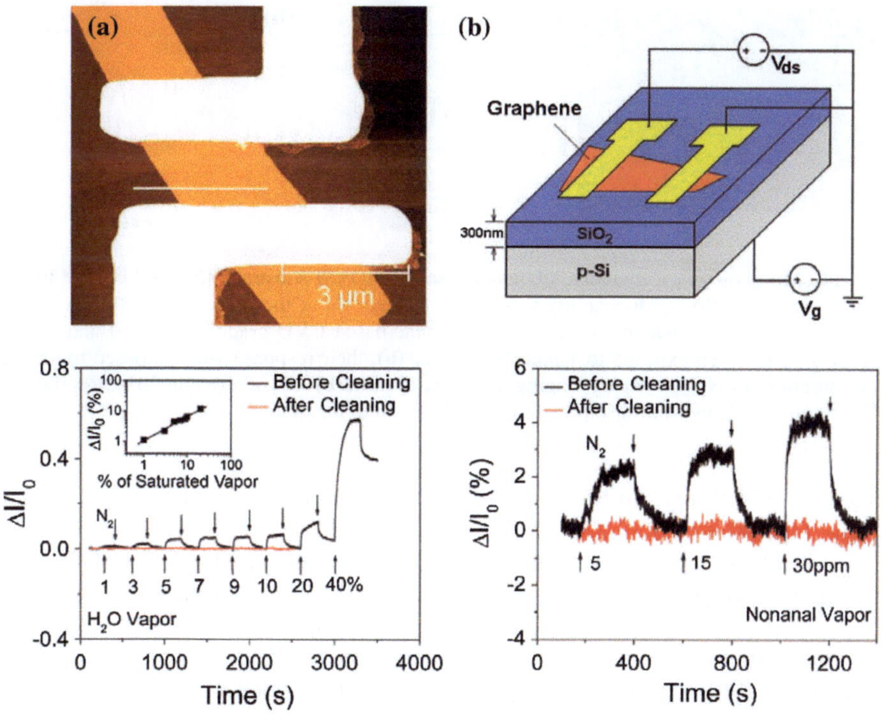

Fig. 13 AFM image of a typical GFET based on the graphene, device schematic of the electrode fabrication by e-beam lithography. Sensing response before (*black*) and after (*red*) sample cleaning to vapors of water (*left*) and nonanal (*right*). The cleaning process removes resist residue from the lithography step that enables the measurement of intrinsic response of the sensors device [75]. Copyright 2009. Reproduced with permission from American Chemical Society

PPr with GO improves the mechanical properties and gave the composite a continuous and porous morphology with an uninterrupted conducting phase. A chemoresistor-type vapor sensor based on the GO/PPr composite film demonstrated a fast, linear and reversible response to toluene with a high normalized sensitivity of 9.87×10^{-4} ppm^{-1}. They claimed that the sensing mechanism of the GO/PPr composite film is due to the PPr layer on rGO sheets interacting with toluene vapor.

Paul et al. [88] reported a graphene nanomesh as highly sensitive chemiresistor gas sensor. In this work, a grown p-type semiconducting large-area monolayer graphene film was grown by ethanol-CVD method and patterned into a nanomesh by the combination of nanosphere lithography and reactive ion etching. The fabricated graphene nanomesh was then tested as a FET and chemiresistor gas sensors. The neck-width of the synthesized nanomesh was about 20 nm corresponding to the gap between polystyrene (PS) spheres. The chemiresistor gas sensor of the ethanol-based graphene nanomesh (gEtOH nanomesh) showed excellent sensitivity toward NO_2 and NH_3. Their sensors exhibited sensitivities of about 4.32 %/ppm in

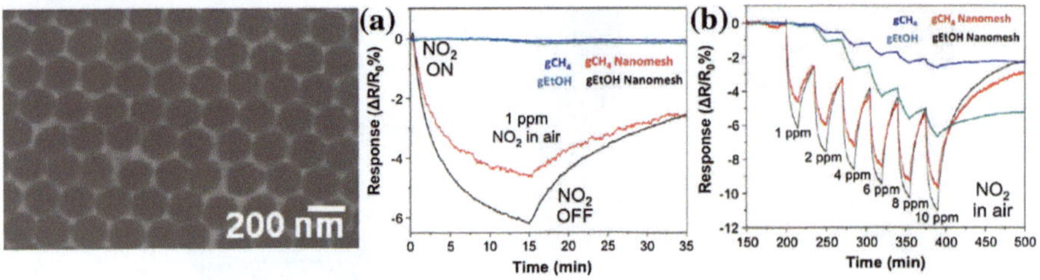

Fig. 14 Morphology of ethanol-CVD nanomesh on SiO_2/Si substrate after etching the unprotected areas of PMMA/graphene films, followed by removal of the Pt mask and PMMA in acetone. Comparison of dynamic responses of methane-based CVD graphene (gCH_4) and gEtOH sensor devices when exposed to 1 ppm NO_2 in air (**a**), their response when exposed to various concentrations of NO_2 in dry air ranging from 1 to 10 ppm (**b**) [77]. Copyright 2014. Reproduced with permission from Multidisciplinary Digital Publishing Institute

NO_2 and 0.71 %/ppm in NH_3 with limits of detection of 15 and 160 ppb, respectively (Fig. 14).

Hu et al. [89] presented a useful gas sensor based on chemically rGO using p-phenylenediamine (PPD) as a reduction agent. A gas sensor based on PPD-rGO showed selective detection of dimethyl methylphosphonate (DMMP). It showed 4.7 times better than hydrazine rGO in response to DMMP at 30 ppm.

Ji et al. [90] reported layer-by-layer films of graphene/ionic liquids (G-IL) composites on quartz crystal microbalances (QCM) for highly selective gas sensing. GO was reduced to form graphene (G) in the presence of different ionic liquids in water. The ILs effectively controls the separation of the graphene layers ranging from 2 to 4 layers with an intercalation of very thin ILs layer, while an increase in alkyl chain length of the imidazolium salt resulted in a longer mean-layer spacing. Composites of graphene/ionic liquids (G–IL) behave like charge-decorated nanosheets and were assembled alternately with poly(sodium styrenesulfonate) (PSS) by electrostatic LbL adsorption on appropriate solid supports. This procedure provides layered assemblies of G–IL composites with PSS on the surface of a QCM resonator, as indicated by regular shifts of the QCM frequency. The different QCM frequency shifts upon gas adsorption of benzene to LbL films of showing their selective toward different chemical vapor at room temperature (Fig. 15).

Very recently, Wang et al. have reported a nanocomposite made of polydiacetylene (PDA) and graphene using stacking film method. As-prepared composite was used as sensors for the detection of environmental hazardous VOCs [91]. The graphene is exploited as a transparent and efficient supporter for the highly ordered PDA monolayer. This colorimetric sensor exhibits a sensitive response to low concentrations of VOCs (~0.01 %), including tetrahydrofuran, chloroform, methanol, and dimethylformamide. The color change caused by relatively high concentrations of VOCs can be perceived by the naked eye, and it is noteworthy that a logarithmic relationship is observed between the chromatic response and the VOC concentration in the range of ~0.01–10 % (Fig. 16).

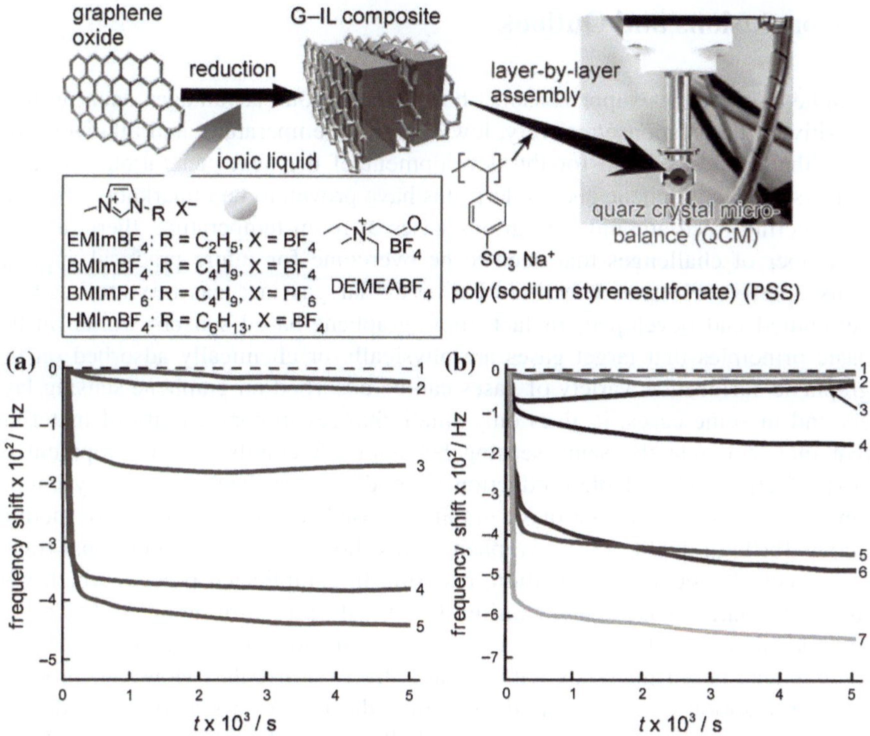

Fig. 15 Schematic illustration of the preparation of graphene/ionic liquid (G–IL) composites and their LbL assembly on a QCM. QCM frequency shifts upon gas adsorption of benzene to LbL films of different G-Ils (**a**), and G–HMImBF$_4$ film prepared by using 10 LbL cycles frequency shifts upon gas adsorption of different vapors of (1) hexane, (2) cyclohexane, (3) ethanol, (4) acetone, (5) benzene, (6) pyridine and (7) toluene (**b**) [79]. Copyright 2013. Reproduced with permission from Elsevier Ltd

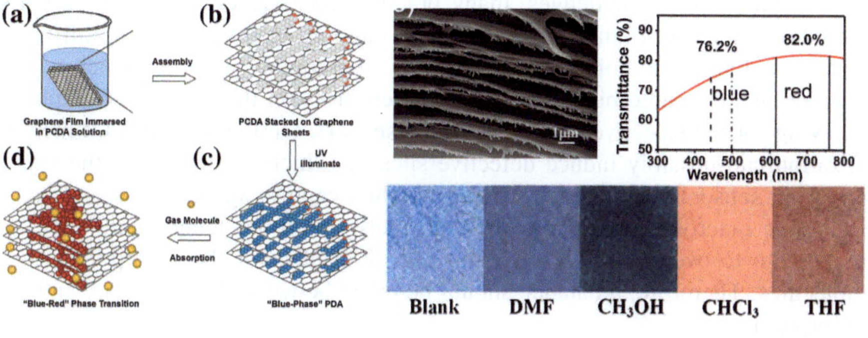

Fig. 16 **a–d** Schematic showing the preparation of the composite films and the molecular arrangement of how the PDA film self-assembles on graphene starting with PCDA: before **b** and after **c** polymerization, and upon exposure to VOC vapors **d**. SEM cross-section images of PDA/graphene film after UV irradiation, and photographs of the PDA/graphene films after exposure to various organic vapors for 2 min [80]. Copyright 2013. Reproduced with permission from Springer

5 Conclusions and Outlook

Graphene-based gas/vapor sensors shows tremendous advantages such as high sensitivity, fast response/recovery, low operation temperature, and relatively low cost, that could be ideal for the development of the next generation gas/vapor sensors. While graphene and its derivates have proven to be remarkably effective in detecting trace amounts of various gases at room temperature, there are still a number of challenges that need to be overcome for future practical applications. Firstly, selective detection of a particular gas in a gas mixture needs to be studied and developed. In fact, most graphene-based sensors works on the basic principles that target gases are physically or chemically adsorbed on the graphene surfaces. A variety of gases can be adsorbed on graphene sensing layers and in some cases, it gives only small changes in conductivity of materials that may generate the same sensing behaviors. Secondly, the development of cost-effective and scalable production methods for graphene is critically important for the commercialization of graphene-based sensors. CVD growth method allows for high quality of 2D graphene films; however it needs additional transfer process for devices fabrication, resulting in complicated process which will hinder for large-scale manufacturing. Chemical exfoliation process shows distinct advantages in terms of yield and cost and is amenable to functionalization of graphene-based materials with organic macro-molecules. However, processing optimization is still required to control the graphene structure and defects for effective sensing. Thirdly, there are challenges in the reliability of graphene-based gas/vapor sensors. As graphene is known as a material sensitive to a vast of external stimuli, graphene-based sensors may be sensitive to changes in environmental conditions such as moisture, temperature, contamination, electromagnetic field and so on [26], raising an issue of reliable detection and lifetime of the sensors. Thus, the environmental effects should be taken into account for the practical applications.

To address these challenges, many practical approaches have been explored including functionalizing graphene with capture agents including different nanocrystalines, biomolecules or organic macromolecules (polymers). The capture agents enable binding specific target gases on the surface of graphene, allowing for the selective detections of the sensors. On the other hand, functionalization can possibly induce defective sites on which desorption of the vapor/gas from sensors is difficult. Particularly when capture agents are organic molecules or macromolecules, the thermal energy at room temperature may not be enough to overcome the activation energy needed for the desorption of gas molecules. Therefore, a balance should be taken between the selectivity and fast desorption.

In short, if the aforementioned issues are fully and addressed, graphene-based gas sensors hold tremendous promise for use in numerous applications and commercialization.

References

1. J. Polleux, A. Gurlo, N. Barsan, U. Weimar, M. Antonietti, M. Niederberger. *Angew. Chem.* 2006, 118, 267.
2. C. K. Ho, A. Robinson, D. R. Miller, M. J. Davis. *Sensors* 2005, 5, 4–37.
3. S. Su, W. Wu, J. Gao, J. Lu, C. Fan. *J. Mater. Chem.* 2012, 22, 18101–18110.
4. I. A. Casalinuovo, D. D. Pierro, M. Coletta, P. D. Francesco. *Sensors* 2006, *6*, 1428.
5. P. Hu, J. Zhang, L. Li, Z. Wang, W. O'Neill, P. Estrela. *Sensors* 2010, *10*, 5133.
6. O. S. Wenger. *Chem. Rev.*113, 2013, 3685–3733.
7. H. J. Dai. 2001 Carbon Nanotubes (*Springer: Berlin*).
8. Y. P. Sun, K. Fu, Y. Lin, W. Huang. *Acc. Chem. Res.* 2002, 35, 1096–104.
9. J. Kong, N. R. Franklin, C. Zhou, M. G. Chapline, S. Peng, K. Cho, H. Dai. *Science* 2000, 287, 622–625.
10. J. Li, Y. Lu, Q. Ye, M. Cinke, J. Han, M. Meyyappan. *Nano lett.* 2003, 3, 929–933.
11. A. Modi, N. Koratkar, E. Lass, B. Wei, P. M. Ajayan. *Nature* 2003, 424, 171–174.
12. J. F. Feller, J. Lu, K. Zhang, B. Kumar, M. Castro, N. Gatt, H. J Choi. *J. Mater. Chem.* 2011, 21, 4142–4149.
13. B. Kumar, M. Castro, J. F. Feller. *J. Mater. Chem.* 2012, 22, 10656–10664.
14. J.C. Bonner. *Expert Rev Respir Med.* 2011, 5, 779–787.
15. K. Kostarelos. *Nat. Biotech.* 2008, 26, 774–776.
16. K. Donaldson, C. A. Poland. *Nat. Nanotech.* 2009, 4, 708–710.
17. H. C. Nerl, C. Cheng, A. E. Goode, S. D. Bergin, B. Lich, M. Gass, and A. E Porter. *Nanomedicine* 2011 6, 849–865.
18. C. Lee, X. Wei, J. W. Kysar, J. Hone. *Science* 2008, 321, 385–388.
19. A. A. Balandin, S. Ghosh, W. Bao, I. Calizo, D. Teweldebrhan, F. Miao, C. N. Lau. *Nano Lett.* 2008, 8, 902–907.
20. K. I. Bolotin, K. J. Sikes, Z. Jiang, M. Klima, G. Fudenberg, J. Hone, P. Kim, H. L. Stormer. *Solid State Commun.* 2008, 146, 351–355.
21. H. G. Park, S. Hwang, J. Lim, D. H. Kim, I. S. Song, J. H. Kim, D. H. Woo, S. Lee, S. C. Jun. *Jpn. J. Appl. Phys.* 2012, 51, 045101.
22. J. Z. Zhang. *J. Phys. Chem. Lett.* 2012, 3, 1806–1807.
23. Q. He, S. Wu, Z. Yin, H. Zhang. *Chem. Sci.* 2012, 3, 1764.
24. S. Mao, G. Lu. *J. Chen. J. Mater. Chem. A* 2014, 2, 5573.
25. W. Yuan, G. Shi. *J. Mater. Chem. A* 2013, 1,10078.
26. F. Yavari, N. Koratkar. *J. Phys. Chem. Lett.* 2012, 3, 1746–1753.
27. Y. Zhang, L. Zhang, C. Zhou. *Acc Chem Res* 2013, 46, 2329–2339.
28. G. H. Lu, S. Park, K. H. Yu, R. S. Ruoff, L. E. Ocola, D. Rosenmann, J. H. Chen. *ACS Nano* 2011, 5, 1154–1164.
29. K. S. Novoselov, A. K. Geim, S. V. Morozov, D. Jiang,Y. Zhang, S. V. Dubonos, et al. *Science* 2004, 306, 666–669.
30. K. S. Kim, Y. Zhao, H. Jang, S. Y. Lee, J. M. Kim, K. S. Kim, et al. *Nature* 2009, 457, 706–710.
31. C. Berger, Z. M. Song, X. B. Li, X. S. Wu, N. Brown, C. Naud, et al. *Science* 2006, **312**, 1191–1198.
32. S. Park, R. S. Ruoff. Nat Nanotechnol. 2009, 4, 217–224.
33. S. Gilje, S. Han, M. S. Wang, K. L. Wang, R. B. Kaner. *Nano Lett.* 2007, 7, 3394–3398.
34. W. Gao, L. B. Alemary, L. Gi, P. M. Ajayan. Nature Chem. 2009, 1, 403–408.
35. F. Schedin, A. K. Geim, S. V. Morozov, E. W. Hill, P. Blake, M. I. Katsnelson, K. S. Novoselov. *Nat Mater.* 2007, 6, 652–655.
36. S. Rumyantsev, G. Liu, M. S. Shur, R. A. Potyrailo, A. A. Balandin. *Nano Lett.* 2012, 12, 2294–2298.

37. R. K. Joshi, H. Gomez, F. Alvi, A. Kumar. *J Phys Chem C* 2010, 114, 6610–6613.
38. J. Hass, W. A. de Heer and E. H. Conrad. *J. Phys.: Condens. Matter*. 2008, 20, 323202.
39. R. Pearce, J. Eriksson, T. Iakimov, L. Hultman, A. L. Spetz, R. Yakimova. *ACS Nano*. 2013, **7**, 4647–4656.
40. Iezhokin, P. Offermans, S. H. Brongersma, A. J. M. Giesbers, C. F. J. Flipse. *Appl. Phys. Lett.* 2013, 103, 053514.
41. X. S. Li, W. W. Cai, A. Jinho, K. Seyoung, N. Junghyo, D. X. Yang, R. Piner, A. Velamakanni, J. Inhwa, E. Tutuc, S. K. Banerjee, L. Colombo, R. S. Ruoff. *Science* 2009, 324, 1312–1314.
42. C. W. Chen, S. C. Hung, M. D. Yang, C. W. Yeh, C. H. Wu, G. C. Chi, F. Ren, S. Pearton. *J. Appl. Phys. Lett.* 2011, 99, 243502.
43. A. Cagliani, D. M. Angus Mackenzie, L. K. Tschammer, F. Pizzocchero, K. Almdal, P. Boggild. *Nano Res*. 2014, 7, 743–754.
44. F. Yavari, Z. Chen, A. V. Thomas, W. Ren, H. M. Cheng, N. Koratkar. *Sci Rep*. 2011, 1, 166.
45. S. Park, R. S. Ruoff. Nat Nanotechnol. 2009, 4, 217–224.
46. S. Gilje, S. Han, M. S. Wang, K. L. Wang, R. B. Kaner. *Nano Lett*. 2007, 7, 3394–3398.
47. W. Gao, L. B. Alemary, L. Gi, P. M. Ajayan. Nature Chem. 2009, 1, 403–408.
48. M. D. Stoller, S. Park, Y. Zhu, J. An, R. S. Ruoff. *Nano Lett*. 2008, 8, 3498–3502.
49. K. K. Sadasivuni, D. Ponnamma, S. Thomas, Y. Grohens. *Prog in Poly Sci*. 2014, 39, 749–780.
50. J. T. Robinson, F. K. Perkins, E. S. Snow, Z. Wei, P. E. Sheehan. *Nano lett*. 2008, 8, 3137–3140.
51. S. Prezioso, F. Perrozzi, L. Giancaterini, C. Cantalini, E. Treossi, V. Palermo, M. Nardone, S. Santucci, L. Ottaviano. *J. Phys. Chem. C*, 2013, 117, 10683–10690.
52. J. D. Fowler, M. J. Allen, V. C. Tung, Y. Yang, R. B. Kaner, B. H. Weiller. *ACS Nano* 2009, 3, 301–306.
53. V. Dua, S. P. Surwade, S. Ammu, S. R. Agnihotra, S. Jain, K. E. Roberts, S. Park, R. S. Ruoff, S. K. Manohar. *Angew. Chem. Int. Ed*. 2010, 49, 2154−2157.
54. H. Song, L. Zhang, C. He, Y. Qu, Y. Tian, Y. Lv. J Mater Chem 2011, 21, 5972–5977.
55. A. Zöpfl, M. M. Lemberger, M. König, G. Ruhl, F. M. Matysik, T. Hirsch. ***Faraday Discuss.***, 2014, **DOI:** 10.1039/C4FD00086B.
56. D. Ponnamma, K. K. Sadasivuni, M. Strankowski, Q. Guo, S. Thomas. *Soft Matter*, 2013, 9, 10343.
57. M. Gautam, A. H. Jayatissa. *Solid State Electron* 2012, 78,159–165.
58. H. Vedala, D. C. Sorescu, G. P. Kotchey, A. Star. *Nano Lett*. 2011, 11, 2342–2347.
59. J. L. Johnson, A. Behnam, S. J. Pearton and A. Ural. *Adv. Mater*., 2010, 22, 4877.
60. T. T. Tung, M. Castro, T. Y. Kim, K. S. Suh, J. F. Feller. *Anal Bioanal Chem* 2014 406, 3995–4004.
61. A. Gutes, B. Hsia, A. Sussman, W. Mickelson, A. Zettl, C. Carraro, R. Maboudian. *Nanoscale*, 2012, 4, 438–440.
62. V. V. Quang, N. V. Dung, N. S. Trong, N. D. Hoa, N. V. Duy, N. V. Hieu. *Appl. Phys. Lett*. 2014, 105, 013107.
63. L. Zhou, F. Shen, X. Tian, D. Wang, T. Zhang, W. Chen. *Nanoscale* 2013, 5, 1564.
64. Y. Wang, L. Zhang, N. Hu, Y. Wang, Y. Zhang, Z. Zhou, Y. Liu, S. Shen, C. Peng. *Naoscale Research Letters* 2014, 9:251.
65. P. A. Russo, N. Donato, S. G. Leonardi, S. Baek, D. E. Conte, G. Neri, N. Pinna. *Angew. Chem., Int. Ed.*, 2012, 51, 11053.
66. W. Yuan, A. Liu, L. Huang, C. Li, G. Shi. *Adv. Mater*. 2013, 25, 766–771.
67. Y. Lu, B. R. Goldsmith, N. J. Kybert, A. T. C. Johnson. *Appl. Phys. Lett*. 2010, 97, 083107.
68. J. Janata,M. Josowicz. *Nat. Mater.* 2003, 2, 19.
69. H. Bai, G. Shi. *Sensors* 2007, 7, 267.
70. S. Virji, J. X. Huang, R. B. Kaner, B. H. Weiller. *Nano Lett*. 2004, 4, 491.
71. P. Hui, Z. Lijuan, C. Soeller, J. Travas-Sejdic. *Biomaterials* 2009, 30, 2132.

72. K. Arshak, V. Velusamy, O. Korostynska, K. Oliwa-Stasiak, C. Adley. *IEEE Sensors J.* 2005, 9, 1942.
73. J. Jang, M. Chang, H. Yoon. *Adv. Mater.* 2005, 17, 1616.
74. H. Yoon, M. Chang, J. Jang. *Adv. Funct. Mater.* 2007, 17, 431.
75. H. S. Yoon. *Nanomaterials* 2013, 3, 524–549.
76. C. M. Hangarter, M. Bangar, A. Mulchandani, N. V. Myung. *J Mater Chem* 2010,20,3131–3140.
77. S. J. Park, O. S. Kwon, J. E. Lee, J. S. Jang , H. S. Yoon. *Sensors* 2014, 14, 3604–3630.
78. L. Al-Mashat, K. Shin, K. Kalantar-zadeh, J. D. Plessis, S. H. Han, R. W. Kojima, R. B. Kaner, D. Li, X. L. Gou, S. J. Ippolito, W. Wlodarski. *J. Phys. Chem. C* 2010, 114, 16168–16173.
79. Z. Wu, X. Chen, S. Zhu, Z. Zhou, Y. Yao, W. Quan, B. Liu. *Sens. Actuators B* 2013, 178, 485–493.
80. W. K. Jang, J. Yun, H. I. Kim, Y. S. Lee. *Colloid Polym Sci* 2013, 291, 1095–1103.
81. N. Hu, Z. Yang, Y. Wang, L. Zhang, Y. Wang, X. Huang, H. Wei, L. Wei, Y. Zhang. *Nanotechnology* 2014, 25, 025502.
82. H. Bai, K. X. Sheng, P. F. Zhang, C. Li, G. Q. Shi. *J. Mater. Chem.* 2011, 21, 18653–18658.
83. T. T. Tung, M. Castro, T. Y. Kim, K. S. Suh, J. F. Feller. ***J. Mater. Chem***. 2012, **22**, 21754–21766.
84. T. T. Tung, M. Castro, J. F. Feller, T. Y. Kim, K.S. Suh. *Org. Electro*. 2013, 14, 2789–279472.
85. T. T. Tung, M. Castro, I. Pillin, T. Y. Kim, K. S. Suh, J. F. Feller. *Carbon* 2014, 74, 104–112.
86. Y. Dan, Y. Lu, N. J. Kybert, Z. Luo, A. T. C. Johnson. *Nano let.* 2009, 9, 1472–1475.
87. L. Zhang, C. Li, A. Liu, G. Q. Shi. ***J. Mater. Chem***. 2012, **22**, 8438–8443.
88. R. K. Paul, S. Badhulika, N. M. Saucedo, A. Mulchandani. *Anal. Chem.* 2012, 84, 8171–8178.
89. N. T. Hu , Y. Y. Wang, J. Chai, R. G. Gao, Z. Yang, E. S. W. Kong, Y. F. Zhang. Sensors and Actuators B 2012, 163,107–114.
90. Q. M. Ji, I. Honma, S. M. Paek, M. Akada, J. P. Hill, A. Vinu, K. Arig. *Angew. Chem. Int. Ed.* 2010, 49, 9737–9739.
91. X. Wang, X. Sun, P. A. Hu, J. Zhang, L. Wang, W. Feng, S. B. Lei, Bi.Yang, and W. W. Cao. *Adv. Funct. Mater.* 2013, *23*, 6044–6050.

Development of Biosensors from Polymer Graphene Composites

Ramendra Sundar Dey

Abstract Graphene has been considered as excellent two dimensional support in recent-times for next-generation graphene-polymer composites towards the development of biosensors. The remarkable properties of polymer and graphene with respect to electrical, mechanical, optical and structural aspect offers an ideal composite support for the development of biosensor. The frontiers of this composites technology is by combining of the polymer and graphene through synergy to achieve the goal of enhanced performance of biosensors with good efficiency and cost effectiveness. Graphene combined with polymer enhances the performance of biosensors in terms of sensitivity, selectivity, response time, and multiplexing capability of biosensors for clinical diagnostics. In this chapter, various methods have been provided to produce the polymer-based graphene composite materials and also discussed the importance of the composite materials to the development of biosensors for clinically important analytes, DNAs, aptamers and immunosensors.

Keywords Graphene · Biosensors · Polymer · Composites · Electrochemistry · Metabolites · DNA · Aptamers · Immunosensors

Abbreviations

Au	Gold
AuNP	Gold nanoparticle(s)
CEA	Carcinobryonic antigen
CHIT	Chitosan
ChOx	Cholesterol oxidase

R.S. Dey (✉)
Department of Chemistry, Technical University of Denmark, 2800 Kgs Lyngby, Denmark
e-mail: rsdey.kgp@gmail.com

K.K. Sadasivuni et al. (eds.), *Graphene-Based Polymer Nanocomposites in Electronics*, Springer Series on Polymer and Composite Materials,
DOI 10.1007/978-3-319-13875-6_11

CNT	Carbon nanotube(s)
Cyt c	Cytochrome C
DET	Direct electron transfer
DMSO	Dimethyl sulfoxide
ECL	Electrochemiluminescence
FET	Field effect transistor
fM	Femtomolar
GCE	Glassy carbon electrode
GO	Graphene oxide
GOx	Glucose oxidase
h	Hour(s)
H_2O_2	Hydrogen peroxide
HRP	Horseradish peroxidase
L	Liter(s)
LbL	Layer-by-layer
min	Minute(s)
mol	Mole(s)
MWCNT	Multi-walled carbon nanotube
Nf	Nafion
nM	Nanomolar
PANi	polyaniline
PAAm	Polyalylamine
PAAc	Polyacrylic acid
PBS	Phosphate buffer solution
PDDA	Poly(diallyldimethylammonium chloride)
PEDOT	Poly-3,4-ethylene dioxythiophene
pM	Picomolar
PPy	Polypyrrole
PSS	Polystrynesulfonate
PVA	Polyvinyl alcohol
rGO	Reduced graphene oxide
rt	Room temperature
RTIL	Room temperature ionic liquid
s	Second(s)
SCE	Saturated calomel electrode
SPCE	Screen-printed carbon electrode
SPE	Screen-printed electrode

1 Introduction

Biosensing of clinically important analytes play a crucial role in human life for the monitoring and diagnosis of various diseases. The development of sensitive, selective and low-cost biosensor device is an important role for early stage screening of disease. Graphene-based composites are extensively used for biosensor development due to its excellent electrical conductivity and high mechanical stiffness and biocompatibility [1, 2]. The high surface area and biocompatibility of graphene-based material help to immobilization of bioreceptor (enzymes, DNA and antibody) on its surface [3, 4]. Additionally, graphene-based material promotes the electron transfer reaction of various redox protein [5, 6]. Moreover, in terms of practical applications, graphene-based material suffers from dispersibility, as it has tendency to irreversible agglomeration due to its planer sp^2-carbon [7].

The combination of the unique properties of polymer and graphene has attracted copious attention in recent times due to the design and creation of a new material in material chemistry and for its enormous application in the field of next-generation sensor devices. Since the discovery of graphene [8], graphene-based polymer nanocomposities is an important material for the nanoscience and nanotechnology. Graphene is a one-atom-thick, planar, sp^2-hybridized carbon nanostructures, possess large specific surface area, high-speed electron mobility, good mechanical strength, high electric and thermal conductivity, room temperature quantum hall effect, good optical transparency, tunable band gap, high elasticity and biocompatibility [9–12]. Because of these excellent characteristics, graphene represents the new growing star at the horizon of materials science, polymer science, condensed matter physics and nanoelectronics [13, 14]. Since 2004 more than 65,000 papers on graphene have materialize (according to scifinder).

Polymer grafted with graphene changes the in-plane properties of the composite material. Polymer help to more complete exfoliation of the graphitic layer towards single sheet and increase the dispersibility of the composites in solution state [15, 16]. It is established that graphene-based polymer composites shows better mechanical, thermal, gas barrier, electrical and flame retardant properties, compared to the straight polymer [16–20]. Although carbon nanotubes (CNT) hold comparable thermal and mechanical properties with graphene, but graphene is still superior material as nano-scale filler material in certain respect such as electrical and thermal conductivity, optical transparency and large surface area [21–25]. It is also described that graphene-based polymer composites have improved mechanical and thermal properties than that of clay and other carbon filler-based composites materials [26, 27]. Therefore, graphene and graphene-based polymer composites triggered scientific interest and impressively used in various field. Graphene-based polymer composites is used in wide range of application like energy storage system (supercapacitor and battery), catalyst for energy conversion system, dye-sensitize solar cell, nanoelectronics, optoelectronics, electrochemical sensors and biosensors [28–34].

As it is clear from the above discussion that graphene-based polymer composites is becoming popular to the scientific communities and current research in the field of nanoscience and bioelectronics demand this material for the development of biosensors. This chapter aims to focus on the recent developments in the field of graphene-based polymer composites and their uses in the biosensors development. This chapter will only describe those polymer-based graphene materials that are used for biosensors application. It includes the method of preparation of

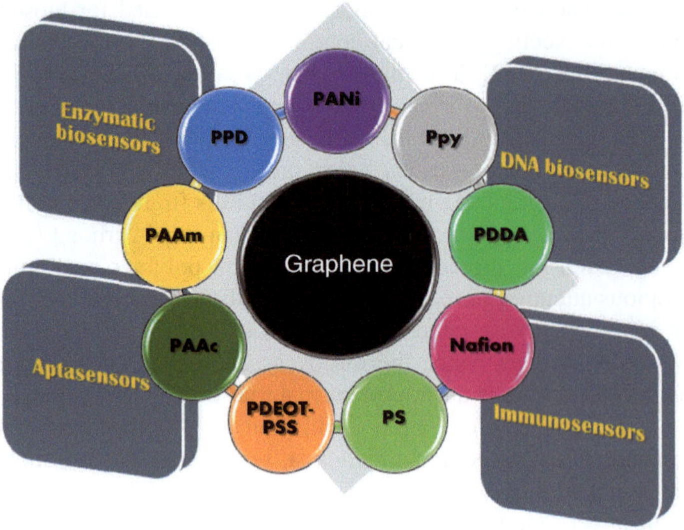

Fig. 1 Figure represents the development of biosensors based on graphene with different polymers

various polymers (PANi, Nf, PDDA, PPy, PS, PDEOT, PSS, PAAc, PAAm, PPD etc.)-based graphene nanocomposites through different route and the application of these materials towards the development of biosensors for biologically important molecules like metabolities, DNAs, aptamers and antigens (Fig. 1).

2 Graphene-Based Polymer Composites and Biosensors

Since the excessive interest generated by the outstanding properties of graphene sheets and the discovery of their various production methods, graphene-based composites materials continued to be an essential research focus around the world. In recent times, a number of processing methods have been conveyed for the dispersion of graphene fillers into polymer matrices. Many of them are similar to those used for other nanocomposite systems, even though some of these techniques have been useful distinctively to graphene-based composites [35].

It is established that graphene oxide (GO)-based fillers exhibit high electrical conductivities (order of thousands S/m) [36], high moduli (ranging from 208 to 650 GPa) [37], and can be functionalized without any trouble to modify their compatibility with the host polymer [38]. Bunnell, in a patent in 1991, proposed the production of polymer nanocomposites incorporating "as thin as possible" as fillers [39]. This study reported tremendous property developments versus conventional polymer composites depend on micron-scale fillers such as untreated flake graphite or carbon black [40, 41].

Graphene is much cheaper than CNTs and other nanocarbon-based material because of its easy synthetic procedure (easily derived from a graphite precursor) and large-scale production. Therefore, graphene-based polymer composites have attracted both academic and industrial importance in recent years. A variety of polymers, such as polyaniline (PANi), Nafion (Nf), Poly(diallyldimethylammonium chloride) (PDDA), Polypyrrole (PPy), PS, Poly-3,4-ethylene dioxythiophene (PDEOT), Polystrynesulfonate (PSS), Polyacrylic acid (PAAc), polyalylamine (PAAm) and PPD (Table 1) have been traditionally used as matrices to the assemble of graphene/polymer composites.

Biosensors are capable of produce electroanalytical information using biological recognition system. According to IUPAC, 'An electrochemical biosensor is a self-contained integrated device, which is capable of providing specific quantitative or semi-quantitative analytical information using a biological recognition element (biochemical receptor) which is retained in direct spatial contact with an electrochemical transduction element' [42]. Recently, graphene family has become promising for the construction of various biosensors with high sensitivity, lower detection limit, selectivity and good stability of the sensors compared to other carbon materials. Its unique structure, extremely high carrier mobility and capacity, prominent electron transfer rate, high surface-to-volume ratio, outstanding flexibility and high mechanical properties [43–45] make it a promising candidate for development of biosensors. Other member of graphene family such as GO and reduced GO (rGO) are more electrochemically active as compared to pristine graphene due to the oxygen-containing reactive sites at edges and in the basal plane, suggesting use for the maturity of biosensors [46]. However, due to higher conductivity of reduced graphene oxide (rGO), it exhibits a higher sensitivity and electrocatalytic activity than GO-based electrode, [45, 47]. But, with the idea of avoiding the irreversible aggregation of graphene, combining with other components like polymer is a promising choice and an encouraged route to stabilize graphene and to extend the purpose of graphene-based materials.

3 Polymer: Functionalization on Graphitic Layers

Functionalization of polymer into graphitic layer is an immense challenge to the researchers in both aqueous and organic solvents. The synthesis of graphene-based polymer composites depends on the polarity, molecular weight, hydrophobicity and reactive groups present in the polymer, graphene and solvents used [45]. There are two different ways for functionalization of the polymer to the basal plane or the edge of graphitic layers.

Table 1 Table demonstrates the method used for graphene-based polymer composites and detection of various bioanalytes

Materials		Preparation method	Analytes	References
Polyaniline	Graphene PANi nanowires	3 Steps electrochemical	DNA	[85]
	Nanocomposite of graphene, polyvinylpyrrolidone and PANi	Mixing and electrospraying	Cholesterol	[52]
	LbL graphene, Fe_3O_4 and PANi	Solution mixing	Glucose	[105]
	Graphene-PANi	Electrochemical	H_2O_2	[89]
	Sulfonated PANi-GO	Solution mixing	DNA	[51]
	GO sheets/PANi/CdSe quantum dots	Solution mixing	Cyt c	[113]
	Graphene/PANi/Au nanoparticles	Solution mixing	Glucose	[104]
	GO nanosheets/PANi nanowires/CdSe quantum dots	Solution mixing	Interleukin-6	[54]
	Graphene-PANi	Solution mixing	Estradiol	[56]
	Graphene/PANi/AuNPs	Solution mixing	Glucose	[57]
	Electrochemically reduced GO-PANi	Chemical-electrochemical	DNA	[58]
Polyphenelenediamine	Graphene/HRP-O-phenylenediamine	Electrochemical and chemical	CEA, AFP, cancer antigen 125, carbohydrate antigen 153	[131]
	Graphene-carried poly(o-phenylenediamine)/gold hybrid nanosheets	Solution mixing	CEA	[78]
	Graphene/aptamers complex-poly(o-phenylenediamine	Chemical-electrochemical	ATP	[92]
PDDA	PDDA functionalized graphene	Autoclave	Glucose	[106]
	PDDA-protected graphene-AuNPs	Solution mixing	Platelet derived Growth factor	[126]
	PDDA-graphene and AuNps	Solution mixing	Human IgG	[130]
	PDDA-capped AuNP functionalized graphene/MWCNTs	Solution mixing	Glucose	[71]
	PDDA-graphene nanosheets-AuNPs	Solution mixing	H_2O_2	[69]
	PDDA-graphene and AuNPs	Solution mixing	H_2O_2	[121]
	PDDA–graphene nanosheets–AuNPs	Solution mixing	glucose	[107]

(continued)

Table 1 (continued)

Materials		Preparation method	Analytes	References
Polypyrrole	Graphene-PPy NT	Ultrasonication	H_2O_2	[34]
	PPy-graphene	Electrochemical	Glucose	[91]
	Poly(styrenesulfonic acid-graphene-pyrrole)	Solution mixing	Hypoxanthine	[81]
	AuNP–PPy–rGO	Electrochemical	Paraoxon-ethyl	[90]
	PPy-converted nitrogen-doped few-layer graphene	Furnace growth	Vascular endothelial growth factor	[80]
Polyallylamine	Graphene-polyallylamine	Plasma polymerization	Thrombin	[74]
PEDOT and/or PSS	Graphene-PEDOT-PSS	Electrolytic exfoliation	Glucose	[93]
	PDEOT-PSS-graphene	Solution mixing	Ethanol	[75]
	Graphene/PEDOT	Solution mixing	H_2O_2	[76]
PVA	PVA/chitosan/GO/Pt	electrospinning	Glucose	[94]
Polydiaminobenzidine	GO-Polydiaminobenzidine	Peroxidase-catalyzed polymerization	Interleukin 5	[134]
Nafion	rGO/Nf	Solution mixing	Organophosphate	[59]
	rGO/Nf	Solution mixing	Glucose	[108]
	GO/Nf	Solution mixing	H_2O_2	[122]
	Graphene sheets–Nf	Solution mixing	H_2O_2	[60]
Polyacrylic acid	Pyrene-grafted polyacrylic acid/graphene	Solution mixing	Glucose and maltose	[109]
Poly(N-butylbenzimidazole)	Poly(N-butyl benzimidazole)-graphene	Solution mixing	Glucose	[110]

3.1 Chemical Polymerization

Chemical functionalization of polymer with graphene is divided by two sub categories; covalent and non-covalent functionalization. Covalent functionalization is generally involves coupling reaction with oxygen functionalities of GO with active group present in the polymer. However, in case of non-covalent polymerization, polymer molecule physically adsorbed on the basal plane of GO. In both cases, first GO is prepared from naturally occurring graphite by modified hummers method [48]. The synthesized GO was then subjected to disperse in either aqueous or organic solvents. Functionalization of polymer with GO/rGO occur (a) monomer, initiator and GO mixed together and follow in situ polymerization to get graphene-based polymer composites [49–51]; (b) this method comprises the synthesis of polymer followed by mixing with GO/rGO to get the desired product [52].

Polyaniline is widely studied polymer among others because of its high conductivity, environment friendly and easy synthetic procedure. Vallés et al. reported the simultaneous reduction GO and PANi followed by the method reported by Jiménez et al. [53] It includes in situ oxidative polymerization of aniline in presence of dispersed GO [49]. However, Chen et al. synthesized PANi nanorod embedded single layer GO nanosheets by in situ polymerization method, where molar ratio of aniline to APS was fixed at 1:1; the weight ratio of GO to aniline was 1:2 [50]. Recently, Yang et al. reported that sulfonated PANi-GO nanocomposite and poly(m-aminobenzenesulfonic acid)-GO hybrid material by mixing all the required component to get the desired product [51]. They claimed that the π–π planar structure of GO and the carboxyl groups on the surface of GO ensured it could act as an excellent substrate for adsorption and polymerization of aniline monomer. Ruecha et al. developed a cholesterol biosensor based ongraphene/polyvinylpyrrolidone/PANi nanocomposite. First, Camphor-10-sulfonic acid doped PANi and polyvinyl alcohol (PVA)/graphene was synthesized separately and then these two solution were mixed together to get the composite material [52]. Liu et al. described the synthesis of GO nanosheets PANi nanowire composite materials by mixing all the component in stepwise manner [54]. Li et al. [55] developed graphene-PANi composites followed by the method of Zhang et al. [56]. Xu et al. [57] developed graphene/PANi/AuNPs nanocomposite, where mass ratio of aniline to graphene was 100:1. In a report, Du et al. [58] described that preparation of graphene/PANI/GCE electrode where first graphene/GCE electrode was prepared and chemically synthesized PANi drop cast on graphene/GCE electrode and then dried in air.

Nafion The hydrophobic backbone of Nf can be used for the functionalization of graphene sheets for the high performance hybrid materials. Choi et al. reported the preparation of free-standing flexible conductive reduced graphene oxide/Nafion (RGON) hybrid films by solution chemistry [59]. They demonstrated that the hydrophobic backbone of Nf could afford well-defined integrated structures (Fig. 2), on micro- and macroscales for the assemblying of hybrid materials

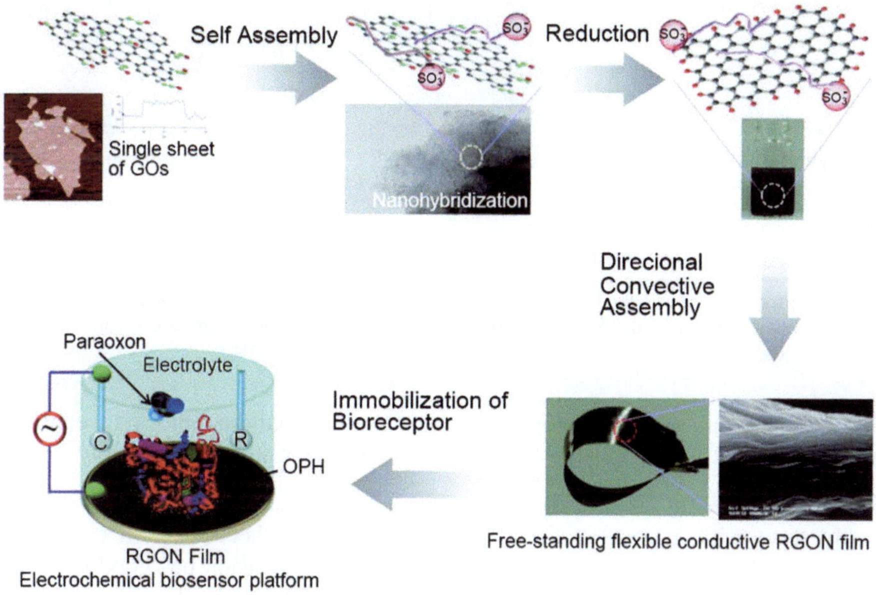

Fig. 2 Figure represents the procedure to design RGON film and development of biosensor [59]. Copyright 2010. Reproduced with permission from American Chemical Society

through self-assembly, while the hydrophilic sulfonate groups enabled highly stable dispersibility and long-term stability for graphene. Homogeneous dispersion of Nf solution with graphene can be made under vigorous ultrasonication for about 1 h and used as an electrode material reported by Xin et al. [60].

It should be noted here that, most of the cases GO/rGO mixed with ethanolic Nf solution either by ultrasonication or by stirring at room temperature to form homogeneous mixture and directly coated on electrode surface for further use [61–66]. Nf also can be used as layer film over the modified electrode to stabilized the electrode material [67].

PDDA is a well-known polymer can act as stabilizing agent to prepare soluble graphene nanosheets from GO. It can also act as reducing agents in some cases. Zhang et al. [68] proposed the synthesis of Pt/graphene composites functionalized with PDDA as shown in Fig. 3, where PDDA induced the reduction of GO and act as stabilizer. In a report, Song et al. was also used PDDA as a reducing agent for the reduction of GO at 90 °C for 5 h [69]. Liu et al. first time synthesized functionalized graphene with PDDA by solution mixing method [70]. First PDDA solution mixed with homogeneous dispersion of GO and reduced by hydrazine hydrate at 90 °C for 24 h. Another work of PDDA functionalized graphene composite material with PDDA-capped AuNP(s) (AuNP) reposted by Yu et al. [71].

PAAc is a nono-conductive polymer, but when PAAc doped with graphene, graphene helps to improve the conductivity of PAAc. Ma et al. [72] reported the

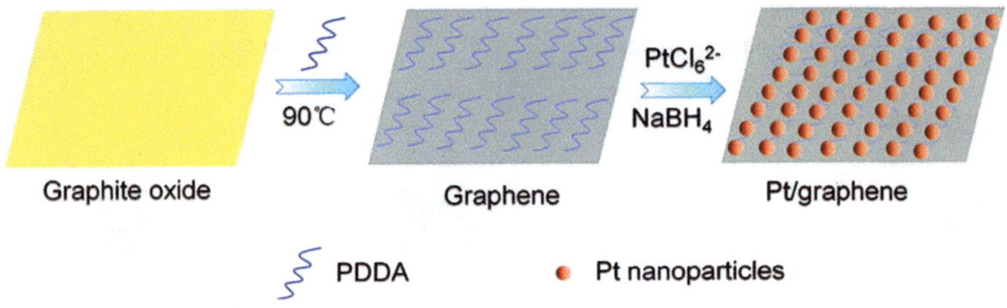

Fig. 3 Scheme represents the PDDA-induced reduction and decoration of GO and in situ growth of Pt nanoparticles on them [68]. Copyright 2011. Reproduced with permission from American Chemical Society

PAAc functionalized graphene surface was used as building block to construct electrochemically functionalized multilayer via electrostatic LbL assembly.

PAAm hold a long alkyl chain with a number of reactive amine groups that can easily react with xygen functional groups of the GO sheets. PAAm chemically cross-linked with GO form homogeneous aqueous suspension has good mechanical properties is recently reported [73]. Zhang et al. [74] reported a nanocomposite based on amine functionalized graphene and plasma polymerized PAAm. PAAm was deposited onto the graphene with self-assembled octadecylamine under plasma input power for various time periods using allylamine as the monomer gas.

PDEOT and/or PSS Poly(3,4 ethylenedioxythiophene):polystyrene sulfonic acid (PEDOT:PSS) is considered as a highly promising conducting polymer matrix. Chemical modification of PDEOT/PSS with graphene can be achieved by solution mixing method recently reported by Gao et al. [75]. Wang et al. [76] reported an improved procedure to fabricate graphene/PDEOT nanocomposites, where they followed the in situ polymerization of EDOT monomer on GO and PSS was used as charge balancing dopant. In another report, Xu et al. [77] proposed a sonochemical method for the preparation of polystyrene functionalized graphene starting from graphite flakes and a reactive monomer styrene. This paper demonstrates that the polystyrene chains are formed from sonochemically initiated radical polymerization of styrene and can make up to $\sim$18 wt% of the functionalized graphene.

Polyphenelinediamine is a highly aromatic polymer containing 1,4-diaminophenazine or quinoraline repeating unit. Chen et al. [78] synthesized PPD by a simple chemical method, where GO and phenelinediamine were mixed together and stirred at 200 rpm for 48 h at rt. During this process, phenelinediamine molecules were conjugated onto GO nanosheets by the π-stacking interactions. Meanwhile, the immobilized phenelinediamine molecules were self-polymerized due to the instability of phenelinediamine monomers.

Polypyrrole is one of the most promising conducting polymers for commercial applications because of its stability and flexibility. In addition, the facile preparation and easy control of the properties of PPy are also attractive features. Recently, Park et al. [79] described a simple route for graphene-PPy nanotube composites synthesis. First one-dimensional PPy nanotube was synthesized and then mixed with GO dispersion followed by reduction by hydrazine to obtain the nanocomposites. Kwon et al. [80] synthesized PPy-converted nitrogen-doped few-layer graphene (PPy-NDFLG) grown on Cu substrate by chemical vapor deposition combined with vapor deposition polymerization. The step-by step procedure is shown in Fig. 4. Zhang et al. [81] synthesized the functionalization of rGO sheets with conducting PPy graft copolymer, poly(styrenesulfonic acid-g-pyrrole), via π-π noncovalent interaction.

3.2 *Electrochemical Polymerization*

Electrochemical polymerization is a promising approach to make polymer layer directly on the electrode surface. During the last 20 years, electronically polymers like PANi, PPy, and polythiophene have received significant attention because of their easy synthetic procedures, less time consuming, remarkable electronic, magnetic, and optical properties and their wide range of potential applications in many areas [82–94].

Polyaniline Among the various electrochemically synthesized polymers, PANi has been mostly studied polymer as a potential candidate with unique doping/dedoping performance, inherent electrical conductivity, facile synthesis and environmental

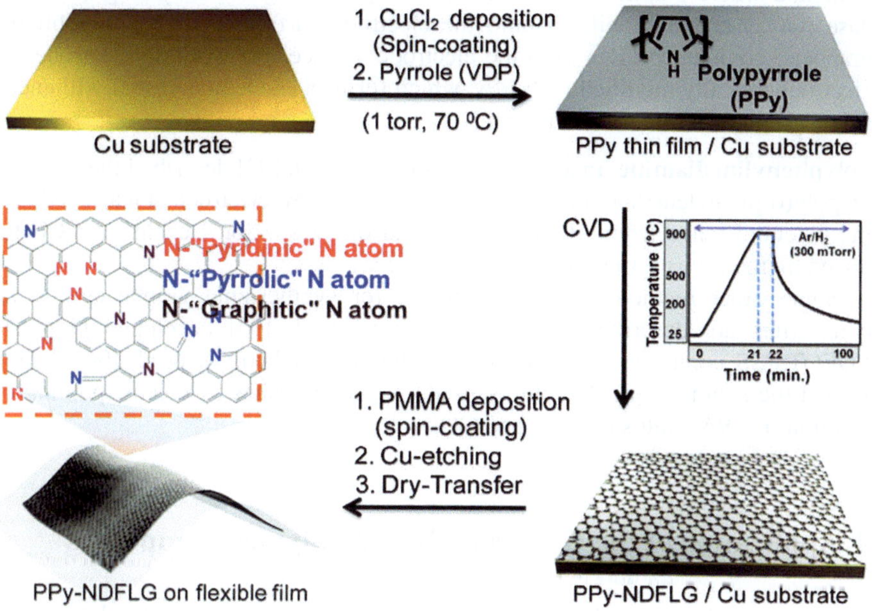

Fig. 4 Scheme represents the protocol of PPy-converted few layer N-doped graphene on flexible substrate [80]. Copyright 2012. Reproduced with permission from American Chemical Society

stability. Recently, Bo et al. [85] developed a synthetic route for oxidized graphene and PANi nanowires modified glassy carbon electrode (GCE) comprise three steps electrochemical method followed by the work reported by Liang et al. [86]. First, a larger current density was applied on the electrode in 1.0 mol L^{-1} HCl solution containing 0.5 mol L^{-1} aniline. Then the initial stage was followed by continuous polymerization with reduced current density. A typical polymerization procedure consist of a current density of 0.06 mA cm^{-2} was applied for 0.5 h, then 0.03 mA cm^{-2} for 3 h, finally followed by another 3 h at 0.015 mA cm^{-2}. Devi et al. [87] reported a method for deposition of chitosan/PANi/GO nanoparticles/PPy on gold (Au) electrode. First, a mixture of GO nanoparticles and PPy was electropolymerized onto Au electrode followed by electrodeposition of a mixture of PANi and CHIT onto GO nanoparticles/PPy/Au to construct CHIT/PANi/ GO nanoparticles/PPy/Au electrode. The nanocomposites of GO–PANi were prepared by electrochemical deposition of a mixture of HCl, aniline and GO solution demonstrated by Radhapyari et al. [88]. The electropolymerization was demonstrated at scan rate of 20 mV s^{-1} in the potential range from 2.0 to 1.1 V. Feng et al. [89] described a large-scale one-step elctrodeposition method using GO and aniline as starting materials via potential scanning in the potential range of −1.3 to 1.0 V in H_2SO_4 solution.

Polypyrrole Yang et al. [90] investigated the polymerization of pyrrole co-deposited with sulfonated rGO. They also demonstrated that due to strong chemisorptions between polymers and carbon materials, rGO was readily incorporated into the PPy to avoid aggregation. In another report, Alwarappan et al. [91] described a method where graphene is covalently conjugated with the enzyme glucose oxidase (GOx) and immobilized onto the electrode surface. Prior to immobilizing graphene-conjugated GOx onto the electrode surface, the working electrode was electrochemically modified with PPy, which forms a stable matrix that will encapsulate graphene-GOx on its surface.

Polyphenylinediamine In a current report, Wen et al. [92] described the synthesis of poly(o-phenylenediamine)/graphene composites by electrochemical cycling in the potential range from −0.50 to 1.00 V at 0.1 V s^{-1} for 4 cycles at GCE followed by the drop casting of graphene solution.

Some other methods are also reported for synthesizing grapheane-polymer composites are described below. Wisitsoraat et al. [93] developed graphene-PDEOT:PSS composite based on electrophoretic method. Su et al. [94] demonstrated the synthesis of PVA-chitosan-GO nanocomposites by electrospinning the solution of PVA, chitosan, GOx and GO.

4 Application of the Graphene-Based Polymer Composites

Even though there are plentiful of challenging works remain in developing with graphene based polymer composites, but the materials are already travel around in the area of energy storage and conversion system, electronic devices, drug delivery and electrochemical sensors and biosensors [95, 96]. Among them electrochemical biosensor is a promising field for the use of graphene-based polymer composites as a matrix because of its superior conductivity, high stability, large surface area and biocompatibility.

4.1 Electrochemical Biosensors

Biosensors are composed of detection technology exploiting enzymes or other biomolecules to sense information relative to biological systems or to mimic biological sensing processes. An electrochemical biosensor comprises of two components; a chemical or biological recognition element and a physical transducer that transduce the analytical signal of the sensing event to an electronic circuit [97]. The actual function of biosensors is the detection of an electroactive analyte involved in the chemical or biological recognition event. The sensing potential of biosensors is evaluated in terms of selectivity, sensitivity, detection limit, response time, stability, etc. The recognition and transduction are the two key factors that determine the overall performance of the biosensor. The environment of recognition elements, that of the transducer, and that of the matrix used for the immobilization of recognition elements generally control the sensing event. The efficacy of the biosensing device can be enhanced by the careful choice of novel functional nanomaterials such as nanocarbon like graphene and polymers. The skillful combination of catalytically active nanoscale materials with polymer and redox enzymes can enhance the overall performance. The large surface-to-volume ratio of nanoscale materials ensures the immobilization of an adequate amount of recognition elements. The high electronic conductivity, admirable catalytic activity, and biocompatibility of nanomaterials make sure the facilitated signal transduction during the sensing experience. The following part of this chapter will describe the biosensors of clinically important analytes such as enzymatic biosensors formetabolites like glucose, cholesterol, uric acid, etc., direct electron transfer (DET) of enzymes, electrochemical immunosensors, aptasensors and DNA biosensors (Table 1).

4.1.1 Enzymatic Biosensors

Enormous attempt has been made to build up new enzyme-based electrochemical biosensors that are suitable for a highly selective, sensitive, and rapid analysis of various biological substances in vivo and in vitro [98]. However, the constancy of enzymes immobilized on electrode surface will not remain intact for a longer period, and as a result, the stability and the sensitivity of the electrode decay with time. DET between redox enzymes and the electrodes surface can be utilized to investigate the enzyme–catalyzed redox reactions in biological systems and to study the electrochemical property for the study of the structure of enzymes, kinetics and thermodynamics of redox transformations of enzyme molecules, and metabolic processes involving redox transformations [99–101]. The graphene-based polymer composites may provide insight to fabricate novel biosensors for practical applications because of its much higher sensitivity resulting from the conductivity and its copious reactive functional groups that can easily bind with the free–NH_2 terminals of the enzyme to result in a strong linkage of enzymes with composites material. Three important strategies have been used for the enzymatic biosensors. These are based on (1) first-generation biosensors: detection of enzymatically generated H_2O_2 or measuring the change in the concentration of dissolved oxygen, (2) second-generation biosensors: redox electrochemistry of the electron transfer mediators, and (3) third generation biosensors: direct electrochemistry of enzyme [102].

Glucose is one of the important metabolites; its normal level in human serum is 80–120 mg/dL. Diabetes mellitus is a life-threatening diseases caused by glucose. Abnormal levels of blood glucose (hyperglycemia and hypoglycemia) are responsible for various life-threatening diseases like stroke, heart disease, kidney failure, blindness, etc. [103].

A number of efforts are being attempted to develop glucose biosensors with improved performance at reasonably priced. Different approaches have been used for the biosensing of glucose. Kong et al. [104] developed disposable screen-printed carbon electrode (SPCE) based third-generation glucose biosensors. The electrode was modified with graphene/PANi/AuNp/GOx biocomposite and then covered by a paper disk impregnated with the sample (Fig. 5). This glucose sensor not only shows the properties of paper (low cost, simplicity, capillary wicking, pre-stored reagent), but also it hold the high specificity of enzyme reaction. They have also demonstrated that this inexpensive, portable, miniaturized, disposable sensor enabled the direct determination of single analyte in complex samples with small volume and without the need of any pre-processing steps. Another glucose biosensor was developed by Xu et al. based on the direct electrochemistry of GOx adsorbed in graphene/PANi/AuNPs nanocomposite [56]. The adsorbed GOx displayed a pair of well-defined *quasi*-reversible redox peaks with a formal potential

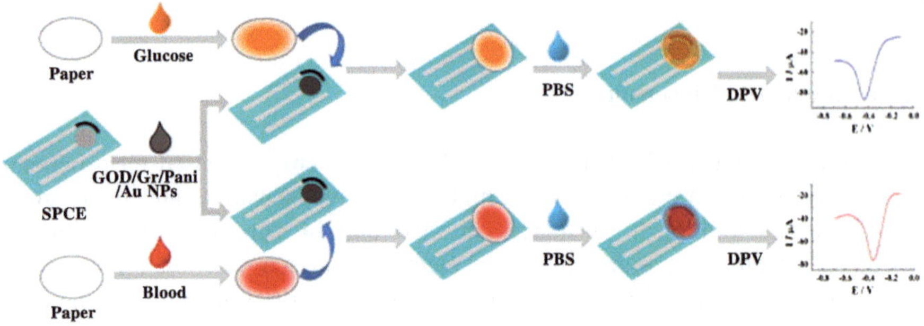

Fig. 5 Schematic representation of the fabrication and assay procedure for the paper-based glucose sensor [104]. Copyright 2014. Reproduced with permission from Elsevier Ltd.

of −0.477 V (vs. SCE). The modified electrode was linearly proportional to the concentration of glucose in the range of 4.0 μM to 1.12 mM with a low detection limit of 0.6 μM. Nguyen et al. describes a glucose electrochemical biosensor, layer-by-layer fabricated from graphene and PANi films [105]. This graphene patterned sensor denoted as graphene/Fe_3O_4/PANi/GOx shows improved glucose sensitivity of 47 μA mM^{-1} cm^{-2}. Jia et al. [106] selected PDDA as an electron acceptor for functionalizing graphene. PDDA-graphene bio-nanocomposite was used as a biosensor for the detection of glucose with a linear range from 0.02 to 1.8 mM and a detection limit of 8 μM. Yu et al. [71] reported PDDA-caped AuNP functionalized graphene nanocomposites for the DET of GOx and biosensing of glucose. The modified electrode shows high sensitivity of 29.72 mA M^{-1} cm^{-2} and low detection limit of 4.8 μM. Song et al. [107] developed a bi-protein bio-interface where GOx and Cytochrome C (Cyt c) were co-entrapped in the PDDA–graphene nanosheets–AuNP hybrid nanocomposites. The optimal molar ratio and total amount of Cyt c and GOx in the bio-interphase for DET of GOx was estimated to be about 3:1 and 1.40 nmol, respectively. The bi-protein bio-interphase was used to detect glucose based on the consumption of O_2 with the oxidation of glucose catalyzed by GOx. Alwarappan et al. [91] reported the enhanced performance of PPy-graphene-GOx based enzymatic biosensors employed for in vitro electrochemical detection of glucose. Prior to immobilizing graphene-conjugated GOx on the electrode surface, the working electrode was electrochemically modified with PPy, which forms a stable matrix that will encapsulate graphene-GOx on its surface (Fig. 6).

Wisitsorrat et al. [93] developed a sensitive glucose biosensor based on graphene–poly(3,4-ethylenedioxythiophene):polystyrene sulfonic acid (graphene-PEDOT:PSS) modified SPCE for electrochemical detection of glucose. Direct electrochemistry of GOx at the graphite/PEDOT:PSS electrode was checked by cyclic voltammetry and graphene plays the main role in a *quasi*-reversible redox process. GOx/GP-PEDOT:PSS electrode exhibits high sensitivity of 7.23 μA mM^{-1} good stability with only 30 % loss of enzyme activity after 30 days.

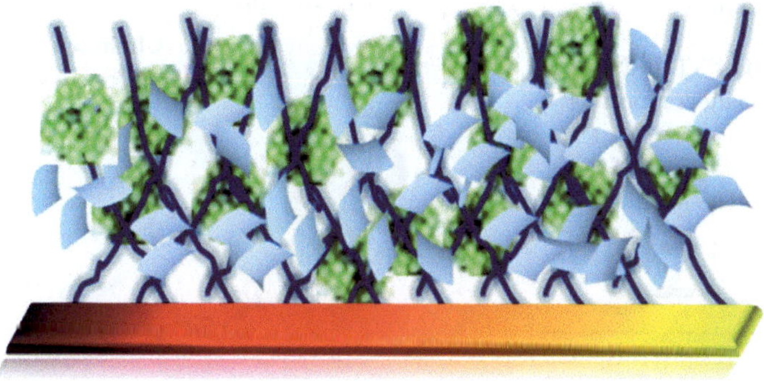

Fig. 6 Scheme represents the graphene-GOx entrapped within a porous PPy matrix [91]. Copyright 2010. Reproduced with permission from American Chemical Society

Su et al. demonstrated a biosensing platform by combining the advantages of electrospun PVA/chitosan nanofibers and GO [94]. The biosensor was successfully applied for the determination of glucose in human serum samples. The mechanism of the efficient biosensing was analyzed in detail and the results show that it can be due to the synergy effects of electrospun nanofibers and GO nanosheets. Flow-injection amperometric glucose biosensors based on rGO/Nf hybrids was described by Choi et al. [108]. The hybrid biosensors showed the fast response time of ~3 s, the sensitivity of 3.8 μA mM^{-1} cm^{-2}, the limit of detection of 170 μM, and the linear detection range of 2–20 mM for the flow-injection amperometric detection of glucose. Zeng et al. demonstrated an unconventional method for the LbL assembly of graphene multilayer films using PAAc and poly(ethyleneimine) [109]. They demonstrated this concept by combining the graphene multilayer with a bienzyme system of GOx and glucoamylase for the fabrication of a maltose amperometric biosensor for glucose and maltose. Chen et al. developed a colloidal suspension of nanostructured poly(N-butyl benzimidazole)-graphene sheets (PBBIns-Gs) for the development of glucose biosensors [110]. These materials have some advantages including a negative working potential, high sensitivity toward H_2O_2, and non-destructive immobilization. Based on the advantages, the glucose biosensor exhibited a fast response time (5.6 s), broad detection range (10 μM to 10 mM), high sensitivity (143.5 μA mM^{-1} cm^{-2}) and selectivity, and excellent stability. They have also developed a choline biosensor by dipping a PBBIns-Gs/Au electrode.

The cholesterol level in blood is an important parameter in the diagnosis and prevention of life threatening coronary heart diseases, cerebral thrombosis, arthrosclerosis, etc. [62, 63, 97]. Ruecha et al. prepared a nanocomposite of graphene/PVP/PANi and used to modify the electrode of a paper-based biosensor via electrospraying [52]. As shown in Fig. 7, the droplet-like structures of the graphene/PVP/PANi nanocomposite-modified electrodes were used for the sensitive determination of H_2O_2 and cholesterol using amperometry technique.

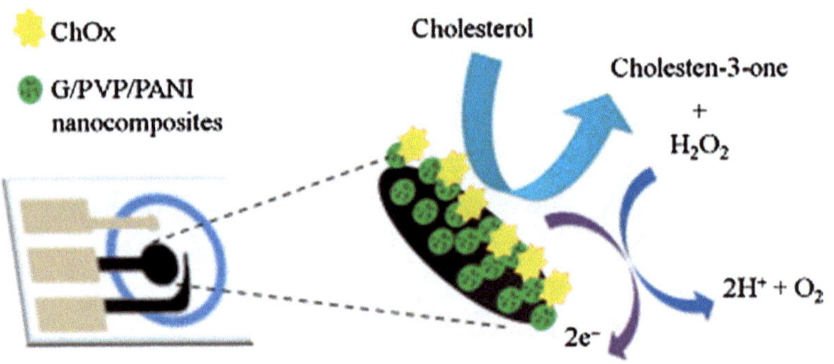

Fig. 7 Scheme represents the enzymatic reaction between cholesterol and ChOx in graphene/PVA/PANi modified electrode [52]. Copyright 2014. Reproduced with permission from Elsevier Ltd.

Hypoxanthine is an essential metabolite to degrade adenine nucleotide mainly accumulated in biological tissues. The determination of hypoxanthine is very essential for the quality control of fish products in food industries [111]. Zhang et al. was designed a biosensing platform for hypoxanthine by the functionalization of rGO sheets with conducting PPy graft copolymer, poly(styrenesulfonic acid-g-pyrrole), via π–π non covalent interaction [81]. The biosensor exhibited a wide linear response ranging from 3.0×10^{-8} to 2.8×10^{-5} M with a high sensitivity of 673 ± 4 μA M^{-1} cm^{-2} and the detection limit of 10 nM. Organophosphates have been commercially used as pesticides, yet their contamination remains a serious public concern for food safety and human health because of their long-term accumulation in the environment. Yang et al. [112] developed a nanohybrid of AuNPs, PPy, and reduced GO sheets (Au–PPy–rGO) by electrochemical deposition of rGO with pyrrole and the introduction of AuNPs. Acetylcholinesterase (AChE) was further encapsulated in a silica matrix and immobilized on the Au–PPy–rGO nanocomposite by co-deposition with $(NH_4)_2SiF_6$. The biosensor led to the rapid and sensitive detection of paraoxon-ethyl from 1.0 nM to 5 mM with a detection limit of 0.5 nM.

Gao et al. developed a sensitive Electrochemiluminescence (ECL) platform for ethanol biosensor based on enzyme and $Ru(bpy)^{3/2+}$ co-immobilized using poly PEDOT and PSS capped graphene (PEDOT-PSS-G) [71]. The biosensor shows enzymatic selectivity with the amplification of PEDOT-PSS-G performed well with a wide linear range, high sensitivity and good stability. The sensing platform also was applied to the determination of amounts of alcohol in real samples. Hu et al. described layer-by-layer assembly of GO sheets/PANi/CdSe quantum dots (GO/PANi/CdSe) nanocomposites layer assembly for the ECL detection of Cyt c [113]. The GO/PANi/CdSe biosensor shows that the ECL intensity decreased linearly with the Cyt c concentrations in the range from 5.0×10^{-8} to 1.0×10^{-4} M with detection limit of 2.0×10^{-8} M.

H_2O_2 is a reactive oxygen species, is connected to several health-disorders such as atherosclerosis, cancer, and Alzheimer's disease [114]. However, H_2O_2 is also a component in the physiological signaling pathways of healthy cells and is indispensable for cell growth, differentiation, migration, and immune system function [115]. Detection of H_2O_2 is a significant challenge for applications of healthcare, food science, pharmaceutical science, and environmental monitoring [116]. Recently, metal nanomaterials, including Ag, Au, Pt, and Pd nanoparticles, have been studied as alternative electrochemical catalysts for nonenzymatic hydrogen peroxide sensors [62, 117–120]. However, rapid response and sensitive methods are required for practical applications.

Feng et al. [89] used graphene/PANi composite film for DET between HRP and the electrode and the enzyme exhibited high bioelectrocatalytic activity toward H_2O_2. A disposable biosensor for determination of H_2O_2 based on Fe_3O_4–Au magnetic nanoparticles coated HRP and graphene sheets–Nf film modified SPCE was fabricated by Xin et al. [60] (Fig. 8).

Under optimized experimental conditions, CV demonstrated that the DET of HRP was realized. The biosensor had an excellent performance in terms of

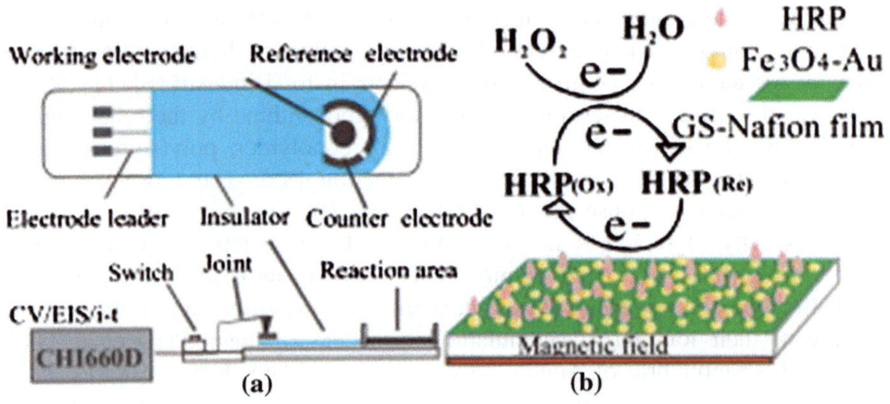

Fig. 8 Schematic diagram of the electrochemical biosensor apparatus (**a**) and the surface of working electrode (**b**) [60]. Copyright 2013. Reproduced with permission from Elsevier Ltd.

electrocatalytic reduction toward H_2O_2. A report by Song et al. [69] describing the entrament of Cyt c in thin films of the RTIL containing nanocomposites of PDDA-graphene nanosheets-gold nanoparticles (PDDA-graphene-AuNPs) at 11-mercaptoundecanoic acid-6-mercapto-1-hexanol modified electrode. They claim that AuNPs/PDDA-graphene nanocomposites could increase the effective surface of the electrode, enhance the fixed amount of Cyt c on the electrode surface, stimulate the electron transfer and facilitate the catalytic activity of Cyt c. Whereas, RTIL could provide a biocompatible microenvironment to keep Cyt c biological activities, act as an effective mediator to immobilize a large number of Cyt c on the electrode and have good conductivity to improve electron transfer. Another report by Feng et al. [121] deals with the AuNPs/PDDA-graphene nanocomposite film exhibited enhanced capability for hemoglobin (Hb) immobilization and realization of its direct electrochemistry. This AuNPs/PDDA-graphene based sensor showed prominent electrocatalytic activity for the detection of H_2O_2 with a wide linear range from 6 to 1,010 μM and a low detection limit of 0.39 μM. Direct electrochemistry of myo globin (Mb) within GO sheets and Nf composite films was fabricated by Guo et al. [122].

They have also showed that Mb in the films displayed good electrocatalytic activities towards various substrates such as hydrogen peroxide, nitrite and oxygen. Park et al. [79] reported a rapid-response and high-sensitivity H_2O_2 biosensor based on a liquid-ion-gated field effect transistor (FET) using graphene-PPy nanotube (rGO-PPy NT) composites as the conductive channel. The rGO-PPy NTs hybrids exhibited enhanced conductivity and surface area that yielded unprecedented sensing performance. The FET sensor provided rapid response (<1 s) in real time and high sensitivity toward H_2O_2 with a limit of detection of 100 pM (Fig. 9). Recently, Wang et al. [76] studied the direct electrochemistry of Cyt c embedded in a novel support matrix of graphene/poly graphene-PEDOT nanocomposite on a GCE. The immobilized Cyt c in G-PEDOT matrix displays admirable

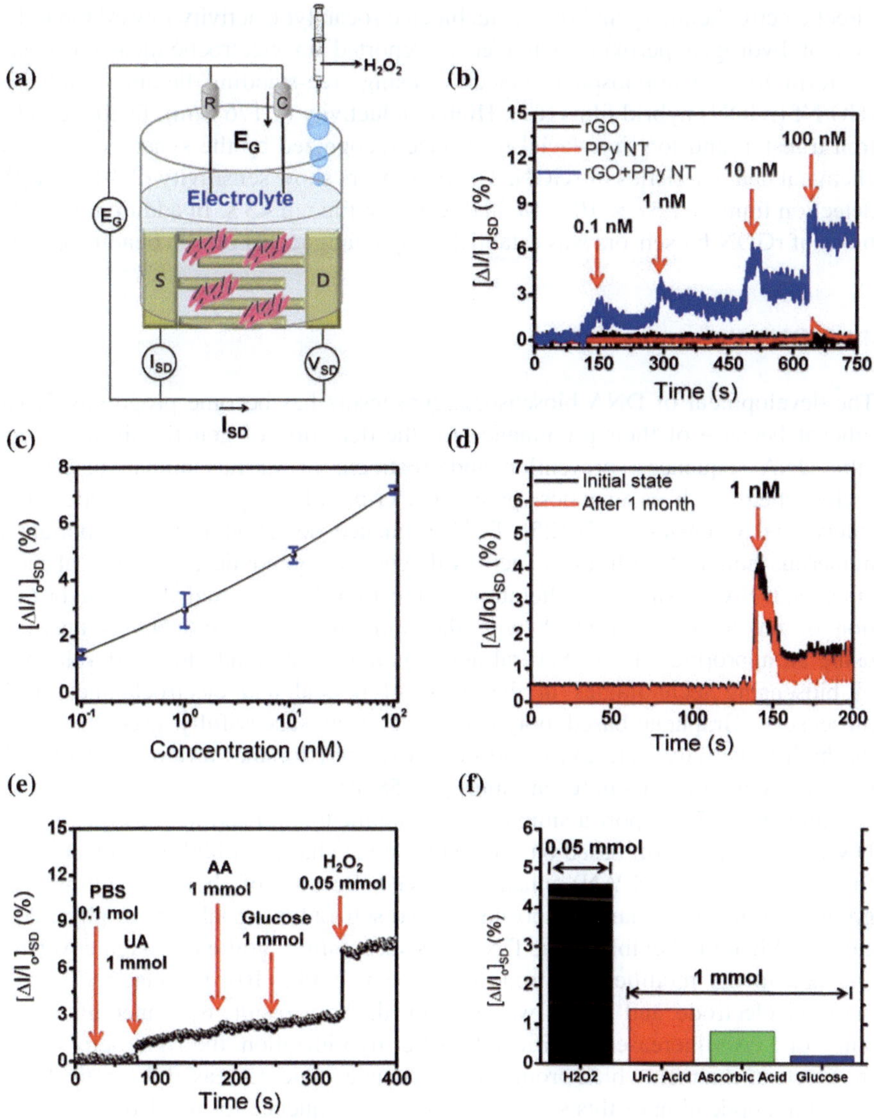

Fig. 9 **a** Schematic diagram shows a liquid-ion-gated FET-type sensor based on rGO/PPy NTs. **b** Real-time responses and **c** a calibration curve for H_2O_2 based on rGO, PPy NTs, and rGO/PPy NTs composites were measured at $V_{sd} = 10$ mV ($V_g = 0.1$ V) with H_2O_2 concentrations of 0.1–100 nM. Storage stability based on rGO/PPy NT nanohybrid sensor performance is shown in part (**d**). Real-time responses to PBS, UA, AA, glucose, and H_2O_2 are shown in part (**e**). **f** A histogram details the sensing performance of the rGO/PPy NTs composites to UA, AA, glucose, and H_2O_2 [79]. Copyright 2014. Reproduced with permission from American Chemical Society

direct electrochemistry and retains its bioelectrocatalytic activity toward the reduction of hydrogen peroxide. Choi et al. reported an electrochemical biosensing platform for organophosphate detection using free-standing flexible conductive rGO/Nf (rGON) hybrid films [61]. High conductivity (1,176 S/m), facilitated electron transfer, and low interfacial resistance recognized by the synergistic electrochemical characteristics of rGON. The biosensors show sensitivity of 10.7 nA/μM, detection limit of 1.37×10^{-7} M, and response time of <3 s. In addition, the reliability of rGON biosensors was established by a fatigue test of 100 bending cycles.

4.1.2 DNA Biosensors

The development of DNA biosensors/genosensors has become progressively significant because of their prominence for the detection of genetic disorders, specific DNA sequences, prevention and treatment of various human diseases in various field such as genetics, pathology, criminology, pharmacogenetics, food safety, and forensics [123–125]. DNA sequence detection is very important in numerous human health areas such as diagnostics, genomics, and clinical medicine. In DNA sensing, the alteration in the transduced signal during hybridization of unknown target DNA with the immobilized probe DNA is measured using an appropriate label. Several attempts have been made to the development of biosensing technologies in the fields of optical and electrochemical DNA biosensors. Graphene-based polymers have been successfully used to achieve the high sensitivity, selectivity and detection limit for the development of DNA biosensors/genosensors in recent times [51, 58, 85].

Yang et al. [51] report a simple and economical approach to construct a direct DNA sensing platform based on self-redox signal change of highly conductive sulfonated polyaniline (SPAN) enhanced by GO. Because of abundant sulfonic acid groups, the resulting nanocomposite showed self-redox signal even at physiological pH. When the flexible probe DNA was successfully grafted through covalently attached to the modified electrode (as shown in Fig. 10), the electron transfer between electrode and buffer was restrained. As a result, the inner impedance value of SPAN increased significantly. After hybridization, the rigid helix opened the electron channel, which prompted impedance value decreased dramatically. As an initial application of this system, fusion gene sequence formed from promyelocytic leukemia and retinoic acid receptor alpha was successfully detected. Bo et al. [85] developed a DNA biosensor based on oxidized graphene and PANi nanowires modified GCE.

The biosensor displayed a fast amperometric response, high sensitivity and good storage stability for the DNA detection. The current response of the sensor increases linearly with the concentration of target from 2.12×10^{-6} to 2.12×10^{-12} mol L^{-1} with a relative coefficient of 0.9938 and the detection limit was 3.25×10^{-13} mol L^{-1}. Du et al. [58] described a DNA electrochemical biosensor for the detection of specific gene sequences by electrochemically reduced GO prepared on PANi nanofibers modified GCE. The hybridization of

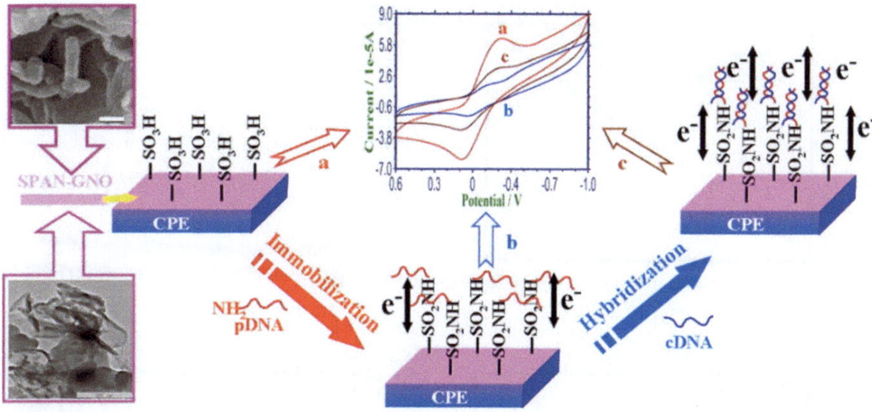

Fig. 10 Schematic diagram of the DNA detection on different electrode modifications [51]. Copyright 2013. Reproduced with permission from American Chemical Society

ssDNA probe with complementary DNA was monitored by the change in the response of surface-bound $[Ru(NH_3)_6]^{3+}$. The dynamic range of the DNA biosensor for detecting the sequence-specific DNA of cauliflower mosaic virus gene was from 1.0×10^{-13} to 1.0×10^{-7} mol L^{-1}, with a detection limit of 3.2×10^{-14} mol L^{-1}.

4.1.3 Aptasensors

Angiogenesis, the formation of new blood vessels developing from the preexisting vasculature, has been well-thought-out at the molecular mechanism level because it is an important factor associated to the metastasis and growth of human tumors [80, 127]. There are various proangiogenic factors such as platelet-derived growth factor (PDGF), basic fibroblast growth factor (bFGF) and vascular endothelial growth factor (VEGF) are well documented in the literature [127, 128]. However, all of these procedures (or recognition processes) still have need of rapid detection and high sensitivity for further improvement.

Kwon et al. [80] developed a flexible FET-type aptasensor based on PPy converted nitrogen-doped few-layer graphene grown on Cu substrate by chemical vapor deposition combined with vapor deposition polymerization to detect VEGF as a cancer biomarker. Field induced high sensitivity was observed for the analyte-binding events, ultimately leading to the recognition of the target molecules at very low concentration of 100 fM. The aptasensor also exhibited excellent reusability, mechanical bendability, and durability in the flexible process (Fig. 11). The developed methodology described the fabrication of N-doped graphene using conducting polymers including heteroatoms in their structures as the carbonization precursor and demonstrates its use in a high-performance aptasensors.

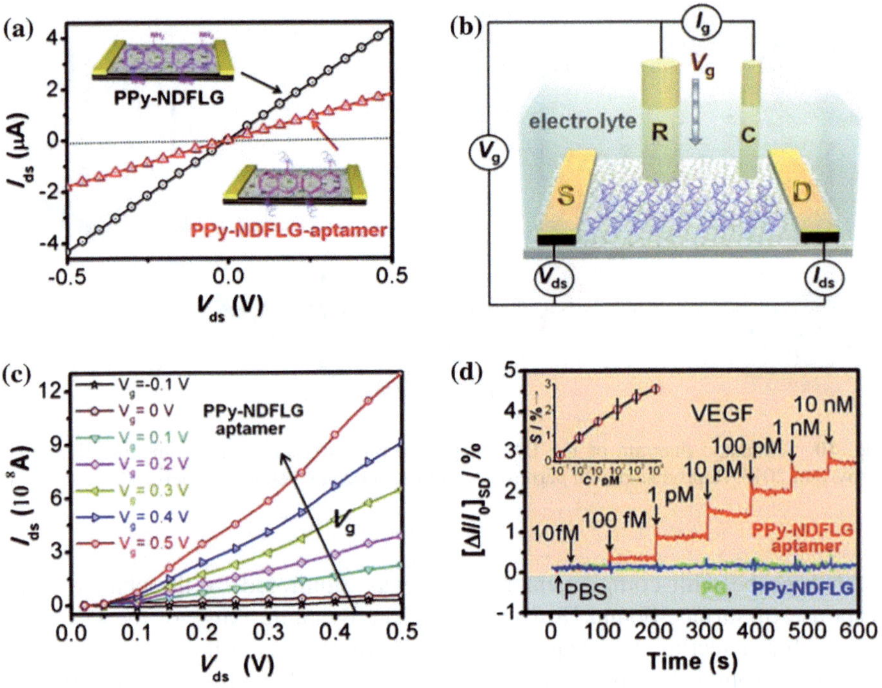

Fig. 11 **a** Current–voltage curves of PPy-NDFLG on the polyethelene naphthalate film before and after aptamer immobilization in air ($V_{ds} = -0.5$ to $+0.5$ V and scan rate was 10 mV s^{-1}). **b** Schematic diagram of a liquid-ion gated FET using aptamer-conjugated PPy-NDFLG. **c** I_{ds}–V_{ds} output characteristics of PPy-NDFLG-aptamer at different V_g from −0.1 to 0.5 V in a step of 0.1 V in phosphate-buffered solution (V_{ds}, 0 to 0.5 V in a step of 50 mV). **d** Real-time responses and a calibration curve (S in the inset indicates ΔI/I0) of aptasensor with various vascular endothelial growth factor concentrations [80]. Copyright 2012. Reproduced with permission from American Chemical Society

Wen et al. [92] demonstrated a unique graphene/aptamers interaction for the construction of electrochemical aptasensors by immobilizing graphene/aptamers complex on the surface of poly(o-phenylenediamine) modified GCE. Adenosine triphosphate (ATP) was chosen as a model analyte and a well-studied 27-mer single-stranded DNA aptamer was tailor-made to have methylene-blue at both ends for use as an electrochemical probe. The biosensors detected ATP down to a detection limit of 0.3 nM with a wide detection range of 10 nM–2 mM and were successfully applied in the field of biological analysis of real samples. Deng et al. [126] constructed a label-free electrochemical aptasensor based on PDDA-protected graphene-AuNPs composite for detection of PDGF based on the direct electrochemistry of GOx. With the DET of double layer GOx membranes, the aptasensor displayed admirable electrochemical response and the peak current decreased linearly with increasing logarithm of PDGF concentration from 0.005 to 60 nM with low detection limit of 1.7 pM. The aptasensor exhibited high specificity, good reproducibility and long-term stability, which provided a new encouraging technique for aptamer-based protein detection. In another report, Zhang et al. [74] developed an aptasensor based on an amino-functionalized nanocomposite of graphene and plasma-polymerized allylamine (PPAA) for thrombin detection. The amounts and kinetics of aptamer immobilization and thrombin detection were recorded using QCM measurements. The aptasensor hold high affinity constant K_a of aptamer bonding on the graphene–PPAA film with the highest saturated adsorbed aptamer amount of 956.76 ng. The limit of thrombin detection was of approximately 5.57 nM.

4.1.4 Immunosensors

The favorable nanolabels for the electrochemical immunoassay generally contain the following features, e.g., high specific surface area, good electron transfer ability and good biocompatibility [129]. The unique properties of graphene based polymer composites make them best fit to fulfill the criteria to the fabrication of sensing platforms for cancer biomarkers. The precise complexation of an antigen with an antibody (Ab) is the key principle in arrears of immunosensing technology. The large surface area of nanocomposites enables the immobilization of sufficient amount of Ab and hence enhancement of the sensitivity can be realized.

GO nanosheets/PANi nanowires/CdSe quantum dots (GO/PANi/CdSe) nanocomposites are successfully used as ECL immunosensor for detection of human-interleukin-6 (IL-6) by Liu et al. [54]. GO/PANi nanocomposites were used to enhance the ECL of CdSe quantum dots. The ECL immunosensor had a sensitive response to IL-6 in a linear range of 0.0005–10 ng mL^{-1} with a detection limit of 0.17 pg mL^{-1}. In another report, Li et al. developed an estradiol immunosensor based on graphene–PANi composites and carboxylated GO GO [56]. The current response of the immunosensor was remarkably improved due to the synergistic effects of GR and PANi. The immunosensor showed a wide linear response to estradiol in the range 0.04–7.00 ng mL^{-1} and a limit of detection of 0.02 ng mL^{-1} ($S/N = 3$) and was y applied to the detection of estradiol in real samples.

Chen et al. designed a electrochemical immunosensing protocol for detection of carcinobryonic antigen (CEA) by using graphene-carried poly(o-phenylenediamine)/Au hybrid nanosheets as signal tags on the hierarchical dendritic Au microstructures modified GCE [78]. The electrochemical immunoassay exhibited a wide dynamic range of 0.005–80 ng mL^{-1} toward CEA standards with a low detection limit of 5.0 pg mL^{-1}. Liu et al. [130] developed fabricated an electrochemical immunosensors by combining PDDA functionalized graphene nanosheets and AuNPs for the detection of human IgG. The immunosensor displayed excellent analytical performance for the detection of human IgG range from 0.1 to 200 ng mL^{-1} with a detection limit of 0.05 ng mL^{-1}. Wu et al. [131] described a signal amplification strategy for ultrasensitive multiplexed detection of cancer biomarkers using a paper-based microfluidic electrochemical immunodevice based on GO-chitosan polymer matrix. Glycidylmethacrylate (GMA) was used as the monomers to provide epoxy groups for immobilization of electrochemical tags HRP. In this work, they have detected four cancer biomarkers, namely, CEA, alpha-fetoprotein (AFP), cancer antigen 125, and carbohydrate antigen 153, as model analytes. These biomarkers were detected using the HRP-O-phenylenediamine-H_2O_2 electrochemical system under optimized environment.

4.2 Optical Biosensors

In recent times, optical/colorimetric biosensors have attracted copious attention because of its simplicity, low cost, and sensibleness. Since color changes can be read by the naked eye, colorimetric biosensors does not require expensive or sophisticated instrumentation and can be applied to field investigation and point-of-care diagnosis [132, 133]. The basic strategy of colorimetric biosensors is converting the detection events into color changes. There are many smart material including graphene-based polymer nanomaterial are used into this events. It is worth mentioned here that there is only handful report published that described the optical biosensor based on graphene and polymer nanocomposites. This field is still open for the young researchers.

In a report, Lim et al. [134] described a GO-based immunosensing system for the detection of interleukin 5 (IL-5) through the quenching of intrinsic GO fluorescence by peroxidase catalyzed polymerization of 3,3′-diaminobenzidine (DAB) naccording to Fig. 12. The fluorescence of GO-based immunosensor diminished by approximately 20 % with 4 ng mL^{-1} IL-5 and the limit of detection of IL-5

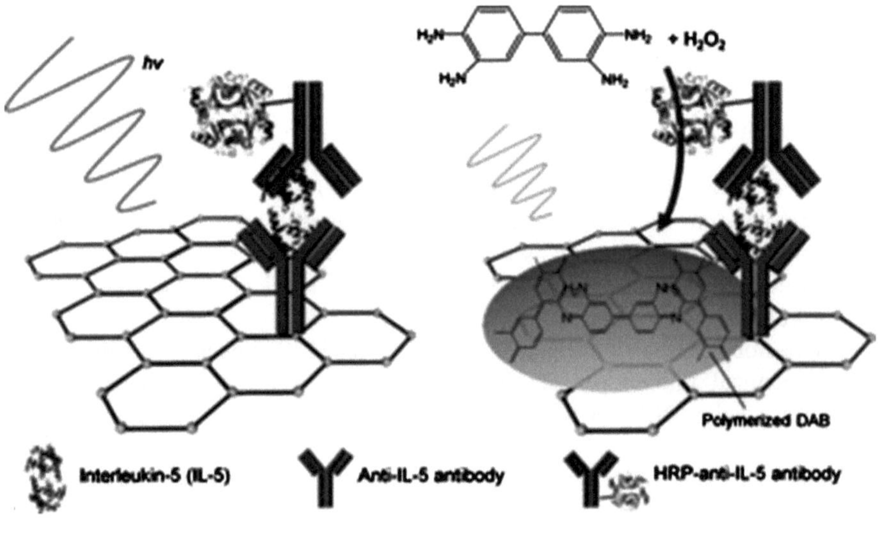

Fig. 12 Scheme represents GO-based immunoassay system employing the fluorescence-quenching of GO by a peroxidase-catalyzed polymerization of DAB [134] Copyright 2012. Reproduced with permission from John Wiley & Sons

was estimated to be approximately 4 pg mL^{-1}. They also reported the lowest level of IL-5 detected in human serum solution was 10 pg mL^{-1}, which is close to the detection limit in phosphate buffered salien, and the degree of fluorescence quenching reached about 50 % at 100 ng mL^{-1} IL-5. Zhao et al. [135] developed a biosensor by exploiting localized plasmons in graphene and biomolecule adsorption on it. The transparent substrate supporting graphene was chosen potentially from a wide range of materials including insulators, semiconductors, polymers, and gels. They have shown that the interaction between sensing medium (containing biomolecules) and incident optical fields is strong. High sensitivity values of up to 1,697 nm/RIU were achieved when the wavelength shift in the plasmon resonance is detected. Moreover, they have also demonstrated that the device can work in a wide-angle range of incident light or on rough substrates. Such a biosensor also works in a reflection structure by using the shift of nearly perfect absorption spectra.

5 Summary, Conclusions, Outlook

This chapter pointed a detailed overview of various type of biosensors based on graphene and polymer nanocomposites. The synthesis and strategy of the material used for the development of such biosensors are demonstrated. The combination of the unique properties of polymer and graphene creats a new material in material chemistry which have enormous application in the field of next-generation sensor devices. Various attempts are being made to further improve and enhance the properties of the graphene-based polymer composites for the development of biosensors. The advantage of graphene-based polymer composite to the development of biosensor are the dispersibility of the material, improve properties of the hybrid material, and analytical performance of the sensing interface. Graphene is a promising nanofiller candidate in the composites materials among the other carbon-based materials, which is likely to replace in the near future. Composite industries are looking for more and more graphene in their product for large scale production and commercialization. The graphene-based polymer composite or hybrid materials can offer a solution for the existing analytical challenges in the fabrication of biosensing devices. There is still some challenges for the production and commercialization of biosensor, this material may be engineered combine with some other nanomaterial for the one-step close to make the biosensing devices.

References

1. Wenrong Y, Kyle RR, Simon PR, Pall T, Gooding JJ, Filip B (2010) Angew Chem Int Ed 49: 2114
2. Potts JR, Dreyer DR, Bielawski CW, Ruoff RS (2011) Polymer 52: 5
3. Menaa F (2013) J Mol Imaging Dynam 2:e103. doi: 10.4172/2155-9937.1000e103

4. Menaa F (2013) Pharmaceut Anal Acta. 5:e161. doi: 10.4172/2153-2435.1000e161
5. Shan C, Yang H, Song J, Han D, Ivaska A, Niu L (2009) Anal Chem 81: 2378
6. Kang X, Wang J, Wu H, Aksay I A, Liu J, Lin Y (2009) Biosens Bioelectron 25: 901
7. Si Y, Samulski ET (2008) Chem Mater 20: 6792
8. Novoselov KS, Geim AK, Morozov SV, Jiang D, Zhang Y, Dubonos SV, Grigorieva IV, Firsov AA (2004) Science 306: 666
9. Geim AK, Novoselov KS (2007) Nat Mater 6: 183
10. Rao CNR, Sood AK, Subrahmanyam KS, Govindaraj A (2009) Angew Chem Int Ed 48: 7752
11. Dey RS, Hajra S, Sahu RK, Raj CR, Panigrahi MK (2012) Chem Commun 48: 1787
12. Menaa F, Abdelghani A, Menaa B, (2014) J Tissue Eng Regen Med. doi: 10.1002/term.1910
13. Geim AK, Novoselov KS (2007) Nat Mater 6:183
14. Steurer P, Wissert R, Thomann R, Mülhaupt R (2009) Macromol Rapid Commun 30:316
15. Ramanathan T, Abdala AA, Stankovich S, Dikin DA, Herrera-Alonso M, Piner RD, Adamson DH, Schniepp HC, Chen X, Ruoff RS, Nguyen ST, Aksay IA, Prud'homme RK, Brinson LC (2008) Nat Nanotechnol 3: 327
16. Stankovich S, Dikin DA, Dommett GHB, Kohlhaas KM, Zimney EJ, Stach EA, et al. (2006) Nature 442: 282
17. Ramanathan T, Abdala AA, Stankovich S, Dikin DA, Alonso, MH, Piner RD, et al. (2008) Nat Nanotechnol 3: 327
18. Lee YR, Raghu AV, Jeong HM, Kim BK (2009) Macromol Chem Phys 210: 1247
19. Xu Y, Wang Y, Jiajie L, Huang Y, Ma Y, Wan X, et al. (2009) Nano Res 2: 343
20. Quan H, Zhang B, Zhao Q, Yuen RKK, Li RKY (2009) Compos Pt A 40: 1506
21. Liang J, Huang Y, Zhang L, Wang Y, Ma Y, Guo T, et al. (2009) Adv Funct Mater 19: 2297
22. Balog R, Jørgensen B, Nilsson L, Andersen M, Rienks E, Bianchi M, et al. (2010) Nat Mater 9: 315
23. Ventura G, Martelli V (2009) Cryogenics 49: 735
24. Woo MW, Wong P, Tang Y, Triacca V, Gloor PE, Hrymak AN, et al. (1995) Polym Eng Sci 35: 151
25. Yuen SM, Ma CCM, Chiang CL, Chang JA, Huang SW, Chen SC, et al. (2007) Compos Pt A 38: 2527
26. Stankovich S, Dikin DA, Dommett GHB, Kohlhaas KM, Zimney EJ, Stach, EA, et al. (2006) Nature 442: 282
27. Quan H, Zhang B, Zhao Q, Yuen RKK, Li RKY (2009) Compos Pt A 40: 1506
28. Hong WJ, Xu YX, Lu GW, Li C, Shi GQ (2008) Electrochem Commun 10: 1555
29. Yao J, Shen XP, Wang B, Liu HK, Wang GX (2009) Electrochem Commun 11:1849
30. Choi DW, Wang DH, Viswanathan VV, Bae IT, Wang W, Nie ZM, Zhang JG, Graff GL, Liu J, Yang ZG, Duong T (2010) Electrochem Commun 12: 378
31. Wang HL, Hao QL, Yang, X J, Lu LD, Wang, X (2009) Electrochem Commun 11: 1158
32. Wu Q, Xu YX, Yao ZY, Liu AR, Shi GQ (2010) ACS Nano 4: 1963
33. Wu JF, Xu MQ, Zhao GC (2010) Electrochem Commun 12: 175
34. Hong WJ, Bai H, Xu YX, Yao ZY, Gu ZZ, Shi GQ (2010) J Phys Chem C 114: 1822
35. Moniruzzaman M, Winey KI, (2006) Macromolecules 39:5194
36. Dreyer DR, Park S, Bielawski CW, Ruoff RS (2010) Chem Soc Rev 39:228
37. Suk JW, Piner RD, An J, Ruoff RS (2010) ACS Nano 4:6557 doi: 10.1021/nnl101781v
38. Gómez-Navarro C, Burghard M, Kern K (2008) Nano Lett 8:2045
39. Bunnell LR (1993) Battelle Memorial Institute 5186919
40. Chen G, Zhao W, editors (2009) Nano- and biocomposites CRCPress, 79
41. Zheng W, Wong S-C (2003) Compos Sci Technol 63:225
42. Thévenot DR, Toth K, Durst RA, Wilson GS, Pure Appl. Chem. 71 (1999) 2333
43. Novoselov KS, Geim AK, Morozov SV, Jiang D, Zhang Y, Dubonos SV, Grigorieva IV, Firsov AA (2004) Science 306:666
44. Pumera M (2009) Chem Rec 9:211
45. Wu S, He Q, Tan C, Wang Y, Zhang H (2013) Small 9:1160

46. Liu Y, Dong X, Chen P (2012) Chem Soc Rev. 41:2283–2307
47. Fan Y, Liu J, Yang C, Yu M, Liu P (2011) Sens. Actuators B Chem 157:669
48. Hummers W, Offeman R (1958) J. Am. Chem. Soc. 80: 1339
49. Cristina V, Pablo J, Edgar, Ana M B, Wolfgang KM (2011) J Phys Chem C, 115: 10468
50. Chen G-L, Shau S-M, Juang T-Y, Lee R-H, Chen C-P, Suen S-Y, Jeng, R-J, (2011) Langmuir, 27: 14563
51. Yang T, Meng L, Wang X, Wang L, Jiao K, (2013) ACS Appl Mater Interfaces, 5: 10889
52. Nipapan R, Rangkupan R, Rodthongkum N, Chailapakul O (2014) Biosens. Bioelectron.52:13
53. Jiménez P, Castell P, Sainz R, Ansón A, Martínez MT, Benito AM, Maser WK (2010) J Phys Chem B 114: 1579
54. Liu P-Z, Hu X-W, Mao C-J, Niu H-L, Song J-M, Jin B-K, Zhang S-Y, (2013) Electrochim. Acta 113: 176
55. Li J, Liua S, Yu J, Lian W, Cui M, Xu W, Huang J, (2013) Sens. Actuators B 188: 99
56. Zhang K, Zhang LL, Zhao XS, Wu J, (2010) Chem. Mat. 22: 1392
57. Xu Q, Gu S-X, Jin L, Zhou Y, Yang Z, Wang W, Hu X, (2014) Sens Actuators B 190: 562
58. Du M, Yang T, Li X, Jiao K, (2012) Talanta 88: 439
59. Choi BG, Park HS, Park TJ, Yang MH, Kim JS, Jang SY, Heo NS, Lee SY, Kong J, Hong WH, (2010) ACS Nano 4: 2910
60. Xin Y, Fu-bing X, Hong-wei L, Feng W, Di-zhao C, Zhao-yang W, (2013) Electrochim. Acta 109: 750
61. Liu Y, Wang G, Li C, Zhou Q, Wang M, Yang L, (2014) Mat. Sci. Eng. C 35 253
62. Dey RS, Raj CR (2010) J Phy. Chem. C 114: 21427
63. Dey RS, Raj CR (2013) ACS Appl. Mater. Interfaces5: 4791
64. Dey RS, Raj CR (2013) RSC Adv. 3: 25858
65. Yang L, Wang G, Liu Y (2013) Anal. Biochem. 437: 144
66. Ni Y, Wang P, Song H, Lin X, Kokot S, (2014) Anal Chim Acta 821: 34
67. Hui J, Cui J, Xu G, Adeloju SB, Wu Y, (2013) Mater. Lett.108: 88
68. Zhang S, Shao Y, Liao H, Engelhard MH, Yin G, Lin Y (2011) ACS Nano, 5: 1785
69. Song Y, Liu H, Wan L, Wang Y, Hou H, Wang L, (2013) Electroanalysis, 25: 1400
70. Liu K, Zhang J, Yang G, Wang C, Zhu J-J, (2010) Electrochem. Commun. 12: 402–405
71. Yu Y, Chen Z, He S, Zhang B, Li X, Yao M (2014) Biosens. Bioelectron. 52: 147
72. Ma J, Cai P, Qi W, Kong D, Wang H, (2013) Colloids Surfaces A: Physicochem Eng Aspects 426: 6
73. Park S, Dikin DA, Nguyen SBT, Ruoff RS, (2009) J Phys Chem C, lett, 113:15801
74. Zhang Z, Liu S, Shi Y, Zhang Y, Peacock D, Yan F, Wang P, He L, Feng X, Fang S (2014) J Mater Chem B, 2: 1530
75. Gao Y, Li J, Yang X, Xiang Q, Wang K, (2014) Electroanalysis, 26: 382.
76. Wang G-X, Qian Y, Cao X-X, Xia X-H, (2012) Electrochem Commun 20:1
77. Xu H, Suslick KS, (2011) J Am Chem Soc, 133: 9148
78. Chen H, Gao Z, Cui Y, Chen G, Tang D, (2013) Biosens. Bioelectron. 44: 108
79. Park JW, Park SJ, Kwon OS, Lee C, Jang J, (2014) Anal Chem, 86: 1822.
80. Kwon OS, Park SJ, Hong J-Y, Han A-R, Lee JS, Lee JS, Oh JH, Jang J, (2012) ACS Nano, 6: 1486
81. Zhang J, Lei J, Pan R, Xue Y, Ju H, (2010) Biosens. Bioelectron. 26: 371
82. Tourillon G (1986) In Handbook of Conducting Polymers, Skotheim, TA, Ed, Marcel Dekker: New York, Vol 1, pp 293-350
83. Chandler GK, Pletcher D (1985) In Electrochemistry, The Royal Society of Chemistry: London, Vol 10, pp 117-150
84. Wei Y, Chan C-C, Tian J, Jang G-W, Hsueh KF (1991) Chem. Mater. 3: 888
85. Bo Y, Yang H, Hu Y, Yao T, Huang S, (2011) Electrochim. Acta 56: 2676
86. Liang L, Liu J, Windisch C, Exarhos G, Lin Y, (2002) Angew Chem Int Ed 41: 3665
87. De*via* R, Relhan S, Pundir CS, (2013) Sens. Actuators B 186: 17

88. Radhapyari K, Kotoky P, Das MR, Khan R, (2013) Talanta 111:47.
89. Feng X-M, Li R-M, Ma Y-W, Chen R-F, Shi N-E, Fan Q-L, Huang W, (2011) Adv Funct Mater.21: 2989
90. Yang Y, Asiri AM, Du D, Lin Y, (2014) Analyst 13: 3055
91. Alwarappan S, Liu C, Kumar A, Li C-Z, (2010) J Phys Chem C, 114:12920
92. Wen W, Bao T, Yang J, Zhang M-Z, Chen W, Xiong H-Y, Zhang X-H, Zhao Y-D, Wang S-F, (2014) Sens. Actuators B 191: 695
93. Wisitsoraat A, Pakapongpan S, Sriprachuabwong C, Phokharatkul D, Sritongkham P, Lomas T, Tuantranont A, (2013) J. Electroanal.Chem. 704: 208
94. Su X, Ren J, Meng X, Rena X, Tang F (2013) Analyst 138: 1459
95. Das TK, Prusty S (2013) Polymer-Plastics Technol. Eng. 52: 319
96. Lei W, Si W, Xu Y, Gu Z, Hao Q, (2014) Microchim Acta 181:7072
97. Dey RS, Bera RK, Raj CR (2013) Anal Bioanal Chem 405:3431
98. Ray PC, (2010) Chem. Rev. 110: 5332
99. Gooding JJ, Wibowo R, Liu JQ, Yang WR, Losic D, Orbons S, Mearns FJ, Shapter JG, Hibbert DB (2003) J Am Chem Soc 125:9006
100. Pulcu GS, Elmore BL, Arciero DM, Hooper AB, Elliott SJ (2007) J Am Chem Soc 129:1838
101. Wang JX, Li MX, Shi ZJ, Li NQ, Gu ZN (2002) Anal. Chem. 74:1993
102. Wang J (2008) Chem. Rev. 108:814
103. Heller A, Feldman B (2010) Acc. Chem. Res. 43:963
104. Kong F-Y, Gu S-X, Li W-W, Chen T-T, Xu Q, Wang W, (2014) Biosens. Bioelectron. 56: 77
105. Nguyen HB, Nguyen VC, Nguyen VT, Ngo TTT, Nguyen NT, Dang TT H, Tran DL, Do PQ, Nguyen XN, Nguyen XP, Phan HK, Phan NM, (2012) Adv Nat Sci: Nanosci Nanotechnol 3:025011
106. Jia L, Liu J, Wang H (2013) Electrochim. Acta 111: 411
107. Song Y, Liu H, Wang Y, Wang L, (2013) Electrochim. Acta 93: 17
108. Choi BG, Im J, Kim HS, Park HS, (2011) Electrochim. Acta 56: 9721
109. Zeng G, Xing Y, Gao J, Wang Z, Zhang X (2010) Langmuir, 26: 15022
110. Chen H-C, Tsaid R-Y, Chen Y-H, Lee R-S, Hua M-Y (2013) Anal. Chim. Acta 792:101
111. Causse E, Pradelles A, Dirat B, Negre-Salvayre A, Salvayre R, Couderc F, (2007) Electrophoresis 28: 381
112. Yang Y, Asiri AM, Du D, Lin Y (2014) Analyst, 139: 3055
113. Hu X-W, Mao C-J, Song J-M, Niu H-L, Zhang S-Y, Huang H-p, (2013) Biosens. Bioelectron. 41: 372
114. Ohshima H, Tatemichi M, Sawa T (2003) Arch Biochem Biophys 417: 3
115. Foreman J, Demidchik V, Bothwell J H F, Mylona P, Miedema H, Torres M A, Linstead P, Costa S, Brownlee C, Jones J D G, Davies J M, Dolan L (2003) Nature 422: 442
116. Vione D, Maurino V, Minero C, Borghesi D, LucchiariM, Pelizzetti E (2003) Environ Sci Technol, 37: 4635
117. Bai J, Jiang X (2013) Anal Chem, 85: 8095
118. Kafi AKM, Ahmadalinezhad A, Wang J, Thomas DF, Chen A (2010) Biosens Bioelectron, 25: 2458
119. Sun X, Guo S, Liu Y, Sun S (2012) Nano Lett, 12: 4859
120. Wang Y, Yang X, Bai J, Jiang X, Fan G (2013) Biosens Bioelectron 43: 180
121. Feng Q, Liu K, Fu J, Zhang Y, Zheng Z, Wang C, Du Y, Ye W, (2012) Electrochim. Acta 60: 304
122. Guo C, Sun H, Zhao XS, (2012) Sens. Actuators B 164: 82
123. Odenthal KJ, Gooding JJ (2007) Analyst 132:603
124. Wang J (2002) Anal Chim Acta 469:63
125. Sassolas A, Leca-Bouvier BD, Blum LJ (2008) Chem. Rev. 108:109
126. Deng K, Xiang Y, Zhang L, Chen Q, Fu W, (2013) Anal. Chim. Acta 759: 61
127. Loureiro R, D'Amore, P (2005) Cytokine Growth Factor Rev. 16: 77

128. Foekens JA, Peters HA, Grebenchtchikov N, Look MP, Gelder ME M-V, G-Moespot A, Kwast THVD, Sweep CGJ, Klijn JGM (2001) Cancer Res. 15: 5407
129. Saha K, Agasti S, Kim C, Li X, Rotello V (2012) Chem. Rev. 112: 2739
130. Liu K, Zhang J-J, Wang C, Zhu J-J (2011) Biosens. Bioelectron. 26: 3627
131. Wu Y, Xue P, Hui KM, Yuejun K, (2014) Biosens. Bioelectron. 52: 180
132. Ho HA, Najari A, Leclerc M, (2008) Acc Chem Res 41: 168
133. Song Y, Wei W, Qu X (2011) Adv. Mater. 23: 4215
134. Lim SY, Ahn J, Lee JS, Kim M-G, Park CB, (2012) small, 8: 1994
135. Zhao Y, Hu X, Chen G, Zhang X, Tan Z, Chen J, Ruoff RS, Zhu Y, Lu Y (2013) Phys Chem Chem Phys 15: 17118

Graphene/Polymer Nanocomposites as Microwave Absorbers

Vadali V.S.S. Srikanth and K.C. James Raju

Abstract A major application identified for graphene/polymer nanocomposites is as electromagnetic (EM) wave absorbers in high frequency electronics which is the backbone of present day communication systems. In this application area, thin and flexible absorbers are essential for ensuring electromagnetic interference (EMI)/EM compatibility standards. Presently, communication modes are primarily mobile in nature and inherently light weight and small in size. In this context, there is a great demand for high performance novel absorbing materials that can offer required solutions. The properties of graphene-filled polymer nanocomposites clearly make them outstanding candidates for microwave absorption. Graphene as a filler is quite unique as it offers the highest surface-to-volume ratio and hence once it is incorporated inside a polymer matrix it offers increased conductive and dielectric loss without a large increase in impedance mismatch. It is possible to disperse graphene in some polymers uniformly and hence their large surface-to-volume ratio becomes advantageous. Once they are well dispersed in the host, the composite can be imagined as a kind of distributed capacitors combining in series and parallel resulting in reduced capacitance but increased dissipation, yielding impedance-matched absorber. Graphene can be functionalized with various functional groups giving an additional degree of freedom to fine-tune its properties. This in turn increases the flexibility in designing novel graphene-based materials. For an absorber, not only its EM response but its mechanical, adhesive, and weatherability characteristics are also important. Since meeting the EM absorption requirement over a range of frequencies by a single material is difficult, the possibility of functionalization of graphene opens up many opportunities and hence graphene/polymer nanocomposites open up scope for a wide

V.V.S.S. Srikanth (✉)
School of Engineering Sciences and Technology (SEST), University of Hyderabad, Hyderabad, Telangana 500 046, India
e-mail: vvsssse@uohyd.ernet.in

K.C.J. Raju (✉)
Centre for Advanced Studies in Electronics Science and Technology, School of Physics, University of Hyderabad, Hyderabad, Telangana 500 046, India
e-mail: kcjrsp@yahoo.com

K.K. Sadasivuni et al. (eds.), *Graphene-Based Polymer Nanocomposites in Electronics*, Springer Series on Polymer and Composite Materials,
DOI 10.1007/978-3-319-13875-6_12

spectrum of combinatorial investigations that are able to give solutions for the emerging scenario where in the usage of microwave spectrum is becoming more widespread, rather than not merely confined to the strategic sector as it used to be.

Keywords Microwave · Shielding · Graphene · Absorbers

1 Microwave Absorbers and Electromagnetic Shielding Materials

1.1 General Background

The need for absorbers of EM radiation was recognized from the time this radiation began to be used in mid 1930s. First known application is a simple resonant absorber used behind a 2 GHz antenna to improve its front-to-back ratio. As their usage proliferated so is the need to remove them from regions where they are deemed unnecessary. The exploration that followed brought out solutions based on materials and design, often requiring a combination of both. The primary material's parameters that help in electromagnetic absorption are the imaginary parts of complex permittivity ($\varepsilon^* = \varepsilon' - j\varepsilon''$), complex permeability ($\mu^* = \mu' - j\mu''$) and complex conductivity ($\sigma^* = \sigma' - j\sigma''$). No material exists that exhibits high values for all these parameters at all frequencies. Hence, there is a need for composites that meet the required set of characteristics. When it comes to applications, apart from the above-mentioned primary characteristics, a number of key material parameters have to be considered depending upon the specific application. A few important requirements that generally need to be met in addition to the EM absorption characteristics are physical strength, weather resistance, light weight, flexibility and resilience, minimum thickness, wide operational bandwidth, low reflectivity, higher power handling capability, stable properties over wide temperature range, low outgassing and more importantly all of these at low costs. Most of the applications involve the following: reduction of undesirable reflections from objects and devices, reduction of radar cross section of target objects (like aircraft, ship, tanks, vehicles, buildings, etc.,), making dummy loads for testing purposes, making test chambers where outside radiation and undesirable radiation are absorbed, and making shielding enclosures to reduce EMI.

1.2 Principles of Absorber Function

A. Material Systems Having Equal Complex Magnetic Permeability and Complex Dielectric Permittivity

If ε^* and μ^* of a medium are equal, there would not be any interface reflection of EM wave (at normal incidence) as such and the medium would offer the same impedance as that of free space. Then depending on medium's thickness and absorption characteristics, the EM wave will be dissipated as it travels through the medium. Only ferrites give similar complex dielectric permittivity and complex magnetic permeability albeit over a limited frequency range. However, the real and imaginary parts of permittivity and permeability of even ferrites are highly dependent on the incident EM wave's frequency and hence to utilize ferrites for absorbing, extensive composition modifications are required. At present, only a limited range of compositions with a thickness of only about 6 mm are available for reflection reductions of 10–25 dB in 30 MHz–1 GHz range. Ferrites are inorganic ceramics that are stable at even high temperatures, are incombustible and possess required outgassing properties. They can dissipate power levels greater than 3 W/cm^2. Ferrites are used in the pure form to make anechoic chambers and absorbing antenna caps that can work up to 1 GHz.

B. Resonant Absorbers Having a Quarter Wavelength Thickness Operating Over a Narrow Frequency Range

For most of the materials, the above-mentioned condition of nearly equal values of permeability and permittivity does not exist for most of the frequencies. Permeability for most of the materials will be too low compared to permittivity. In such situations, the impedance mismatch between free space and the material is so high that the energy will be reflected back from the material. There are composites that contain metal nanoparticles which exhibit dielectric constant in the range of 10^5 and loss tangent of 10^2. However, such high values of loss tangent are not useful because of the reason that the resulting impedance mismatch will make most of the energy reflected back. Therefore, it is necessary to look for ways by which an extended interaction between the material and the incident EM energy is possible. Achieving this over short range of frequencies is relatively easy by using "Resonant" absorbers. In this approach, the high interface reflection is canceled by another reflection from the back surface. Interface reflection will take place if the material is backed by either a metal or by free space. Now the condition for destructive interference is that the path length should be an integral multiple of half of the wavelength in the medium. That means the material/medium thickness should be a quarter of wavelength. By adjusting the thickness and complex magnetic permeability and permittivity of the medium, a condition of low reflection is achieved at the resonant frequency for angles close to normal incidence. The condition is slightly different for other angles of incidence. If magnetic permeability is nonzero, a wider band width would be used. Then absorption would take place over the full thickness of the material but over a limited range of frequencies.

It is possible to achieve an absorber function by making use of a solution from transmission line theory. A quarter wavelength ahead of a short (conductive surface), the line acts as open and if a resistive sheet with sheet resistance of free space (377 Ohm per square) is kept there, full dissipation of energy will take place in the resistive sheet at the frequency where the quarter wave condition is matched. This is called as "Salisbury Screen." Composite flexible materials can be used to make such screens and depending on the amount of resistive material added to them, their sheet resistance can be varied.

If the bandwidth of an application is narrow, a resonant absorber will offer a thin absorber. They are generally used as flexible absorbers over metallic surfaces to absorb power from selected regions at narrow frequency ranges like in parts of antennas or to mask reflecting objects. Another variant of it is a "Jaumann" absorber which extends the "Salisbury Screen" concept to multiple layers. Resistive sheets separated by low-loss dielectrics enable broadband performance. In general, the resistivity of the sheets decreases from front to back in a Jaumann absorber.

C. Broadband Absorbers

In a metal-backed resonant absorber, the path length becomes doubled because of the reflection from the metal surface, thereby increasing the absorption. If a metal layer cannot be used, the attenuation offered by the material should be quite high so that any wave leaving the material should have sufficiently low amplitude. Increasing the imaginary part of permittivity of the material medium alone is not the correct solution as it will also increase the real part of permittivity making the impedance of the medium different from free space and thereby making it difficult for the energy to freely enter the absorber. Even though high values of permittivity of the medium can miniaturize the size of the medium, it will not always result in reduction in absorber size. Under such circumstances, the method employed is to adopt a composite approach wherein at the front face the impedance offered by the medium will be close to that of free space while the medium changes its characteristics slowly so that both propagation and absorption will progress. Hence at the front face the permittivity will be very low and as the wave advances inside, it will see that the medium is offering increasing values of permittivity with higher loss characteristics. It is achieved either by a graded inclusion of the absorber material in the host or by using multiple layers of the absorbers with changing impedance and loss characteristics or by using geometric tapering of the absorbers. Since it is not a resonance type of absorption, such absorbers can support broadband absorption. Since multiple layers of absorbers with stepped characteristics or concentration gradient of inclusion of absorbers is required, it is possible to use materials with different values of permittivity, loss, and conductivity to achieve broadband absorption. Under best conditions, absorption levels less than 60 dB below incident energy levels are reported by such approaches.

D. Thin Absorbers to Attenuate Induced Surface Currents on Metal Structures

Most of the absorbers are designed for minimum specular reflection which is the case with the main lobe of the radiation where the angles of incidence are not so large. To attenuate the side lobes, surface current absorbers are required. Surface current absorbers attenuate the side lobe energy by reducing the currents flowing along reflecting surfaces. The surface currents on a metallic surface give rise to reflections when the current encounters discontinuities like gaps, steps, and sharp bends or edges. The absorbers perform by reducing the amplitude of surface currents before they encounter such discontinuities. It is the surface currents that act as the source for the secondary radiation which is the reflection from a metallic surface. Hence a reduction in surface current leads to a reduction in radiated energy. They can be made with thin sheets or coatings with high permeability or a material that contains absorbers with large values of σ'' such as carbon or potentially graphene.

2 Basic Theory of Absorbers

2.1 General Idea

Absorbers are in general composites with a matrix material containing one or more filler materials which do most of the absorbing. The host matrix is generally chosen for its physical properties like adhesion, temperature stability, water absorption, weatherability, etc. It is the complex permittivity, permeability, and conductivity that determine the absorption characteristics of a material. Permittivity dictates the interaction of the electric field of the EM radiation with the material while permeability governs the interaction between the magnetic field of the radiation and conductivity determines the fate of the induced oscillating current [1].

The permittivity is complex and is generally written as

$$\varepsilon^* = \varepsilon' - j\varepsilon''$$

The permittivity arises from the dielectric polarization of the material. The quantity ε'' is a measure of the attenuation of the electric field caused by the material. The electric loss tangent of a material is defined as

$$tan\delta_e = \frac{\varepsilon''}{\varepsilon'}$$

The greater the loss tangent of the material, the greater the attenuation as the wave travels through the material. Analogous to the electric permittivity is the magnetic permeability which is written as

$$\mu^* = \mu' - j\mu''$$

With magnetic loss tangent defined as

$$tan\delta_m = \frac{\mu''}{\mu'}$$

The free space has a permittivity value of $\varepsilon_0 = 8.854 \times 10^{-12}$ (F/m) and permeability value of $\mu_0 = 4\pi \times 10^{-7}$ ($H \cdot m^{-1}$). At present, they are termed as "electric constant" and "magnetic constant", respectively. In general, permittivity and permeability of a medium are expressed relative to that of vacuum and they are termed as relative permittivity (ϵ_r) and relative permeability (μ_r), respectively. The relative permittivity is used to be known as dielectric constant and that is now replaced with the term "relative static permittivity".

The permeability is a measure of the material's influence on the incident magnetic field. Both components contribute to wavelength compression inside the material. Additionally, due to the coupled EM wave, loss in either the magnetic or electric field will attenuate the energy in the wave. In most absorbers, both permittivity and permeability are functions of frequency [2]. If the complex permittivity and permeability are known over a frequency range then the material's effect on the wave is completely known. Hence, it is these quantities that are primarily measured for a material medium as a function of frequency. In terms of these material parameters, the other parameters of engineering design importance like reflection, and transmission coefficients are to be calculated and hence the corresponding relations are given below. Most absorber filler materials are spherical in shape leading to isotropic EM parameters, i.e., propagation and attenuation in the material is independent of direction.

2.2 Attenuation

Attenuation is a measure of how much a wave propagating through a material is attenuated. It is not a direct measurement but is calculated from the material's complex permittivity and permeability. The definition is that if all space is filled with the material, a wave will attenuate at this rate per unit distance. Attenuation is usually expressed in dB/cm. Attenuation values do not relate directly to any particular measurement and they cannot be used to predict reflectivity [3]. It is used to compare the relative absorption of different materials when their material impedance is also taken into account. Attenuation in dB/cm is given by

$$\text{Attenuation}\left(\frac{dB}{cm}\right) = \frac{2\pi(8.686)}{\lambda_0}\sqrt{\frac{\mu'\varepsilon'}{2}\left(\sqrt{(1+tan^2\delta_e)(1+tan^2\delta_m)} - (1 - tan\delta_e tan\delta_m)\right)}.$$

2.3 Reflection and Transmission of Waves at a Material Boundary

Absorbers are used to eliminate unwanted EM energy. In free space they do so by presenting an impedance to an incoming wave equal to the impedance of free space (377 Ω). At a material interface, the incident, reflected, and refracted waves must obey the boundary condition that the sum of E and H fields of the waves must be continuous. Requiring continuity of the amplitudes leads to Fresnel's equations. Continuity of phase leads to Snell's Law. Reflection from a dielectric interface depends on the polarization. There are two polarization states defined. Parallel polarization occurs when the electric field vector is parallel to the plane of incidence. The plane of incidence is defined by the vector normal to the material and the propagation direction of the incident wave. Perpendicular polarization occurs when the electric field vector is perpendicular to the plane of incidence.

The phase delay experienced by the wave in propagating a distance d is given by

$$\phi = \frac{2\pi d}{\lambda}\sqrt{\varepsilon^*\mu^* - sin^2\theta}$$

where λ is the free space wavelength. For a nonmagnetic material, these equations are simplified by $\mu^* = 1$.

The interface reflection coefficients are only half the story. Eventually, the wave will reach the other side of the absorber and reflect. The total reflection is then derived from the sum of the reflected waves [4].

The voltage reflection coefficient for a thickness d of a material is

$$R = \frac{-r^*(1 - e^{-j2\phi})}{1 - r^2e^{-j2\phi}}$$

where r is the appropriate interface reflection coefficient given by

$$r_{par} = \frac{\sqrt{\varepsilon^*\mu^* - sin^2\theta} - \mu^*\varepsilon^* cos\theta}{\sqrt{\left(\varepsilon^*\mu^* - sin^2\theta\right)} + \mu^*\varepsilon^* cos\theta}$$

$$r_{per} = \frac{cos\theta - \sqrt{\varepsilon^*\mu^* - sin^2\theta}}{cos\theta + \sqrt{\left(\varepsilon^*\mu^* - sin^2\theta\right)}}$$

where θ is the angle of incidence.

Reflection coefficients is usually expressed in dB.

$$\text{Reflection Coefficient (dB)} = 10 \log\left(\frac{1}{|R|^2}\right)$$

The voltage transmission coefficient is given by

$$\mathrm{T} = \frac{(1 - r^2)e^{-j\phi}}{1 - r^2e^{-j2\phi}}$$

Transmission coefficient in dB is given by

$$\text{Transmission Coefficient (dB)} = 10 \log\left(\frac{1}{|T|^2}\right)$$

In most of the cases with absorbers, the material is backed by metal [5]. The total reflection coefficient (now called the reflectivity) becomes

$$\mathrm{R_{MB}} = \frac{r - e^{-j2\phi}}{1 - re^{-j2\phi}}$$

and the reflectivity in dB is given by

$$\text{Reflectivity(dB)} = 10 \log\left(\frac{1}{R_{MB}^2}\right)$$

Once the complex permittivity and permeability of a material medium are measured, these quantities can be calculated. Hence, some of the methods available for measuring these material parameters over a wide frequency range will be briefly reviewed in the following sections.

2.4 Material Parameter Measurement Methods

Accurate measurement of EM parameters is critical for modeling the performance of microwave absorbers. ε^* and μ^* are in general functions of the frequency, so swept frequency methods are desired. At low frequencies (<1 GHz), parameters can be measured using an impedance analyzer with custom test fixtures. The permittivity test fixture measures the capacitance of two parallel test heads both with and without the material under test. The permittivity is then derived from the capacitance. The permeability is determined from the change in inductance of a cylindrical cavity by insertion of a doughnut-shaped material under test. Very good results can be found down to 1 MHz. At higher frequencies, measuring the capacitance or inductance no longer yields satisfactory results and hence field theory must be used. Material parameter measurement testing at these frequencies requires sending a wave into the material and measuring the material response. Since four results are needed (real and imaginary values of ε^* and μ^*), four measurements need to be taken on a sample which are usually the magnitude and phase of S_{11} (reflection) and S_{21} (transmission) through the sample. These are scattering parameters which are generally measured using a vector network analyzer. If it is known that the material has no magnetic components ($\mu = 1$), then the electric permittivity can be determined with 2 measurements, S_{11} magnitude and phase or S_{21} magnitude and phase. While free space measurements of reflection/transmission amplitude and phase can yield good results, the best results are seen in closed systems using coaxial lines or waveguides. In such measurements, the calibration of the network analyzer is critical. Full 2 port calibration or TRL calibration is needed for accurate phase measurements, particularly of reflection. Fitting of the sample inside the coax or waveguide is very important as a poorly fit sample will not yield good results. As frequencies extend into millimeter waves, calibration and sample fit become even more critical due to the short wavelength. For nonmagnetic materials, free space techniques can yield excellent results all the way up through millimeter waves. It is much more straightforward to measure transmission phase in free-space. In addition, good measurements can be made in free space without the sample fit problem.

2.5 Insertion Loss Measurement

Insertion loss is a measure of how much microwave energy traveling from Point A to Point B is reduced by the introduction (or insertion) of a microwave absorbent material in the path. An insertion loss measurement does not differentiate between all the factors which will affect the reduction in power including reflection from the material and loss as the wave transits through the material. A well designed setup for testing insertion loss would include two antennas oriented so that their maximum directivity is towards each other. They will be separated sufficiently to satisfy far field requirements though the greater the separation, the larger the sample size must be to minimize errors caused by energy leaking around the edges of the sample under test. In practice insertion loss measurement is straightforward. A signal is transmitted through one antenna and the response is measured at the second antenna. This establishes the reference or 0 dB level and is usually measured as a function of frequency. The material under test is then placed between the antennas and a measurement is performed. The insertion loss is expressed in dB as a function of frequency.

3 Material Measurements in the Microwave Frequency Range

3.1 General Description

To measure material parameters like complex permittivity and permeability there are many methods available. For a wide band frequency dependent measurement, the instrument that is widely used now is a vector network analyzer (VNA). However, it measures only the magnitude and phase of the complex scattering parameters between 2 ports. Deducing the material parameters from the measured S-parameters requires careful modeling of the test structure in which except the property of the material under test everything else should be known. This demands not only a careful calibration but also taking up a test structure with well-defined boundary conditions so that it is easier to arrive at exact relations for material parameters in terms of the measured scattering parameters and the dimensions of the sample. The sample size also matters along with the value of the material property parameters. For example, the techniques that are to be used for bulk samples and that for thin films are different. Also the techniques that work well for low loss samples are not suitable for lossy samples. Similar selection exists for dielectric measurements and magnetic measurements. Again there are techniques well suited for narrow band measurements which are different from broadband measurements. Three techniques that can be used in an academic laboratory for material parameter measurement is briefly introduced next. However, it must be noted that in the case of a heterostructured medium, what can be experimentally measured is the "effective permittivity" and "effective magnetic permeability" of the medium which represents the averaged out response of a the heterostructured medium in presence of an electric or magnetic field of the measurement frequency.

3.2 Waveguide Based Measurement Technique

Advantage in using a waveguide is that the boundary conditions are exact and hence well-defined field patterns are available leading to relatively simple method to calculate the material parameters. Rectangular or circular waveguides can be used. The samples should have the same size and cross section as that of the waveguide used. Hence the measurements have to be done band by band as waveguides of each band differ in their dimensions. There should not be any air gaps between the sample and the waveguide walls. In case of any small gaps, especially if the gap is of irregular shape, it needs to be filled with conducting silver paste. A small length of the waveguide attached with its flange becomes the sample holder. To connect it with the measurement ports of VNA that are coaxial cables, a pair of coaxial to waveguide adapters are required. The measurement setup is shown in Fig. 1. The calibration is to be carried out using the Thru-Reflect-Line (TRL) method first at the VNA ports and then at the waveguide adapters and the interface between the adapters become the calibration plane. Calibration at the VNA ports are to be done with coaxial standards and the calibration at the waveguide calibration plane is to be done with a waveguide calibration kit comprising a short, a standard waveguide as a through transmission line and a quarter wavelength waveguide section are to be used. The waveguide sample holder should be connected in between the waveguide adapters (calibration plane). The calibration plane from the adapter surface is to be extended to the sample surfaces by a port extension procedure which adds the phase factor that is equivalent to the electrical length of the distance from the sample surface to calibration plane. This is required if the sample length is less than the sample holding waveguide length. The microwave scattering parameters were measured after port extension from the surface of the samples. The material parameters can be derived using the scattering parameters S_{11} and S_{12} in magnitude and phase.

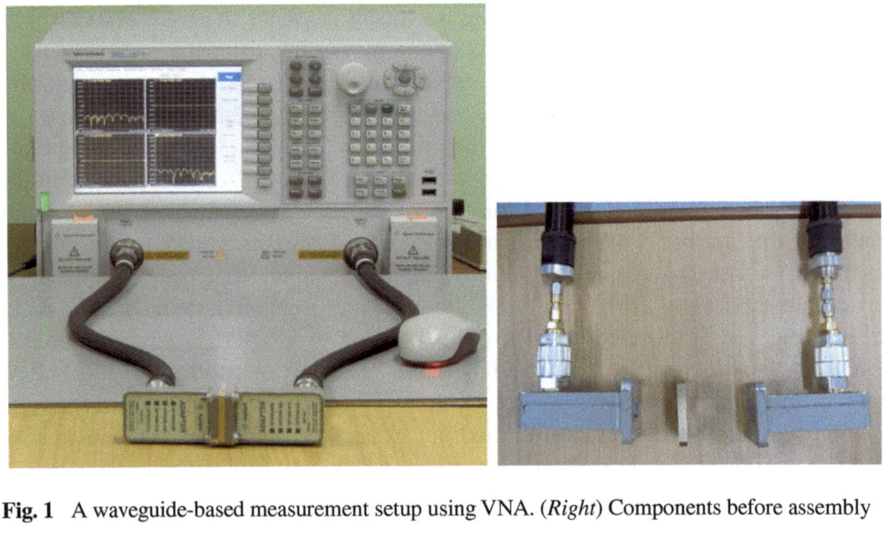

Fig. 1 A waveguide-based measurement setup using VNA. (*Right*) Components before assembly

There are different approaches available for the calculation of material parameters from the measured S-parameters. They include Thru-Reflect-Line algorithm (NRW) [6], NIST algorithm [7], and the calibration comparison technique [8] for which the details are given in the references. A detailed discussion on these methods and procedures heavily relies on microwave engineering principles.

3.3 *Coaxial Line Probe Based Technique*

In this procedure, an open-ended coaxial line probe is attached to the VNA probe and terminated in the material under test. Since the probe is a coaxial transmission line, the bandwidth of operation is quite wide. The port of the network analyzer used in the experiment is to be calibrated with open, short, and matched 50 Ω loads in sequence followed by calibrating the coaxial probe with open (air), short circuit (with a flat-polished metal surface), and deionized water at 25 °C. In this method, flat samples of thickness higher than $20 \div \sqrt{|\epsilon r * |}$mm and diameter >20 mm with uniform composition are to be used. This method can be used for materials that give relative permittivity below 100 and a tan δ higher than 0.05. The seamless contact between the sample and the flat face of the coaxial probe is quite important. Good contacts can be achieved if the sample surface is little pliable. The probe transmits a signal into the material under test. The measured reflected response from the material is then related to its dielectric properties. During the measurement, the rigid coaxial cable which is connected to the analyzer and the probe is to be fixed to avoid the effect of cable position and shape changes on the measurement accuracy. Probes that withstand temperatures from −40 to +200 °C are available. After calibration, the S_{11} parameter is measured using VNA in magnitude and phase. With this data the material parameters can be derived using the procedure described in [9]. The setup is shown in Fig. 2.

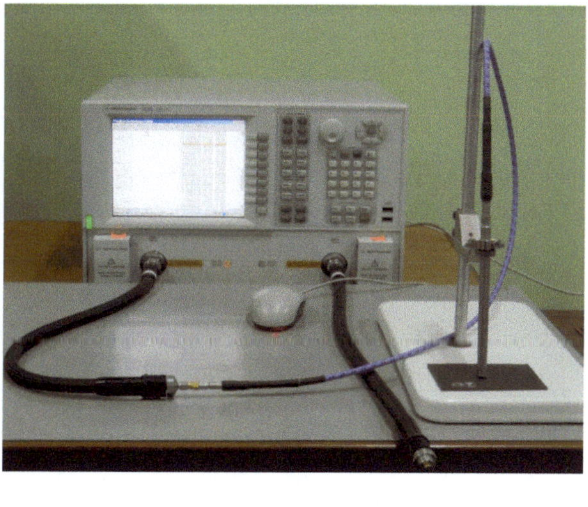

Fig. 2 Measurement of material properties using a VNA and open-ended coaxial line probe

3.4 Free Space Measurement Technique Using Antennas

The free space method of measurement has found a special place in the field of research and industry because of its varied attractive features like its ability to do the measurement over a wide bandwidth, ability to handle large samples and that too at elevated temperatures [10]. A free space measurement setup usually contain two horn antennas facing each other at the open end either directly or through reflecting mirrors and connected to a VNA from the back end. A sample holder is placed in between the antennas where a large planar sample is placed to be analyzed as shown below (Fig. 3).

The microwave signal, generated in the VNA is transmitted from one of the antenna. It strikes the sample, gets partly reflected and partly transmitted from it and comes back to VNA either through the transmitting or reflecting antenna. The VNA uses the magnitude and phases of these waves to calculate the four complex S-parameters. Using one of the available algorithms, the material parameters can be elucidated from the measured S-parameters. NIST algorithm [7] is one of them mainly used to measure the dielectric properties.

Instead of a simple horn antenna, corrugated circular horn antennas can be used for the following reasons. A horn antenna may be regarded as a flared out or opened out waveguide. The function of the horn is to produce a uniform wave front with a larger aperture than that of the waveguide and hence greater directivity. The circular horn antennas are excited by circular waveguides. Corrugated horns provide reduced edge diffraction, improved pattern symmetry, and reduced cross-polarization. Corrugations on the horn walls acting as $\lambda/4$ chokes are used to reduce E field to very low values at horn edges at all polarizations. These prevent waves from diffracting around the edges of the horn. This results in high aperture efficiency as reflections at the circumference of the aperture can lead to a distorted result. In order to minimize the effect of the reflection at the edges of the antenna aperture, the antennas can be made long which also result in enhanced directivity of the wave.

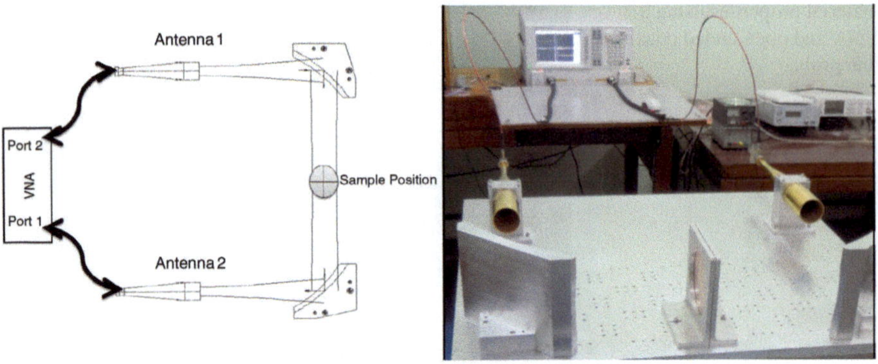

Fig. 3 Free Space Measurement setup using VNA, corrugated horn antenna, and ellipsoidal mirrors

However, such reflections at discontinuities cannot be avoided completely. Hence an important antenna parameter known as the "Effective Aperture" of the antenna can be used. For horn antennas, the effective aperture is 0.5–0.8 times the actual physical aperture. Ellipsoidal mirrors can be used to reduce the size of the free space measurement setup. Two ellipsoidal mirrors are shown in the Fig. 3. These mirrors/reflectors have two conjugate foci. Waves from one focus passes through the other, after reflection. Ellipsoids collect a much higher fraction of total emitted energy than a spherical mirror or conventional lens system. Wider bandwidth and lower absorption loss are the two features of ellipsoidal mirrors that make them so attractive. The EM wave will come out of the antenna in the form of a Gaussian Beam. For a significant analysis, the sample is to be kept at the location of the beam waist. At this location, i.e., the location of the narrowest part of the beam waist, will be the sample holder. The free space measurement technique usually involves large samples but powdered samples kept on the beam waist at fixtures can also be effectively analyzed. If the measurements are at quasi-optic frequencies, the sample thickness would be comparable with the wavelength. As the sample is held in air, the free space measurement technique is contactless and nondestructive. This is an added advantage of this technique.

The experiment is to be carried out by first performing the calibration of VNA using gated-reflected-line (GRL) technique. Calibration is the mathematical procedure of removing imperfections in the test system and is performed before the actual measurement of the sample. Known quantities called standards are measured instead of the sample. Error models are used for corrections. Errors are classified as systematic, random, and drift errors. Systematic errors are due to hardware limitations of the VNA itself and can be removed by calibration. To perform GRL calibration we need two calibration standards, an empty air fixture and a metal plate of known thickness. First, the measurement of reflection coefficient of the empty air fixture is done. With the help of the reflection coefficient and time domain analysis, the transmission coefficients can be determined. The same procedure is applied for the metal plate also. After the calibration, the sample can be placed in the sample holder and the measurements are to be made. The obtained values of the S-parameters are to be subjected to one of the algorithms that determine the values of complex permittivity and complex permeability of the different samples [11].

NIST ALGORITHM [7]: NIST method performs the calculation using a Newton-Raphsons root finding method and is suitable for permittivity calculation only. It utilizes all four S-parameters or a pair of S-parameters (S_{11}, S_{21}) of samples to calculate the reflection and transmission coefficients. It works well if a good initial guess is available. The method bypasses the inaccuracy peaks that exist in NRW method when the sample thickness is an integer multiple of one half wavelength ($n\lambda/2$). By using this method, a stable permittivity over the frequency spectrum can be obtained from the S-parameters and it allows measurements to be taken on samples of arbitrary length. This method is robust for low-loss and high-loss materials. However, this iterative method got the disadvantages that it is applicable for permittivity measurement only and it need an initial guess value for permittivity.

4 Absorber Applications in the Microwave Frequencies

4.1 General Description

Virtually every component of a microwave system uses absorbers. However, some of the applications are in confined spaces where the objective is to prevent the radiation generated in one compartment to go to any other compartments and thereby interfere with the field there. Also in such confined spaces, one would like to prevent the outside radiation from entering into the compartment which is being isolated. With the rampant usage of high frequency radiation for communication purposes, need for such isolations in electronic systems are becoming increasingly important to achieve EMI/EMC standards. Once radiating circuits are used, the best way to achieve required levels of EMI/EMC in adjacent blocks is to cover the inner walls of the circuit blocks of a system with suitable absorbers. Since most of the circuit blocks in high frequency circuits that use relatively higher power are housed in metallic housings, they act as a cavity resonator for EM radiation. A cavity sustains radiation in different modes governed by its size and geometry which can seriously impact the circuit performance. The field pattern of the electric and magnetic fields in each mode is different and it can be computed. The correct absorber material when introduced to the cavity can damp the resonance, enabling proper operation of the circuit. Even a physically small cavity or enclosure could encompass several wavelengths at millimeter wave frequencies.

4.2 Absorbers in Enclosed Space

The physics governing absorber performance in an enclosed space is different from that of in a free space volume. In an enclosed space, there are no propagating waves, only standing waves. In standing waves, the E field and H field are 90° out of phase with each other. Material thickness is not as crucial as it is with free space absorbers since material resonance is not the goal [3]. In cavity resonance damping the absorber is a high permittivity/permeability material that will attract the energy and absorb it. Since the impedance at a given point is proportional to the E field divided by the H field, the cavity resonance can cause wild swings in impedance across a cavity. Hence where the electric field is strong, the magnetic field is weak and vice versa. When an absorber is inserted into the cavity, the high permittivity/permeability of the absorber causes the energy to move into the absorber. The field equations are too complex to solve directly for a partially filled cavity but using Finite Element Method (FEM) software solutions for the fields can be found. The higher the permittivity and permeability, the more the energy "wants" to go into the material. The tangential electric field is zero on a conducting wall while the magnetic field is maximum. For this reason, magnetically loaded absorbers or surface current absorbers are the most effective in damping cavity resonances.

4.3 Absorbers for Near Field

A radiation pattern emanating from a radiator will have a far field and near field that are different in their radiation characteristics. The energy in the near field will be predominantly magnetic. Near-field absorbers are a class of absorbers that are placed near or directly upon a radiating element and they will have high magnetic permeability and high magnetic loss. Also, since they are often in direct contact with circuit elements, they must have very low conductivity. Even the best designed circuit will contain elements that will resonate and radiate at particular frequencies [3]. These radiators could be inductors or capacitors or connecting wires that will behave differently at certain frequencies. Magnetic energy dies off very quickly with distance but can still interfere with nearby circuit components and hence near-field absorbers can enhance the EMI/EMC performance of high frequency circuit blocks.

4.4 Free Space Reflection Reduction

Any system that transmits energy can experience interference from reflections back to the transmitter. Also, unwanted reflections can interfere with other systems. Often there would not be any control over the source of reflections and hence absorbers will have to be used to reduce the reflection level. Typical reflectivity reduction for weather resistant outdoor absorber material is −20 dB which will eliminate 99 % of the reflection. It is necessary that the chosen absorber is designed to absorb at the transmit frequency.

4.5 Radar Cross Section Reduction (RCSR)

Absorbers can also be used to reduce the radar cross section of a target object. By reducing the reflection level, the object will present a smaller cross section. However, due to the narrow-banded performance of thin radar absorbent material (RAM) and the thickness and weight of broadband RAM, it is difficult to achieve effective radar cross section reduction using absorber alone over broad range of frequencies.

4.6 Absorbers in Anechoic Chambers

Anechoic chambers are used to create a free space condition in an enclosed room Very high performance absorber material is to be used on the walls, ceiling, and floor to eliminate reflections. Reflectivity of absorbers used in anechoic chambers can be −50 dB or better. Such absorption levels are needed to do an effective antenna or radar cross section measurement in a controlled environment.

5 Why Graphene as a Filler?

Nanosized materials of many types have been tried as fillers in polymers and they offer varying degrees of absorption at different frequencies. However, the opportunities thrown open by graphene is of a different kind. Being atomically thin, it offers the highest surface-to-volume ratio with good in-plane conductivity. High specific surface area of graphene gives an ample opportunity to attach numerous functional groups or anchoring groups that can alter its properties in a particular host polymer. The high aspect ratio of graphene allows its uniform dispersion in polymer matrices. There are ways by which graphene can be dispersed in a polymer matrix such that they can form parallel layers in the matrix resulting in the formation of capacitor-like structures of the nanometer scale size. The way the charge carriers in graphene interact with the host matrix depends a lot on the chemical properties of the host. These factors throw open a large variety of opportunities to consider different compositions yielding hitherto unexplored properties and the already published work reflects only a glimpse of that potential. The requirement for shielding materials in high frequency electronics is also becoming ubiquitous rendering the available materials inadequate. For example: in mobile handsets the available dimensions and weight margins are so narrow that a thin layer of coatable polymer layer is expected to provide sufficient EMI shielding over many frequency bands to isolate different circuit compartments.

5.1 General Background

Graphene is a 2D monolayer of densely packed sp^2-bonded carbon atoms in a hexagonal or honeycomb lattice [12, 13]. Graphene has shown a great promise as a material for electronic and energy applications owing to its unique electronic band structure [14–17]. Graphene has been touted as an excellent microwave absorber owing to its high dielectric loss and low density. But its poor impedance matching characteristic has hindered its application as a microwave absorbing material. In recent past, graphene-related materials have attracted a great attention as microwave absorbers. First of such materials is reduced graphene oxide (r-GO), which showed a microwave absorption (MA) of ~6.9 dB (in other words a reflection loss (RL) of −6.9 dB) at 7 GHz [18–20]. Impedance matching characteristic was improved by preparing r-GO by chemically reducing graphite such that the process leaves residual defects and functional groups (like C–O and C = O) in the material which in turn will prompt energy transition from contiguous states to Fermi level and introduce defects' related polarization relaxation and functional groups' related electronic dipole relaxation, which will favor microwave penetration and absorption. Figure 4 shows RL in r-GO in comparison to graphite (matching thickness of ~2 mm).

The most popular graphene-related microwave absorbing materials are r-GO or few-layered graphene (FLG) filled polymer composites. These composites have gained popularity owing mainly to ease of their preparation, their practical

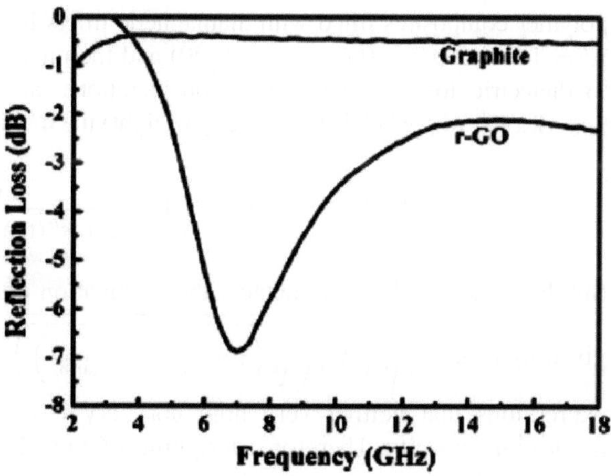

Fig. 4 Reflection Loss of r-GO in comparison to graphite [18]. Copyright 2011. Reproduced with permission from American Institute of Physics

applicability and viability of their industrial production. However, fillers' dimensions, surface area, surface functionality, and aggregate structure effect the overall interaction between fillers and suitable polymer matrices, which in turn affects the overall physical properties and applicability of the composites as microwave absorbers. Thorough understanding of the fillers' intrinsic characteristics, which depend on their synthesis procedures and/or postsynthesis surface modifications, is also important to have a control on the required properties of the composite to be used as a microwave absorber. All in all, suitable fillers have to be homogeneously dispersed in suitable polymer matrices without compromising on the intrinsic properties of both the fillers and matrices. Development of novel synthesis procedures has indeed made this possible to an extent. r-GO or FLG particles are being filled into suitable polymer matrices by i) mechanically mixing them with molten polymers, or ii) their in situ formation in the presence of a suitable polymer matrix, or iii) dispersing them in a suitable monomer solution which subsequently polymerizes to form the host matrix.

5.2 *Effective Microwave Absorption*

Let an EM wave (possessing an angular frequency $\omega = 2\pi f$, "f" is measured in Hz) be incident perpendicularly on the surface of a monolayer (of thickness "t") of an absorber material in contact with a perfect electric conductor. In this scenario, the real parts of dielectric permittivity and magnetic permeability of the material indicate the measurethrough magnetic of the material's ability to store external EM energy whilst the imaginary parts govern the dissipation (or loss) of EM energy within the material. Moreover, out of the total EM power that enters the material, a part is used in increasing the field energy stored in the material whilst the rest is lost as heat [21]. Dissipation results through conduction, relaxation, and resonance mechanisms, i.e., both conductive and dielectric losses contribute to the heat. However, in the case of polymer composites filled with diamagnetic fillers like r-GO or FLG (or graphene), $\mu' = 1$ whilst $\mu'' = 0$ (i.e., $\mu = 1 - j0$) and the microwave absorption is purely due to dielectric losses. Under such considerations, at the matching frequency (i.e., $f = c/2\pi t\varepsilon''$, where "c" is the velocity of light) the maximum RL is given by

$$[RL]_{Max} = 20\log_{10}\left(1 - \frac{4}{2 + \varepsilon''(\tan\delta_e + \frac{1}{\tan\delta_e})}\right)$$

Similarly, at matching frequency, the attenuation constant "α" of the material is given by $\alpha = \frac{\omega}{\sqrt{2}t}\sqrt{\left(\frac{1}{\varepsilon''}\right)\left(\sqrt{\left(1 + \frac{1}{\tan^2\delta_e} - \frac{1}{\tan\delta_e}\right)}\right)}$. It is evident from the above expressions that neither very high nor very low ε'' values will result in high absorption capacity. Therefore, designing r-GO or FLG filled polymer composites with an optimal ε'' is very much needed. It is also evident from the formulae that an optimal value for loss tangent is required to achieve high absorption capacity.

RL in terms of input impedance Z_{in} of the absorber material and air impedance Z_0 is given by

$$RL = 20\log_{10}\left(\frac{Z_{in} - Z_0}{Z_{in} + Z_0}\right)$$

where $Z_{in} = Z_0\sqrt{\frac{\mu_r}{\varepsilon_r}}\tanh\left[j\left(\frac{2\pi ft}{c}\right)(\mu_r\varepsilon_r)^{1/2}\right]$. ε_r and μ_r are relative permittivity and permeability of the material, respectively. In the present case however, $\mu_r = 1$.

Additionally, MA through magnetic losses by incorporation of magnetic graphene (mostly achieved by decorating r-GO or FLG with magnetic nanoparticles) into the design is also possible. Moreover, as the microwave traverses in a composite material it can come across a variety of interfaces leading to local electric field variations (it should be noted that there is a quadratic relation between absorption and the interacting electric field intensity) which will strongly influence the overall MA. Three different mechanisms namely reflection (R), absorption (A), and multiple internal reflections (M) contribute to the overall attenuation of an EM field as it interacts with a material. If AE_R, AE_A and AE_M are attenuation effectiveness components corresponding to R, A, and M, respectively, the total attenuation effectiveness AE_{total} can be expressed in logarithmic scales as the ratio of transmitted to incident electric (or magnetic) powers. AE is conveniently measured in decibels (dB) and is given by:

$$AE_{total} = AE_R + AE_A + AE_M = -10\log_{10}\left(\frac{P_T}{P_I}\right) = -20\log_{10}\left(\frac{E_T}{E_I}\right) = -20\log_{10}\left(\frac{H_T}{H_I}\right)$$

where P_I (E_I or H_I) and P_T (E_T or H_T) are the powers (electric or magnetic field intensity) of incident and transmitted EM waves, respectively. As mentioned previously VNA is a popular experimental setup that is used to measure AE. Using this setup reflected and transmitted power with respect to the power incident on the sample (sheet/monolayer) surface is generally measured. The sum of absorption coefficient (A), reflection coefficient (R), and transmission coefficient (T) is equal to 1. R, T, and A can be calculated using the below given formulae:

$$\mathrm{R} = \left(\mathrm{E_R}/\mathrm{E_T}\right)^2$$
$$\mathrm{T} = \left(\mathrm{E_T}/\mathrm{E_I}\right)^2$$
$$\mathrm{A} = 1 - \mathrm{R} - \mathrm{T}$$

The reflection is related to the impedance mismatch between air and absorber. Absorption is regarded as the energy dissipation of EM wave in the material over multiple reflections at the interfaces and scattering from inhomogenieties inside the material while the multiple reflections are the consequence of impedance mismatch at the two sample-air interfaces. When $\mathrm{AE_{total}} \geq 15\,\mathrm{dB}$, it is usually assumed that $\mathrm{E_M}$ is negligible and thus, $\mathrm{AE_{total}} \approx \mathrm{AE_R} + \mathrm{AE_A}$. The effective absorbance A_{eff} can be therefore expressed as $\mathrm{A}_{eff}=\frac{(1 - \mathrm{R} - \mathrm{T})}{(1 - \mathrm{R})}$. AE due to reflection and absorption of the sheet material with respect to power of the effective incident EM wave inside the material can now be expressed as

$$\mathrm{AE_R} = -10 \log (1 - \mathrm{R})$$
$$\mathrm{AE_A} = -10 \log \left(\mathrm{T}/(1 - \mathrm{R})\right)$$
$$\mathrm{AE_{total}} = \mathrm{AE_R} + \mathrm{AE_A} = -10 \log (\mathrm{T})$$

5.3 Graphene/Polymer Composites as Microwave Absorbers

5.3.1 General Description

Even though there are a variety of graphene-filled polymer composites [22, 23], only few have been explored as microwave absorbing materials. Polymer matrices explored are also limited. They are epoxy resin, poly vinyl alcohol (PVA), poly (methyl methacrylate) (PMMA), polyvinyledene fluoride (PVDF), polyaniline (PANI), poly (ethylene oxide) (PEO), polyetherimide (PEI), polystyrene (PS) and praffin wax. It can be clearly observed that both electrically conductive and non-conductive polymer matrices have been considered. Microwave absorbing composites should have a reasonable (at least 1 S/m) electrical conductivity. This can be achieved either by filling high amounts of electrically conductive fillers into a non-conductive polymer matrix or by using electrically conductive matrix and not so high amount of electrically conductive filler. The former may have adverse effects on formability of the composite while the latter is difficult to produce. Nonetheless there are a good number of carefully designed and performed experiments that have overcome such difficulties. In all of these works, the composite absorber materials are in the form of sheets of different thicknesses. Sheets are the most convenient to fabricate, handle and are industrially viable. The formulae presented earlier are valid for all the cases discussed in this chapter. In the following paragraphs, the graphene-filled polymer composites in which the filler has no magnetic properties are discussed. It is important to bring to the notice of the readers that several works include discussion only on MA of nonmagnetic and magnetic graphene-related materials. These studies are made by incorporating nonmagnetic and

Fig. 5 Cross-sectional scanning electron micrograph of graphene/epoxy composite with 7 wt% loading of SPFG [24]. Copyright 2009. Reproduced with permission from Elsevier Ltd

magnetic graphene-related materials as fillers in wax matrices. Examples of such material systems with magnetic graphene as filler are given under the sub-heading "Graphene/wax composites".

5.3.2 Epoxy Composites

The first report on graphene-filled polymer is on solution-processable functionalized graphene (SPFG) filled epoxy composite [24]. The composite was used for EM wave shielding in the X-band (8.2–12. GHz). The thickness of each SPFG sheet is 0.8–1.1 nm corresponding to 2–3 graphene layers stacked along the c-axis. The lateral dimensions of each SPFG sheet are in the submicron (100–400 nm) range. Epoxy solution was mixed with the suspension containing partially reduced (intentionally retaining oxygen-containing functional groups) SPFG sheets. Simple solution casting was used to obtain the composite. Electron micrograph (please see Fig. 5) of the composite's cross section showed homogeneous dispersion of SPFG sheets in the epoxy matrix. Different SPFG loadings were used and the composites showed a low percolation threshold of only 0.52 vol.% owing to the high aspect ratio of the sheets and their homogeneous dispersion as revealed by electron microscopy. DC electrical conductivity could be varied with SPFG loadings.

A high overall AE (or commonly known as shielding effectiveness) of ~21 dB was measured for a loading of 8.8 vol% (~15 wt%). This was attributed to the conducting interconnected graphene-based sheets' networks in the insulating epoxy matrix. However, the mechanism (reflection or absorption or any other) leading to the overall AE was not discussed. But based on the arrangement of fillers in the matrix and the measured electrical conductivity it can be speculated that the maximum contribution to AE is from absorption (dielectric loss). However, in a similar work [25] which is discussed below elucidates the contribution of RL to the overall AE of graphene-filled epoxy composites.

Fillers named as graphene nanoplatelet (GNP) particles (these are similar to SPFG particles) are used to prepare microwave absorbing epoxy composites [25].

Each GNP particle was ~8 nm thick (i.e., 4–5 layers of graphene) and had lateral dimensions of ~100 nm. Acetone with dispersed GNP particles was mixed with epoxy resin and sonicated for 1 h. A curing agent EPIKURE 3234 was added to the mixture and manually stirred for just 3 min. This mixture was then casted similarly to the work [24] reported earlier.Before casting the acetone evaporated. Different wt% of GNP particles were added to obtain composites with different filler loadings. After casting, the samples were cured at 80 °C for 1 h in an oven. The thickness of all the samples was maintained at 3 mm.

Electron microscopy has shown that for filler loadings of greater than 15 wt%, a continuous conductive network of GNP particles throughout the epoxy matrix has formed. DC conductivity measurements on the samples with respect to the filler loading corroborated the observation of conductive network of GNP particles [25]. It was argued that the edges of GNP particles offer anchor sites that would enable their optimal dispersion in epoxy. The DC conductivity of 15 wt% GNP particles loaded epoxy composite was measured as 1.3×10^{-6} S/m which is far less than 1 S/m. Nonetheless, this value is 3 orders higher than the value measured for 10 wt% GNP particles loaded epoxy composite. Further increase in the loading of GNP particles the conductivity could only be increased to ~2 orders (for epoxy filled with 30 wt% GNP particles). The improved DC conductivity values with increase in the loading of GNP particles in correlation with the corresponding cross-sectional morphology (as observed using electron microscopy) clearly indicate the formation of conductive networks in epoxy matrix for loadings greater than 15 wt% of GNP particles. Interestingly, in the previously discussed SPFG filled epoxy composites the optimum AE was observed in the case of 15 wt% SPFG loaded epoxy [13]. As discussed in the section "*Affective Microwave Absorption*" the necessity of controlling both ε'' and loss tangent (tanδ) of the composite for optimum MA has been clearly demonstrated in reference [25].

The variation of ε' and ε'' with frequency (in the range 8–20 GHz) and different GNP loadings is given in Fig. 6. ε' of the GNP filled epoxy composites slightly increased from ~3.6 to ~5.4 as the loading of GNP particles in epoxy increased from 1 to 10 wt%. This is owing to the absence of long range interaction among the filler particles even though interfacial polarization (between epoxy and GNP) in the composite is in existence during the composite's interaction with the external EM wave. Further increase in the loading of GNP particles in epoxy resulted in sharp increase in ε' values attributed to Maxwell–Wagner interfacial polarization. This can be attributed to increase in the concentration of free electrons loosely bound with the systemwith a increase in loading of GNP particles. In the high frequency regime, ε' values depended on frequency. In this regime ε' decreased with increase in frequency for a particular loading of GNP particles. This is expected when induced charges in the composite lag in response to the reversal of the interacting EM wave. Similar to the behavior of ε', ε'' of the GNP filled epoxy composites slightly increased from 0 to ~0.4 as the loading of GNP particles in epoxy increased from 1 to 10 wt%. Very low electrical conductivity of the composites for loadings in the range 1–10 wt% is the reason for such low ε'' values. Similar to the behaviour of ε', as the loading of GNP particles in epoxyincreased sharp increase in ε'' values of the composites was observed owing to the onset of the formation of a suitable conductive network of

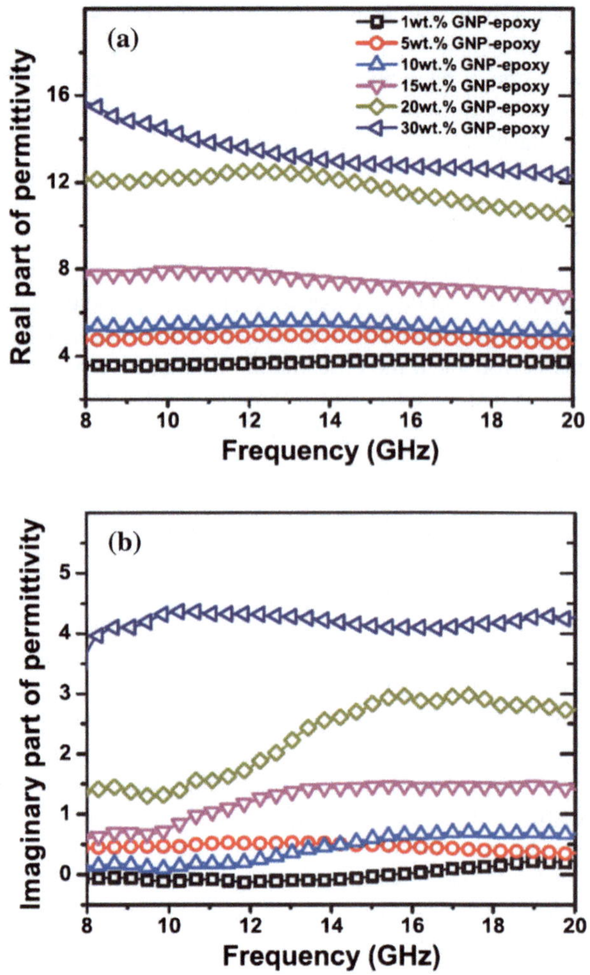

Fig. 6 **a** ε′ (real) and **b** ε″ (imaginary) parts of the permittivity of different GNP filled epoxy composites in the frequency range 8–20 GHz [25]. Copyright 2014. Reproduced with permission from American Institute of Physics

GNP particles. ε″ values also strongly depended on frequency. In the case of 15 wt% GNP particles loaded epoxy, ε″ increased from ~0.6 at 8 GHz to a saturation value of ~1.5 for frequencies ≥13.5 GHz. Saturation ε″ values of 20 and 30 wt% loaded epoxy composites are ~3 and ~4, respectively.

RL (in the frequency range 8–20 GHz) of the GNP/epoxy composites as a function of loadings of GNP particles is shown in Fig. 7. RL of −5 to −10 dB was measured for the composites with 10 wt% or less of GNP particles in epoxy. The minimum RL was measured as −14.5 dB at 18.9 GHz for 15 wt% GNP particles loaded epoxy. This is mainly attributed to dielectric loss due to optimal electric conductivity and charge multipoles at the polarized interfaces in the composite. For loadings greater than 15 wt%RL minimum decreased while the frequency of RL minimum shifted to lower values. The minimum RL was −11.2 dB at 17.2 GHz and −9 dB at 13.2 GHz for 20 and 30 wt% GNP loadings, respectively.

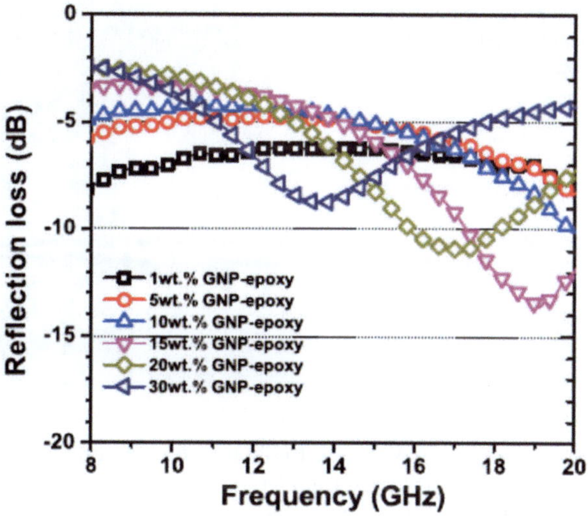

Fig. 7 Reflection loss of GNP/epoxy composites in the frequency range 8–20 GHz [25]. Copyright 2014. Reproduced with permission from American Institute of Physics

In the present case, a minimum reflection requires a perfect matching between permittivity, frequency, and thickness of the epoxy composite with the free space. Here, all the samples have the same thickness. Therefore, minimum RL for GNP/epoxy composites with GNP loadings >15 wt% is mainly determined by the matching condition of permittivity of these composites for a particular frequency. The observed decrease in RL minimum for composites with high GNP loadings (in other words as the loading of GNP particles increased beyond 15 wt%) is due to impedance mismatch at the interface of air with the composites. The impedance mismatch is owing to the high-dielectric permittivity values for the composites with high loadings of GNP particles. However, this can be avoided if the thickness of the GNP/epoxy composites could be reduced (please see the expressions for RL and Z_{in}) while maintaining the electrical conductivity and dielectric permittivity for a particular GNP loading.

5.3.3 Poly Vinyl Alcohol (PVA) Composites

PVA is a synthetic polymer and it is well known for its water solubility, high transparency, very good flexibility, and wide commercial availability. It has excellent film forming, emulsifying, and adhesion properties. These characteristics make PVA an attractive matrix material to host graphene-related fillers. In a recent work [26], very thin (0.3–0.6 mm) graphene nanoribbon/polyvinyl alcohol (GNR/PVA) composite films with different amounts of GNR particles have been developed for shielding EM waves in the X-band. GNR particles were obtained by unzipping multiwall carbon nanotubes [27]. GNR particles were first dispersed in water using sonication and then PVA was added to the solution. The solution was stirred until PVA completely dissolved in water to yield a uniform solution containing GNR particles and PVA.

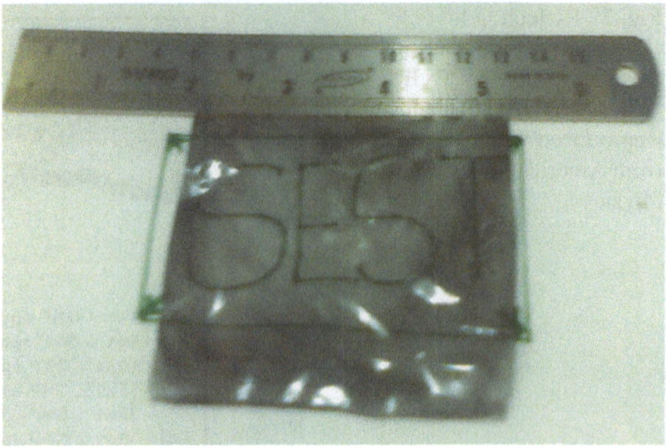

Fig. 8 Photograph of a semitransparent FLG/PVA composite sheet

Solution casting was used to obtain free-standing composite films which were subsequently allowed to dry for 12 h in a vacuum oven at 40 °C. EMI shielding of as high as 60 dB was achieved for a particular GNR/PVA composite. However in this work, the contributions to overall AE were not elucidated.

In another recent work [28], 1 mm thick flexible FLG/PVA composite sheets containing different amounts of FLG are synthesized using an easy eco-friendly solution mixing process followed by a simple casting process. 2.5 g of PVA powder was mixed with 50 ml of distilled water and stirred at 333 K for 3 h to form a homogeneous solution. 1 wt% (~0.5 vol.%) of the filler material (named as graphene worms [29]) was dispersed in 50 ml of a solution (distilled water+small volume of ethanol AR) by using probe sonicator. This solution was added drop by drop to PVA solution and stirred for 2 h to form a homogeneous solution which was then casted in a borosilicate glass Petri dish and was left overnight under room conditions. Next day, the cast was dried at 333 K to evaporate any residual water and ethanol content. This resulted in free-standing, flexible, and semitransparent (to visible light) FLG/PVA sheet (Fig. 8). The composite contained uniformly distributed but a network of FLG particles as revealed by the cross-sectional scanning electron micrograph (Fig. 9). A maximum AE of ~19.5 dB was obtained in

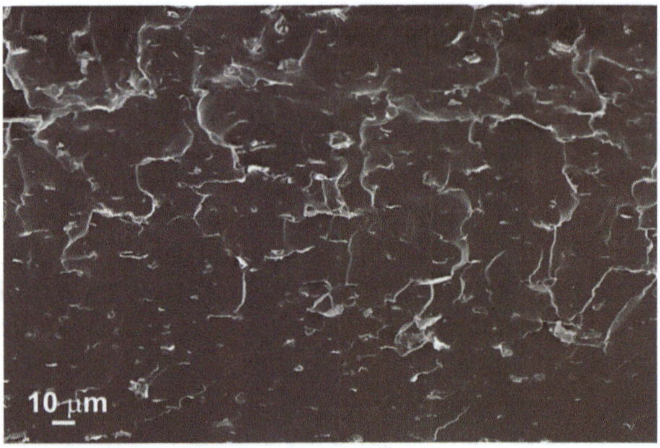

Fig. 9 Secondary electron micrograph of the cross section of 1 wt% FLG filled PVA composite

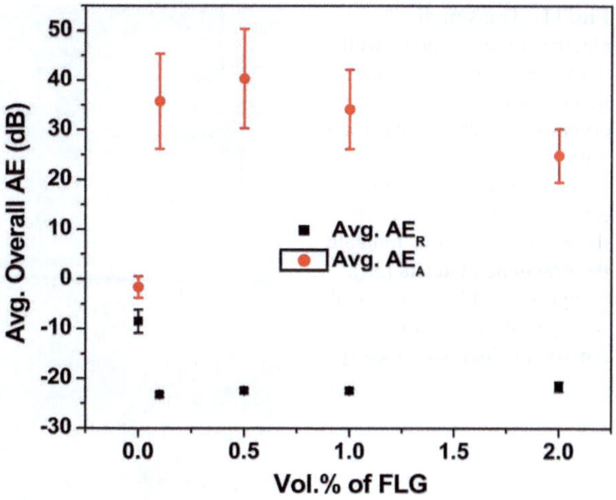

Fig. 10 Average AE vs. vol.% of FLG in PVA matrix

X-band range (8.2–12.4 GHz) for this composite. AE value decreased as vol% of FLG in the composite increased above 0.5 vol%. This is attributed to the dispersion instability of FLG in PVA matrix. It is evident from Fig. 10 that the major contribution to the overall AE is from the absorption process. Unfortunately, the dielectric behavior of the composites and its correlation to MA is still unknown.

5.3.4 Poly (Methyl Methacrylate) (PMMA) Composites

Similar to epoxy and PVA, PMMA is another popular and inexpensive polymer. PMMA is also known for its transparency and machinability. Graphene/PMMA composite microcellular foams for MA were synthesized by solution blending (followed by melt compounding) of PMMA with graphene sheets (GS) followed by foaming with subcritical CO_2 as a foaming agent [30]. The graphene platelets used in this work are constituted by 3–4 graphene layers and these platelets are similar to GNP particles used in preparation of epoxy composites [25]. Figure 11 shows the dispersion of graphene in a cell wall of the graphene/PMMA composite microcellular foam. Formation of microcellular cells has greatly improved the ductility and tensile toughness of the foam in comparison to the normal composite. The composite foam with graphene platelets loading of 1.8 vol.% exhibited a superior electrical conductivity (of 3.11 S/m) when compared to other compositions. This is typically due to the formation of an appropriate conductive percolating network among graphene platelets. This particular sample exhibited an overall AE of 13-19 dB in the frequency range 8–12 GHz. At the matching thickness of 1.5 mm 1.8 vol.% graphene platelets loaded composite foam exhibited a maximum absorption of 18 dB at ~9 GHz (Fig. 12) and a broad band width ranging from 9.5 to 14.6 GHz in which the maximum RL is more than 10 dB. Similar to the case of PVA composites, the MA of PMMA composite foam was not discussed in correlation with its dielectric behavior.

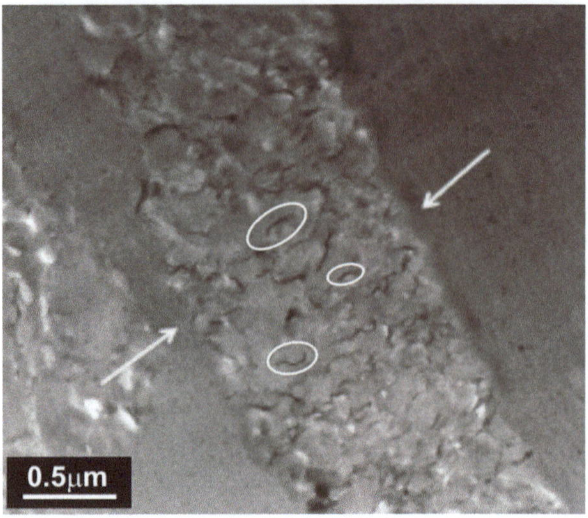

Fig. 11 Transmission electron micrograph of well-dispersed graphene platelets in a cellwall of the PMMA composite microcellular foam with 0.8 vol.% graphene sheets. The *arrows point* to the cell wall surfaces while the encircled regions indicate the graphene platelets [30]. Copyright 2011. Reproduced with permission from American Chemical Society

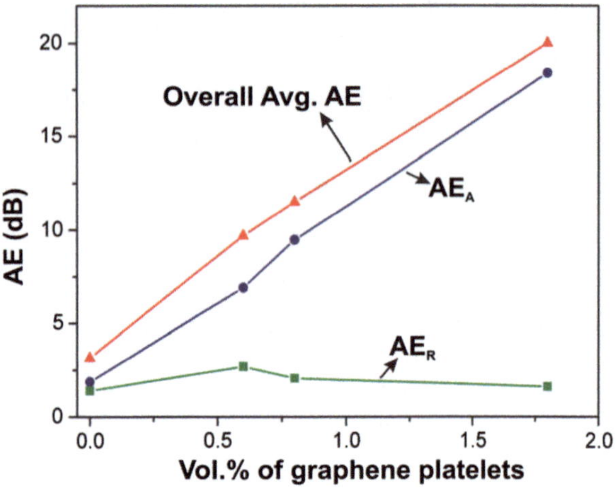

Fig. 12 Comparison of overall AE, microwave absorption (AE_A), and microwave reflection (AE_R) measured from 1.8 vol.% graphene loaded PMMA composite foam at 9 GHz [30]. Copyright 2011. Reproduced with permission from American Chemical Society

5.3.5 Poly Vinyledene Fluoride (PVDF) Composites

PVDF is resistant to chemicals and can sustain high temperatures unlike the polymers discussed so far. To avail the advantages of PVDF including its excellent dielectric characteristics, it has also been tried as a matrix material in preparing functionalized graphene (f-G) filled PVDF composite foam [31]. The thickness of each FLG particle used as filler in this work is 2–5 nm corresponding to 6–15 graphene layers stacked along the c-axis. The lateral dimensions of each FLG particle were in the range of 20–40 μm. FLG particles were first functionalized to obtain f-G particles which were then used in solution mixing and traditional foaming process to obtain f-G filled PVDF composite foam. The schematic of the synthesis is shown in Fig. 13.

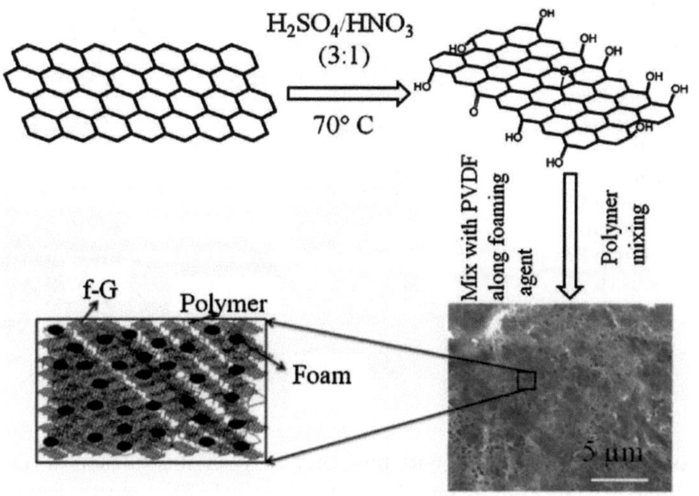

Fig. 13 Schematic representation of the preparation of f-G and f-G/PVDF composite foam [31]. Copyright 2011. Reproduced with permission from John Wiley & Sons

Different f-G loadings were used and the composites showed a low percolation threshold of 0.5 wt% owing to the high aspect ratio of the sheets and their homogeneous dispersion. Each cell in the foam had a uniform distribution of graphene sheets as revealed by high magnification scanning electron micrograph. DC electrical conductivity could be varied with f-G loadings similar to the case of SPFG filled epoxy composites. DC conductivity was measured as 10^{-16} S/m for pure PVDF while it was measured as 10^{-3} S/m for 0.5 wt% f-G filled PVDF composite foam. In the calculations, it was assumed that mass fraction and volume fraction of PVDF and f-G are almost the same. AE of 20 dB is measured in X-band (8–12 GHz) region and 18 dB in broadband (1–8 GHz) region for 5 wt% f-G filled PVDF composite foam. The increase in AE is attributed to the increase in conductivity of the composite foam owing the formation of conducting network of f-G fillers in PVDF matrix. As the loading of f-G increased the number of conducting f-G interconnections increased resulting in more interaction between the fillers as evidenced from the DC conductivity measurements. In turn this has improved the AE of the composite foam. The coefficients R, T, and A are measured as 0.78, 0.01, and 0.21, respectively for 5 wt% f-G loaded PVDF composite foam. This shows that AE in the frequency range 1–12 GHz of composite foam is mainly due to reflection but not absorption. This result is quite different from epoxy composites which showed stronger MA (more RL). It should be noted that the sizes of FLG particles used in the case of preparing PVDF composites are greater than those used in preparing epoxy composites. It FLG particles of even higher aspect ratio are used, the percolation threshold could have been even less leading to an improvement of the required properties. Availability of mechanistic studies on the dielectric behavior of different PVDF composites and on matching thickness

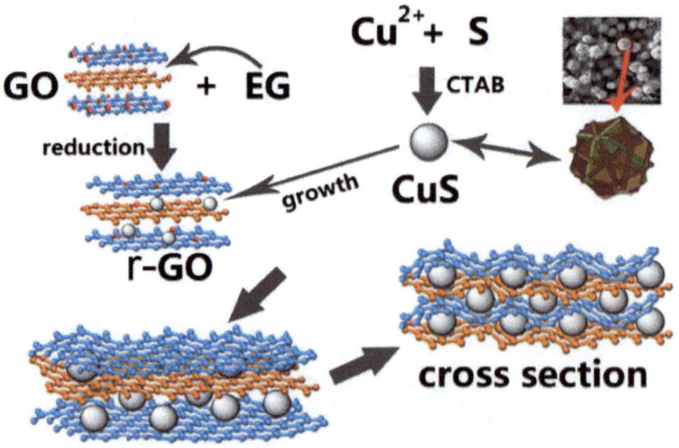

Fig. 14 Schematic representation of the preparation of CuS microspheres embedded in r-GO. EG is ethylene glycol and CTAB is cexadecyltrimethyl ammonium bromide [33]. Copyright 2013. Reproduced with permission from Royal Society of Chemistry

contribution to the overall AE will probably help in the further development of graphene-filled PVDF composites for MA. Here it is also worth mentioning that modified graphite filled PVDF composite [32] is far better than the composite foam discussed above for microwave absorption. Microwave absorption intensity of −64.4 dB was measured for modified graphite filled PVDF composite (it is also named as graphite intercalated composite) at 10.59 GHz and it was claimed that the composite was as flexible as PVDF even for 9 wt% modified graphite loading. It was elucidated in [32] that the microwave absorption capacity of the modified graphite filled composites is governed by their conductivity and dielectric behavior which in turn are related to the phenomena like Debye relaxation, Maxwell–Wagner relaxation, and electronic polarization taking place in the composites during their interaction with EM waves. Improvement in microwave absorption was obtained by incorporating microwave absorbing CuS particles along with r-GO in PVDF [33]. Wet chemical method was used to obtain CuS microspheres embedded in r-GO layers as shown in Fig. 14. This product was used as the filler by blending it homogeneously with PVDF solution before hot pressing to obtain CuS microspheres embedded in r-GO in PVDF matrix.

At a filler loading of 5 wt% (of CuS microspheres embedded in r-GO), 2.5 mm thick composite exhibited good microwave absorption (i.e., high RL values are measured) with a maximum RL of 32.7 dB at 10.7 GHz. RL values for different compositions at a fixed thickness of 2.5 mm and for the composite with a filler loading of 5 wt% are shown in Fig. 15a, b, respectively.

Higher filler (CuS microspheres embedded in r-GO) loadings lead to increase in the dielectric constant but did not result an optimal loss tangent which is very important to increase microwave absorption capacity. The variation in the loss tangent for a filler loading of 5 wt% can be accessed from Fig. 16. 4 semicircles (resembling Cole–Cole plot) are identified in Fig. 16 indicating the contribution of the Debye relaxation process. For convenience, the variation of dielectric constant with frequency for different compositions is shown in Fig. 17. It was observed that at higher filler loadings, microwave absorption capacity decreased and this was attributed to the contributions from leakage current, impedance match and skin effect. The

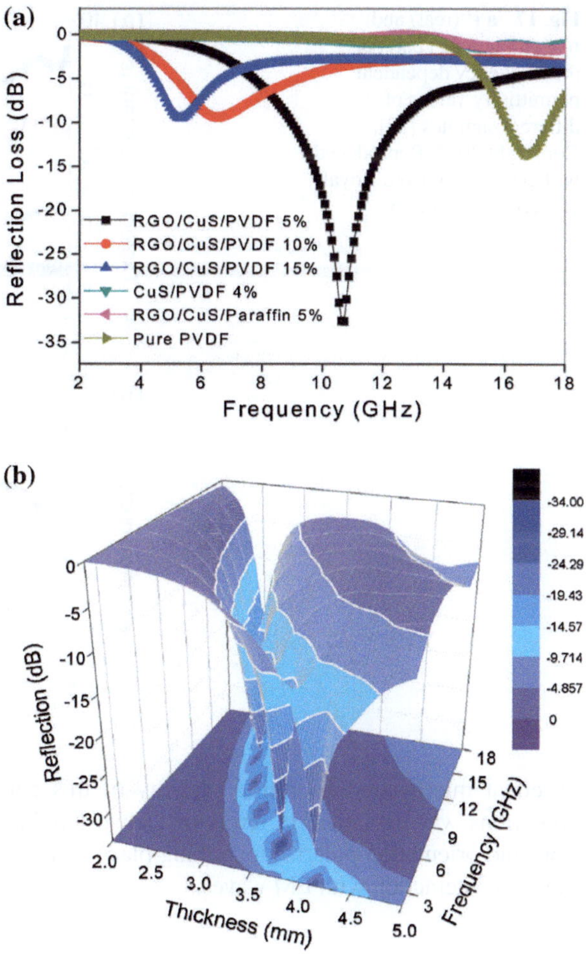

Fig. 15 **a** Reflection loss values in the frequency range of 2–18 GHz for different samples at a thickness of 2.5 mm and **b** reflection loss values of 5 wt% filler (CuS microspheres embedded in r-GO) loaded PVDF composite [33]. Copyright 2013. Reproduced with permission from Royal Society of Chemistry

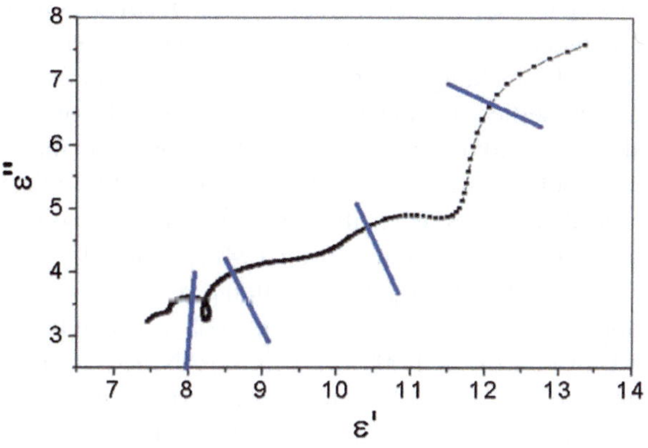

Fig. 16 ε″ (imaginary part) versus ε′ (real part) of the permittivity for the composite with 5 wt% filler loading [33]. Copyright 2013. Reproduced with permission from Royal Society of Chemistry

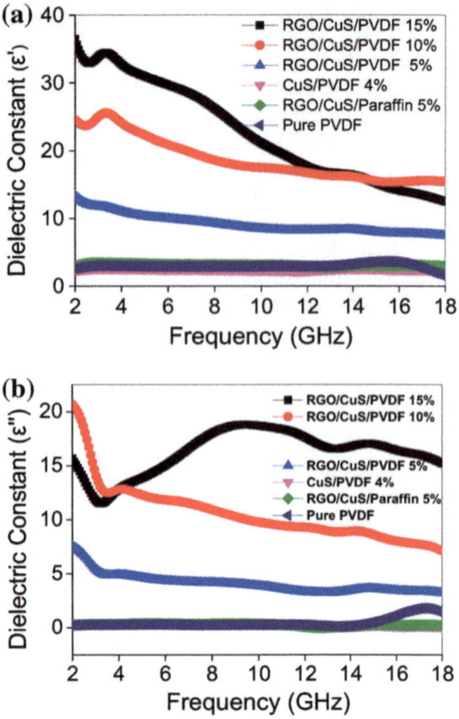

Fig. 17 **a** ε′ (real) and **b** ε″ (imaginary) parts of the frequency dependent permittivity values of different samples [33]. Copyright 2013. Reproduced with permission from Royal Society of Chemistry

excellent microwave absorption capacity of this composite under certain optimal conditions was explained in terms of Debye dipolar relaxation, interface polarization, and orientation polarization taking place simultaneously in the composites during their interaction with EM waves.

5.3.6 Graphene/Conducting Polymer Composites

Owing to their ease of preparation and electrical conductivity, intrinsically conducting polymers are probably the best candidates as matrix materials to host r-GO or FLG for EMI shielding applications. Even though poor- (i) formability and (ii) mechanical properties are the general drawbacks of conducting polymers, excellent EMI shielding and/or microwave absorbing materials that are constituted with polyacetylene, polyaniline (PANi), and polypyrrole (PPy) related components have been fabricated. Polyacetylene film when exposed to iodine attained metallic conductivity and was found to be useful for EMI shielding [34]. PANi grafted glass fabrics were synthesized using *p*-toluene sulfonic acid or camphor-10-sulfonic acid as primary dopant and 4-chloro-3-methyl phenol as secondary dopant. These fabrics showed AE of ~54 dB at 1,000 MHz [35]. PANi-silver and PANI-graphite composites [36] and PANI coated nickel carbon black composite in co-poly (ethylene-propylene) [37] have also been found suitable for EMI shielding. AE of fabric complexes coated with

PPy and thermally evaporated silver was found to be ~80 dB and was found to be useful in military applications [38]. It was also reported that the far field AE of PPy impregnated microporous polyethylene tapes was ~40–50 dB [39]. Different graphene (r-GO or FLG like materials) filled conducting polymer composites are discussed in the following paragraphs.

A. Polyethylene Oxide (PEO) Composites

A green approach to synthesize chemically reduced graphene (CR-G)-poly (ethylene oxide) (PEO) composite through simple aqueous mixing method was reported [40]. GO was prepared by Hummers method and then it was in situ reduced in PEO by eco-friendly L-ascorbic acid. The thickness of each CR-G sheet is ~1 nm while its lateral size is ~1.5 μm. In situ reduction of GO to r-GO or in this case CR-G generally overcomes the problem of aggregation of the fillers as commonly observed while using completely reduced GS as the fillers. Electron microscopy has shown excellent dispersion of the fillers in matrix. X-ray diffraction analysis corroborated this observation. GO with its enhanced interlayer spacing allows polymers to intercalate while the polymers like PVA and PEO can interact with the filler through hydrogen bonding resulting in the presence of exfoliated GO (i.e., monolayers after in situ reduction) in final composite. In other words, polymer matrix materials like PVA and PEO act as barriers for any agglomeration of the graphene fillers. It is noteworthy that such cases have shown better microwave absorption capacity.

PEO composites (0.54 and 2.6 vol.% CR-G filled PEO) were found to exhibit very high ε′ and ε″ values in comparison to PEO. The dielectric loss tangent versus frequency for the composites in comparison to PEO is shown in Fig. 18. The composites exhibited higher loss tangent values than PEO. The composites exhibited similar loss tangent values up to ~7.6 GHz. For frequencies greater than 7.6 GHz, 2.6 vol.% CR-G filled PEO composite exhibited greater loss tangent values than those exhibited by 0.54 vol.% CR-G filled PEO composite.

2.6 vol.% CR-G filled PEO composite exhibited an excellent MA capacity. RL values pertaining to this composite with different thicknesses are shown in Fig. 19.

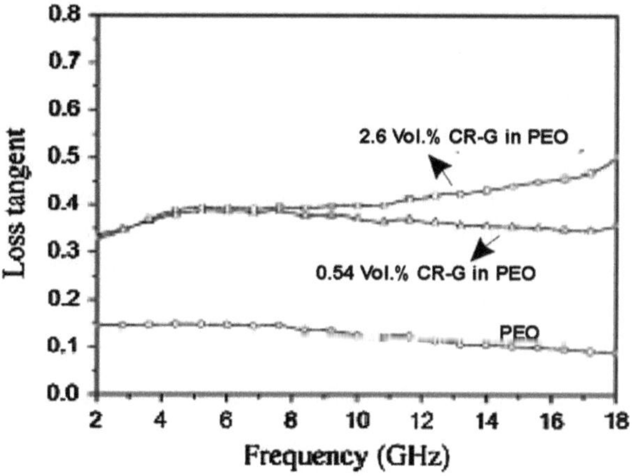

Fig. 18 The variation of loss tangent values with frequency pertaining to PEO composites in comparison with those of PEO in the frequency range 2–18 GHz [40]. Copyright 2011. Reproduced with permission from American Chemical Society

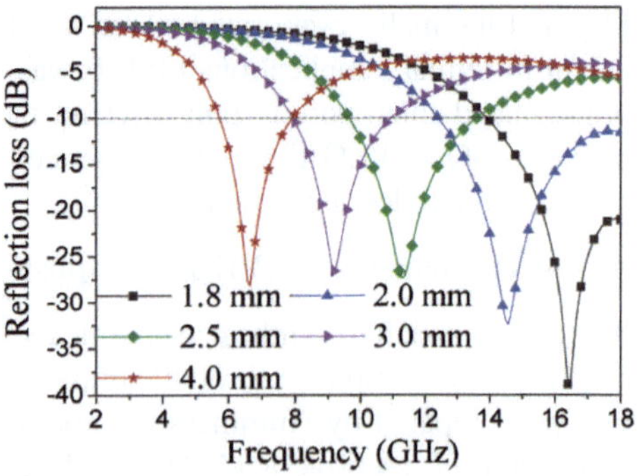

Fig. 19 Reflection loss curves pertaining to 2.6 vol.% CR-G filled PEO composite with different thickness in the frequency range of 2–18 GHz [40]. Copyright 2011. Reproduced with permission from American Chemical Society

For all the thicknesses (ranging from 1.8–4 mm) considered, the RL values were less than −25 dB in the considered frequency range. For the thickness of 1.8 mm, the least RL of −38.8 dB was measured. 0.54 vol.% CR-G filled PEO composite can also be considered as a good microwave absorbing material as it exhibited RL values of less than −10 dB for sample thicknesses ranging from 3–5 mm. The superiority of 2.6 vol.% CR-G filled PEO composite in absorbing microwaves over 0.54 vol.% CR-G filled PEO composite can be correlated to its higher loss tangent values. Importantly for frequencies greater than 7.6 GHz, the imaginary part of the dielectric permittivity has attained an optimal value (neither too high nor too low) whilst the real part of the dielectric permittivity decreased with increase in the frequency.

B. Polyaniline (PANi) Composites

Yet another composite which is found to be useful for MA is graphene sheets (GS) filled PANi composite [41]. GS were prepared by using hydrogen arc-discharge method. Each sheet had a stack of 2–10 graphene layers. GS and PANi powders in different proportions are first mixed and grounded. The mixed powders are smashed in the presence of an alcohol by using ultrasonic cell crusher and further ultrasonicated to obtain suspensions of composite powders. The dried composite powders are molded into sheets by applying a compressive pressure of 3 MPa for 1 min [42]. The thickness of the composite sheets was 2.4 mm. The morphology analysis showed that GS are not only coated with PANi but are uniformly dispersed in PANi matrix. The best overall AE for 33 wt% GS filled PANi was measured as 34.2 dB. AE_A was found to be greater than AE_R for this composite. AE_A values against the overall AE are found to be 66, 68, 75, 79, and 81 % at 2, 6, 10, 14 and 18 GHz, respectively. The reason for good MA was attributed to improved electrically conductive network in the composite. RL values lower than −40 dB are measured for r-GO-PANi composite systems [43, 44]; but these values are obtained by incorporating PANi coated r-GO in wax and not by directly making PANi composites.

C. Polystyrene (PS) Composites

High pressure (1,000 MPa) compression molding followed by salt-leaching was used to prepare an excellent microwave absorbing functionalized graphene sheets (FGS) filled PS porous composite [45]. The schematic of the synthesis procedure is given in Fig. 20.

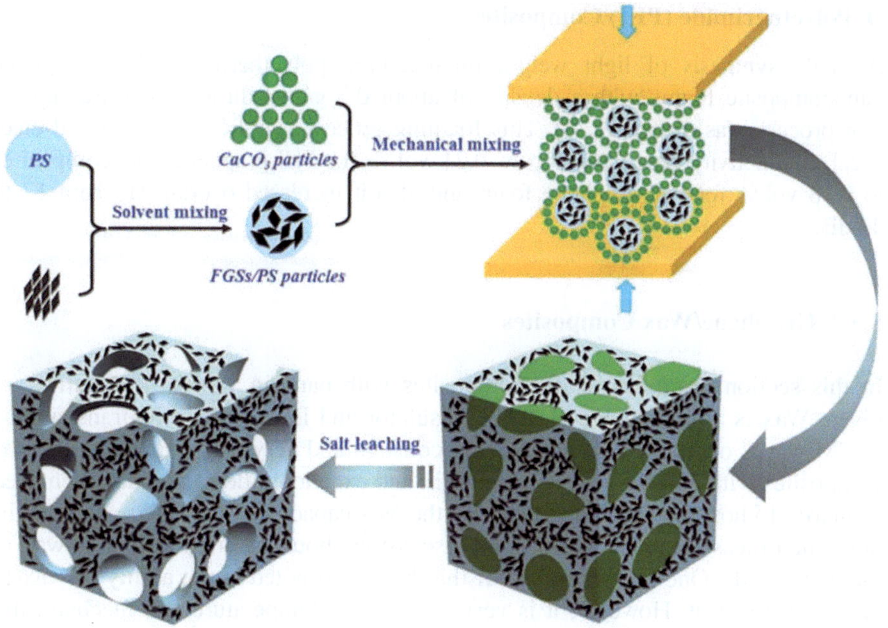

Fig. 20 Schematic for the fabrication of porous FGS/PS composite [45]. Copyright 2012. Reproduced with permission from American Chemical Society

5.6 vol.% (~30 wt%) FGS filled PS composite exhibited an excellent MA capacity. The thickness of the sheet was 2.5 mm and the density of the composite was only 0.45 g/cm^3. For this composite the average overall AE, AE_A, and AE_R (Fig. 21) are measured as 29, 27.7, and 1.3 dB, respectively. Interestingly the conductivity of the sample was 1.25 S/m which is greater than the required value of 1 S/m for optimal attenuation.

Fig. 21 Comparison of overall attenuation effectiveness (AE), microwave absorption (AE_A), and microwave reflection (AE_R) [45]. Copyright 2012. Reproduced with permission from American Chemical Society

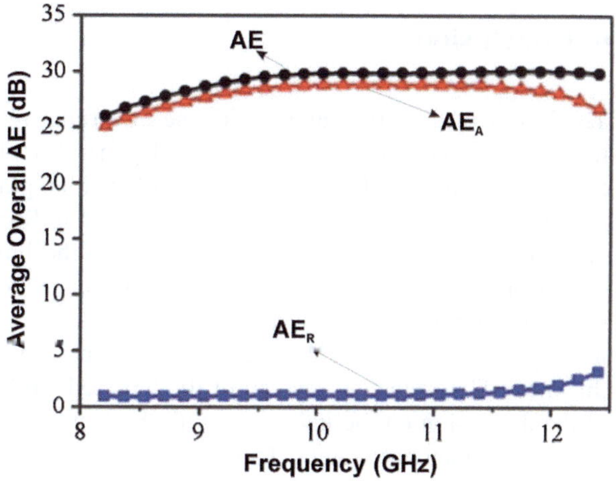

D. Polyetherimide (PEI) Composites

A facile synthesis of light weight microcellular polyetherimide (PEI)-graphene nanocomposite foams with a density of about 0.3 g/cm^3 through a phase separation process was reported [46]. This foaming process significantly decreased electrical conductivity percolation from 0.21 vol.% for PEI-graphene nanocomposites to 0.18 vol.% for PEI-graphene foam and also it increased overall AE from 17 to 44 dB.

5.3.7 Graphene/Wax Composites

In this section examples of the composites with paraffin wax as the matrix are given. Wax is a well-known electrical insulator and is known for its transparency to EM waves owing to its low dielectric constant and very low tangent loss. In the composites with paraffin wax as the matrix and excellent fillers, wax has been used primarily to host the fillers and measure the MA capacity of the fillers. Wax facilitates such measurements. There is no discussion about the contribution of wax to the overall AE. One important aspect that has to be noted is the ability of wax to be a sink for heat. However, it is very sensitive to temperature and mechanically not flexible when compared to other polymers and therefore may not be viable as a matrix material for preparing graphene-based composites for real applications. Hexgonal Ni nanocrystals decorated monolayer graphene [47], polycrystalline Fe_3O_4 nanoparticles decked r-GO/FLG [48–50], and submicron sized $CoFe_2O_4$ hollow spheres decorated on single sheet r-GO [51] filled wax composites showed an excellent MA capacity (RL less than -10 dB corresponding to 90 % MA for different thicknesses of the wax composite samples). These composites have shown far superior microwave absorption capacity than the typical nonmagnetic r-GO/FLG particles filled wax composites [52, 53].

6 Conclusion

The basics of absorbers along with the governing relationships and a review of a few primary measurement methods and general applications are discussed under Sects. 1–4. The details of the behavior of various graphene/polymer nanocomposites at the microwave frequencies are presented in Sect. 5. The discussion was only on absorbers in the form of a sheet owing to the limited literature available. It is clear from the discussion that dielectric and magnetic losses are primarily responsible for microwave absorption. These loses can be controlled by adjusting factors namely nature, distribution and content of fillers, nature of the matrix, and matching thickness. Unfortunately several, of the reported works are incomplete with respect to the discussion on the influence of all the factors on the microwave absorption of a particular material system. This is probably owing to the difficulties in preparing the samples for testing and hence the field is open for research. It is well known that design of the absorber can also influence the absorption characteristics. In the case of graphene/polymer nanocomposites, there are no research works yet that discuss about designs other than the sheet form. This is one aspect that is open for research and development. In the case of discussed examples, wherein there is no problem of formability and mechanical stability, there was inadequacy in the required dielectric properties of the composite material. In the cases wherein the properties were extremely good, formability and mechanical stability are a matter of concern. Nonetheless, there are examples of extremely good microwave absorbers. It is well known that metallic magnetic materials and ferrites have strong EM attenuation property owing to their high complex permeability. In this context, researchers have recently tried to decorate graphene (and related forms) with such magnetic materials and then use these structures as fillers in polymers in order to generate both dielectric and magnetic losses in the composite. For example, high dielectric and magnetic losses have been observed to contribute to the MA value of as high as 26 dB (attenuation >99.7 %) in PANi filled with Fe_3O_4 decorated r-GO [54]. In another recent work, γ-Fe_2O_3 decorated r-GO-PANi core–shell structures containing composite sheet exhibited AE_A as high as 43 dB [55].

The use of graphene/polymer nanocomposites for MA can still be considered at infancy owing to the inadequacy of the available information. This mainly owes to the difficulties in synthesizing controlled samples for testing. This in turn has not helped in fully understanding the science governing the absorption mechanism. For example, the presence of any cross-coupling among Debye relaxation, Maxwell–Wagner relaxation, and electronic polarization has to be ascertained. Local characterization (graphene/polymer interface) of charge distribution is experimentally challenging. This forms an interesting area of research. Probing the large shape anisotropy (which in turn increase the microwave absorption) that the magnetic graphene structures can bring-in is also worth pursuit. Since high specific surface area helps in functionalization of graphene, a great variety of functionalized graphene filled polymers can be synthesized. Similarly, the high aspect ratio of graphene allows designing composites with varied filler contents closely following the percolation theory. Last but not the least, industrial production of graphene/polymer composites which are already well-understood demands immediate attention.

References

1. Arthur R Von Hippel, Dielectric Materials and Applications, Artech House, 1995
2. L. F. Chen, C. K. Ong, C. P. Neo, V. V. Varadan, and Vijay K. Varadan Microwave Electronics: Measurement and Materials Characterization, 2004, Wiley
3. Paul Dixon, IEEE Microwave magazine, 2005, 6 (2), 74
4. Y. J. An, K. Nishida, T. Yamamoto, S. Ueda and T. Deguchi, Journal of Ceramic Processing Research, 2008, 9(4), 430
5. S.-H. Park, P.Theilmann, P. Asbeck, and P. R. Bandaru, The IEEE Transactionson Nanotechnology, 2009, 13, 1
6. W. B Weir, Proceedings of the IEEE, 1974, 62, 1, 33

7. M. D. Janezic and J. A. Jargon, IEEE Microwave and Guided Wave Letters, 1999, 9(2), 76
8. K. Sudheendran, K. C. James Raju, M. Ghanashyam Krishna, and Anil. K Bhatnagar, URSI Proceedings, 2006, ProcGA05/pdf/D06b.4(0873)
9. D. V. Blackham and R. D. Pollard, IEEE Transactions on Instrumentation and Measurement, 1997, 46(5), 1093
10. P. F. Goldsmith, "Quasi-Optical Techniques", Proceedings of The IEEE, 1992, 80, 11
11. J. Baker-Jarvis, E. J. Vanzura, and W. A. Kissick, IEEE Transactions on Microwave Theory and Technology, 1990, 38, 1096
12. K. S. Novoselov, A. K. Geim, S. V. Morozov, D. Jiang, Y. Zhang, S. V. Dubonos, I. V. Grigorieva, and A. A. Firsov, Science, 2004, 306, 666
13. A. K. Geim and K. S. Novoselov, Nature Materials, 2007, 6 (3), 183
14. A. A. Balandin, S. Ghosh, W. Bao, I. Calizo, D. Teweldebrhan, F. Miao, and C. N. Lau, Nano Letters, 2008, 8, 902
15. K. I. Bolotin K. J. Sikes, Z. Jiang, M. Klima, G. Fudenberg, J. Hone, P. Kim, and H. L. Stormer, Solid State Communications, 2008, 146, 351
16. M. D. Stoller, S. Park, Y. Zhu, J. An, and R. S. Ruoff, Nano Letters, 2008, 8, 3498
17. A. K. Geim, Science, 2009, 324, 1530
18. C. Wang, X. Han, P. Xu, X. Zhang, Y. Du, S. Hu, J. Wang, and X. Wang, Applied Physics Letters, 2011, 98, 072906
19. F. Qin and C. Brosseau, Applied Physics Letters, 2012, 100, 046101
20. C. Wang, X. Han, P. Xu, X. Zhang, Y. Du, S. Hu, J. Wang, and X. Wang, Applied Physics Letters, 2012, 100, 046102
21. R. E. Colin, Foundations of Microwave Engineering (McGraw Hill, NewYork, 1966)
22. J. Du and H.-M. Cheng,Macromolecular Chemistry and Physics, 2012, 213, 1060
23. H. Kim, A. A. Abdala, and C. W. Macosko, Macromolecules, 2010, 43, 6515
24. J. Liang, Y. Wang, Y. Huang, Y. Ma, Z. Liu, J. Cai, C. Zhang, H. Gao, and Y. Chen, Carbon, 2009, 47, 922
25. Z. Wang, J. Luo, and G.–L. Zhao, AIP Advances, 2014, 4, 0171239
26. A. Joshi, A. Bajaj, R. Singh, P. S. Alegaonkar, K. Balasubramanian, and S. Datar, Nanotechnology, 2013, 24, 455705
27. D. V.Kosynkin, A. L. Higginbotham, A. Sinitskit, J. R. Lomeda, A. Dimiev, B. K. Price, and J. M. Tour, Nature, 2009, 458, 872
28. S. K. Marka, M. Tech. Thesis, University of Hyderabad, 2013
29. R. N. Kumar, P. Shaikshavali, V. V. S. S. Srikanth, and K. B. S. Rao, AIP Conference Proceedings, 2013, 1538, 262
30. H.-B. Zhang, Q. Yan, W.-G. Zheng, Z. He, and Z.-Z. Yu, ACS Applied Materials & Interfaces 2011, 3, 918
31. V. Eswaraiah, V. Sankaranarayanan, and S. Ramaprabhu, Macromolecular Materials and Engineering 2011, 296, 894
32. G.-S. Wang, X.-J. Zhang, Y.-Z. Wei, S. He, L. Guo, and M.-S. Cao, Journal of Materials Chemistry A, 2013, 1,7031
33. X.-J. Zhang, G.-S. Wang, Y.-Z. Wei, L. Guo, and M.-S. Cao, Journal of Materials Chemistry A, 2013, 1, 12115
34. H. Shirakawa, E.J. Louis, A.G. MacDiarmid, C.K. Chiang, and A.J. Heeger, Journal of Chemical Society, Chemical Communications, 1977, 578
35. S. Geetha, K.K. Satheesh, and D.C. Trivedi, Composites Science and Technology, 2005, 65, 973
36. C.Y. Lee, H.G. Song, K.S. Jang, E.J. Oh, A.J. Epstein, and J. Joo, Synthetic Metals, 1999, 102, 1346
37. P. Kathirgamanathan, Journal of Materials Chemistry, 1993, 3, 259
38. Y.K. Hong, C.Y. Lee, C.K. Jeong, J.H. Sim, K. Kim, J. Joo, M.S. Kim, J.Y. Lee, S.H. Jeong, and S.W. Byun, Current Applied Physics, 2001, 1, 439
39. P. Kathirgamanathan, Advanced Materials, 1993, 5, 281
40. X. Bai, Y. Zhai, and Y. Zhang, Journal of Physical Chemistry C, 2011, 115, 11673
41. B. Yuan, L. Yu, L. Sheng, K An, and X. Zhao, Journal of Physics D: Applied Physics, 2012, 45, 235108

42. J. H. Du, L. Zhao,Y. Zeng, L. L. Zhang, F. Li, P. F. Liu, and C. Liu, Carbon, 2011,49, 1094
43. H. Yu, T. Wang, B. Wen, M. Lu, Z. Xu, C. Zhu, Y. Chen, X. Xue, C. Sun, and M. Cao, Journal of Materials Chemistry, 2012, 22, 21679
44. P. Liu and Y. Huang, Journal of Polymer Research, 2014, 21, 430
45. D.-X. Yan, P.-G. Ren, H. Pang, Q. Fu, M.-B. Yang, and Z.-M. Li, Journal of Materials Chemistry, 2012, 22, 18772
46. J. Ling, W. Zhai, W. Feng, B. Shen, J. Zhang, and W. Zheng, ACS Applied Materials & Interfaces, 2013, 5, 2677
47. T. Chen, F. Deng, J. Zhu, C. Chen, G. Sun, S. Ma, and X. Yang, Journal of Materials Chemistry, 2012, 22, 15190
48. X. Sun, J. He, G. Li, J. Tang, T. Wang, Y. Guo, and H. Xue, Journal of Materials Chemistry C, 2013, 1, 765
49. X. Li, H. Yi, J. Zhang, J. Feng, F. Li, D. Xue, H. Zhang, Y. Peng, and N. J. Mellors, Journal of Nanoparticle Research, 2013, 15, 1472
50. G. Wang, Z. Gao, G. Wan, S. Lin, P. Yang, and Y. Qin, Nano Research 2014, 7(5), 704
51. M. Fu, Q. Jiao, Y. Zhao, and H. Lia, Journal of Materials Chemistry A, 2014, 2, 735
52. W.-L. Song, M.-S. Cao, M.-M. Lu, J. Liu, J. Yuan, and L.-Z. Fan, Journal of Materials Chemistry C, 2013, 1, 1846
53. W.-L. Song, M.-S. Cao, M.-M. Lu, J. Yang, H.-F. Ju, Z.-L. Hou, J. Liu, J. Yuan, and L.-Z. Fan, Nanotechnology, 2013, 24, 115708
54. K. Singh, A. Ohlan, V.H. Pham, R. Balasubramaniyan, S. Varshney, J. Jang, S.H. Hur, W.M. Choi, M. Kumar, S.K. Dhawan, B.-S. Kong, and J.S. Chung, Nanoscale, 2013, 5, 2411
55. A.P. Singh, M. Mishra, P. Sambyal, B.K. Gupta, B.P. Singh, A. Chandra, and S.K. Dhawan, Journal of Materials Chemistry A, 2014, 2, 3581

Graphene Nanocomposites for Electromagnetic Induction Shielding

Yang Li and Wentao Zhai

Abstract The unique properties of graphene, such as high specific surface area, aspect ratio and electrical conductivity, make it very promising to fabricate electromagnetic induction (EMI) shielding materials. In this chapter, we first made a brief introduction about the development of EMI shielding materials as well as the preparation of graphene and polymer/graphene nanocomposites (PGNs). Typical surface modification of graphene to optimize its dispersion within polymer matrix was reviewed later. After that, we presented critical factors for the EMI shielding effectiveness (SE) of PGNs in detail. Meanwhile, the EMI shielding mechanism was introduced associated with corresponding examples.

Keywords Electromagnetic interference · Shielding effectiveness · Graphene · Polymer · Nanocomposites · Electrical conductivity

Abbreviations

PGNs	Polymer/graphene nanocomposites
EMI	Electromagnetic interference
SE	Shielding effectiveness
CNF	Carbon nanofiber
MWCNTs	Multiwalled carbon nanotubes
G–O	Graphite oxide
GO	Graphene oxide
RGO	Reduced graphene oxide
GNSs	Graphene nanosheets
G@Fe_3O_4	Fe_3O_4-decorated graphene

Y. Li · W. T. Zhai (✉)
Ningbo Institute of Material Technology and Engineering,
Chinese Academy of Sciences, Ningbo 315201, Zhejiang Province, China
e-mail: wtzhai@nimte.ac.cn

K.K. Sadasivuni et al. (eds.), *Graphene-Based Polymer Nanocomposites in Electronics*, Springer Series on Polymer and Composite Materials,
DOI 10.1007/978-3-319-13875-6_13

FGS	Functionalized graphene sheets
Fe_2O_3	Ferric oxide
Fe_3O_4	Ferroferric oxide
PMMA	Polymethylmethacrylate
PS	Polystyrene
PVA	Poly(vinyl alcohol)
PEI	Polyetherimide
PC	Polycarbonate
PSFG	Polystyrene-functional graphene
P-GNS	Pristine graphene nanosheets
S-GNS	STAC-absorbed-GNSs
PANI	Polyaniline
PVDF	Polyinylidene fluoride
PU	Polyurethane
WPU	Water-borne polyurethane
PDA	Polydopamine
EVA	Ethylene-vinyl acetate
STAC	Stearyl trimethyl ammonium chloride
WVIPS	Water vapor-induced phase separation

1 Introduction

The electromagnetic induction (EMI) is defined as electromagnetic radiation emitted by electrical circuits under current operation [1]. To date, electronic products have been widely used in many fields ranging from commercial to military electronic devices [2, 3], leading enormous electromagnetic radiation in our surrounding environment to a high level that has never achieved before. As we know, electromagnetic radiation at high frequencies (e.g., radio waves) may disturb the proper operation of electrical equipment nearby and do dreadful radiative damage to human beings [4]. Herein, the development of EMI shields becomes a significantly meaningful work and exerts a peculiar fascination on a great many researchers.

In 1821, Michael Faraday first introduced the concept that an enclosed conductive house has zero electric field, which has been well known to us all as "Faraday cage." From then on, it is the principle that exactly established the foundation of today's electromagnetic shielding technology. Metals like copper and aluminum are the most common materials for EMI shielding due to their high electrical conductivity, and the shielding mechanism is reflection which stems from the interaction between free electrons in metals and external fields [4]. However, properties of metals, such as high density and easily corrosion, restrict their usage to a large extent as EMI shields in aerospace industries. In order to solve these problems, polymer-based composites with effective EMI Shielding effectiveness (SE) have developed and gradually come into fashion. There are various range of composite filler, including metallic and magnetic nanoparticles [5–7], intrinsically conductive polymers like polyaniline (PANI) [8–10], and carbon particles (carbon black, carbon fibers, carbon nanotubes, etc.) [11–23].

Graphene, a flat monolayer of carbon atoms tightly packed into a two-dimensional honeycomb lattice, rose to fame swiftly since it was reported by Novoselov et al. [24]. Because of its unique properties: excellent thermal conductivity (~5,000 W m^{-1} K^{-1}) [25], remarkable mobility of charge carriers (200,000 cm^2 V^{-1} s^{-1}) [26], high specific surface area (2,630 m^2 g^{-1}) [27], great values of Young's modulus (~1,100 GPa) [28], graphene is regarded as an ideal candidate to take over carbon nanotubes (CNT) in various fields, such as electrode, supercapacitor, oxidation resistance, transistor, Lithium ion batteries, fuel cells, solar cells, photocatalysis, Raman enhancement, and the EMI shielding applications [29–44].

In this chapter, we will initially introduce something about the preparation of graphene and polymer/graphene nanocomposites (PGNs) as EMI shields, together with surface modification of graphene. Then the EMI shielding theory will be briefly summarized and critical factors for the EMI SE of PGNs will be discussed in detail associated with corresponding experimental results, including the electrical conductivity, magnetic permeability, thickness and structures of PGNs. Meanwhile, the EMI shielding mechanism will be discussed.

2 Preparation of PGNs

Prior to preparation of PGNs, synthesis of graphene will be given first. Then to acquire a good dispersion of graphene within polymer matrix, chemical modification of graphene containing both covalent and noncovalent methods is presented. Finally, typically processing methods including in situ polymerization, solution and melt blending will be summarized briefly.

2.1 Synthesis of Graphene

For years, tremendous efforts have been made in the preparation of graphene, such as chemical vapor deposition (CVD) [45–47], micromechanical exfoliation of graphite [24], epitaxial growth on the surface of SiC [48], and the exfoliation and reduction of graphite oxide (G–O) [49–52]. As we know, synthesis routes like CVD or epitaxial growth are too expensive for large-scale preparation of graphene and the micromechanical exfoliation of graphite does hardly control. Therefore, the exfoliation and reduction of G–O seems to be the only one left. Generally, the exfoliation and reduction of G–O either can be realized by chemical or thermal treatment. Some researchers prefer thermal techniques to chemical ones for the simple reason that the enormous usage of reluctant (hydrazine [50] and $NaBH_4$ [53]) for chemical methods does harm to environment. However, a high temperature of 1,050 °C and heating rate of about 2,000 °C min^{-1} are required for thermal method to ensure the complete exfoliation of G–O [52].

As we know, high temperature always means huge consumption of energy. By taking advantage of vacuum condition, our group managed to synthesis graphene from G–O at a low temperature of 135 °C [54]. The expanded force during the exfoliation of G–O originates from not only the release of gas like H_2O and CO_2 by the decomposition of oxygen-containing functional groups in G–O, but also the negative pressure perpendicular to the surface of G–O produced by vacuum condition. We also substituted volatile substance for vacuum condition in the exfoliation of G–O [55]. The force resulting from the evaporation of volatile substance like HCl at 135–145 °C may surpass the van der Waals force within graphite interlayers, resulting in the full exfoliation of G–O under ambient atmosphere. Additionally, the as-prepared graphene contains a number of functional groups, which is helpful for further chemical modification of graphene.

2.2 Chemical Modification of Graphene

When graphene sheets are dispersed in water or some organic solvents, they tend to irreversibly agglomerate due to its hydrophobic nature. To address this issue, people proposed surface modification of graphene, which consists of covalent and noncovalent methods.

2.2.1 Covalent Modification of Graphene

With covalent modification of graphene, it is inevitable to mention the well-known "grafting from" and "grafting to" methods. The "grafting from" relies on immobilization of initiators at surface of graphene, followed by in situ surface initiated polymerization. Fang et al., for example, functionalized graphene with polystyrene (PS) via atom transfer radical polymerization [56]. The other method, "grafting to," is performed by bonding the terminated-functionalized polymer chains to surface of graphene. Considering the existence of two terminated epoxide groups in epoxy chains, our group coupled successfully epoxy chains onto graphene oxide (GO) via the "grafting to" technique [57]. To further simplify the preparing process, we also readily functionalized graphene sheets with poly(vinyl alcohol) (PVA) chains by ultrasonication [58]. In addition, the oxygen-containing groups remaining in GO plane like hydroxyl and carboxyl groups afford tremendous opportunities to the occurrence of amidation and esterification reaction. Stankovich et al. explored the addition of isocyanates in GO aqueous solution [59]. The formation of homogeneous colloidal suspension in polar solvent like *N*, *N*-dimethylformamide confirmed that the isocyanate compound had been attached to the surface of GO via amidation and esterification reaction.

2.2.2 Noncovalent Modification of Graphene

Though the grafting methods may improve the dispersion of graphene in polymer matrix, the covalent modification of graphene does certainly disrupt the conjugated graphene sheets and exert a negative impact on intrinsic electrical conductivity of graphene, and consequently EMI SE of PGNs. In contrast, noncovalent methods are more promising as the conjugated structure of graphene remains unaffected. Especially, when excellent electrical conductivity and effective EMI SE are required, noncovalent methods receive more attention.

Generally, traditional noncovalent methods are composed of π-π stacking interactions, van der Waals force, hydrogen bonding, electrostatic, and coordination bonding, among which the π-π stacking interactions draw more attention owing to its strong strength comparable to covalent bonding. With the formation of π-π stacking, two relatively large nonpolar aromatic rings having overlapping π orbitals are always needed. Su et al., for instance, verified that molecules with large aromatic structures and extra negative charges are noncovalently immobilized on the basal plane of graphene sheets via strong π-π interactions [60]. By melt blending, we realized the formation of π-π stacking between PS/polycarbonate (PC) and graphene [61, 62]. Both the UV-vis absorption spectroscopy of PS-functional graphene and that of PC-functional graphene exhibited an obvious red shift, indicating the formation of π-π stacking.

2.3 *Preparation Methods of PGNs*

Strategies for preparing PGNs are commonly composed of three segments: in situ polymerization, solution, and melt blending. Each of them has advantages and disadvantages alike. Table 1 has listed different preparation techniques of PGNs, as well as its corresponding EMI SE. Details are as follows:

2.3.1 In Situ Polymerization

In situ polymerization is initially performed with mixture between graphene and GO with monomers. Then initiator is added and polymerization begins. Homogeneous dispersion of incorporated fillers, as well as strong interactions between composite fillers and polymer matrix can be attained in this way. PGNs as EMI shields made in this way mainly concentrate on epoxy/graphene [63], PANI/graphene [64], PANI/graphene@Fe_3O_4 [65], and polyurethane (PU)/graphene [66]. The decrease of the polymerization rate at the later stages of polymerization is its drawback [67].

Table 1 A comparison for the EMI SE of different polymer/graphene systems

Matrix/filler	Preparation methods	Percolation threshold	Filler loading	Thickness (mm)	DC EC (S m^{-1})	EMW FR (GHz)	EMI SE (dB)	TDEMI SM	Ref.
Epoxy/SPFG	In situ	0.52 vol%	15 wt%	–	–	8.2–12.4	21	–	[63]
			8.8 vol%						
PS/FGS	Solution	–	30 wt%	2.5	0.22	8.2–12.4	17.3	Absorption	[68]
			3.2 vol%						
PS/FGS	Solution	–	30 wt%	2.5	1.25	8.2–12.4	29.3	Absorption	[68]
			5.6 vol%						
PS/G-MWCNT	In situ solution	–	3.5 wt%	5.6	–	8.2–12.4	20.2	Reflection	[72]
PMMA/G-5.0	Solution	0.92 vol%	2.67 vol%	3.4	–	8–12	15	Absorption	[69]
PMMA/G-9.6	Solution	0.77 vol%	2.67 vol%	3.4	–	8–12	17	Absorption	[69]
PMMA/G-13.2	Solution	0.62 vol%	2.67 vol%	3.4	10	8–12	25	Absorption	[69]
PMMA/G	Solution	–	5.0 wt%	2.4	3.11	8–12	13–19	Absorption	[74]
			1.8 vol%						
PVDF/graphene	Solution	0.5 wt%	7 wt%	–	–	1–8	28	Reflection	[71]
						8-12	20		
PEI/G	Solution	0.21 vol%	10 wt%	2.3	4.8×10^{-6}	8–12	20	Absorption	[70]
			5.87 vol%						
PEI/G	Solution	0.18 vol%	10 wt%	2.3	2.2×10^{-5}	8–12	11	Absorption	[70]
			1.38 vol%						
PEI/G@Fe_3O_4	Solution	–	10 wt%	2.5	–	8–12	14.3–18.2	Absorption	[89]
PVA/G	Solution	0.0075 wt%	0.025 wt%	0.6	–	8.2–12.4	45	–	[106]
PVA/MG	Solution	–	–	0.36	–	8.2-1.4	20.3	Absorption	[90]
PU/FGS	In situ	0.1 vol%	4.75 vol%	2	–	0.03–2	17.6	Absorption	[81]
WPU/P-GNS	Solution	–	5 vol%	2.0	0.97	8.2–12.4	21	Absorption	[80]

(continued)

Table 1 (continued)

Matrix/filler	Preparation methods	Percolation threshold	Filler loading	Thickness (mm)	DC EC ($S\ m^{-1}$)	EMW FR (GHz)	EMI SE (dB)	TDEMI SM	Ref.
WPU/S-GNS	Solution	–	5 vol%	2.0	5.1	8.2–12.4	32	Absorption	[80]
EVA/G	Solution	–	60 vol%	1	2.5×10^{-2}	8.2–12.4	27	Reflection	[76]
PANI/G	Solution	–	33 wt%	2.4	–	2–18	34.2	Absorption	[85]
PANI/G@Au	In situ	–	–	–	865	2–12	87–108	–	[79]
PANI/G@Au	In situ	–	–	2	–	2–12	90–120	–	[107]
PANI/G@Fe_3O_4	In situ	–	66.7 wt%	2.5	–	12.4–18	26.7–32.3	Absorption	[65]
PANI/G@γ-Fe_2O_3	In situ	–	75 wt%	2.5	–	8.2–12.4	51	Absorption	[97]
PANI/G-MWCNT	In situ Solution	–	14.5 wt%	2.5	2,950	12.4–18	98	Absorption	[108]
PIL-PEDOT/G@Fe_3O_4	Solution In situ	–	1 wt%	0.01	–	0.02–1	22	–	[94]
Phenolic/G@γ-Fe_2O_3-CNF	In situ Solution	–	50 wt%	0.2–0.4	–	8.2–12.4	42.83	Absorption	[91]
PDMS/G	Others	–	0.7 wt%	1	180	0.03–1.5	30	Absorption	[3]
			0.8 wt%		200	8–12	20		
PE/G	Solution	0.063 vol%	1.50 wt%	2.5	3.4	8.2–12.4	28.3–32.4	Absorption	[105]
			0.660 vol%						
EVA/G	Solution	–	60 vol%	1	2.5×10^{-2}	8.2–12.4	27	Reflection	[76]

G graphene, G@Au graphene decorated with gold particles, G@γ-Fe_2O_3 γ-Fe_2O_3-decorated graphene, *PIL-PEDOT* poly(ionic liquid)-modified poly (3,4-ethylenedioxythiophene), G@γ-Fe_2O_3-CF γ-Fe_2O_3-decorated both graphene and carbon fiber, *EC* electrical conductivity, *EMW* electromagnetic wave, *FR* frequency range, *TDEMI* the dominant electromagnetic interference, *SM* shielding mechanism

2.3.2 Solution Blending

During solution blending process, graphene or modified graphene can be readily dispersed in a suitable solvent, sometimes, with the help of ultrasonication. Then the solvent is evaporated or distilled, following the addition of polymer. A diverse range of PGNs like PS/graphene [68], polymethyl methacrylate (PMMA)/graphene [69], polyetherimide (PEI)/graphene [70] and polyvinylidene fluoride (PVDF)/graphene [71] have been prepared for EMI shielding. The difficulty in removing residual solvents limits its application [44].

2.3.3 Melt Blending

In contrast, melt blending needs no solvent and graphene or modified graphene can be easily mixed with polymer matrices in molten state. Due to the simplicity of this method, it is much more commercially attractive than two the others. The drawback of this technique is the low degree of graphene dispersion so that researchers rarely utilize it in the preparation of EMI shielding materials as shown in Table 1.

3 A Brief Introduction of the EMI Shielding Theory

EMI SE is defined as a function of the logarithm of the ratio of incoming power (P_i) to outgoing power (P_o) of radiation and normally expressed in decibels (dB). With any kind of EMI waves, there are three basic mechanisms for the EMI shielding: reflection, absorption, and multiple reflections. Therefore, the total EMI SE (SE_T) equals the sum of the absorption (SE_A), the reflection (SE_R), and the multiple reflections (SE_M) as given by:

$$SE_T = SE_A + SE_R + SE_M = 10\log_{10}(P_i/P_o) \tag{1}$$

From this equation, we can infer that the higher the EMI SE, the less the transmitted electromagnetic wave through shielding materials. For example, a shielding efficiency with 20 dB, the value required for commercial applications, leaves only 1 % of electromagnetic waves to be transmitted. Furthermore, when we measure the scattering parameters with Vector Network Analyzer, the power coefficients of reflectivity (*R*), tansmissivity (*T*), and absorptivity (*A*) can be calculated using *S*-parameters as follows [72]:

$$P_i = P_a + P_r + P_o \tag{2}$$

$$R = P_r / P_i = |S_{11}|^2 = |S_{22}|^2 \tag{3}$$

$$T = P_o / P_i = |S_{12}|^2 = |S_{21}|^2 \tag{4}$$

$$R + T + A = (P_r + P_o + P_a) / P_i = 1 \tag{5}$$

where, P_a is the power absorbed in slabs; P_r is the power reflected at input interface. In the case of $SE_T \geq 15$ dB, SE_M is usually neglected and, we can have here:

$$SE_T \approx SE_A + SE_R \tag{6}$$

The effective absorbance (A_{eff}) can be described as [23]:

$$A_{eff} = (1 - R - T) / (1 - R) \tag{7}$$

$$SE_R = -10 \log_{10} (1 - R) \tag{8}$$

$$SE_A = -10 \log_{10} \left(1 - A_{eff} = -10 \log_{10} \left[T / (1 - R)\right]\right) \tag{9}$$

Generally, the basic component of electromagnetic energy is a kind of electromagnetic wave where any point contains two fields: E-field and H-field, which always keep perpendicular to each other along the direction of propagation. When placed in an electric field (E-field), a thin, conductive, and spherical shell will shield the E-field by inducing electronic charges along the shell and generate an induced E-field with its direction opposite to that of the original field. In the case of a magnetic field (H-field), however, Faraday's effect does not work. The issue can be addressed effectively by magnetic materials with high permeability (relative permeability $\mu \geq 1$). Additionally, sufficient thickness of magnetic shields is requested in magnetic absorption [73]. As a matter of fact, it is the electrical and magnetic properties, together with thickness that have a profound impact on the EMI SE of a shield [4, 73].

4 Influence Factors on the EMI SE of PGNs

In this section, items that have a critical impact on the EMI SE of PGNs will be listed, such as electrical and magnetic properties, sample thickness, and structures. Meanwhile, EMI shielding mechanism will be discussed associated with corresponding examples.

4.1 Electrical Conductivity of PGNs

Previous studies revealed that electrical conductivity is of utmost importance for EMI SE [14, 74, 75]. Specifically, the electrical conductivity about 1 S m^{-1} is said to be the lowest value to obtain the EMI SE of 20 dB [14, 68]. From the

EMI shielding mechanism point of view, electrical conductivity plays the most significant role in the reflection and absorption contribution alike [76]. With respect to the electrical conductivity of PGNs, there are various influence factors covering not only the type of polymer matrix but the properties of the graphene nanosheets (GNSs), such as their concentration, dispersion, aspect ratio, orientation, etc. [43, 44, 77].

4.1.1 Concentration, Dispersion, and Conductivity of Graphene

A critical concentration of graphene is always necessary for the transition of PGNs from insulative to conductive state [14, 18, 63]. This critical concentration, known as percolation threshold, increases the electrical conductivity of PGNs by several orders of magnitude by forming a continuous network within polymer matrix. The direct current (DC) conductivity of PGNs can be depicted using a power law as follows [78]:

$$\sigma = \sigma_0 (P - P_c)^t \tag{10}$$

where, σ is the conductivity of PGNs; σ_o is a saling factor; P represents the content of graphene; P_c represents the percolation threshold of PGNs, thus the critical concentration of graphene; t stands for the critical exponent describing the lattice dimensionality.

Liang et al. was the first group to employ G–O as a conductive filler blending with epoxy resin [63]. The G–O was first exfoliated with ultrasonication and referred to as solution-processable functionalized graphene (SPFG). Then the epoxy/SPFG composites were subjected to both chemical and thermal reduction to increase its conductivity. As shown in Fig. 1a, the DC conductivity of the epoxy/SPFC composites agrees well with the power law and exhibits a nonlinear sharp increase with the increase of SPFG concentration. This low percolation threshold (0.52 vol%) was attributed to the high aspect ratio of the graphene and its homogeneous dispersion in the epoxy, which were confirmed later by atomic force microscopy and scanning electron microscopy, respectively. Above the percolation threshold, further increase in SPFG loading has few influences on the DC conductivity of the epoxy/SPFG composites. The EMI SE of the epoxy/SPFG composites is shown in Fig. 1b, from which the EMI SE increases with the graphene loading increasing over the frequency range of 8.2–12.4 GHz (X band). This phenomenon was ascribed to the formation of conducting interconnected graphene-based sheets networks in the insulating epoxy matrix.

In order to improve the dispersibility of GNSs in water or organic solvents, different sorts of surfactants have been used in the fabrication of PGNs [66, 79, 80]. Ma et al., for instance, used a cationic surfactant: stearyl trimethyl ammonium chloride (STAC) to produce the STAC-absorbed-graphene nanosheets (S-GNS), effectively preventing the aggregation and restacking of GNSs by

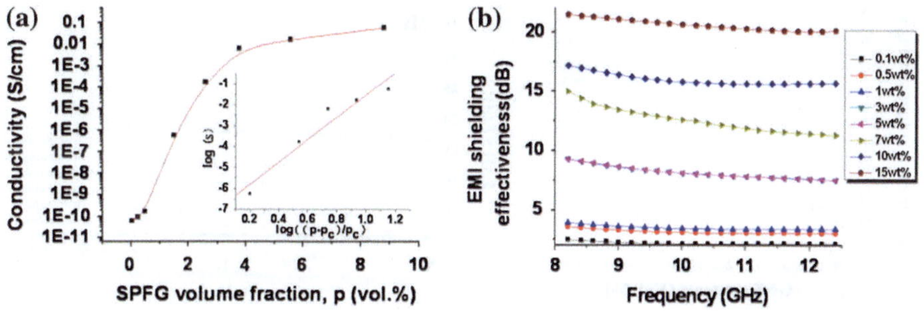

Fig. 1 **a** Log_{10} DC conductivity (σ) versus volume fraction (P) of epoxy/SPFG composites measured at room temperature. *Inset*: log–log plot for σ versus $((P-P_c)/P_c)$ for the same composites. The best fit gave values $P_c = 0.52$ vol% and $t = 5.37$ with a correlation factor of 0.97; **b** EMI SE of epoxy/SPFG composites with various SPFG loadings as a function of frequency in the X band. The epoxy/SPFG composites exhibited SE ~21 dB in the X-band for a 15 wt% loading. [63] Copyright 2009. Reproduced with permission from Elsevier Ltd.

electrostatic attraction [80]. In contrast, GNSs without the STAC was also prepared and named as P-GNS. Then both the S-GNS and P-GNS were blended with one kind of water-borne polyurethane (WPU) via solution compounding, respectively. Similar to that of the epoxy/SPFG composites, both the conductivity of the WPU/S-GNS and the WPU/P-GNS in Fig. 2a present a nonlinear increase. When loaded with the same content of GNS (5 vol%), the electrical conductivity of WPU/S-GNS is up to 5.1 S m^{-1}, much higher than that of P-GNS/WPU (0.97 S m^{-1}). What is more, the WPU/S-GNS composite shows a lower percolation threshold than the WPU/P-GNS. In Fig. 2b, the highest EMI SE of the S-GNS/WPU is 32 dB in X band, much higher than 21 dB for the P-GNS/WPU composite with the same graphene loading of 5 vol% (Fig. 2c). These phenomena were ascribed to the better dispersion of the S-GNS within the WPU matrix, that is, better graphene dispersion makes easier formation of interconnected conducting network within polymer matrix and higher EMI SE. Additionally, Fig. 2d says that the absorption loss (SE_A) of the WPU/S-GNS composites contributes more than 90 % contribution to the total EMI SE (SE_{total}) over the whole S-GNS loadings and varies according to the change in SE_{total}. While the reflection contribution (SE_R) seems to be independent with the change of S-GNS loading, indicating that the EMI shielding for the WPU/S-GNS composite was absorption dominant.

In another case, Yang et al. modified graphene sheets with polydopamine (PDA) [81]. The as-obtained PDA-coated graphene sheets (D-Graphene) was then used to blend with PU by solution blending. The low percolation threshold of the as-obtained PU/D-Graphene nanocomposites (0.1 vol%) was ascribed to the good dispersion of graphene and the decreased sheet–sheet junction resistance resulting from the PDA-functionalized conductive interface. The EMI SE of the

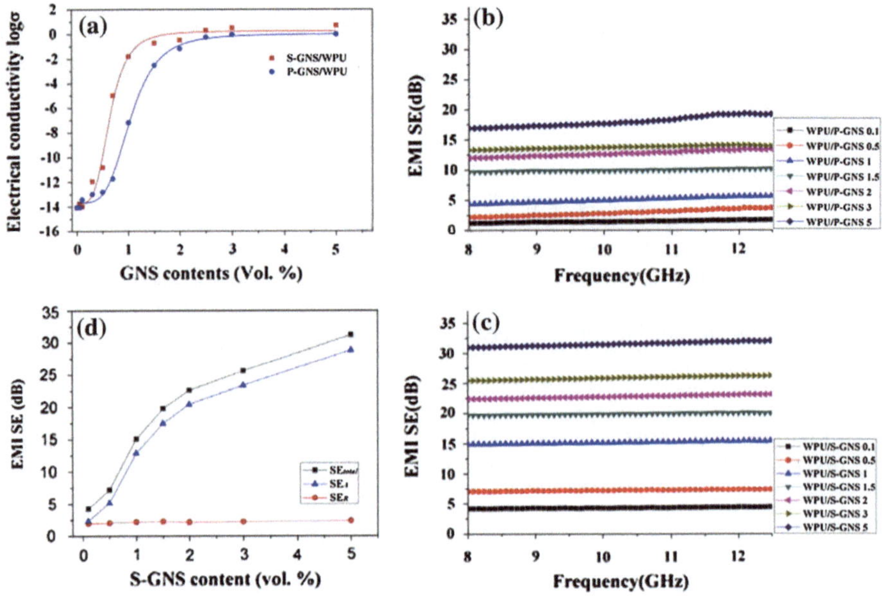

Fig. 2 **a** The electrical conductivity (σ) of WPU/P-GNS and WPU/S-GNS composites measured at room temperature as a function of GNS content; EMI SE of **b** WPU/P-GNS and **c** WPU/S-GNS composites with various GNS loadings as a function of frequency in the range of 8.2–12.4 GHz. At 5 vol% loading, the WPU/P-GNS exhibited EMI SE of approximately 21 dB, and the EMI SE of WPU/S-GNS showed approximately 32 dB in the range of 8.2–12.4 GHz; **d** Comparison of SE_{total}, SE_A and SE_R of S-GNS/WPU composite at 9 GHz as a function of S-GNS content [80]. Copyright 2013. Reproduced with permission from Elsevier Ltd.

nanocomposites tends to increase with the increasing loading of the D-Graphene in the frequency range of 0.03–1.8 GHz and reaches to 17.6 dB at graphene loading of 4.75 vol%. Moreover, its EMI SE is dominated by absorption. The explanation, however, was not given.

The electrical conductivity of PGNs depends not only on the concentration and dispersion of graphene, but also on its intrinsic conductivity [44]. Motivated by this, our group prepared three kinds of graphene with C/O ratios being 5.0, 9.6, and 13.2 and noted them as graphene-5.0, graphene-9.6, and graphene-13.2, respectively [69]. After incorporated into PMMA via solution process, each of the PMMA/graphene nanocomposites shows an insulator–conductor transition as shown in Fig. 3a. Moreover, The PMMA/graphene-13.2 composite exhibits the lowest percolation threshold of 0.40 vol% for the simple reason that the oxygen-containing groups within graphene disrupt the conjugate graphitic sp^2 structure and decrease its intrinsic conductivity. As shown in Fig. 3b, the EMI SE of the PMMA/graphene-13.2 tends to increase with the graphene loading increasing and the highest EMI SE for the PMMA/graphene-13.2 is about 30 dB. In Fig. 3c, we can see clearly that when loaded with the same graphene content of 2.67 vol%, the EMI SE of the PMMA/graphene composites is proportional to the value of the C/O ratio of graphene. This phenomenon is in good accordance with the electrical conductivity of

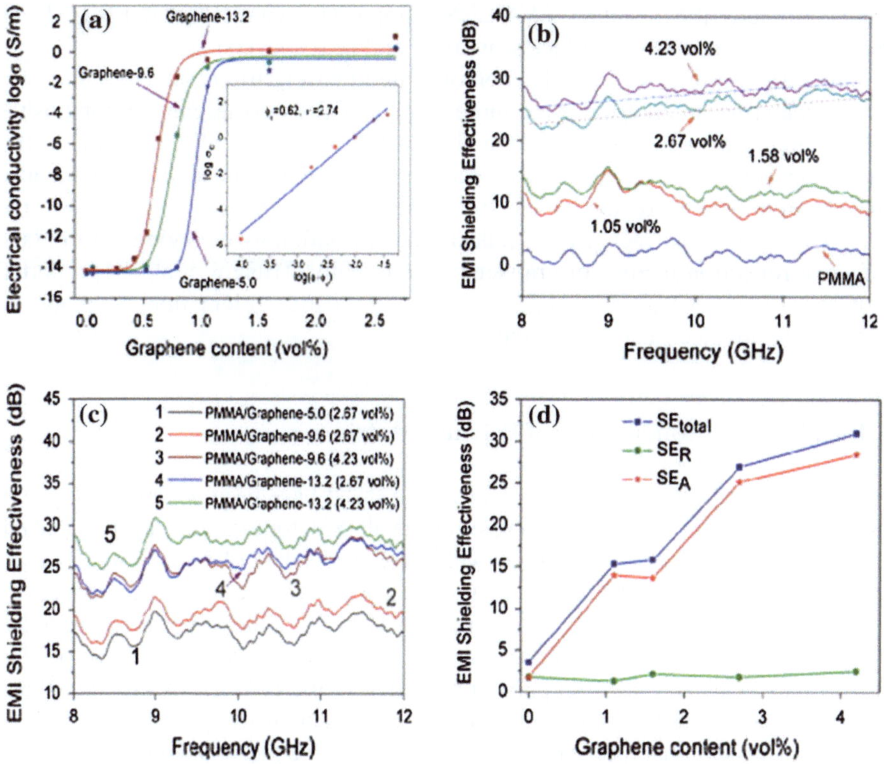

Fig. 3 **a** Plots of electrical conductivity of PMMA/graphene composites as a function of graphene content; **b** EMI SE of PMMA/graphene-13.2 composites; **c** comparison of EMI SE of PMMA composites with graphene-13.2, graphene-9.6 and graphene-5.0; **d** comparison of SE_{total}, SE_A and SE_R at 9 GHz as a function of graphene content [69]. Copyright 2012. Reproduced with permission from Elsevier Ltd.

PMMA/graphene in Fig. 3a. Furthermore, the primary EMI shielding mechanism of PMMA/graphene composites is absorption rather than reflection as shown in Fig. 3d, which is related closely to the conductive networks created by graphene.

4.1.2 Aspect Ratio and Orientation of Graphene

To prepare PGNs with enhanced electromagnetic attenuation, Song et al. [82] enlarged the aspect ratio of commercial expandable graphite (EG) through a direct chemical exfoliation approach and named the resulting product as carbon nanosheets (CNS). Then both the EG and CNS were mixed with ethylene-vinyl acetate (EVA) resin via a solution process, respectively. The results revealed that the percolation threshold of the EVA/CNS composites is about 20 vol% (36 wt%), much lower than that of EG/EVA 30 vol% (50 wt%). The reason was mainly attributed to the larger aspect ratio of the CNS. Moreover, the increase of CNS fillers provides the resultant EVA/CNS composites with efficient EMI SE by means of dielectric loss, interface scattering, and heat conversion.

Similarly, Song et al. [83] also demonstrated that electric conductive interconnections can be much established more easily for the post-alignment composites at the same graphene loading for the simple reason that the alignment may offer more opportunities for graphene to form conducting interconnections and enhance the absorption contribution of electromagnetic waves. Specifically, few-layered graphene (FLG) may increase the reflection contribution by enlarging the in-plane effective reflection regions of graphene. Additionally, Tripathi et al. [84] reported that the PMMA/graphene nanocomposite with better orientation of graphene shows a higher electrical conductivity.

4.1.3 Electrical Conductivity of Polymer Matrix

Recently, intrinsically conductive polymers like PANI have been widely employed in the preparation of PGNs as EMI shields. Yuan et al. [85], for example, reported that the EMI SE of PANI/graphene composite with graphene loading of 33 wt% was 34.2 dB in X band, higher than 17.3 dB for the PS/graphene composite [68]. The conductivity of PANI matrix partially leads to the remarkable EMI SE of the composites. In another case, Chen et al. [64] functionalized graphene by growing PANI nanorods perpendicular to its surface and the absorption contribution in the EMI SE of PANI/graphene composites was significantly improved because of the conductive property of PANI.

It has been said that there are two sorts of conductive mechanism for PGNs: tunneling and direct contact. Tunneling is the dominant mechanism for PGNs whose *I*–*V* curves can be characterized by power law, while the linear *I*–*V* relationship following Ohm's law means the dominant mechanism of direct contact [18]. For example, the conductivity of forementioned PU/D-Graphene was supposed to be tunneling dominant [81]. Yang et al. [14] established that there is a close relationship between the EMI SE of conductive polymer-based composites and its electrical conductivity. A plot of the EMI SE of PS/CNT composite foam at 10 GHz versus its electrical conductivity had been given, in which the SE varies slightly at first with the great increase of conductivity and then increase dramatically above the percolation threshold even though a small change in conductivity. Similar phenomenon was also observed by Thomassin et al. [43] after plenty of statistics.

4.2 Magnetic Permeability of PGNs

In terms of electromagnetic theory, there are two possible contributions for microwave absorption: dielectric loss and magnetic loss [86–88]. Pristine graphene is nonmagnetic and its contribution to microwave absorption mainly derives from dielectric loss. To further improve the EMI SE of PGNs, a number of researchers prefer to introduce magnetic particles into graphene sheets [89].

Our group functionalized graphene with Fe_3O_4 particles (G@Fe_3O_4) initially, and then blended the G@Fe_3O_4 with PEI through a water vapor-induced phase separation (WVIPS) [89]. For comparison, PEI/Fe_3O_4 foams were also prepared. The results revealed that the specific EMI SE of PEI/Fe_3O_4 foams is 15.2 dB cm^3 g^{-1}, lower than 41.5 dB cm^3 g^{-1} for the PEI/G@Fe_3O_4 foam. The specific EMI SE (EMI SE divided by density) proposed by Gupta et al. [14], is supposed to be more appropriate for comparing the shielding performance between lightweight conductive polymer composites. So the decrease in specific EMI SE of the PEI/Fe_3O_4 composite was assigned to the high resistivity of Fe_3O_4 nanoparticles as well as their poor dispersion in PEI. When compared with the PEI/graphene foams with the same graphene loading (10 wt%) and sample thickness (almost 2.4 mm), the contribution of absorption to total EMI SE is 97.2 %, much higher than 90.6 % for the PEI/graphene foam. As shown in Fig. 4b, the G@Fe_3O_4 retains their superparamagnetism when mixed with PEI resin and the magnetization of PEI/G@Fe_3O_4 foams significantly increases with the reduction of

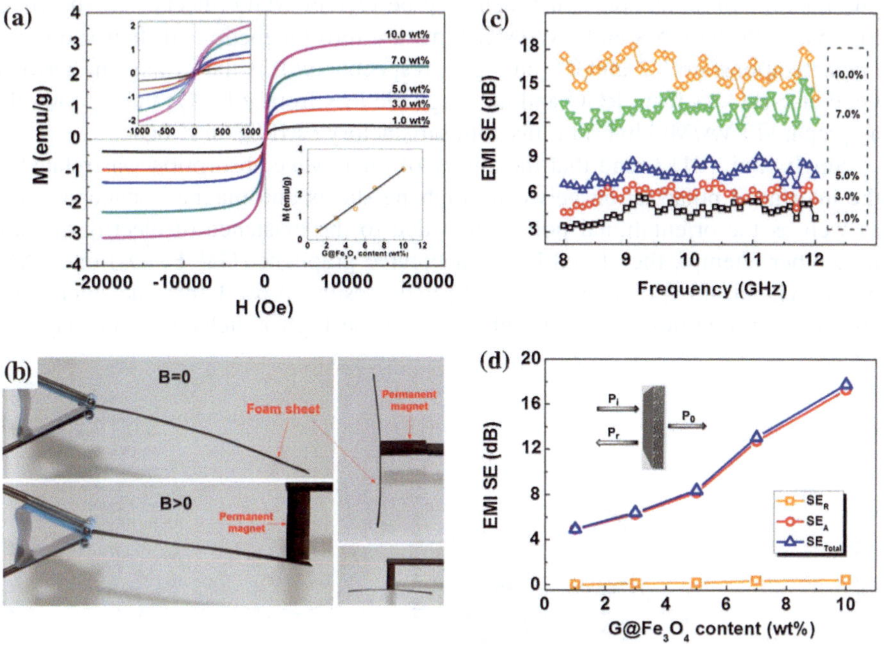

Fig. 4 Superaramagnetic microcellular PEI/G@Fe_3O_4 foam sheets. **a** magnetization of the composite foams with different G@Fe_3O_4 loading. The Ms of the foams is in the range of 0.38–3.09 emu g^{-1} and tended to linearly increase with the G@Fe_3O_4 content (*inset* at the *lower right*). Moreover, the PEI/G@Fe_3O_4 foams are superaramagnetic as there is no significant hysteresis in the Ms (*inset* at the *top left*); **b** this free-standing composite foam can be actuated by an external magnetic field; **c** EMI SE as a function of frequency measured in the 8–12 GHz range of the PEI/G@Fe_3O_4 foams with various G@Fe_3O_4 concentrations; **d** SE_{total}, SE_A and SE_R of microcellular foams at 9.6 GHz [89]. Copyright 2013. Reproduced with permission from American Chemical Society

$G@Fe_3O_4$ concentration (Fig. 4a). Therefore, the introduction of magnetic particles within graphene plain is supposed to be a good way to enhance the microwave absorption of PGNs due to the introduction of magnetic loss by Fe_3O_4 nanoparticles. From Fig. 4c, we can see that the EMI SE of the PEI/graphene foam is proportional to the $G@Fe_3O_4$ content and the highest value of SE is 17.8 dB in X band. Moreover, Fig. 4d demonstrates that the primary EMI shielding mechanism of PEI/$G@Fe_3O_4$ foams is absorption.

In addition, Yuan et al., synthesized γ-Fe_2O_3 incorporated graphene nanosheets (GNS), then used PVA to blend with GNS via evaporation-induced assembly and obtained PVA/MG hybrid films [90]. In Fig. 5, we can see clearly that the average EMI SE of the PVA/MG hybrid film is bigger than 20 dB with sample thickness only of 0.36 mm. For most PGNs, at least 2 mm thick is always required to meet this value. In Fig. 5b, the PVA/MG hybrid film has a higher SE_A than the PVA/reduced graphene oxide (RGO) film and the EMI shielding mechanism for the PVA/MG film is absorption rather than reflection. Because of RGO being nonmagnetic, the enhancement in SE_A for the PVA/MG film was ascribed to the introduction of magnetic constituents (γ-Fe_2O_3) in RGO sheets. As shown in Fig. 5a, both the GNS and PVA/MG films are ferromagnetic and their saturation magnetizations are 5.2 and 0.9 emu g^{-1}, respectively. It is the synergism between the dielectric loss from RGO and the magnetic loss from γ-Fe_2O_3 that endows the as-prepared PVA/MG hybrid films with satisfactory EMI SE in X band.

Singh et al. [91] found that the ability of microwave absorption was related to the interfacial polarization between graphene sheets and magnetic nanoparticles, as well as the orientational polarization due to the existence of electric dipoles. In another attempt, they filled Fe_3O_4-decorated graphene (G@ Fe_3O_4) into PANI matrix via in situ polymerization [65]. The results proved that the microwave absorption properties had been enhanced by the higher dielectric and magnetic

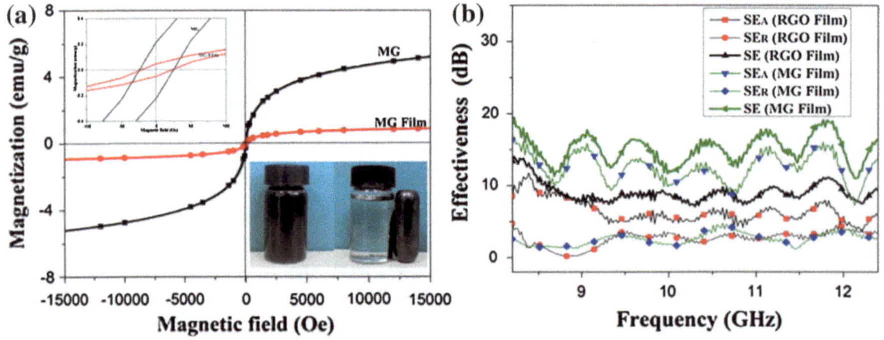

Fig. 5 **a** Magnetic hysteresis loops of MG and MG hybrid film. The *top left inset* shows an enlarged view of the magnetic hysteresis loops. Digital photo of MG aqueous dispersion in the *bottom right inset* shows the magnetic response under an external magnetic; **b** EMI SE_A and SE_R of RGO and MG hybrid films with thickness of 0.23 mm as a function of frequency [90]. Copyright 2014. Reproduced with permission from Elsevier Ltd.

loss because of the synergistic effect of graphene and Fe_3O_4. Similar researches have also been conducted [92–95] and the results highlighted the important role of magnetic particles playing in microwave absorption.

In summary, the ability of microwave absorption of PGNs can be enhanced by introducing magnetic particles in graphene sheets due to the association of dielectric loss from graphene and magnetic loss from magnetic particles like Fe_3O_4. Because of the insulative nature of most magnetic fillers, too much of them may cause a decrease in electrical conductivity, and consequently weak microwave absorption even though the magnetic loss is relatively high.

4.3 Sample Thickness of PGNs

From the absorption loss point of view, a sufficient thickness is one of the crucial factors for a shield to acquire the satisfactory EMI SE [73]. The sufficient sample thickness for efficient EMI SE has two different meanings: (a) to eliminate multiple reflections; (b) to enhance absorption contribution. Details are given as follows.

Once an electric wave entering a shield, its amplitude will decrease exponentially. The distance at which the field drops to 1/e or 37 % of the incident value is defined as the skin depth δ, which can be expressed as follows:

$$\delta = \frac{1}{\pi\sqrt{f\mu\sigma}} \tag{7.11}$$

where, f is frequency in Hz; μ is magnetic permeability in H m^{-1}; σ is electrical conductivity in S m^{-1}. In general, the skin depth decreases with increasing frequency, conductivity as well as permeability. As one of the EMI shieling mechanisms, multiple reflections occur frequently among interfaces and surfaces of shielding materials, particularly in fillers with large surface and interface area [4]. While, it can be ignored when the thickness of a shield is larger than the skin depth as the internal reflection will be attenuated by the shield. On the contrary, if the thickness of a shield is smaller than the skin depth, multiple reflections should be concerned. However, multiple reflections may induce decrement in overall EMI SE and is always expected to be eliminated effectively by handling shielding thickness larger than the skin depth [76].

On the other hand, Song et al. prepared EVA/graphene (G-E) nanocomposite films with different thickness ranging from 50 to 350 μm [76]. In Fig. 6a, the total SE of G-E films are proportional to the sample thickness. When in 350 μm thick, the G-E film containing 60 vol% graphene has an average SE more than 23 dB, meeting the commercial level of 20 dB. The absorption contribution (SE_{abs}) in Fig. 6b indicates that thicker shields give more microwave absorption. Yuan et al. [90] found that the EMI SE of the PVA/MG film in 0.24 mm thick has an average EMI SE of 15.1 dB in X band, much lower than 20.3 dB for the same film with 0.36 mm thick (Fig. 7). Chen et al. [11] also demonstrated the significant role

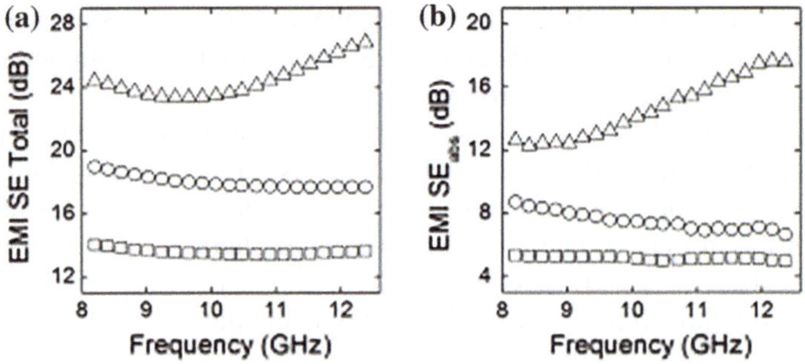

Fig. 6 EMI **a** SE and **b** SE_{abs} of the G-E films with 60 vol% filler loading in thickness of ~50 μm (*square*), ~120 μm (*circle*) and ~350 μm (*triangle*). [76] Copyright 2013. Reproduced with permission from Elsevier Ltd.

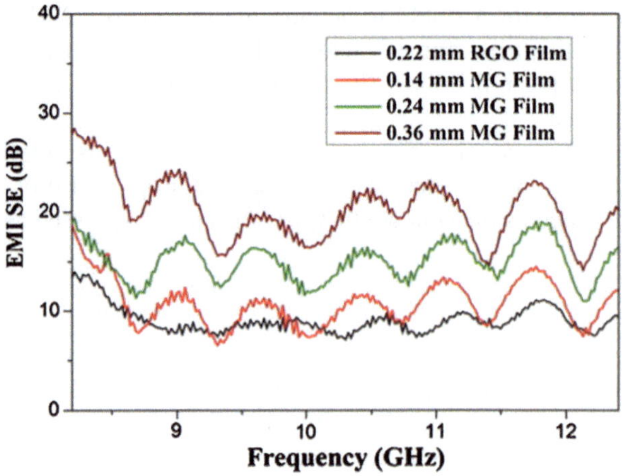

Fig. 7 EMI SE of RGO and MG hybrid films in X band [90]. Copyright 2014. Reproduced with permission from Elsevier Ltd.

of materials' shielding thickness in microwave absorption. Besides, Gupta et al. [96] investigated the EMI SE of the manganese oxide (MnO_2) decorated graphene nanoribbons (GNRs) in X band tended to increase as the MnO_2–GNRs pellets' thickness increasing from 0.5 to 3.0 mm.

4.4 *Structures of PGNs*

In this section, EMI shields with cells, multilayers, and core-shell structures will be reviewed, as well as their influences on EMI SE [76, 89, 97]. Meanwhile, more emphasis will be given to the shields with cellular structure because of their wonderful EMI shielding performance.

4.4.1 Cellular Structure

In 2010, our group first employed graphene sheets to mix with PMMA and the resultant PMMA/graphene foams have a cell size distribution ranging from 1 to 10 μm. When loaded with 1.8 vol% graphene, the PMMA/graphene foam exhibited an EMI SE of 13–19 dB in X band and its specific EMI SE is 17–25 dB cm^3 g^{-1}. Specifically, the dominant EMI shielding mechanism for this foam is absorption, instead of reflection for the PS/CNT foam [14]. After that, Ramaprabhu et al. prepared the PVDF/graphene foams via a chemical foaming process for EMI shielding [71]. The EMI SE of PVDF/functional graphene (FGS) foam containing 7 wt% graphene is 28 dB in X band. Yan et al. [68] also prepared PS/FGS foams as EMI shielding material using a compression molding plus salt-leaching method. The lowest density of the PS/FGS foam with FGS loading of 30 wt% is 0.27 g cm^{-3} and its' EMI SE in X band approaches 29.3 dB. Furthermore, the corresponding specific EMI SE was 64.4 dB cm^3 g^{-1}, much higher than that for the PMMA/graphene foams mentioned above. Table 2 has listed a comparison of properties of different PGNs.

In order to broaden the applications of PGNs in areas of aircraft, spacecraft and automobiles, higher performance polymer PEI has been utilized by us to prepare lightweight microcellular PEI/graphene composite foams as EMI shields [70]. However, there are plenty of challenges lying in the way of foaming PEI/graphene composite solid via traditional physical foaming, including high

Table 2 A comparison of EMI SE of different polymer/graphene foams

Matrix	PMMA	PVDF	PS	PDMS	PEI	PEI
Filler	Graphene	Graphene	Graphene	Graphene	Graphene	Graphene @Fe_3O_4
Filler loading (vol%[a], wt%[b])	1.80[a], 5.00[b]	7.00[b]	30.00[b]	0.80[b]	1.38[a], 10.00[b]	10.00[b]
Pore size (μm)	5.0	0.5–2.0	20.0–40.0	–	9.0	–
Density (g cm^{-3})	0.79	–	0.45	0.06	0.28	0.40
Thickness (mm)	2.4	–	2.5	1.0	2.3	2.5
DC EC (S m^{-1})	3.11	–	1.25	2.00	2.20×10^{-3}	6.50×10^{-5}
EMI SE (dB) (8.2–12.4 GHz)	13.0–19.0	28.0	29.3	20.0	11.0	17.8
Specific EMI SE (dB cm^3 g^{-1})	17.0–25.0	–	64.4	333.0	44.1	41.5
TDEMI SE	Absorption	Reflection	Absorption	Absorption	Absorption	Absorption
References	[74]	[71]	[68]	[33]	[70]	[89]

DC direct current, *EC* electrical conductivity, *TDEMI* the dominant electromagnetic interference, *SM* shielding mechanism, *a* stands for the volume fraction of filler loading in polymer matrix, while the *b* is the weight fraction of fillers processing temperature (340–360 °C), extremely long saturation time (280–520 h) as well as the low expansion ratio (≤2). To solve these problems, the WVIPS was applied in the foaming process. The density of the resultant PEI/graphene foams is 0.28 g cm^{-3} at graphene loading of 10 wt% and its average cell size is 9.0 μm. The EMI SE of the composite foams in X band increased gradually up to 20 dB at 10 wt% graphene loading. The specific EMI SE of microcellular PEI/graphene foams is 44.1 dB cm^3 g^{-1}, about 2.5 times higher than that of the corresponding solid. As about 98.9 % electromagnetic energy being absorbed, the primary EMI shielding mechanism for the PEI/graphene foam is absorption in X band.

To analyze the effect of porous structure on EMI SE, we need to compare the same sort of PGN composites with porous and nonporous structure under the same condition. After series of analysis, we consider that the power efficient of absorptivity A (equals SE_A/SE_T) and the specific EMI SE are more suitable for the comparison of EMI SE of PGNs solid and foam. Take the PMMA/graphene composite for an example, the power efficient of absorptivity A for the PMMA/graphene foam is 94.7 % at graphene loading of 1.8 vol% [74]. The value for the PMMA/graphene solid with graphene loading of 4.23 vol% is supposed to be much bigger than 94.7 % due to its higher graphene content and electrical conductivity, together with its absorption mechanism. However, the truth is that only 93.3 % incident energy can be attenuated by this PMMA/graphene solid [69], which is lower than the 94.7 % for the PMMA/graphene foam with the same sample thick. So we may assume that porous shields may perform better in microwave absorption than bulk ones. When it comes to the PEI/graphene system, the A for PEI/graphene foam is 98.9 %, still much higher than 90.8 % for the PEI/graphene solid with the same graphene loading of 10 wt% and thickness of 2.3 mm [70]. This phenomenon is fully convinced that porous materials possess a higher ability in microwave absorption.

According to the EMI shielding theory, reflection results from the impedance mismatch between the incident wave and shielding materials [21, 73, 98]. The wave impedance is defined as the ratio of E to H (for air, E/H is 377 Ω), in which the E and H are the intensity of the electric and magnetic field within electromagnetic wave [73, 99]. The higher the degree of impedance mismatch, the more the contribution of reflection, and vice versa. Krause et al. claimed that the air bubbles in polymer matrix can decrease the dielectric constant without any effect on the component of polymer matrix [100]. As shown in Fig. 8, the existence of the spherical microscale air bubbles in foams can attenuate the incident electromagnetic microwaves by reflecting and scattering between the cell wall and nanofillers. Microwaves can hardly escape from the foams before being absorbed and transferred to heat. In addition, Thomassin et al. also prepared multiwalled carbon nanotubes (MWCNTs) filled polycaprolactone (PCL) composite foams via supercritical CO_2. Compared to MWCNT/PCL solids, MWCNT/PCL foams own higher shielding efficiencies and lower reflectivities [21]. Thereby the porous structure can certainly help to match the wave impedances between foams and ambient atmosphere, leading to the enhancement in microwave absorption. It makes foaming shields very promising in the fabrication of lightweight, effective microwave absorbers.

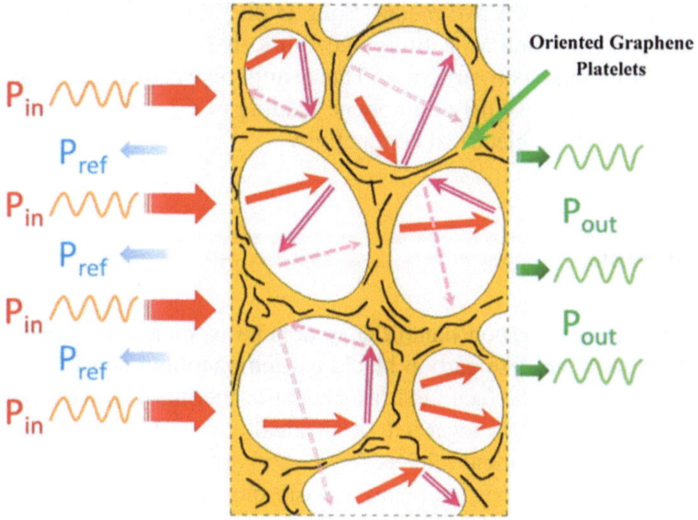

Fig. 8 Schematic description of the microwave transfer across PEI/graphene nanocomposite foam [70]. Copyright 203. Reproduced with permission from American Chemical Society

4.4.2 Core-Shell Structure

Additionally, graphene-based composites with conductive shells and magnetic cores also exhibited unique properties when used in EMI shielding [101, 102]. For instance, Singh et al. [97] synthesized γ-Fe_2O_3 decorated graphene (RF) and fabricated PANI-RF shell-core tubes for EMI shielding. High EMI SE around 51 dB was obtained with sample thickness of 2.5 mm, which was regarded as the result of interfacial polarization and effective anisotropy energy from the existence of RF. Additionally, Chen et al. [11] prepared hematite/graphene composites with a unique quasi shell-core nanostructures. They found that the microwave absorption of hematite/graphene composites could be enhanced by the unique quasi shell-core nanostructure, which may yield interfacial polarization and associated relaxation and make the dielectric loss much bigger.

4.4.3 Multilayered Structure

Fabricating shields with multilayered structure is another way to fabricate effective EMI shields [43]. One case is to stack several layers of polymer composites with the same content of fillers and the improvement in EMI SE results mainly from multiple reflections. This strategy has been conducted effectively in PMMA/MWCNT composites [103]. In another case, the multilayered structure consists of layers with CNTs concentration increasing from one to the next [104]. The contribution from reflection can be decreased by reducing mismatch of the dielectric constant between two media. Song et al. [76] synthesized free-standing graphene/EVA (G-E) composite films and fabricated wax |PVA|G - E|PVA| wax hybrid films with sandwich structure. The results revealed that the highest EMI SE of the hybrid film is 14 dB at graphene loading of 60 vol% and it is reflection dominant due to its high electrical conductivity.

5 Conclusions

As a two-dimensional carbon nanofiller, graphene is somehow superior to other conventional nanofillers like carbon black, carbon nanofiber and CNTs, due to its unique properties: high surface area, aspect ratio, thermal and electrical conductivity. Because of this, it is regarded as an ideal carbon filler to blend with lots of polymer resins and fabricate PGNs for all sorts of applications including EMI shields. However, traditional strategies for the preparation of graphene, such as CVD, epitaxial growth, and micromechanical exfoliation, either are too expensive or hardly control. With respect to the relatively convenient method: thermal exfoliation and reduction of G–O, the high temperature of 1,050 °C used to exfoliate G–O is another challenge for the large-scale fabrication of graphene. With the assistance of vacuum condition or volatile substance like HCl, the temperature has been reduced successfully to 135 °C. As we know, composites with good properties always need a good dispersion of fillers within matrix, so do PGNs. But the truth is that the hydrophobic nature of graphene makes it incompatible with most organic polymers and tend to irreversibly agglomerate itself. To solve this problem, graphene must be modified by either noncovalent or covalent methods. Generally, the PGNs made by covalent modification possesses stronger interfacial interactions between polymer matrix and graphene due to the formation of chemical bonding. In the case of noncovalent modification of graphene, more emphasis has been given to the π-π stacking interactions owing to its strong strength comparable to covalent bonding. In fact, this technique is much more preferential on the conditions where excellent electrical conductivity and EMI SE are required, for the conjugated structure of graphene can be saved in this way.

There are a number of influence factors for PGNs conductivity, including the intrinsic conductivity, dispersion, aspect ratio and orientation of graphene, coupled with the type of polymer matrix. Additionally, the idea of decorating graphene with magnetic nanoparticles has been proved to be very promising. Nevertheless, too much insulative magnetic particles may lead to a reduction in the electrical conductivity of PGNs, and finally the EMI SE of PGNs. From the EMI shielding point of view, the EMI SE mechanism is composed of reflection, absorption and multiple reflections. The absorption contribution stems mainly from dielectric and magnetic loss. Specifically, the dielectric loss results from complex phenomena such as electronic polarization and relaxation, dipolar relaxation and natural resonances. For PGNs, graphene nanosheets may induce electronic polarization and dipolar relaxation and the interface between graphene and polymer matrix is able to cause interfacial polarization and relaxation. It is these polarization and relaxation that make up of the dielectric loss and finally lead to microwave absorption. In fact, most PGNs are dominated by absorption mechanism, which is related infinitely to the high electrical conductivity and aspect ratio of graphene. As to reflection mechanism, it is related to the impedance mismatch between PGNs and ambient atmosphere. The introduction of air bubbles into PGN solids to form PGN foams is a good choice for reducing the degree mismatch and reflection loss, as well as the mass production of microwave absorbers. However, multiple reflections are frequently considered by the PGNs with multilayered structures due to the additional reflecting surfaces. Generally, an increase in the thickness of PGNs does also help to improve their EMI SE. In summary, an effective EMI shielding material should possess high electrical conductivity and magnetic permeability, together with sufficient thickness to meet the need of skin depth at low frequency.

Though PGNs have been studied as EMI shields for many years, there are still some challenges as follows: (a) a homogeneous graphene dispersion within polymer matrix is hard to get due to its hydrophobic nature; (b) the PGNs with satisfactory EMI SE for practical applications always need large graphene loading, make the preparation of PGNs very expensive and difficult and severely restrict the applications of PGNs as EMI shields; (c) the magnetic nanoparticles functionalized graphene sheets like graphene@ Fe_3O_4 may have a bad dispersion within polymer matrix even though effective EMI SE can be achieved by this way and the interfacial interactions between them are supposed to be week due to the lake of chemical bondings; and (d) thin EMI shields will be eagerly needed with modern electronic devices getting smaller and lighter, while the thickness of current PGN shields is always more than 2 mm. However, good news is that the absorption mechanism for the PGN foam as well as its extremely high power absorptivity makes it very promising for the microwave absorber.

References

1. Liu J W, Che R C, Chen H J et al (2012) Microwave absorption enhancement of multifunctional composite microspheres with spinel Fe_3O_4 cores and anatase TiO_2 shells. Small 8: 1214-1221
2. Watts P C P, Hsu W K, Barnes, A et al (2003) High permittivity from defective multiwalled carbon nanotubes in the X-band. Adv Mater 15: 600-603
3. Chen Z P, Xu C, Ma C et al (2013) Lightweight and flexible graphene foam composites for high-performance electromagnetic interference shielding. Adv Mater 25: 1296-1300
4. Chung D D L (2001) Electromagnetic interference shielding effectiveness of carbon materials. Carbon 39: 279-285
5. Nguyen M T, Diaz A F (1994) A novel method for the preparation of magnetic nanoparticles in a polypyrrole powder. Adv Mater 6: 858-860
6. Lee H C, Kim J Y, Noh C H et al (2006) Selective metal pattern formation and its EMI shielding efficiency. Appl Surf Sci 252: 2665-2672
7. Huang X H, Neretina S, Ei S et al (2009) Gold nanorods: from synthesis and properties to biological and biomedical applications. Adv Mater 21: 4880-4910

8. Wang Y Y, Jing X L (2005) Intrinsically conducting polymers for electromagnetic interference shielding. Polym. Advan Technol 16: 344-351
9. Olmedo L, Hourquebie P, Jousse F (1993) Microwave absorbing materials based on conducting polymers. Adv Mater 5: 373-377
10. Joo J, Epstein A J (1994) Electromagnetic radiation shielding by intrinsically conducting polymers. Appl Phys Lett 65: 2278-2280
11. Chen D Z, Wang G S, He S et al (2013) Controllable fabrication of mono-dispersed RGO-hematite nanocomposites and their enhanced wave absorption properties. J Mater Chem A 1: 5996-6003
12. Rahaman M, Chaki T K, Khastgir D (2011) Development of high performance EMI shielding material from EVA, NBR, and their blends: effect of carbon black structure. J Mater Sci 46: 3989-3999
13. Yang Y, Gupta M C, Dudley K L et al (2005) Conductive carbon nanofiber–polymer foam structures. Adv Mater 17: 1999-2003
14. Yang Y, Gupta M C, Dudley K L et al (2005) Novel carbon nanotube−polystyrene foam composites for electromagnetic interference shielding. Nano Lett 5: 2131-2134.
15. Brosseau C, Boulic F, Queffelec P et al (1997) Dielectric and microstructure properties of polymer carbon black composites. J Appl Phys 81: 882-890
16. Brosseau C, Molinie P, Boulic F et al (2001) Mesostructure, electron paramagnetic resonance, and magnetic properties of polymer carbon black composites. J Appl Phys 89: 8297-8310
17. Brosseau C, Queffelec P, Talbot P (2001) Microwave characterization of filled polymers. J Appl Phys 89: 4532-4540
18. Al-Saleh M H, Sundararaj U (2009) A review of vapor grown carbon nanofiber/polymer conductive composites. Carbon 47: 2-22
19. Jou W S, Wu T L, Chiu S K et al (2002) The influence of fiber orientation on electromagnetic shielding in liquid-crystal polymers. J Electron Mater 31: 178-184
20. Kim H M, Kim K, Lee C Y et al (2004) Electrical conductivity and electromagnetic interference shielding of multiwalled carbon nanotube composites containing Fe catalyst. Appl Phys Lett 84: 589-591
21. Thomassin J M, Pagnoulle C, Bednarz L et al (2008) Foams of polycaprolactone/MWNT nanocomposites for efficient EMI reduction. J Mater Chem 18: 792-796
22. Che R C, Peng L M, Duan X F et al (2004) Microwave absorption enhancement and complex permittivity and permeability of Fe encapsulated within carbon nanotubes. Adv Mater 16: 401-405
23. Li N, Huang Y, Du F et al (2006) Electromagnetic interference (EMI) shielding of single-walled carbon nanotube epoxy composites. Nano Lett 6: 1141-1145
24. Novoselov K S, Geim A K, Morozov S V et al (2004) Electric Field Effect in Atomically Thin Carbon Films. Science 306: 666-669
25. Balandin A A, Ghosh S, Bao W Z et al (2008) Superior thermal conductivity of single-layer graphene. Nano Lett 8: 902-907
26. Bolotin K I, Sikes K J, Jang Z et al (2008) Ultrahigh electron mobility in suspended graphene. Solid State Commun 146: 351-355
27. Stoller M D, Park S J, Zhu Y W et al (2008) Graphene-based ultracapacitors. Nano Lett 8: 3498-3502
28. Lee C, Wei X D, Kysar J W et al (2008) Measurement of the elastic properties and intrinsic strength of monolayer graphene. Science 321: 385-388
29. Geim A K (2009) Graphene: status and prospects. Science 324: 1530-1534.
30. Li X S, Zhu Y W, Cai W W et al (2009) Transfer of large-area graphene films for high-performance transparent conductive electrodes. Nano Lett 9: 4359-4363
31. Fan Z J, Yan J, Zhi L J et al (2010) A three-dimensional carbon nanotube/graphene sandwich and its application as electrode in supercapacitors. Adv Mater 22: 3723-3728
32. Chen S S, Brown L, Levendorf M et al (2011) Oxidation resistance of graphene-coated Cu and Cu/Ni alloy. Acs Nano 5: 1321-1327

33. Schwierz F (2010) Graphene transistors. Nat Nanotechnol. 5: 487-496.
34. Wang D H, Kou R, Choi D et al. (2010) Ternary self-assembly of ordered metal oxide-graphene nanocomposites for electrochemical energy storage. Acs Nano 4: 1587-1595
35. Seger B, Kamat P V (2009) Fuel cell geared in reverse: photocatalytic hydrogen production using a TiO2/Nafion/Pt membrane assembly with no applied bias. J Phys Chem C 113: 18946-18952
36. Shao Y Y, Zhang S, Wang C M et al (2010) Highly durable graphene nanoplatelets supported Pt nanocatalysts for oxygen reduction. J Power Sources 195: 4600-4605
37. Li X M, Zhu H W, Wang K L et al (2010) Graphene-on-silicon schottky junction solar cells. Adv Mater 22: 2743-2748
38. Yin Z Y, Sun S Y, Salim T et al (2010) Organic photovoltaic devices using highly flexible reduced graphene oxide films as transparent electrodes. Acs Nano 4: 5263-5268
39. Zhang X Y, Li H P, Cui X L et al (2010) Graphene/TiO2 nanocomposites: synthesis, characterization and application in hydrogen evolution from water photocatalytic splitting. J Mater Chem 20: 2801-2806
40. Ling X, Xie L M, Fang Y et al (2010) Can graphene be used as a substrate for Raman enhancement? Nano Lett 10: 553-561
41. Thrall E S, Crowther A C, Yu Z H et al (2012) R6G on graphene: high Raman detection sensitivity, yet decreased Raman cross-section. Nano Lett 12: 1571-1577
42. Hong S K, Kim K Y, Kim T Y et al (2012) Electromagnetic interference shielding effectiveness of monolayer graphene. Nanotechnology. doi: 10.1088/0957-4484/23/45/455704
43. Thomassin J M, Jérôme C, Pardoen T et al (2013) Polymer/carbon based composites as electromagnetic interference (EMI) shielding materials. Mat Sci Eng R 74: 211-232.
44. Kuilla T, Bhadra S, Yao D H et al (2010) Recent advances in graphene based polymer composites. Prog Polym Sci 35: 1350-1375.
45. Cai W W, Moore A L, Zhu Y W et al (2010) Thermal transport in suspended and supported monolayer graphene grown by chemical vapor deposition. Nano Lett 10: 1645-1651
46. Li X S, Cai W W, An J H et al (2009) Large-area synthesis of high-quality and uniform graphene films on copper foils. Science 324: 1312-1314
47. Kim K S, Zhao Y, Jang H et al (2009) Large-scale pattern growth of graphene films for stretchable transparent electrodes. Nature 457: 706-710
48. Berger C, Song Z M, Li X B et al (2006) Electronic confinement and coherence in patterned epitaxial graphene. Science 312: 1191-1196
49. Li D, Muller M B, Gilje S et al (2008) Processable aqueous dispersions of graphene nanosheets. Nat Nanotechnol 3: 101-105
50. Tung V C, Allen M J, Yang Y et al (2009) High-throughput solution processing of large-scale graphene. Nat nanotechnol 4: 25-29
51. Schniepp H C, Li J L, McAllister M J et al (2006) Functionalized single graphene sheets derived from splitting graphite oxide. J Phys Chem B 110: 8535-8539
52. McAllister M J, Li J L, Adamson D H et al (2007) Single sheet functionalized graphene by oxidation and thermal expansion of graphite. Chem Mater 19: 4396-4404
53. Si Y C, Samulski E T (2008) Synthesis of water soluble graphene. Nano Lett 8: 1679-1682.
54. Zhang H B, Wang J W, Yan Q et al (2011) Vacuum-assited synthesis of graphene from thermal exfoliation and reduction of graphite oxide. J Mater Chem 21: 5392-5397
55. Shen B, Lu D D, Zhai W T et al (2013) Synthesis of graphene by low-temperature exfoliation and reduction of graphite oxide under ambient atmosphere. J Mater Chem 1: 50-53
56. Fang M, Wang K G, Lu H B et al (2009) Covalent polymer functionalization of graphene nanosheets and mechanical properties of composites. J Mater Chem 19: 7098-7105
57. Shen B, Zhai W T, Tao M M et al (2013) Chemical functionalization of graphene oxide toward the tailoring of the interface in polymer composites. Compos Sci Technol 77: 87-94
58. Shen B, Zhai W T, Lu D D et al (2012) Ultrasonication-assisted direct functionalization of graphene with macromolecules. Rsc Adv 2: 4713-4719
59. Stankovich S, Piner R D, Nguyen S T et al (2006) Synthesis and exfoliation of isocyanate-treated graphene oxide nanoplatelets. Carbon 44: 3342-3347

60. Su Q, Pang S P, Alijani V et al (2009) Composites of graphene with large aromatic molecules. Adv Mater 21: 3191-3195
61. Shen B, Zhai W T, Chen C et al (2011) Melt blending in situ enhances the interaction between polystyrene and graphene through pi-pi stacking. Acs Appl Mater Inter 3: 3103-3109
62. Shen B, Zhai W T, Tao M M et al (2013) Enhanced interfacial interaction between polycarbonate and thermally reduced graphene induced by melt blending. Compos Sci Technol 86: 109-116
63. Liang J J, Wang Y, Huang Y et al (2009) Electromagnetic interference shielding of graphene/epoxy composites. Carbon 47: 922-925
64. Yu H L, Wang T S, Wen B et al (2012) Graphene/polyaniline nanorod arrays: synthesis and excellent electromagnetic absorption properties. J Mater Chem 22: 21679-21685
65. Singh K, Ohlan A, Pham V H et al (2013) Nanostructured graphene/Fe3O4 incorporated polyaniline as a high performance shield against electromagnetic pollution. Nanoscale 5: 2411-2420
66. Bernal M M, Martin G M, Molenberg I et al (2014) Influence of carbon nanoparticles on the polymerization and EMI shielding properties of PU nanocomposite foams. Rsc Adv 4: 7911-7918
67. Liu J Q, Tang J G, Gooding J J (2012) Strategies for chemical modification of graphene and applications of chemically modified graphene. J Mater Chem 22: 12435-12452
68. Yan D X, Ren P G, Pang H et al (2012) Efficient electromagnetic interference shielding of lightweight graphene/polystyrene composite. J Mater Chem 22: 18772-18774
69. Zhang H B, Zheng W G, Yan Q et al (2012) The effect of surface chemistry of graphene on rheological and electrical properties of polymethylmethacrylate composites. Carbon 50: 5117-5125
70. Ling J Q, Zhai W T, Feng W W et al (2013) Facile preparation of lightweight microcellular polyetherimide/graphene composite foams for electromagnetic interference shielding. Acs Appl Mater Inter 5: 2677-2684
71. Eswaraiah V, Sankaranarayanan V, Ramaprabhu S (2011) Functionalized graphene-PVDF foam composites for EMI shielding. Macromol Mater Eng 296: 894-898
72. Maiti S, Shrivastava N K, Suin S et al (2013) Polystyrene/MWCNT/graphite nanoplate nanocomposites: efficient electromagnetic interference shielding material through graphite nanoplate-MWCNT-graphite nanoplate networking. Acs Appl Mater Inter 5: 4712-4724
73. Tong X C (2008) Advanced materials and design for electromagnetic interference shielding. CRC, Boca Raton.
74. Zhang H B, Yan Q, Zheng W G et al (2011) Tough graphene-polymer microcellular foams for electromagnetic interference shielding. Acs Appl Mater Inter 3: 918-924
75. Antunes M, Velasco J I (2014) Multifunctional polymer foams with carbon nanoparticles. Prog Polym Sci 39: 486-509
76. Song W L, Cao M S, Lu M M et al (2014) Flexible graphene/polymer composite films in sandwich structures for effective electromagnetic interference shielding. Carbon 66: 67-76
77. Gelves G A, Al-Saleh M H, Sundararaj U et al (2011) Highly electrically conductive and high performance EMI shielding nanowire/polymer nanocomposites by miscible mixing and precipitation. J Mater Chem 21: 829-836
78. Shante V K, Kirkpatrick S (1971) An introduction to percolation theory. Adv Phys 20: 325-357
79. Basavaraja C, Kim W J, Kim D G et al (2012) Microwave absorption studies of polyaniline nanocomposites encapsulating gold nanoparticles on the surface of reduced graphene oxide in the presence of 2-naphthalene sulfonic acid. Colloid Polym Sci 290: 829-838
80. Hsiao S T, Ma C C M, Tien H W et al (2013) Using a non-covalent modification to prepare a high electromagnetic interference shielding performance graphene nanosheet/water-borne polyurethane composite. Carbon 60: 57-66
81. Yang L P, Phua S L, Toh C L et al (2013) Polydopamine-coated graphene as multifunctional nanofillers in polyurethane. Rsc Adv 3: 6377-6385

82. Song W L, Cao M S, Lu M M et al (2013) Improved dielectric properties and highly efficient and broadened bandwidth electromagnetic attenuation of thickness-decreased carbon nanosheet/wax composites. J Mater Chem C 1: 1846-1854
83. Song W L, Cao M S, Lu M M et al (2013) Alignment of graphene sheets in wax composites for electromagnetic interference shielding improvement. Nanotechnology. doi: 10.1088/0957-4484/24/11/115708
84. Tripathi S N, Saini P, Gupta D et al (2013) Electrical and mechanical properties of PMMA/reduced graphene oxide nanocomposites prepared via in situ polymerization. J Mater Sci 48: 6223-6232
85. Yuan B Q, Yu L M, Sheng L M et al (2012) Comparison of electromagnetic interference shielding properties between single-wall carbon nanotube and graphene sheet/polyaniline composites. J Phys Appl Phys. doi: 10.1088/0022-3727/45/23/235108
86. Chen T T, Deng F, Zhu J et al (2012) Hexagonal and cubic Ni nanocrystals grown on graphene: phase-controlled synthesis, characterization and their enhanced microwave absorption properties. J Mater Chem 22: 15190-15197
87. Hu C G, Mou Z Y, Lu G W et al (2013) 3D graphene-Fe_3O_4 nanocomposites with high-performance microwave absorption. Phys Chem Chem Phys 15: 13038-13043
88. Chen Y J, Xiao G, Wang T S et al (2011) Porous Fe_3O_4/carbon core/shell nanorods: synthesis and electromagnetic properties. J Phys Chem C 115: 13603-13608
89. Shen B, Zhai W T, Tao M M et al (2013) Lightweight, multifunctional polyetherimide/graphene@Fe_3O_4 composite foams for shielding of electromagnetic pollution. Acs Appl Mater Inter 5: 11383-11391
90. Yuan B H, Bao C L, Qian X D et al Design of artificial nacre-like hybrid films as shielding to mitigate electromagnetic pollution. Carbon 75: 178-189
91. Singh A P, Garg P, Alam F et al (2012) Phenolic resin-based composite sheets filled with mixtures of reduced graphene oxide, gamma-Fe_2O_3 and carbon fibers for excellent electromagnetic interference shielding in the X-band. Carbon 50: 3868-3875
92. Liu P B, Huang Y, Wang L et al (2013) Synthesis and excellent electromagnetic absorption properties of polypyrrole-reduced graphene oxide-Co_3O_4 nanocomposites. J Alloy Compd 573: 151-156
93. Zong M, Huang Y, Zhao Yang et al (2013) Facile preparation, high microwave absorption and microwave absorbing mechanism of RGO-Fe_3O_4 composites. Rsc Adv 3: 23638-23648
94. Tung T T, Feller J F, Kim H et al (2012) Electromagnetic properties of Fe_3O_4-functionalized graphene and its composites with a conducting polymer. J Polym Sci Pol Chem 50: 927-935
95. He H K, Gao C (2010) Supraparamagnetic, conductive, and processable multifunctional graphene nanosheets coated with high-density Fe_3O_4 nanoparticles. Acs Appl Mater Inter 2: 3201-3210.
96. Gupta T K, Singh B P, Singh V N et al (2014) MnO_2 decorated graphene nanoribbons with superior permittivity and excellent microwave shielding properties. J Mater Chem A 2: 4256-4263
97. Singh A P, Mishra M, Sambyal P et al (2014) Encapsulation of gamma-Fe_2O_3 decorated reduced graphene oxide in polyaniline core-shell tubes as an exceptional tracker for electromagnetic environmental pollution. J Mater Chem A 2: 3581-3593
98. Wang C, Han X J, Xu P et al (2011) The electromagnetic property of chemically reduced graphene oxide and its application as microwave absorbing material. Appl Phys Lett. doi: 10.1063/1.3555436
99. Geetha S, Satheesh Kumar K K, Rao C R K et al (2009) EMI shielding: methods and materials—a review. J Appl Polym Sci 112: 2073-2086
100. Krause B, Koops G H, van der Vegt N F A et al (2002) Ultralow-k dielectrics made by supercritical foaming of thin polymer films. Adv Mater 14: 1041-1046
101. Che R C, Peng L M, Duan X F et al (2004) Microwave absorption enhancement and complex permittivity and permeability of Fe encapsulated within carbon nanotubes. Adv Mater 16: 401-405
102. Lee C C, Chen D H (2007) Ag nanoshell-induced dual-frequency electromagnetic wave absorption of Ni nanoparticles. Appl Phys Lett. doi: 10.1063/1.2731706

103. Yuen S M, Ma C C M, Chuang C Y et al (2008) Effect of processing method on the shielding effectiveness of electromagnetic interference of MWCNT/PMMA composites. Compos Sci Technol 68: 963-968
104. Chen M X, Zhu Y, Pan Y B et al (2011) Gradient multilayer structural design of CNTs/SiO_2 composites for improving microwave absorbing properties. Mater Design 32: 3013-3016
105. Yan D X, Pang H, Xu L et al (2014) Electromagnetic interference shielding of segregated polymer composite with an ultralow loading of in situ thermally reduced graphene oxide. Nanotechnology. doi: 10.1088/0957-4484/25/14/145705
106. Joshi A, Bajaj A, Singh R et al (2013) Graphene nanoribbon-PVA composite as EMI shielding material in the X band. Nanotechnology. doi: 10.1088/0957-4484/24/45/455705
107. Basavaraja C, Kim W J, Kim Y D et al (2011) Synthesis of polyaniline-gold/graphene oxide composite and microwave absorption characteristics of the composite films. Mater Lett 65: 3120-3123
108. Gupta T K, Singh B P, Singh V N et al (2014) Multi-walled carbon nanotube-graphene-polyaniline multiphase nanocomposite with superior electromagnetic shielding effectiveness. Nanoscale 6: 842-851

Index

K.K. Sadasivuni et al. (eds.), *Graphene-Based Polymer Nanocomposites in Electronics*, Springer Series on Polymer and Composite Materials,
DOI 10.1007/978-3-319-13875-6

Q

R

S

T

MIX
Papier aus verantwortungsvollen Quellen
Paper from responsible sources
FSC® C105338
FSC
www.fsc.org

If you have any concerns about our products,
you can contact us on
ProductSafety@springernature.com

In case Publisher is established outside the EU,
the EU authorized representative is:
Springer Nature Customer Service Center GmbH
Europaplatz 3, 69115 Heidelberg, Germany

Printed by Libri Plureos GmbH
in Hamburg, Germany